MARKOV PROCESSES AND POTENTIAL THEORY

(ROBERT McCALLUM)

Robert M. Blumenthal, 1931 -
Department of Mathematics
University of Washington
Seattle, Washington

Ronald K. Getoor
Department of Mathematics
University of California, San Diego
La Jolla, California

DOVER PUBLICATIONS, INC.
Mineola, New York

Copyright

Bibliographical Note

This Dover edition, first published in 2007, is an unabridged republication of the work originally published by Academic Press, New York, in 1968.

Library of Congress Cataloging-in-Publication Data

Blumenthal, R. M. (Robert McCallum), 1931–
Markov processes and potential theory / Robert M. Blumenthal and Ronald K. Getoor.
p. cm.
"This Dover edition, first published in 2007, is an unabridged republication of the work originally published by Academic Press, New York, in 1968."
ISBN-13: 978-0-486-46263-9
ISBN-10: 0-486-46263-3
1. Markov processes. 2. Potential theory (Mathematics). I. Getoor, R. K. (Ronald Kay), 1929– II. Title.

QA274.7.B589 2007
519.2'33—dc22

2007024512

Manufactured in the United States of America
Dover Publications, Inc., 31 East 2nd Street, Mineola, N.Y. 11501

PREFACE

The study of the relationship between Markov processes and potential theory began in the early 1930's. Initially attention was directed toward analytic aspects such as differential equations satisfied by the transition probabilities and questions concerning existence and uniqueness of solutions of these equations. Then, through the work of Doob, Feller, Kac, and Kakutani, among others, it emerged that this relationship has a deep probabilistic basis. An extensive study of the relationship between Brownian motion and classical potential theory was given by Doob [4] in 1954. This paper marks the beginning of the modern era in the subject. Then in 1956 and 1957 Hunt [2–4] laid bare a large part of the entire subject: he gave a rather general definition of "potential theory" and associated with each such theory a Markov process in terms of which potential-theoretic objects and operations (superharmonic functions, balayage, etc.) have probabilistic interpretations. He used this relationship to generalize and reinterpret (often in a more illuminating form) many facts from classical potential theory.

The purpose of this book is to collect within one cover most of the contents of Hunt's fundamental papers, portions of the theory of additive functionals, and some closely related matters. We hope that we have presented the material in such a way that it is accessible to a diligent advanced graduate student. The reader is assumed to be familiar with general measure theory at the level of a basic graduate course. Aside from this, the book is very nearly self-contained. Still it is unrealistic to recommend it to a reader who does not have some background in probability theory or at least a bit of intuition concerning random variables, independence, conditioning, and the like.

Unfortunately the material is somewhat top-heavy; that is, one must start with some extensive (and perhaps dull) measure-theoretic preliminaries. Frequently the need for a particular level of generality (e.g., taking as sample space a general measure space rather than some specific function space) is not immediately apparent. However, we have tried to select the lowest level of abstraction that allows one firstly to operate freely and rigorously in subsequent developments, and secondly to avoid unnecessary restrictions on the transformations of processes one is willing to consider. Most of the measure-theoretic preliminaries are presented in Chapter I, while the more interesting developments begin in Chapter II; we ask the reader to go at least this far before passing judgement on the entire presentation. The reader especially interested in potential theory can go directly from the end of Chapter II to Section 6 of Chapter III and from there to Section 1 of Chapter V and Sections 1, 2, and 4 of Chapter VI. These sections are more or less independent of the rest of the book, and they contain the basic potential theoretic facts.

Most sections are followed by exercises designed to further the theory as well as give practice. We do not hesitate to use the results of an exercise in subsequent sections.

Historical references and credits are collected in the "Notes and Comments." We hope that we always have told the truth, but realize that it is seldom the whole truth. It is not our intention to give anyone less than his full measure of credit and we apologize in advance to anyone who may feel slighted.

This book does not cover all of the theory of Markov processes or even all of probabilistic potential theory. We have not said anything about infinitesimal generators (which, however, receive excellent treatment in Dynkin [2] and in Ito-McKean [1]). Neither have we said anything about the problem of constructing a Markov semigroup starting from a potential operator (or a class of excessive functions) nor about the boundary theory for the representation of harmonic functions—to mention only a few omissions.

Finally we have the pleasure of acknowledging a few of our debts. We benefited greatly from the interest which P. A. Meyer took in this project; he read most of the manuscript and set us straight on a number of important matters. Harry Dym and Frank Knight read portions of the manuscript and made useful comments. We received financial support from the National Science Foundation and from the Air Force Office of Scientific Research during part of the writing. We are particularly indebted to our typists Donna Thompson and Olive Lee for the capable manner in which they handled the job. They lightened considerably our work in preparing the manuscript.

February, 1968

R. M. Blumenthal
R. K. Getoor

CONTENTS

0

PRELIMINARIES

This chapter contains preliminary material that will be used in the main text. Section 1 establishes the basic notation that will be used without further explanation in the sequel. As a result this section should be read before the main text is attempted. The remaining sections of this chapter need not be read by everyone. It will suffice to refer to them whenever necessary. Section 2 contains the so-called "monotone class" theorems, which are perhaps the most useful results from set theory for us. They will be used extensively in the sequel. In Section 2 we also give a number of typical applications of these theorems in order to illustrate their use to readers who are unfamiliar with this type of argument. Often we will omit such arguments in the body of the text. Finally in Section 3 we collect some results connecting topology and measure theory.

1. Notation

Let $\mathbf{R}$ denote the real numbers, $\mathbf{Z}$ the integers, and $\overline{\mathbf{R}}$ the extended real numbers $\{-\infty\} \cup \mathbf{R} \cup \{+\infty\}$ with the usual topology. If a and b are in $\overline{\mathbf{R}}$, then $a \vee b = \max(a, b)$, $a \wedge b = \min(a, b)$, and $a^+ = a \vee 0$. If E is a set, a numerical (real-valued) function on E is a function from E to $\overline{\mathbf{R}}(\mathbf{R})$. For any numerical function f on E, $\|f\| = \sup_{x \in E} |f(x)|$. A σ-algebra $\mathscr{E}$ on E is a collection of subsets of E such that $E \in \mathscr{E}$ and $\mathscr{E}$ is closed under complements and countable unions. We will use the usual notation "$\cap$" and "$\cup$" to denote set intersection and union. If $A \subset E$, then A^c will denote the complement of A. Finally $A - B = A \cap B^c$ and $A \,\Delta\, B = (A - B) \cup (B - A)$ will denote set difference and symmetric difference, respectively. If $\mathscr{M}$ is any collection of subsets of E, $\sigma(\mathscr{M})$ denotes the σ-algebra generated by $\mathscr{M}$, that is, the smallest σ-algebra on E containing $\mathscr{M}$. Clearly $\sigma(\mathscr{M})$ is the intersection of all σ-algebras on E containing $\mathcal{M}$. A pair $(E, \mathcal{E})$ consisting of a set E and a σ-algebra on E is called a *measurable space*.

Let $(E, \mathcal{E})$ and $(F, \mathcal{F})$ be measurable spaces. A function $f: E \to F$ will be called *measurable relative to* $\mathcal{E}$ and $\mathcal{F}$ provided that for every $B \in \mathcal{F}$, $f^{-1}(B) \in \mathcal{E}$. Following Chung and Doob [1] we will write simply $f \in \mathcal{E}/\mathcal{F}$ if f is measurable relative to $\mathcal{E}$ and $\mathcal{F}$. In the case where $F = \overline{\mathbf{R}}$ and $\mathcal{F}$ is the class of all Borel subsets of $\overline{\mathbf{R}}$ we write simply $f \in \mathcal{E}$ for a numerical function $f \in \mathcal{E}/\mathcal{F}$. If A is a subset of E we denote the indicator function of A by I_A, so that for example, the statements $I_A \in \mathcal{E}$ and $A \in \mathcal{E}$ have the same meaning. We denote the collection of all real-valued $\mathcal{E}$ measurable functions by $r\mathcal{E}$. If $\mathcal{H}$ is any collection of numerical functions, then $\mathcal{H}_+$ denotes the set of nonnegative functions in $\mathcal{H}$ and $b\mathcal{H}$ denotes the set of bounded functions in $\mathcal{H}$. If f is a numerical function on E, then $f = 0$, $f \geq 0$, and $f > 0$ mean $f(x) = 0$ for all $x \in E$, $f(x) \geq 0$ for all $x \in E$, and $f(x) > 0$ for all $x \in E$, respectively. If E, F, and G are three sets and if $f: E \to F$ and $g: F \to G$, then we write $g \circ f$ for the composition map from E to G, $g \circ f(x) = g[f(x)]$. In particular if $(E, \mathcal{E})$, $(F, \mathcal{F})$, and $(G, \mathcal{G})$ are measurable spaces and if $f \in \mathcal{E}/\mathcal{F}$ and $g \in \mathcal{F}/\mathcal{G}$, then $g \circ f \in \mathcal{E}/\mathcal{G}$.

Let Ω be a set and let $(E_i, \mathcal{E}_i)_{i \in I}$ be a collection of measurable spaces indexed by some set I. For each i let f_i be a map from Ω to E_i; then $\sigma(f_i, \mathcal{E}_i; i \in I)$ denotes the σ-algebra on Ω generated by the sets $\{f_i^{-1}(A_i): A_i \in \mathcal{E}_i\}$ as i ranges over I. Clearly this is the smallest σ-algebra on Ω relative to which all the f_i are measurable. We will write merely $\sigma(f_i; i \in I)$ when the E_i and $\mathcal{E}_i$ are understood clearly from the context.

By a measure on $(E, \mathcal{E})$ we always will mean a *nonnegative* measure. A *measure space* is a triple $(E, \mathcal{E}, \mu)$ where $(E, \mathcal{E})$ is a measurable space and μ is a measure on $(E, \mathcal{E})$. We recall that μ is finite if $\mu(E) < \infty$ and μ is σ-finite if there exists a sequence $\{E_n\}$ of elements of $\mathcal{E}$ such that $E = \bigcup E_n$ and $\mu(E_n) < \infty$ for all n. We assume the reader is familiar with the theory of integration on a measure space. Often we write $\mu(f)$ instead of $\int f\, d\mu$ for those $f \in \mathcal{E}$ for which the integral exists. A measure space $(E, \mathcal{E}, \mu)$ is called *complete* if $B \in \mathcal{E}$ and $\mu(B) = 0$ implies that every subset of B is in $\mathcal{E}$. Given any measure space $(E, \mathcal{E}, \mu)$ we define $\mathcal{E}^\mu$ to consist of all sets $B \subset E$ for which there exist $B_1, B_2 \in \mathcal{E}$ such that $B_1 \subset B \subset B_2$ and $\mu(B_2 - B_1) = 0$. It is evident that $\mathcal{E}^\mu$ is a σ-algebra and that the measure μ has a unique extension to $\mathcal{E}^\mu$. We denote the extension again by μ. $\mathcal{E}^\mu$ is called the *completion* of $\mathcal{E}$ with respect to μ. One verifies easily that $\mathcal{E} = \mathcal{E}^\mu$ if and only if $(E, \mathcal{E}, \mu)$ is complete. Define $\mathcal{E}^* = \bigcap_\mu \mathcal{E}^\mu$ where the intersection is over all *finite* μ on $(E, \mathcal{E})$. It is easy to see that $\mathcal{E}^*$ is a σ-algebra and every finite measure on $\mathcal{E}$ has a unique extension to $\mathcal{E}^*$. $\mathcal{E}^*$ is called the σ-algebra of *universally measurable sets* over $(E, \mathcal{E})$. If μ is a measure on $(E, \mathcal{E})$ we say that μ *does not charge* a set A provided $A \in \mathcal{E}^\mu$ and $\mu(A) = 0$.

Let $(E, \mathscr{E}, \mu)$ be a measure space and let $(F, \mathscr{F})$ be a measurable space. If $f \in \mathscr{E}/\mathscr{F}$, then the formula $\nu(A) = \mu[f^{-1}(A)]$ for $A \in \mathscr{F}$ defines a measure ν on $(F, \mathscr{F})$. We will write $\nu = \mu f^{-1}$. If $h \in \mathscr{F}$, then $h \circ f \in \mathscr{E}$ and

$$\text{(1.1)} \qquad \int_E (h \circ f)\, d\mu = \int_F h\, d\mu f^{-1}$$

in the sense that if either integral exists then so does the other and they are equal.

If $(E_i, \mathscr{E}_i)_{i \in I}$ is a family of measurable spaces, then the product $\prod_{i \in I}(E_i, \mathscr{E}_i) = (E, \mathscr{E})$ is a measurable space where E is the ordinary Cartesian product of the E_i's and $\mathscr{E}$ is the σ-algebra on E generated by sets of the form $\prod_i A_i$ with $A_i \in \mathscr{E}_i$ for all i and only finitely many A_i different from E_i. We will write $\mathscr{E} = \prod_i \mathscr{E}_i$ and call $\mathscr{E}$ the product of the σ-algebras $\mathscr{E}_i$. We assume the reader to be familiar with the theory of product measures and the various forms of the Fubini theorem.

A probability space $(\Omega, \mathscr{F}, P)$ is a measure space with $P(\Omega) = 1$. If $(E, \mathscr{E})$ is a measurable space then a map $X : \Omega \to E$ is called an $(E, \mathscr{E})$ random variable provided $X \in \mathscr{F}/\mathscr{E}$. The distribution of X is the probability measure $\mu_X = PX^{-1}$ on $(E, \mathscr{E})$ and we often write $P(X \in B)$ for $\mu_X(B)$. A numerical random variable is by definition an element of $\mathscr{F}$. If X is a numerical random variable, then the *expectation* of X, $E(X)$, is defined by $E(X) = \int X\, dP$ provided the integral in question exists. If X is an $(E, \mathscr{E})$ random variable and $f \in b\mathscr{E}$, then (1.1) becomes

$$\text{(1.2)} \qquad E[f(X)] = \int_E f\, d\mu_X,$$

and this extends to any $f \in \mathscr{E}$ for which either side exists, in particular to $f \in \mathscr{E}_+$. Often we will establish formulas such as (1.2) for $f \in b\mathscr{E}$ and then use these formulas for nonnegative $f \in \mathscr{E}$ without special mention. Obviously this is permissible in view of the monotone convergence theorem. Frequently when X is a function in $\mathscr{F}$ and Λ is a set in $\mathscr{F}$ we write $E(X; \Lambda)$ in place of $\int_\Lambda X\, dP$.

We assume that the reader is familiar with the concepts of independence and conditional expectation as set forth, for example, in Loève [1] or Neveu [1]. In particular if $\mathscr{G}$ is a σ-algebra contained in $\mathscr{F}$ and $X \in \mathscr{F}$ is P-integrable we use $E(X \mid \mathscr{G})$ to denote any function f such that: (1) $f \in \mathscr{G}$ and (2) $E(f; \Lambda) = E(X; \Lambda)$ for every $\Lambda \in \mathscr{G}$. $E(X \mid \mathscr{G})$ is called the *conditional expectation* of X given $\mathscr{G}$. When $X = I_A$, $A \in \mathscr{F}$, we write $P(A \mid \mathscr{G})$ in place of $E(I_A \mid \mathscr{G})$. Of course conditions (1) and (2) determine f up to a set in $\mathscr{G}$ of P measure 0. On the other hand, when we write $f = E(X \mid \mathscr{G})$ we mean simply that f is a function satisfying (1) and (2); consequently, relationships involving conditional expectations are not enhanced by the qualifying phrase "almost everywhere relative to P." When $\mathscr{G}$ is of the form $\sigma(f_i, \mathscr{E}_i; i \in I)$ we sometimes write $E(X \mid f_i; i \in I)$ instead of $E(X \mid \mathscr{G})$.

Finally we summarize the basic results on supermartingales that will be needed in later chapters. Let $(\Omega, \mathscr{F}, P)$ be a probability space and let $\mathbf{T}$ be a subset of the extended real numbers. For each $t \in \mathbf{T}$, let X_t be a numerical random variable over $(\Omega, \mathscr{F}, P)$ and let $\mathscr{F}_t$ be a sub-σ-algebra of $\mathscr{F}$ such that $\mathscr{F}_t \subset \mathscr{F}_s$ whenever $t \le s$, t and s in $\mathbf{T}$. Then $\{X_t, \mathscr{F}_t; t \in \mathbf{T}\}$ is a *supermartingale* (over $(\Omega, \mathscr{F}, P)$) provided

$$\text{(1.3)} \qquad \begin{aligned} &\text{(i)} \quad X_t \in \mathscr{F}_t, && t \in \mathbf{T}, \\ &\text{(ii)} \quad E(|X_t|) < \infty, && t \in \mathbf{T}, \\ &\text{(iii)} \quad E\{X_s \mid \mathscr{F}_t\} \le X_t; && s, t \in \mathbf{T}, \quad s > t. \end{aligned}$$

If the σ-algebras $\mathscr{F}_t$ are omitted from the definition, then it is understood that $\mathscr{F}_t = \sigma(X_s; s \in \mathbf{T}, s \le t)$. We say that $\{X_t, \mathscr{F}_t\}$ is a *submartingale* provided $\{-X_t, \mathscr{F}_t\}$ is a supermartingale, and that $\{X_t, \mathscr{F}_t\}$ is a *martingale* provided that it is both a supermartingale and a submartingale. If I is an arbitrary index set, then a family $\{Y_i; i \in I\}$ of numerical random variables over $(\Omega, \mathscr{F}, P)$ is said to be *uniformly integrable* provided that $\lim_n E\{|Y_i|; |Y_i| > n\} = 0$ uniformly on I. We refer the reader to Meyer [1] or Neveu [1] for the properties of uniformly integrable families. The next two theorems contain the basic results which will be needed in later chapters. Proofs may be found in Doob [1], Loève [1], Meyer [1], or Neveu [1].

(1.4) THEOREM. Let $\mathbf{N} = \{1, 2, \ldots\}$ and $\overline{\mathbf{N}} = \mathbf{N} \cup \{\infty\}$. Let $\{X_n, \mathscr{F}_n; n \in \mathbf{N}\}$ be a supermartingale such that $\sup_n E(|X_n|) < \infty$. Then $X_\infty = \lim_n X_n$ exists and is finite almost surely. Moreover if $\{X_n; n \in \mathbf{N}\}$ is uniformly integrable then $\{X_n, \mathscr{F}_n; n \in \overline{\mathbf{N}}\}$ is a supermartingale where $\mathscr{F}_\infty = \sigma(\bigcup \mathscr{F}_n)$.

Note that if $X_n \ge 0$ for all n, then (1.3) implies that $E(|X_n|) = E(X_n) \le E(X_1)$, and so a nonnegative supermartingale $\{X_n, \mathscr{F}_n; n \in \mathbf{N}\}$ always satisfies the hypothesis of Theorem 1.4.

(1.5) THEOREM:

(a) If $\{X_t, \mathscr{F}_t; t \in (0, \infty)\}$ is a supermartingale such that $t \to X_t(\omega)$ is right continuous on $(0, \infty)$ for almost all ω, then $t \to X_t(\omega)$ has left-hand limits on $(0, \infty)$ for almost all ω, that is, for almost all ω, $\lim_{s \uparrow t} X_s(\omega)$ exists (and is finite) for all $t \in (0, \infty)$.

(b) Let $\{X_t, \mathscr{F}_t; t \in [0, \infty)\}$ be a supermartingale, and let $\mathbf{D}$ be a countable dense subset of $[0, \infty]$. Then for almost all ω, $\lim_{s \uparrow t, s \in \mathbf{D}} X_s(\omega)$ and $\lim_{s \downarrow t, s \in \mathbf{D}} X_s(\omega)$ exist (and are finite) for all $t \in [0, \infty)$.

The following fact will also be needed in Chapter I. We leave its proof to the reader as an exercise.

(1.6) PROPOSITION. Let $\{X_t, \mathscr{F}_t; t \in [0, \infty)\}$ be a nonnegative supermartingale and let $\mathbf{D}$ be a countable dense subset of $[0, \infty)$. Then for each $t \in [0, \infty)$, $P\{X_t > 0, \inf_{s<t, s\in\mathbf{D}} X_s = 0\} = 0$.

2. The Monotone Class Theorem

In this section we will discuss the monotone class theorem in the form we find most useful for application in probability theory.

(2.1) DEFINITION. Let Ω be a set and $\mathscr{S}$ a collection of subsets of Ω; then

(i) $\mathscr{S}$ is a π-system (on Ω) if $\mathscr{S}$ is closed under finite intersections;

(ii) $\mathscr{S}$ is a d-system (on Ω) if

(a) $\Omega \in \mathscr{S}$,

(b) if $A, B \in \mathscr{S}$, $A \subset B$, then $B - A \in \mathscr{S}$,

(c) if $\{A_n\}$ is an increasing sequence of elements of $\mathscr{S}$, then $\bigcup A_n \in \mathscr{S}$.

Obviously any σ-algebra is both a d-system and π-system. Conversely, one easily verifies that if $\mathscr{S}$ is both a d-system and a π-system, then $\mathscr{S}$ is a σ-algebra. If $\mathscr{M}$ is any collection of subsets of Ω, then we define $d(\mathscr{M})$ to be the smallest d-system on Ω containing $\mathscr{M}$. The existence of $d(\mathscr{M})$ is clear since the intersection of an arbitrary number of d-systems is again a d-system. We will say that $d(\mathscr{M})$ is the d-system generated by $\mathscr{M}$.

We come now to the main result on d-systems.

(2.2) THEOREM. Let $\mathscr{S}$ be a π-system on Ω; then $d(\mathscr{S}) = \sigma(\mathscr{S})$.

Proof. Since $d(\mathscr{S}) \subset \sigma(\mathscr{S})$ it suffices to show that $d(\mathscr{S})$ is a σ-algebra, and for this it suffices to show that $d(\mathscr{S})$ is a π-system.

To this end define

$$\mathscr{D}_1 = \{B \in d(\mathscr{S}) : B \cap A \in d(\mathscr{S}) \text{ for all } A \in \mathscr{S}\}.$$

The reader will easily verify that $\mathscr{D}_1$ is a d-system and that $\mathscr{D}_1 \supset \mathscr{S}$ since $\mathscr{S}$ is a π-system. Hence $\mathscr{D}_1 \supset d(\mathscr{S})$. But by definition $\mathscr{D}_1 \subset d(\mathscr{S})$ and so $\mathscr{D}_1 = d(\mathscr{S})$. Next define

$$\mathscr{D}_2 = \{B \in d(\mathscr{S}) : B \cap A \in d(\mathscr{S}) \text{ for all } A \in d(\mathscr{S})\}.$$

Again one shows without difficulty that $\mathscr{D}_2$ is a d-system. If $A \in \mathscr{S}$, then $B \cap A \in d(\mathscr{S})$ for all $B \in \mathscr{D}_1 = d(\mathscr{S})$, and consequently $\mathscr{S} \subset \mathscr{D}_2$. Hence

$d(\mathscr{S}) = \mathscr{D}_2$, and this is just the statement that $d(\mathscr{S})$ is closed under finite intersections. Thus Theorem 2.2 is established.

We next give a version of Theorem 2.2 that deals with functions rather than sets.

(2.3) THEOREM. Let Ω be a set and $\mathscr{S}$ a π-system on Ω. Let $\mathscr{H}$ be a vector space of real-valued functions on Ω satisfying:

(i) $1 \in \mathscr{H}$ and $I_A \in \mathscr{H}$ for all $A \in \mathscr{S}$;

(ii) if $\{f_n\}$ is an increasing sequence of nonnegative functions in $\mathscr{H}$ such that $f = \sup_n f_n$ is finite (bounded), then $f \in \mathscr{H}$.

Under these assumptions $\mathscr{H}$ contains all real-valued (bounded) functions on Ω that are $\sigma(\mathscr{S})$ measurable.

Proof. Let $\mathscr{D} = \{A\colon I_A \in \mathscr{H}\}$. According to Assumption (i) $\Omega \in \mathscr{D}$ and $\mathscr{S} \subset \mathscr{D}$. If $A_1 \subset A_2$ are in $\mathscr{D}$, then $I_{A_2 - A_1} = I_{A_2} - I_{A_1} \in \mathscr{H}$ since $\mathscr{H}$ is a vector space, and so $A_2 - A_1 \in \mathscr{D}$. Finally if $\{A_n\}$ is an increasing sequence of sets in $\mathscr{D}$, then $I_{\cup A_n} = \sup I_{A_n}$ which is in $\mathscr{H}$ by (ii). Thus $\mathscr{D}$ is a d-system on Ω containing $\mathscr{S}$ and hence $\mathscr{D} \supset \sigma(\mathscr{S})$ by Theorem 2.2.

If $f \in \sigma(\mathscr{S})$ is real-valued then $f = f^+ - f^-$ with f^+ and f^- being nonnegative, real-valued, and $\sigma(\mathscr{S})$ measurable. On the other hand, if $f \in \sigma(\mathscr{S})$ is nonnegative then f is an increasing limit of simple functions $f_n = \sum_{i=1}^{n} a_i^n I_{A_i^n}$ with each $A_i^n \in \sigma(\mathscr{S})$. Hence each $f_n \in \mathscr{H}$ and using (ii) the result is now immediate.

Theorems 2.2 and 2.3 will be used in the following form: Let Ω be a set and $(E_i, \mathscr{E}_i)_{i \in I}$ be a family of measurable spaces indexed by an arbitrary set I. For each $i \in I$, let $\mathscr{S}_i$ be a π-system generating $\mathscr{E}_i$ and let f_i be a map from Ω to E_i. Using this setup we state the following two propositions.

(2.4) PROPOSITION. Let $\mathscr{S}$ consist of all sets of the form $\bigcap_{i \in J} f_i^{-1}(A_i)$ where $A_i \in \mathscr{S}_i$ for $i \in J$ and J ranges over all finite subsets of I. Then $\mathscr{S}$ is a π-system on Ω and $\sigma(\mathscr{S}) = \sigma(f_i;\, i \in I)$.

(2.5) PROPOSITION. Let $\mathscr{H}$ be a vector space of real-valued functions on Ω such that

(i) $1 \in \mathscr{H}$;

(ii) if $\{h_n\}$ is an increasing sequence of elements of $\mathscr{H}_+$ such that $h = \sup h_n$ is finite (bounded), then $h \in \mathscr{H}$;

(iii) $\mathscr{H}$ contains all products of the form $\prod_{i \in J} I_{A_i} \circ f_i$ where J is a finite subset of I and $A_i \in \mathscr{S}_i$ for $i \in J$.

Under these assumptions $\mathscr{H}$ contains all real-valued (bounded) functions in $\sigma(f_i; i \in I)$.

Proposition 2.4 is immediate and (2.5) follows from (2.4) and (2.3). In the sequel any of the results, (2.2)–(2.5) will be referred to as the Monotone Class Theorem (MCT). We now give some applications of these results. We begin with the following uniqueness theorem for finite measures. (See Exercise 3.2 for an extension to σ-finite measures.)

(2.6) PROPOSITION. Let μ and ν be finite measures on a measurable space $(E, \mathscr{E})$. Let $\mathscr{S} \subset \mathscr{E}$ be a π-system containing E and generating $\mathscr{E}$. If μ and ν agree on $\mathscr{S}$, then they agree on $\mathscr{E}$.

This is an immediate consequence of (2.2) since $\{A \in \mathscr{E}: \mu(A) = \nu(A)\}$ is clearly a d-system containing $\mathscr{S}$.

We close this section with two useful facts.

(2.7) PROPOSITION. Let Ω be a set and $(E, \mathscr{E})$ a measurable space. Let $f: \Omega \to E$; then $\varphi: \Omega \to \overline{\mathbf{R}}$ is $\sigma(f)$ measurable if and only if there exists $h \in \mathscr{E}$ such that $\varphi = h \circ f$.

Proof. If $\varphi = h \circ f$ with $h \in \mathscr{E}$ then obviously $\varphi \in \sigma(f)$. To prove the converse let $\mathscr{H}$ be all real functions on Ω of the form $h \circ f$ with $h \in r\mathscr{E}$. Clearly $\mathscr{H}$ is a vector space containing the constants. Suppose $\{h_n \circ f\}$ is an increasing sequence in $\mathscr{H}_+$ such that $\psi = \sup(h_n \circ f)$ is finite. Let $A = \{x \in E: \sup h_n(x) < \infty\}$; then $A \in \mathscr{E}$ and $f(\Omega) \subset A$. Define $h = \sup h_n$ on A and $h = 0$ on A^c. Then $h \in r\mathscr{E}$ and $\psi = h \circ f$. If $C \in \sigma(f)$ then $C = f^{-1}(A)$ for some $A \in \mathscr{E}$ and so $I_C = I_A \circ f \in \mathscr{H}$. Thus $\mathscr{H}$ contains all the real-valued functions in $\sigma(f)$ according to (2.5). If $\varphi \in \sigma(f)$ is numerical valued then $\varphi' = \arctan \varphi \in \sigma(f)$ and is real valued. So $\varphi' = h' \circ f$ for some $h' \in \mathscr{E}$. Clearly we may assume that $h'(E) \subset [-\pi/2, \pi/2]$ because φ' takes values only in this interval. Then $\varphi = h \circ f$ with $h = \tan h'$ and of course $h \in \mathscr{E}$.

We should point out that if φ is real valued (bounded) we may assume that h is real valued (bounded).

(2.8) PROPOSITION. Let Ω be a set, $\{\mathscr{G}_i; i \in I\}$ a family of σ-algebras on Ω, and $\mathscr{G} = \sigma(\mathscr{G}_i; i \in I\}$. Then for every $\Lambda \in \mathscr{G}$ there is a countable subset $J \subset I$ such that $\Lambda \in \sigma(\mathscr{G}_i; i \in J)$.

Proof. The class of sets Λ having this property is clearly a σ-algebra and it contains each $\mathscr{G}_i$.

The following corollary is an easy consequence of (2.7) and (2.8). We leave the details to the reader as an exercise.

(2.9) COROLLARY. Let Ω be a set and let $(E_i, \mathscr{E}_i)_{i \in I}$ be a family of measurable spaces. For each i let $f_i : \Omega \to E_i$ and set $\mathscr{F} = \sigma(f_i;\, i \in I)$. If $\varphi : \Omega \to \overline{\mathbf{R}}$ is in $\mathscr{F}$, then φ depends on at most countably many coordinates in the sense that there exist a countable subset J of I and a function $h : \prod_{i \in J} E_i \to \overline{\mathbf{R}}$ which is $\prod_{i \in J} \mathscr{E}_i$ measurable and such that $\varphi = h \circ f_J$ where f_J is the map $\omega \to (f_i(\omega))_{i \in J}$ from Ω to $\prod_{i \in J} E_i$.

3. Topological Spaces

A topological space E will always mean a *Hausdorff* space unless explicitly stated otherwise. If E is a topological space the *Borel sets* of E is the smallest σ-algebra containing the open sets of E. We write $\mathscr{B}(E)$ (or just $\mathscr{B}$) for the Borel sets of E. If E is locally compact, then $\mathbf{C}(E)$, $\mathbf{C}_0(E)$, $\mathbf{C}_K(E)$ denote, respectively, the bounded real-valued continuous functions on E, the real-valued continuous functions vanishing at ∞, the real-valued continuous functions with compact support. Clearly $\mathbf{C}_K \subset \mathbf{C}_0 \subset \mathbf{C}$ and $\mathbf{C}_0$ and $\mathbf{C}$ are Banach spaces under the supremum norm $\|\cdot\|$. Moreover $\mathbf{C}_K$ is dense in $\mathbf{C}_0$. The *Baire sets* of E is the σ-algebra $\sigma(\mathbf{C}_K) = \mathscr{B}_0(E)$. Obviously $\mathscr{B}_0 \subset \mathscr{B}$.

In most of our work we will be concerned with locally compact spaces with a countable base (LCCB). We summarize some of the relevant properties of such a space.

(3.1) Let E be an LCCB; then

(i) $\mathscr{B} = \mathscr{B}_0$;

(ii) E is metrizable, and one can choose a metric d compatible with the topology, such that (E, d) is a separable complete metric space in which every closed and d bounded set is compact;

(iii) if $\mathscr{U}$ is a countable base for the topology of E, then $\sigma(\mathscr{U}) = \mathscr{B}$;

(iv) $\mathbf{C}_0$ is separable (as a Banach space).

Let E be an LCCB. A measure μ on $(E, \mathscr{B}(E))$ is called a *Radon measure* on E if $\mu(K) < \infty$ for all compact K. In particular a Radon measure on an LCCB is σ-finite. If L is a nonnegative linear functional on $\mathbf{C}_K$, then there exists a unique Radon measure μ on E such that $L(f) = \mu(f)$ for all $f \in \mathbf{C}_K$. Every Radon measure is regular; that is, if $B \in \mathscr{B}$ then

$$\mu(B) = \sup\{\mu(K) \colon K \subset B,\ K \text{ compact}\}$$
$$= \inf\{\mu(G) \colon G \supset B,\ G \text{ open}\}.$$

Let $\mathbf{M}_+$ be the set of all Radon measures μ on E such that $\mu(E) < \infty$ and let $\mathbf{M} = \mathbf{M}_+ - \mathbf{M}_+$ be the vector space of all real-valued set functions λ on E which have the form $\lambda = \mu - \nu$ with $\mu, \nu \in \mathbf{M}_+$. For $\lambda \in \mathbf{M}$ we define $\|\lambda\| = \sup \sum |\lambda(A_i)|$ where the supremum is taken over all finite partitions $\{A_1, \ldots, A_n\}$ of E into Borel sets. Under this norm $\mathbf{M}$ is a Banach space and, in fact, $\mathbf{M} = \mathbf{C}_0^*$. We will call the topology induced on $\mathbf{M}$ by $\mathbf{C}_0$ (the weak*-topology on $\mathbf{M}$) the weak (or vague) topology on $\mathbf{M}$.

Exercises

(3.2) Let μ and ν be two σ-finite measures on the measurable space $(E, \mathscr{E})$. Let $\mathscr{S} \subset \mathscr{E}$ be a π-system generating $\mathscr{E}$ and suppose $\mathscr{S}$ contains a sequence $\{B_n\}$ with $\bigcup B_n = E$. If μ and ν agree on $\mathscr{S}$ and $\mu(B_n) = \nu(B_n) < \infty$ for all n, then μ and ν agree on $\mathscr{E}$.

(3.3) Let $(E_i, \mathscr{E}_i)$, $i = 1, 2$, be measurable spaces. If $f \in \mathscr{E}_1/\mathscr{E}_2$, then $f \in \mathscr{E}_1^*/\mathscr{E}_2^*$.

(3.4) Let $(E, \mathscr{E}, \mu)$ be a measure space. Show that $f \in \mathscr{E}^\mu$ if and only if there exists f_1 and $f_2 \in \mathscr{E}$ such that $f_1 \le f \le f_2$ and $\mu(\{f_1 < f_2\}) = 0$.

(3.5) Prove that $\mathscr{B}(E) = \mathscr{B}_0(E)$ when E is LCCB.

(3.6) Let Ω be a set and E an LCCB. Let $\{X_i;\, i \in I\}$ be a family of functions from Ω to E. Let $\mathscr{H}$ be a vector space of real-valued functions on Ω satisfying:

(i) $1 \in \mathscr{H}$;

(ii) If $\{F_n\}$ is an increasing sequence in $\mathscr{H}_+$ such that $F = \sup F_n$ is bounded, then $F \in \mathscr{H}$;

(iii) $\mathscr{H}$ contains all functions of the form $\prod_{i \in J} f_i \circ X_i$ where J is a finite subset of I and $f_i \in \mathbf{C}_K$ for $i \in J$.

Show that $\mathscr{H}$ contains all bounded functions in $\sigma(X_i;\, i \in I)$.

(3.7) Let $(\Omega, \mathscr{F})$ be a measurable space and E be a metric space. Let $\{f_n\}$ be a sequence in $\mathscr{F}/\mathscr{B}(E)$. Prove the following statements:

(i) if $f_n(\omega) \to f(\omega)$ for all ω, then $f \in \mathscr{F}/\mathscr{B}(E)$.

(ii) If E is complete and x_0 a point in E, then

$$
\begin{aligned}
f(\omega) &= \lim_n f_n(\omega) && \text{when the limit exists} \\
&= x_0 && \text{when the limit does not exist}
\end{aligned}
$$

is in $\mathscr{F}/\mathscr{B}(E)$. The completeness hypothesis cannot be entirely eliminated.

What happens if $\mathscr{B}$ is replaced by $\mathscr{B}^*$ (the universally measurable sets over $(E, \mathscr{B})$) throughout?

(3.8) Let $(E, \mathscr{E})$ be a measurable space and let T be a positive linear map from $b\mathscr{E}$ to $b\mathscr{E}$ such that whenever $\{f_n\}$ is a decreasing sequence in $b\mathscr{E}$ with $\inf f_n = 0$ then $\inf Tf_n = 0$. Show that there exists a function $P(x, A)$ defined for $x \in E$ and $A \in \mathscr{E}$ such that

(i) $A \to P(x, A)$ is a finite measure on $\mathscr{E}$ for each $x \in E$;

(ii) $x \to P(x, A)$ is in $b\mathscr{E}$ for each $A \in \mathscr{E}$;

(iii) $Tf(x) = \int P(x, dy) f(y)$ for each $f \in b\mathscr{E}$ and $x \in E$.

[Hint: let $P(x, A) = T I_A(x)$.]

I

MARKOV PROCESSES

1. General Definitions

In this section we will give the general definition of a Markov process following more or less the definition in Doob [1]. However this definition is too general for the purposes of this book, and so very shortly we will restrict ourselves to a more tractable class of processes.

Let $(\Omega, \mathscr{F}, P)$ be a probability space, $(E, \mathscr{E})$ a measurable space, and $\mathbf{T}$ an arbitrary index set. Then a *stochastic process with values in* $(E, \mathscr{E})$ *and parameter set* $\mathbf{T}$ (over $(\Omega, \mathscr{F}, P)$) is a collection $X = \{X_t : t \in \mathbf{T}\}$ of maps X_t from Ω to E each of which belongs to $\mathscr{F}/\mathscr{E}$. (See Section 1 of Chapter 0 for notation.) For each $\omega \in \Omega$ the map $t \to X_t(\omega)$ from $\mathbf{T}$ to E is called the path or trajectory corresponding to ω. It will sometimes be convenient to write $X(t, \omega)$ for $X_t(\omega)$ and also $X(t)$ for X_t. The measure P clearly plays no role in this definition and so one can define in exactly the same way a stochastic process over a measurable space $(\Omega, \mathscr{F})$. This will sometimes be useful.

From now on we assume that $\mathbf{T}$ is a subset of $\overline{\mathbf{R}}$ (the extended real numbers). If $\{\mathscr{F}_t; t \in \mathbf{T}\}$ is an increasing family of sub-σ-algebras of $\mathscr{F}$, that is, $\mathscr{F}_t \subset \mathscr{F}_s$ whenever t and s are in $\mathbf{T}$ and $t < s$, we say that the process X is *adapted to* $\{\mathscr{F}_t\}$ provided $X_t \in \mathscr{F}_t/\mathscr{E}$ for each t in $\mathbf{T}$. In particular X is always adapted to the family of sub-σ-algebras $\mathscr{F}_t = \sigma(X_s : s \leq t)$. Of course in expressions such as this the letters t and s refer to elements of $\mathbf{T}$.

(1.1) DEFINITION. Let $X = \{X_t : t \in \mathbf{T}\}$ be a stochastic process with values in $(E, \mathscr{E})$. One says that X is a Markov process with respect to an increasing family $\{\mathscr{F}_t;\ t \in \mathbf{T}\}$ of sub-σ-algebras of $\mathscr{F}$ provided (1) X is adapted to $\{\mathscr{F}_t\}$ and (2) for each $t \in \mathbf{T}$ the σ-algebras $\mathscr{F}_t$ and $\sigma(X_s : s \geq t)$ are conditionally independent given X_t, that is,

(1.2) $$P(A \cap B \mid X_t) = P(A \mid X_t)P(B \mid X_t)$$

whenever A is in $\mathscr{F}_t$ and B is in $\sigma(X_s;\ s \geq t)$. One says that X is a Markov process (without specifying the σ-algebras $\mathscr{F}_t$) if it is a Markov process with respect to the family of σ-algebras $\mathscr{G}_t = \sigma(X_s;\ s \leq t)$.

Notice that if X is a Markov process with respect to some family $\{\mathscr{F}_t\}$ of σ-algebras, then it is also a Markov process with respect to the family $\{\mathscr{G}_t = \sigma(X_s;\ s \leq t)\}$. If X is a Markov process with respect to $\mathscr{G}_t = \sigma(X_s;\ s \leq t)$, then the σ-algebras $\sigma(X_s;\ s \leq t)$ and $\sigma(X_s;\ s \geq t)$ enter symmetrically in the definition. Consequently the Markov property is preserved if one reverses the order in $\mathbf{T}$. The intuitive meaning of this condition should be clear: namely, given the present, X_t, the past, $\sigma(X_s;\ s \leq t)$, and the future, $\sigma(X_s;\ s \geq t)$ are independent.

In discussing conditional expectations the following situation will often arise. Let $(\Omega, \mathscr{F}, P)$ be a probability space, and $\mathscr{G}_1$ and $\mathscr{G}_2$ σ-algebras contained in $\mathscr{F}$ with $\mathscr{G}_1 \subset \mathscr{G}_2$. In this situation if $f \in \mathscr{F}$, then the equality

$$E(f \mid \mathscr{G}_1) = E(f \mid \mathscr{G}_2)$$

should be interpreted as meaning that there exists a function $g \in \mathscr{G}_1$ such that $E(f;\ \Lambda) = E(g;\ \Lambda)$ for all $\Lambda \in \mathscr{G}_2$.

We now give some equivalent formulations of Definition (1.1).

(1.3) THEOREM:

(a) Let $X = \{X_t;\ t \in \mathbf{T}\}$ be a stochastic process adapted to the family $\{\mathscr{F}_t\}$. Then the following statements are equivalent:

(i) X is a Markov process with respect to $\{\mathscr{F}_t\}$;

(ii) for each $t \in \mathbf{T}$ and Y in $b\sigma(X_s\colon s \geq t)$ one has

(1.4) $$E(Y \mid \mathscr{F}_t) = E(Y \mid X_t);$$

(iii) if t and s are in $\mathbf{T}$ and $t \leq s$ then

(1.5) $$E(f \circ X_s \mid \mathscr{F}_t) = E(f \circ X_s \mid X_t)$$

for all $f \in b\mathscr{E}$.

(b) X is a Markov process (with respect to $\sigma(X_s;\ s \leq t)$) if for each finite collection $t_1 \leq \ldots \leq t_n \leq t$ from $\mathbf{T}$ and $f \in b\mathscr{E}$ one has

(1.6) $$E(f \circ X_t \mid X_{t_1}, \ldots, X_{t_n}) = E(f \circ X_t \mid X_{t_n}).$$

Proof. We will first establish the equivalence of (i) and (ii). Suppose (i) holds. It is immediate from MCT that it suffices to prove (1.4) when $Y = I_B$ with $B \in \sigma(X_s\colon s \geq t)$. If $A \in \mathscr{F}_t$, then

$$\begin{aligned} P(A \cap B) &= E\{P(A \cap B \mid X_t)\} \\ &= E\{P(A \mid X_t)\, P(B \mid X_t)\} \\ &= E[E\{I_A\, P(B \mid X_t) \mid X_t\}] \\ &= E[I_A\, P(B \mid X_t)]. \end{aligned}$$

Hence (i) implies (ii). Conversely suppose (ii) holds. If $A \in \mathscr{F}_t$ and $B \in \sigma(X_s;\, s \geq t)$, then

$$\begin{aligned} P(A \cap B \mid X_t) &= E\{P(A \cap B \mid \mathscr{F}_t) \mid X_t\} \\ &= E\{I_A\, P(B \mid \mathscr{F}_t) \mid X_t\} \\ &= E\{I_A\, P(B \mid X_t) \mid X_t\} \\ &= P(A \mid X_t)\, P(B \mid X_t), \end{aligned}$$

and so (ii) implies (i).

We next establish the equivalence of (ii) and (iii). Clearly (iii) is a special case of (ii) and so we need only show that (iii) implies (ii). Therefore assume (iii) holds. First observe that if $\mathbf{H}$ denotes those elements $Y \in b\sigma(X_s;\, s \geq t)$ for which (1.4) holds, then $\mathbf{H}$ is a vector space containing the constants and closed under bounded monotone limits. Thus by MCT in order to show that $\mathbf{H} = b\sigma(X_s;\, s \geq t)$ it suffices to show that $\mathbf{H}$ contains all Y of the form $Y = \prod_{i=1}^{n} f_i(X_{s_i})$ where $t \leq s_1 < \ldots < s_n$ and $f_i \in b\mathscr{E}$ for $1 \leq i \leq n$. We will establish this last statement by induction on n. When $n = 1$ this reduces to (iii). If $n > 1$, we may write for such a Y

$$\begin{aligned} E(Y \mid \mathscr{F}_t) &= E(E(Y \mid \mathscr{F}_{s_{n-1}}) \mid \mathscr{F}_t) \\ &= E\left\{\prod_{i=1}^{n-1} f_i(X_{s_i})\, E[f_n(X_{s_n}) \mid \mathscr{F}_{s_{n-1}}] \mid \mathscr{F}_t\right\}. \end{aligned}$$

From (iii) and (2.7) of Chapter 0 we have $E(f_n(X_{s_n}) \mid \mathscr{F}_{s_{n-1}}) = E(f_n(X_{s_n}) \mid X_{s_{n-1}}) = g(X_{s_{n-1}})$ where $g \in \mathscr{E}$ may be assumed bounded since f_n is. Consequently applying the induction hypothesis to the functions $f_1, \ldots, f_{n-2}, f_{n-1} g$ evaluated at $X_{s_1}, \ldots, X_{s_{n-1}}$ we obtain

$$\begin{aligned} E(Y \mid \mathscr{F}_t) &= E\left\{\prod_{i=1}^{n-1} f_i(X_{s_i})\, g(X_{s_{n-1}}) \mid X_t\right\} \\ &= E\left\{\prod_{i=1}^{n-1} f_i(X_{s_i})\, E(f_n(X_{s_n}) \mid \mathscr{F}_{s_{n-1}}) \mid X_t\right\} \\ &= E(Y \mid X_t). \end{aligned}$$

Thus Part (a) of Theorem 1.3 is established.

If X is a Markov process then clearly (1.6) follows from (1.5), and so in order to establish (b) it will suffice to show that (1.6) implies $E(f \circ X_s \mid \mathscr{G}_t) =$

$E(f \circ X_s \mid X_t)$ for $t \leq s$ and $f \in b\mathscr{E}$ where $\mathscr{G}_t = \sigma(X_s \colon s \leq t)$. This amounts to showing that for all $\Lambda \in \mathscr{G}_t$ one has

$$\int_\Lambda f \circ X_s \, dP = \int_\Lambda E(f \circ X_s \mid X_t) \, dP. \tag{1.7}$$

Let $\mathscr{D}$ be the collection of all sets $\Lambda \in \mathscr{G}_t$ for which (1.7) holds. Plainly $\mathscr{D}$ is a d-system. On the other hand it follows from (1.6) that if $\Lambda = \bigcap_{j=1}^n \{X_{t_j} \in A_j\}$ where $0 \leq t_1 < \ldots < t_n = t$ and $A_j \in \mathscr{E}$ for $1 \leq j \leq n$, then $\Lambda \in \mathscr{D}$. But Λ's of this form are a π-system generating $\mathscr{G}_t$ and hence $\mathscr{D} = \mathscr{G}_t$ completing the proof of Theorem 1.3.

2. Transition Functions

In this section we will assume that the parameter set $\mathbf{T}$ of any stochastic process under discussion is $\mathbf{R}_+ = [0, \infty)$ unless explicitly stated otherwise. This will be the case of most interest to us and the reader should have no difficulty in adapting the definitions to other cases of interest: for example, $\mathbf{T} = \mathbf{Z}_+ = \{0, 1, 2, \ldots\}$.

(2.1) Definition. Let $(E, \mathscr{E})$ be a measurable space; then a function $P_{t,s}(x, A)$ defined for $0 \leq t < s < \infty$, x in E, and A in $\mathscr{E}$, is called a *Markov transition function* on $(E, \mathscr{E})$ provided

(i) $A \to P_{t,s}(x, A)$ is a probability measure on $\mathscr{E}$ for each t, s, and x;
(ii) $x \to P_{t,s}(x, A)$ is in $\mathscr{E}$ for each t, s, and A;
(iii) if $0 \leq t < s < u$, then

$$P_{t,u}(x, A) = \int P_{t,s}(x, dy) \, P_{s,u}(y, A) \tag{2.2}$$

for all x and A.

The relationship (2.2) is called the Chapman–Kolmogorov equation.

(2.3) Definition. A Markov transition function $P_{t,s}(x, A)$ on $(E, \mathscr{E})$ is said to be *temporally homogeneous* provided there exists a function $P_t(x, A)$ defined for $t > 0$, $x \in E$, and $A \in \mathscr{E}$ such that $P_{t,s}(x, A) = P_{s-t}(x, A)$ for all t, s, x, and A.

In this case $P_t(x, A)$ is called a *temporally homogeneous Markov transition function* over $(E, \mathscr{E})$ and the Chapman–Kolmogorov equation becomes

$$P_{t+s}(x, A) = \int P_t(x, dy) \, P_s(y, A) \tag{2.4}$$

for all t, s, x, and A. In actual fact the transition functions that we will deal with in this book are the temporally homogeneous ones and we will very shortly restrict ourselves to consideration of these only.

(2.5) DEFINITION. Let X be a stochastic process with values in $(E, \mathscr{E})$ that is adapted to $\{\mathscr{F}_t\}$ and let $P_{t,s}(x, A)$ be a Markov transition function on $(E, \mathscr{E})$. One says that X is a Markov process with respect to $\{\mathscr{F}_t\}$ having $P_{t,s}(x, A)$ as transition function provided

$$E(f \circ X_s \mid \mathscr{F}_t) = P_{t,s}(X_t, f), \tag{2.6}$$

for all $0 \leq t < s$ and f in $b\mathscr{E}$. ($P_{t,s}(x, f) = \int P_{t,s}(x, dy) f(y)$.)

Taking conditional expectations with respect to $\sigma(X_t)$ in (2.6) one obtains

$$E(f \circ X_s \mid X_t) = P_{t,s}(X_t, f) \tag{2.7}$$

and upon combining (2.6) and (2.7) with Theorem 1.3 we see that the above definition is consistent with Definition 1.1. We remark in passing that there exist processes satisfying (2.7) but not (2.6). Moreover there exist Markov processes (in the sense of Definition 1.1) which do not possess transition functions. The fact that a Markov process has a transition function means that it is possible to define "nice" conditional probability distributions for the conditional probabilities $P(X_s \in A \mid \mathscr{F}_t)$, $0 \leq t < s$, $A \in \mathscr{E}$. As usual if we say that X is a Markov process with transition function $P_{t,s}(x, A)$ without specifying the σ-algebras $\mathscr{F}_t$, it is understood that $\mathscr{F}_t = \sigma(X_s;\ s \leq t)$.

The intuitive meaning of the transition function should be clear: $P_{t,s}(x, A)$ is "the conditional probability that $X_s \in A$ given that $X_t = x$ when $0 \leq t < s$." Although this statement is very attractive, one must beware of such statements since conditional probabilities are determined only almost surely. Perhaps we should remark that without further restrictions a Markov process may have more than one transition function. But of course (2.6) implies that if P^1 and P^2 are transition functions for X then for each fixed t, s, and A, $P^1_{t,s}(x, A) = P^2_{t,s}(x, A)$ for almost all x relative to the distribution in E of X_t.

A Markov process X with respect to $\{\mathscr{F}_t\}$ is called *temporally homogeneous* provided it possesses a temporally homogeneous transition function. In particular we will apply the adjective temporally homogeneous only to those Markov processes which have transition functions.

Let X be a Markov process with values in $(E, \mathscr{E})$ and transition function $P_{t,s}(x, A)$. Let μ be the distribution of X_0; that is, μ is the probability measure on $\mathscr{E}$ defined by $\mu(A) = P(X_0 \in A)$. The measure μ is called the *initial distribution* of X. Using the Markov property repeatedly one easily obtains the following formula for the *finite-dimensional distributions* of X: if $0 \leq t_1 < t_2 < \ldots < t_n$ and $f \in b\mathscr{E}^n$ where $\mathscr{E}^n = \mathscr{E} \times \ldots \times \mathscr{E}$ (n factors), then

(2.8) $$E[f(X_{t_1}, \ldots, X_{t_n})]$$
$$= \int \mu(dx_0) \int P_{0,t_1}(x_0, dx_1) \int \ldots \int P_{t_{n-1},t_n}(x_{n-1}, dx_n)\, f(x_1, \ldots, x_n)$$

where one integrates first on x_n, then on $x_{n-1}, \ldots,$ and finally on x_0, and if $t_1 = 0, P_{0,0}(x, \cdot)$ is taken to be unit mass at x. Thus all of the finite-dimensional distributions of X are expressible in terms of its initial measure and transition function.

It is now natural to ask the following question: Suppose we are given a probability measure μ and a transition function $P_{t,s}(x, A)$ on a measurable space $(E, \mathscr{E})$. Then does there exist a Markov process X with values in $(E, \mathscr{E})$ which has μ for its initial measure and $P_{t,s}(x, A)$ as transition function? If $\mathbf{T} = \mathbf{Z}_+$, then there is always such a process, but if $\mathbf{T} = \mathbf{R}_+$ the most general conditions on $(E, \mathscr{E})$ for which the answer is affirmative are unknown. However, if E is a σ-compact Hausdorff space and $\mathscr{E}$ the σ-algebra of topological Borel sets, then the celebrated theorem of Kolmogorov guarantees the existence of a Markov process with the desired properties. The proof will be outlined in the following paragraphs.

Let $(E, \mathscr{E})$ be a measurable space and let $\mathbf{T}$ be an arbitrary index set. Define $\Omega = E^{\mathbf{T}}$ and $\mathscr{F} = \mathscr{E}^{\mathbf{T}}$ so that $(\Omega, \mathscr{F})$ is the usual product measurable space. For each $t \in \mathbf{T}$, let $X_t : \Omega \to E$ be the coordinate map $X_t(\omega) = \omega(t)$. In particular $\mathscr{F} = \sigma(X_t : t \in \mathbf{T})$. If J is a finite subset of $\mathbf{T}$ we will write $(E^J, \mathscr{E}^J)$ for the appropriate product space, and we will denote the natural projection of Ω on E^J by π_J; that is, if $J = (t_1, \ldots, t_n)$

$$\pi_J \omega = (\omega(t_1), \ldots, \omega(t_n)) \in E^J.$$

If $J = \{t\}$ then $\pi_J = X_t$. Finally if I and J are finite subsets of $\mathbf{T}$ with $I \subset J$ we will denote the natural projection of E^J on E^I by π_I^J. Note that $\pi_I^J \in \mathscr{E}^J/\mathscr{E}^I$ and that if $K \subset J \subset I$ then $\pi_K^I = \pi_K^J \pi_J^I$.

(2.9) Definition. Let $\varphi(\mathbf{T})$ be the class of all finite subsets of $\mathbf{T}$ and suppose that for each $J \in \varphi(\mathbf{T})$ we are given a probability measure P_J on $(E^J, \mathscr{E}^J)$. Then the system $\{P_J : J \in \varphi(\mathbf{T})\}$ is called a *projective system* over $(E, \mathscr{E})$ provided

(2.10) $$P_J(\pi_I^J)^{-1} = P_I \quad \text{for} \quad I \subset J \in \varphi(\mathbf{T}).$$

Let $\mathscr{E}_0^{\mathbf{T}}$ be the *algebra* of all *finite-dimensional* cylinder sets in Ω, that is, all subsets Λ of Ω for which $J \in \varphi(\mathbf{T})$ and $B \in \mathscr{E}^J$ exist such that $\Lambda = \pi_J^{-1}B$. It is immediate that $\sigma(\mathscr{E}_0^{\mathbf{T}}) = \mathscr{E}^{\mathbf{T}}$. Moreover using condition (2.10) it is easy to construct a *finitely additive* probability measure P on $\mathscr{E}_0^{\mathbf{T}}$ such that $P\pi_J^{-1} = P_J$ for all $J \in \varphi(\mathbf{T})$. If P is countably additive on $\mathscr{E}_0^{\mathbf{T}}$, then it can be extended uniquely to a probability measure on $\mathscr{E}^{\mathbf{T}} = \mathscr{F}$, which we again denote by P, so that $(\Omega, \mathscr{F}, P)$ is a probability space. Finally $X = \{X_t; t \in \mathbf{T}\}$ is a stochastic process over $(\Omega, \mathscr{F}, P)$ with values in $(E, \mathscr{E})$ and the finite-dimensional distributions of X are the P_J's we started with; that is, if $J = (t_1, \ldots, t_n)$ and $B \in \mathscr{E}^J$ then

$$P[(X_{t_1}, \ldots, X_{t_n}) \in B] = P_J(B).$$

The following theorem of Kolmogorov gives sufficient conditions under which P is countably additive on $\mathscr{E}_0^{\mathbf{T}}$. We omit the proof, which may be found in any standard text on probability theory.

(2.11) THEOREM (*Kolmogorov*). Let E be a σ-compact Hausdorff space and $\mathscr{E}$ be the topological Borel sets of E. If $\{P_J : J \in \varphi(\mathbf{T})\}$ is a projective system over $(E, \mathscr{E})$, then the finitely additive measure P on $\mathscr{E}_0^{\mathbf{T}}$ constructed above is actually countably additive and hence can be extended to $\mathscr{E}^{\mathbf{T}}$.

Let us now apply this result to the Markov process situation. Let $\mathbf{T} = \mathbf{R}_+$ and suppose that a probability measure μ and a transition function $P_{t,s}(x, A)$ are given on a measurable space $(E, \mathscr{E})$. If $J = \{t_1 < t_2 < \ldots < t_n\}$ is a finite subset of $\mathbf{T}$, then it is not difficult to see that (2.8) defines a probability measure P_J on $(E^J, \mathscr{E}^J)$, and that $\{P_J\colon J \in \varphi(\mathbf{T})\}$ is a projective system over $(E, \mathscr{E})$. Thus if $(E, \mathscr{E})$ satisfies the conditions of Kolmogorov's theorem we obtain a probability measure P on $(\Omega, \mathscr{F})$ where $\Omega = E^{\mathbf{T}}$ and $\mathscr{F} = \mathscr{E}^{\mathbf{T}}$. The reader should now verify that the coordinate mappings $\{X_t\}$ form a Markov process over $(\Omega, \mathscr{F}, P)$ with values in $(E, \mathscr{E})$ which has μ as initial measure and $P_{t,s}(x, A)$ as transition function.

We will illustrate this result with an important class of examples. Let $E = \mathbf{R}^n$ be Euclidean n-space and $\mathscr{E}$ be the Borel sets of E. Let $\{\mu_t; t > 0\}$ be a semigroup (under convolution) of probability measures on $(E, \mathscr{E})$ such that $\mu_t \to \varepsilon_0$ weakly as $t \to 0$ where ε_0 is unit mass at the origin. If $f \in b\mathscr{E}$ and $t > 0$ define

$$P_t(x, f) = \int f(x + y)\, \mu_t(dy). \tag{2.12}$$

It is then immediate that $P_t(x, A)$ is a temporally homogeneous transition function on $(E, \mathscr{E})$. Consequently if μ is any probability measure on $(E, \mathscr{E})$ it follows from the above discussion that there exists a (temporally homogeneous) Markov process X with values in $(E, \mathscr{E})$ which has μ as initial measure and $P_t(x, A)$ as transition function. Note that $P_t(x, A)$ is *translation invariant*; that is, $P_t(x + y, A + y) = P_t(x, A)$ for all $y \in E$. The reader should check for himself that the translation invariance of the transition function implies that the process X has *independent increments*; that is, if $t_0 < t_1 < \ldots < t_n$, then the $(E, \mathscr{E})$ random variables X_{t_0}, $X_{t_1} - X_{t_0}, \ldots,$ $X_{t_n} - X_{t_{n-1}}$ are independent.

It is well known (for example, Lévy [1] or Bochner [1]) that such a semigroup $\{\mu_t\}$ may be characterized as follows: Let $\varphi_t(x) = \int e^{i(x,y)}\, \mu_t(dy)$ be the Fourier transform of μ_t. Then $\varphi_t(x) = \exp[-t\,\psi(x)]$ where

$$\text{(2.13)}\qquad \psi(x) = i(a, x) + \tfrac{1}{2}(Sx, x) + \int \left[1 - e^{i(x,y)} + \frac{i(x, y)}{1 + |y|^2}\right] \nu(dy)$$

with $a \in \mathbf{R}^n$, S a nonnegative definite symmetric operator on $\mathbf{R}^n$, (Sx, x) the usual quadratic form associated with S, and ν a measure on $\mathbf{R}^n$ satisfying $\int |x|^2(1 + |x|^2)^{-1}\, \nu(dx) < \infty$. Here (x, y) is the usual inner product in $\mathbf{R}^n$ and $|x| = (x, x)^{1/2}$ is the distance between x and 0. Conversely if ψ is defined by (2.13) with a, S, and ν as described, then there exists a semigroup $\{\mu_t\}$ of probability measures on $\mathbf{R}^n$ such that

$$\int e^{i(x,y)}\, \mu_t(dy) = \exp[-t\,\psi(x)]$$

for all $t > 0$ and $\mu_t \to \varepsilon_0$ weakly as $t \to 0$. Some important special cases of such transition functions will be described in the exercises at the end of this section. Other important examples of Markov processes will be introduced later.

Exercises

(2.14) Let X be a Markov process over $(\Omega, \mathscr{F}, P)$ with respect to $\{\mathscr{F}_t\}$. Suppose $\mathscr{M}$ is a σ-algebra contained in $\mathscr{F}$ and that for every t, $\mathscr{M}$ and $\mathscr{F}_t$ are independent, $(P(A \cap B) = P(A)\,P(B)$ for all $A \in \mathscr{M}$ and $B \in \mathscr{F}_t)$. Prove that X is a Markov process relative to the family $\{\sigma(\mathscr{M} \cup \mathscr{F}_t)\}$.

(2.15) Let X be a Markov process with values in $(E, \mathscr{E})$ and having a transition function $P_{t,s}(x, A)$, which for each A is jointly measurable in (t, s, x). Prove that the process Y given by $Y_t(\omega) = (X_t(\omega), t)$ with values in $(E \times \mathbf{R}_+, \mathscr{E} \times \mathscr{B}(\mathbf{R}_+))$ is a temporally homogeneous Markov process.

(2.16) Let $\{\eta_t;\, t > 0\}$ be a convolution semigroup of probability measures on $(0, \infty)$. Let $P_t(x, A)$ be a transition function on $(E, \mathscr{E})$ such that for all A the mapping $(t, x) \to P_t(x, A)$ is jointly measurable. Prove that $Q_t(x, A) = \int_0^\infty P_u(x, A)\, \eta_t(du)$ defines a transition function on $(E, \mathscr{E})$. Prove that if $P_t(x, A)$ is of the form (2.12) then $Q_t(x, A)$ is also of this form.

(2.17) Let $E = \mathbf{R}^n$, $g_t(x) = (4\pi t)^{-n/2} \exp(-|x|^2/4t)$ for $t > 0$, $x \in \mathbf{R}^n$. Verify that $P_t(x, A) = \int_A g_t(y - x)\, dy$ is a transition function (the *Brownian motion* transition function in $\mathbf{R}^n$) of the form (2.12) and that in the representation (2.13), $a = 0$, $\nu = 0$, and $\frac{1}{2}(Sx, x) = |x|^2$. The dy denotes the element of Lebesgue measure in $\mathbf{R}^n$. The function $g_t(x)$ is called the *Gauss kernel* for $\mathbf{R}^n$.

(2.18) Given positive numbers d and λ let X be a (Poisson) random variable with the distribution $P(X = nd) = \exp(-\lambda)[\lambda^n/n!]$, $n = 0, 1, \ldots$. Prove that the Laplace transform $E(\exp(-uX)) = \exp[-\lambda(1 - \exp(-du))]$. Prove that if v is a measure on $\mathscr{B}(0, \infty)$ such that $\int_0^\infty [x/(1+x)]\, v(dx) < \infty$, then $\theta(u) = \int_0^\infty \{1 - \exp(-xu)\}\, v(dx)$, $u \geq 0$, defines a function θ such that $\exp(-\theta(u))$ is the Laplace transform of a probability measure η on $(0, \infty)$.
[Hint: approximate the integral by a sum and use the result of the first part together with the convolution and continuity theorems for Laplace transforms.]

(2.19) Prove that the measure η in (2.18) is η_1 in a convolution semigroup $\{\eta_t\}$ of probability measures on $(0, \infty)$, and that the measure v is the same one appearing in the representation (2.13). Prove that for $\beta \in (0, 1)$ the function $\theta(u) = u^\beta$ is a special case of (2.18). (If we denote the corresponding semigroup of measures by $\{\eta_t^\beta\}$ then the transition function defined by (2.12) with η_t^β in place of μ_t is called the *one-sided stable* transition function *of index* β.) Prove that the measures η_t^β have continuous density functions.

(2.20) Let $E = \mathbf{R}^n$. Given $\alpha \in (0, 2)$ define a transition function P^α by $P_t^\alpha(x, A) = \int P_u(x, A)\eta_t^{\alpha/2}(du)$, as in (2.16), where $P_u(x, A)$ is the transition function in (2.17) and the semigroup $\eta_t^{\alpha/2}$ is the one in (2.19). (P^α is called the *symmetric stable* transition function *of index* α *in* $\mathbf{R}^n$.) By (2.16) P^α is of the form (2.12). Verify that in this example $\psi(x) = |x|^\alpha$ and that in the representation (2.13) of $\psi(x)$ the first two terms are absent while $v(dx) = c\,|x|^{-\alpha-n}\,dx$ with

$$c = \alpha 2^{\alpha-1}\Gamma\left(\frac{\alpha+n}{2}\right) \Big/ \pi^{n/2}\Gamma\left(1 - \frac{\alpha}{2}\right).$$

(2.21) In the first part of (2.18) take $d = 1$. Prove that the distribution of X there is μ_1 in a convolution semigroup on $\mathbf{R}$. (The transition function (2.12) in this case is called the *Poisson* transition function *with parameter* λ.) Find the quantities a, S, and v in the representation (2.13).

(2.22) Let $(\Omega, \mathscr{F}, P)$ be a probability space. Let λ be a positive number and let $\{U_n\}$ be a sequence of independent identically distributed random variables on Ω with $P(U_n > x) = \exp(-\lambda x)$, $x > 0$. Define a stochastic process $\{X_t;\ t \geq 0\}$ by

$$\begin{aligned} X_t(\omega) &= 0 \qquad \text{if} \quad t < U_1(\omega), \\ &= n \qquad \text{if} \quad U_1(\omega) + \ldots + U_n(\omega) \leq t < U_1(\omega) + \ldots + U_{n+1}(\omega). \end{aligned}$$

Prove directly that X is a Markov process and that the transition function arising in (2.21) is a transition function for this process.

(2.23) Let $P_{t,s}(x, A)$ be a transition function on a measurable space $(E, \mathscr{E})$. Let $\{X_t; t \in \mathbf{R}_+\}$ be a stochastic process with values in $(E, \mathscr{E})$ and suppose that for all $f \in b\mathscr{E}^n$ and $0 \leq t_1 < \ldots < t_n$ Eq. (2.8) holds. Prove that $\{X_t\}$ is a Markov process with transition function $P_{t,s}(x, A)$ and initial distribution μ.

3. General Definitions Continued

In this section we will introduce the type of Markov process that we will deal with in this book. Roughly speaking it will be a temporarily homogeneous Markov process defined on a "suitable" Ω. In this section the parameter set $\mathbf{T}$ will be $[0, \infty]$ and as in Section 2 $\mathbf{R}_+$ will denote the interval $[0, \infty)$. As will become clear, our point of view in this section is somewhat different from that in the previous sections.

Consider the following objects:

(i) A measurable space $(E, \mathscr{E})$ and point Δ not in E. We write $E_\Delta = E \cup \{\Delta\}$ and let $\mathscr{E}_\Delta$ be the σ-algebra in E_Δ generated by $\mathscr{E}$. Note that $\{\Delta\} \in \mathscr{E}_\Delta$.

(ii) A measurable space $(\Omega, \mathscr{M})$ and an increasing family $\{\mathscr{M}_t \colon t \in \mathbf{T}\}$ of sub-σ-algebras of $\mathscr{M}$. Also a distinguished point ω_Δ of Ω.

(iii) For each $t \in \mathbf{T}$ a map $X_t \colon \Omega \to E_\Delta$ such that if $X_t(\omega) = \Delta$ then $X_s(\omega) = \Delta$ for all $s \geq t$, $X_\infty(\omega) = \Delta$ for all $\omega \in \Omega$, and $X_0(\omega_\Delta) = \Delta$. We will sometimes write $X(t, \omega)$ for $X_t(\omega)$ and $X(t)$ for X_t.

(iv) For each $t \in \mathbf{T}$ a map $\theta_t \colon \Omega \to \Omega$ such that $\theta_\infty \, \omega = \omega_\Delta$ for all ω.

(v) For each x in E_Δ a probability measure P^x on $(\Omega, \mathscr{M})$.

(3.1) Definition. The collection $X = (\Omega, \mathscr{M}, \mathscr{M}_t, X_t, \theta_t, P^x)$ is called a (temporally homogeneous) *Markov process* (*with translation operators*) and with state space $(E, \mathscr{E})$ (augmented by Δ) provided the following axioms hold:

Axiom R (Regularity Conditions)

(a) For each $t \in \mathbf{R}_+$, $X_t \in \mathscr{M}_t/\mathscr{E}_\Delta$.

(b) The map $x \to P^x(X_t \in B)$ from E to $[0, 1]$ is in $\mathscr{E}$ for each $t \in \mathbf{R}_+$ and $B \in \mathscr{E}$.

(c) $P^\Delta(X_0 = \Delta) = 1$.

Axiom H (Homogeneity)

For all $t, h \in \mathbf{T}$, $X_t \circ \theta_h = X_{t+h}$. Note that this is consistent if either t or h is ∞.

Axiom M (Markov Property)

$$P^x(X_{t+s} \in B \mid \mathcal{M}_t) = P^{X(t)}(X_s \in B), \tag{3.2}$$

for all $x \in E_\Delta$, $B \in \mathscr{E}_\Delta$, and $t, s \in \mathbf{T}$.

Comments on the axioms. One may check immediately that if Axiom R holds then (b) remains valid when we replace E, $\mathscr{E}$, and $\mathbf{R}_+$ by E_Δ, $\mathscr{E}_\Delta$, and $\mathbf{T}$. Define $\mathscr{F}_t^0 = \sigma(X_s: s \le t)$ and $\mathscr{F}^0 = \sigma(X_s: s \in \mathbf{T})$. These are σ-algebras in Ω and it follows from Axiom R(a) that $\mathscr{F}_t^0 \subset \mathcal{M}_t$ and hence that $\mathscr{F}^0 \subset \mathcal{M}$. Clearly $\{\mathscr{F}_t^0\}$ is an increasing family. It will sometimes be convenient to write $\mathscr{F}_\infty^0$ for $\mathscr{F}^0$.

It follows from Axiom H that if B is in $\mathscr{E}_\Delta$ then $\theta_h^{-1} X_s^{-1}(B) = X_{s+h}^{-1}(B) \in \mathscr{F}_{s+h}^0$, and consequently

$$\theta_h \in \mathscr{F}_{t+h}^0 / \mathscr{F}_t^0 \tag{3.3}$$

for all $t, h \in \mathbf{T}$. In particular $\theta_h \in \mathscr{F}^0/\mathscr{F}^0$.

If $f(x) = P^x(X_s \in B)$, then $f \in \mathscr{E}_\Delta$ and so $P^{X(t)}(X_s \in B) = f \circ X_t$ is in $\mathscr{F}_t^0$. In addition the reader should check that (3.2) is consistent with Axiom R(c) if $x = \Delta$ and with (iii) and (iv) if either t or s is equal to ∞. Equation (3.2) displays the conditional probability on the left as a measurable function of X_t, and so Axiom M does indeed imply that for each $x \in \mathscr{E}_\Delta$ the family $\{X_t;\ t \in \mathbf{T}\}$ is a Markov process (in the sense of Definition (1.1)) over $(\Omega, \mathcal{M}, P^x)$ with respect to $\{\mathcal{M}_t;\ t \in \mathbf{T}\}$ taking values in $(E_\Delta, \mathscr{E}_\Delta)$. In particular Definition 3.1 requires not one Markov process but a family of them, one for each of the measures P^x. We will shortly show (Proposition 3.5) that each of them has the same (temporally homogeneous) transition function.

Usually we will omit the phrases in parentheses in Definition 3.1; thus the collection $X = (\Omega, \mathcal{M}, \mathcal{M}_t, X_t, \theta_t, P^x)$ will be called simply a Markov process with state space $(E, \mathscr{E})$. If the σ-algebras $\mathcal{M}$ or $\mathcal{M}_t$ are omitted, then it is understood that we are taking $\mathcal{M}$ to be $\mathscr{F}^0$ or $\mathcal{M}_t$ to be $\mathscr{F}_t^0$.

Here is an intuitive way of thinking about Definition 3.1. If we regard $t \to X_t(\omega)$ as the path (or trajectory) of a particle moving in the space E, then P^x should be thought of as the probability law of the particle assuming it starts from x at time $t = 0$. The particle moves in E until it "dies" at which time it is transported to Δ where it remains forever. (The point Δ may be thought of as a "cemetery" or "heaven" depending on one's point of view.) Roughly speaking Axiom M states that if we know the history of the particle up to time t, then probabilistically its future behavior is exactly the same as that of a particle starting at $X_t(\omega)$. Finally the existence of the translation operators, θ_t, says that the underlying space Ω is "sufficiently rich."

With this interpretation in mind we define

$$\zeta(\omega) = \inf\{t : X_t(\omega) = \Delta\}$$

provided the set in braces is not empty and $\zeta(\omega) = \infty$ if it is empty. Since ($\mathbf{Q}$ denotes the rationals)

(3.4) $$\{\zeta < t\} = \bigcup_{r<t,\, r\in\mathbf{Q}} \{X_r = \Delta\} \in \mathscr{F}^0_{t},$$

ζ is a numerical random variable. It is called the *lifetime* of the process X.

(3.5) PROPOSITION. Define $N_t(x, A) = P^x(X_t \in A)$ for $t \in \mathbf{T}$, $x \in E_\Delta$, and $A \in \mathscr{E}_\Delta$. Then $N_t(x, A)$, $0 < t < \infty$, is a (temporally homogeneous) transition function on $(E_\Delta, \mathscr{E}_\Delta)$. Moreover for each $x \in E_\Delta$, N is a transition function for the Markov process $\{X_t; t \in \mathbf{T}\}$ over $(\Omega, \mathscr{M}, P^x)$ with values in $(E_\Delta, \mathscr{E}_\Delta)$.

Proof. The fact that P^x is a probability measure on $\mathscr{M}$ implies that $N_t(x, A)$ is a probability measure in A. Axiom R(b) and the comments following the axioms imply that $N_t(x, A)$ is in $\mathscr{E}_\Delta$ as a function of x. As to the Chapman–Kolmogorov equation we have

$$\begin{aligned} N_{t+s}(x, A) &= P^x(X_{t+s} \in A) \\ &= E^x\{P^{X_t}(X_s \in A)\} \\ &= E^x\{N_s(X_t, A)\} \\ &= \int N_s(y, A)\, N_t(x, dy). \end{aligned}$$

Finally Eq. (3.2) with the right side replaced by $N_s(X_t, B)$ is just the statement that N is the required transition function.

It is immediate from Axiom R(c) that $N_t(\Delta, \{\Delta\}) = 1$ for all t and that $N_t(x, A)$ is completely determined by its restriction to $(E, \mathscr{E})$. We will denote the restriction of $N_t(x, A)$ to $(E, \mathscr{E})$ by $P_t(x, A)$ and call $P_t(x, A)$ the (sub-Markov) *transition function of* X. It satisfies all the conditions of a temporally homogeneous Markov transition function on $(E, \mathscr{E})$ except that $A \to P_t(x, A)$ need not be a probability measure on $\mathscr{E}$; one can assert only that $P_t(x, E) \leq 1$. From now on the term transition function with no qualifying adjectives will mean such an object. If $f \in b\mathscr{E}$ we write

$$P_t f(x) = \int P_t(x, dy)\, f(y) = E^x\{f(X_t); X_t \in E\}.$$

Clearly P_t is a positive linear operator from $b\mathscr{E}$ to $b\mathscr{E}$.

The following result is analagous to Theorem 1.3 and may be proved in a similar manner.

(3.6) THEOREM.

(a) $x \to E^x(Y)$ is $\mathscr{E}_\Delta$ measurable for all $Y \in b\mathscr{F}^0$.

(b) Under Axioms R and H, Axiom M is equivalent to each of the following:

$$(\mathrm{M}_1) \qquad E^x\{f \circ X_{t+s} \mid \mathcal{M}_t\} = E^{X(t)} f \circ X_s$$

for all x, t, s, and $f \in b\mathcal{E}_\Delta$;

$$(\mathrm{M}_2) \qquad E^x\{Y \circ \theta_t \mid \mathcal{M}_t\} = E^{X(t)} Y$$

for all x, t, and $Y \in b\mathcal{F}^0$.

(c) If $\mathcal{M}_t = \mathcal{F}_t^0$, then under Axioms R and H, Axiom M is equivalent to:

(M_3) Given $0 \le t_1 < \ldots < t_n$ and $f_1, \ldots, f_n \in b\mathcal{E}_\Delta$ then

$$E^x \prod_{j=1}^n f_j(X_{t_j})$$

$$= \int N_{t_1}(x, dx_1)\, f_1(x_1) \int \ldots \int N_{t_n - t_{n-1}}(x_{n-1}, dx_n)\, f_n(x_n).$$

In the remainder of this book a Markov process is always understood in the sense of Definition 3.1 unless explicitly stated otherwise.

Exercises

(3.7) Let $\Omega = \mathbf{R} \cup \{\Delta\}$ where Δ is the usual point at ∞. Let $\mathcal{M} = \mathcal{M}_t = \mathcal{B}(\Omega)$, the usual σ-algebra of Borel sets on $\mathbf{R} \cup \{\Delta\}$. Let P^x denote unit mass at x and for $t < \infty$ and $\omega \in \mathbf{R}$ define $X_t(\omega) = \omega + t$, $\theta_t \omega = \omega + t$. Complete the definitions of X_t and θ_t as required for (3.1) and prove that the resulting process is a Markov process with state space $(\mathbf{R}, \mathcal{B}(\mathbf{R}))$ (called *uniform motion to the right*). Show that the resulting transition function, when restricted to $\mathbf{R}$, is of the form (2.12).

(3.8) Let $X = (\Omega, \mathcal{M}, \mathcal{M}_t, X_t, \theta_t, P^x)$ be a Markov process, with state space $(E, \mathcal{E})$ and $\tilde{\Omega} = \Omega \times [0, \infty)$ (with points $\tilde{\omega} = (\omega, r)$, $\omega \in \Omega$, $r \ge 0$). Let $\tilde{\mathcal{M}} = \mathcal{M} \times \mathcal{B}[0, \infty)$ and $\tilde{\mathcal{M}}_t$ consist of all sets $\tilde{\Lambda} \in \tilde{\mathcal{M}}$ such that $\tilde{\Lambda} \cap \{\Omega \times (t, \infty)\}$ is of the form $\Lambda \times (t, \infty)$ for a suitable $\Lambda \in \mathcal{M}_t$. Let $\tilde{P}^x = P^x \times \lambda\, e^{-\lambda r}\, dr$ (λ a positive constant), and finally if $\tilde{\omega} = (\omega, r)$

$$\tilde{X}_t(\tilde{\omega}) = X_t(\omega), \qquad t < r,$$
$$= \Delta, \qquad t \ge r.$$

Define translation operators $\tilde{\theta}_h$ and a point $\tilde{\omega}_\Delta$ as required for Definition 3.1 and prove that the resulting $\tilde{X}$ is a Markov process with state space $(E, \mathcal{E})$.

Prove that its transition function is $e^{-\lambda t} P_t(x, A)$ if $P_t(x, A)$ is the transition function for X defined following the proof of (3.5). (A more general construction along these lines will be given in Chapter III.)

4. Equivalent Processes

Again in this section all process have $\mathbf{T} = [0, \infty]$ as parameter set.

(4.1) DEFINITION. Two Markov processes with the same state space $(E, \mathscr{E})$ are *equivalent* if they have the same transition function.

In this section we will single out from a given equivalence class of Markov processes a particularly "nice" representative defined over a function space where the meaning of the translation operators θ_t will be transparent. Let $(E, \mathscr{E})$ be a measurable space which we extend to $(E_\Delta, \mathscr{E}_\Delta)$ as in Section 3. Consider the following objects:

(i) W: the space of all maps $w: \mathbf{T} \to E_\Delta$ such that $w(\infty) = \Delta$ and if $w(t) = \Delta$ then $w(s) = \Delta$ for all $s \geq t$. Let w_Δ be the constant map $w_\Delta(t) = \Delta$ for all $t \geq 0$.

(ii) Let Y_t, $t \in \mathbf{T}$, be the coordinate maps $Y_t(w) = w(t)$, and define in W the σ-algebras $\mathscr{G}^0 = \sigma(Y_t: t \in \mathbf{T})$, $\mathscr{G}_t^0 = \sigma(Y_s: s \leq t)$.

(iii) Let $\varphi_t: W \to W$ be defined by $\varphi_t w(s) = w(t+s)$. Note that $\varphi_\infty w = w_\Delta$ and that $Y_t \circ \varphi_s = Y_{t+s}$ for all $t, s \in \mathbf{T}$.

(4.2) DEFINITION. A Markov process $X = (\Omega, \mathscr{M}, \mathscr{M}_t, X_t, \theta_t, P^x)$ with state space $(E, \mathscr{E})$ is said to be of *function space type* provided $\Omega = W$, $\mathscr{M} \supset \mathscr{G}^0$, $\mathscr{M}_t \supset \mathscr{G}_t^0$, $X_t = Y_t$, and $\theta_t = \varphi_t$.

(4.3) THEOREM. Any Markov process $X = (\Omega, \mathscr{M}, \mathscr{M}_t, X_t, \theta_t, P^x)$ with state space $(E, \mathscr{E})$ is equivalent to a Markov process Y of function space type with $(E, \mathscr{E})$ as state space.

Proof. Using the notation developed above (4.2) we define a map $\pi: \Omega \to W$ as follows

$$(\pi\omega)(t) = X_t(\omega).$$

Hence $Y_t \circ \pi = X_t$, and consequently

$$\pi^{-1}(Y_t \in A) = \pi^{-1} Y_t^{-1}(A) = X_t^{-1}(A)$$

if $A \in \mathscr{E}_\Delta$. Therefore $\pi^{-1}\mathscr{G}_t^0 \subset \mathscr{F}_t^0 \subset \mathscr{M}_t$; that is, $\pi \in \mathscr{F}_t^0/\mathscr{G}_t^0$ for all $t \in \mathbf{T}$ where as usual $\mathscr{F}_\infty^0 = \mathscr{F}^0$ and $\mathscr{G}_\infty^0 = \mathscr{G}^0$. We now define measures $\hat{P}^x$ on $(W, \mathscr{G}^0)$ by $\hat{P}^x = P^x \pi^{-1}$ and we claim that $Y = (W, \mathscr{G}^0, \mathscr{G}_t^0, Y_t, \varphi_t, \hat{P}^x)$ is a

Markov process equivalent to X. Once this is established Theorem 4.3 will be proved since Y is clearly of function space type. Now

$$\hat{P}^x(Y_t \in A) = P^x \pi^{-1} Y_t^{-1}(A) = P^x X_t^{-1}(A) = P^x(X_t \in A)$$

and so it only remains to show that Y is a Markov process. It is clear that Y satisfies Axioms R and H, and so only Axiom M needs to be checked. First note that

$$(\pi\theta_t \omega)(s) = X_s(\theta_t \omega) = X_{t+s}(\omega) = (\pi\omega)(t+s) = (\varphi_t \pi\omega)(s);$$

that is, $\pi\theta_t = \varphi_t \pi$. We next claim that if $H : W \to \mathbf{R}$ is in $b\mathscr{G}^0$, then $H \circ \pi \in b\mathscr{F}^0$ and $\hat{E}^x(H) = E^x(H \circ \pi)$. The first statement is obvious, while if $H = I_\Lambda$ for $\Lambda \in \mathscr{G}^0$, then

$$\hat{E}^x(H) = \hat{P}^x(\Lambda) = P^x \pi^{-1}\Lambda = E^x(I_{\pi^{-1}\Lambda}) = E^x(I_\Lambda \circ \pi),$$

and so the result follows for general $H \in b\mathscr{G}^0$. To check Axiom M for the process Y we must show that

$$\textbf{(4.4)} \qquad \hat{P}^x[\{Y_{t+s} \in A\} \cap \Lambda] = \hat{E}^x\{\hat{P}^{Y(t)}(Y_s \in A); \Lambda\}$$

for all $\Lambda \in \mathscr{G}_t^0$ and t, s, x, and $A \in \mathscr{E}_\Delta$. Using the facts that $\pi^{-1}\Lambda \in \mathscr{F}_t^0$ and that X is a Markov process we see that the left side of (4.4) equals

$$\begin{aligned} P^x[\pi^{-1} Y_{t+s}^{-1}(A) \cap \pi^{-1}\Lambda] &= P^x[X_{t+s}^{-1}(A) \cap \pi^{-1}\Lambda] \\ &= E^x\{P^{X(t)}(X_s \in A); \pi^{-1}\Lambda\}, \end{aligned}$$

while according to the above remark the right side of (4.4) becomes

$$\begin{aligned} E^x\{\hat{P}^{Y_t \circ \pi}(Y_s \in A); \pi^{-1}\Lambda\} &= E^x\{\hat{P}^{X(t)}(Y_s \in A); \pi^{-1}\Lambda\} \\ &= E^x\{P^{X(t)}(X_s \in A); \pi^{-1}\Lambda\}. \end{aligned}$$

Thus (4.4) is established and so the proof of Theorem 4.3 is completed.

5. The Measures P^μ

In this section we will extend the basic σ-algebras $\mathscr{F}^0$ and $\mathscr{F}_t^0$ by completing them in an appropriate manner. We assume throughout this section that $X = (\Omega, \mathscr{M}, \mathscr{M}_t, X_t, \theta_t, P^x)$ is a given Markov process with state space $(E, \mathscr{E})$ and that $\mathbf{T} = [0, \infty]$.

We have already remarked that $x \to P^x(\Lambda)$ is $\mathscr{E}_\Delta$ measurable whenever Λ is in $\mathscr{F}^0$ (Theorem 3.6). Therefore, given a finite measure μ on $\mathscr{E}_\Delta$, we can define for any Λ in $\mathscr{F}^0$

$$\textbf{(5.1)} \qquad P^\mu(\Lambda) = \int P^x(\Lambda)\, \mu(dx).$$

It is immediate that P^μ is a finite measure on $\mathcal{F}^0$ (P^μ is a probability on $\mathcal{F}^0$ if and only if μ is a probability on $\mathcal{E}_\Delta$). Moreover if ε_x denotes unit mass at x then $P^{\varepsilon_x} = P^x$ on $\mathcal{F}^0$. In general the measures P^μ can *not* be defined on $\mathcal{M}$ unless we make some measurability assumption on the functions $x \to P^x(\Lambda)$ for Λ in $\mathcal{M}$.

We now pause from our main development in order to introduce some terminology that is essential in the sequel. Let $(E, \mathcal{E})$ be a measurable space. If μ is a measure on $\mathcal{E}$ and $\mathcal{F}$ is a sub-σ-algebra of $\mathcal{E}$, then $\mathcal{F}^\mu$, as in Chapter 0, denotes the completion of $\mathcal{F}$ with respect to μ. Of course $\mathcal{F}^\mu \subset \mathcal{E}^\mu$ and the measure μ extends uniquely to $\mathcal{E}^\mu$. This extension will again be denoted by μ. Suppose now that $\mathbf{U}$ is a family of finite measures on $\mathcal{E}$; then the *completion*, $\mathcal{F}^\mathbf{U}$, of $\mathcal{F}$ *with respect to* $\mathbf{U}$ is defined by

$$(5.2) \qquad \mathcal{F}^\mathbf{U} = \bigcap_{\mu \in \mathbf{U}} \mathcal{F}^\mu.$$

The reader should verify that $(\mathcal{F}^\mathbf{U})^\mathbf{U} = \mathcal{F}^\mathbf{U}$, and that $f \in \mathcal{F}^\mathbf{U}$ if and only if for each $\mu \in \mathbf{U}$ there exist f_1 and f_2 in $\mathcal{F}$ such that $f_1 \leq f \leq f_2$ and $\mu(f_1 < f_2) = 0$. If $\mathbf{U}$ is the family of all finite measures on $\mathcal{E}$, then $\mathcal{E}^\mathbf{U} = \mathcal{E}^*$, the σ-algebra of universally measurable sets over $(E, \mathcal{E})$. See Chapter 0.

(5.3) DEFINITION. Let $(E, \mathcal{E})$ and $\mathbf{U}$ be as above. If $\mathcal{G}$ is any σ-algebra contained in $\mathcal{E}^\mathbf{U}$ we define the *completion of* $\mathcal{G}$ *in* $\mathcal{E}^\mathbf{U}$ *with respect to* $\mathbf{U}$, which we denote for the moment by $\tilde{\mathcal{G}}$, as follows: $A \in \tilde{\mathcal{G}}$ if and only if for each $\mu \in \mathbf{U}$ there exists $A_\mu \in \mathcal{G}$ such that $A - A_\mu$ and $A_\mu - A$ are in $\mathcal{E}^\mathbf{U}$ and $\mu(A - A_\mu) = \mu(A_\mu - A) = 0$.

The following characterization of $\tilde{\mathcal{G}}$ is useful in some proofs. The notation is that of (5.3).

(5.4) PROPOSITION. A is in $\tilde{\mathcal{G}}$ if and only if for each $\mu \in \mathbf{U}$ there exist D_μ in $\mathcal{G}$, A_μ and B_μ in $\mathcal{E}$ such that $D_\mu - A_\mu \subset A \subset D_\mu \cup B_\mu$ and $\mu(A_\mu) = \mu(B_\mu) = 0$.

Proof. Let A have the property of Proposition 5.4. Since $D_\mu - A_\mu$ and $D_\mu \cup B_\mu$ are in $\mathcal{E}^\mathbf{U}$ and $\mu[(D_\mu \cup B_\mu) - (D_\mu - A_\mu)] \leq \mu(A_\mu \cup B_\mu) = 0$, it follows that A is in $\mathcal{E}^\mathbf{U}$. Hence $A - D_\mu$ and $D_\mu - A$ are in $\mathcal{E}^\mathbf{U}$. But $\mu(A - D_\mu) \leq \mu(B_\mu) = 0$ and $\mu(D_\mu - A) \leq \mu(A_\mu) = 0$. Consequently A is in $\tilde{\mathcal{G}}$. Conversely if A is in $\tilde{\mathcal{G}}$, then given μ there exists D_μ in $\mathcal{G}$ such that $A - D_\mu$ and $D_\mu - A$ are in $\mathcal{E}^\mathbf{U}$ and are μ null. Hence there exist A_μ and B_μ in $\mathcal{E}$ such that $D_\mu - A \subset A_\mu$ and $A - D_\mu \subset B_\mu$ with $\mu(A_\mu) = \mu(B_\mu) = 0$. But

$$D_\mu - (D_\mu - A) \subset A \subset D_\mu \cup (A - D_\mu)$$

and so $D_\mu - A_\mu \subset A \subset D_\mu \cup B_\mu$. Thus A meets the requirements of Definition 5.3.

The reader should now have no difficulty in checking the following elementary properties:

$$\text{(5.5)} \qquad \begin{aligned} &\text{(i)} \quad \tilde{\mathscr{G}} \text{ is a } \sigma\text{-algebra}, \mathscr{G} \subset \tilde{\mathscr{G}} \subset \mathscr{E}^{\mathbf{U}};\\ &\text{(ii)} \quad \mathscr{G}^{\mathbf{U}} \subset \tilde{\mathscr{G}};\\ &\text{(iii)} \quad (\tilde{\mathscr{G}})^{\sim} = \tilde{\mathscr{G}}. \end{aligned}$$

The following measurability lemma will be of frequent use in the sequel. (This generalizes Exercise 3.3 of Chapter 0.)

(5.6) LEMMA. Let $(E_i, \mathscr{E}_i)$ be measurable spaces and $\mathbf{U}_i$ be families of finite measures on $\mathscr{E}_i$, $i = 1, 2$. Let $\mathscr{G}_i$ be σ-algebras contained in $\mathscr{E}_i^{\mathbf{U}_i}$, $i = 1, 2$. Suppose f is in $\mathscr{G}_1/\mathscr{G}_2$ and in $\mathscr{E}_1/\mathscr{E}_2$. If $\mu f^{-1} \in \mathbf{U}_2$ for all $\mu \in \mathbf{U}_1$, then $f \in \tilde{\mathscr{G}}_1/\tilde{\mathscr{G}}_2$.

Proof. If $A \in \tilde{\mathscr{G}}_2$ we must show that $f^{-1}(A)$ is in $\tilde{\mathscr{G}}_1$. Given $\mu \in \mathbf{U}_1$ then $\nu = \mu f^{-1} \in \mathbf{U}_2$ and so there exist $D_\nu \in \mathscr{G}_2$, A_ν and B_ν in $\mathscr{E}_2$ such that $D_\nu - A_\nu \subset A \subset D_\nu \cup B_\nu$ and $\nu(A_\nu) = \nu(B_\nu) = 0$. Defining $D_\mu = f^{-1}(D_\nu)$, $A_\mu = f^{-1}(A_\nu)$, and $B_\mu = f^{-1}(B_\nu)$ it is clear that $D_\mu - A_\mu \subset f^{-1}(A) \subset D_\mu \cup B_\mu$ and that $\mu(A_\mu) = \mu(B_\mu) = 0$. Consequently $f^{-1}(A) \in \tilde{\mathscr{G}}_1$.

In the special case that $\mathscr{G}_1 = \mathscr{E}_1$ and $\mathscr{G}_2 = \mathscr{E}_2$ the lemma states that if $f \in \mathscr{E}_1/\mathscr{E}_2$ and $\mathbf{U}_1 f^{-1} \subset \mathbf{U}_2$ then $f \in \mathscr{E}_1^{\mathbf{U}_1}/\mathscr{E}_2^{\mathbf{U}_2}$. In particular $\mathscr{E}_1/\mathscr{E}_2 \subset \mathscr{E}_1^*/\mathscr{E}_2^*$. If we assume only $\mathscr{G}_2 = \mathscr{E}_2$, then the above argument actually shows that $f \in \mathscr{G}_1^{\mathbf{U}_1}/\mathscr{E}_2^{\mathbf{U}_2}$.

Let us return now to the consideration of a Markov process $X = (\Omega, \mathscr{M}, \mathscr{M}_t, X_t, \theta_t, P^x)$ with state space $(E, \mathscr{E})$. We define $\bar{\mathscr{M}}$ to be the completion of $\mathscr{M}$ with respect to the family of measures $\{P^x, x \in E_\Delta\}$, and $\mathscr{F}$ to be the completion of $\mathscr{F}^0$ with respect to the family $\{P^\mu;\ \mu$ a finite measure on $\mathscr{E}_\Delta\}$. We next define $\bar{\mathscr{M}}_t$ to be the completion of $\mathscr{M}_t$ in $\bar{\mathscr{M}}$ with respect to $\{P^x; x \in E_\Delta\}$ and $\mathscr{F}_t$ to be the completion of $\mathscr{F}_t^0$ in $\mathscr{F}$ with respect to $\{P^\mu;\ \mu$ a finite measure on $\mathscr{E}_\Delta\}$. Since $\mathscr{M} \supset \mathscr{F}^0$ and $\mathscr{M}_t \supset \mathscr{F}_t^0$ it is immediate that $\bar{\mathscr{M}} \supset \mathscr{F}$ and $\bar{\mathscr{M}}_t \supset \mathscr{F}_t$. These completions turn out to be the appropriate ones for our theory. See, for example, Proposition 8.12.

The next definition is of central importance.

(5.7) DEFINITION. Let $q(\omega)$ be a property of ω; then q is said to hold *almost surely* (a.s.) *on* $\Lambda \in \mathscr{F}$ if the set Λ_0 of ω in Λ for which $q(\omega)$ fails to hold is in $\mathscr{F}$ and $P^x(\Lambda_0) = 0$ for all x. If $\Lambda = \Omega$, we simply say "almost surely."

The following propositions, (5.10)–(5.12), are designed to show that, roughly speaking, one can replace the σ-algebras $\mathscr{M}_t$ and $\mathscr{F}_t^0$ by $\bar{\mathscr{M}}_t$ and $\mathscr{F}_t$ provided one replaces $\mathscr{E}_\Delta$ by $\mathscr{E}_\Delta^*$. (Clearly $\mathscr{E}_\Delta^*$, the universally measurable sets over $(E_\Delta, \mathscr{E}_\Delta)$, is just the σ-algebra in E_Δ generated by $\mathscr{E}^*$.) In particular it will follow from our discussion that $(\Omega, \bar{\mathscr{M}}, \bar{\mathscr{M}}_t, X_t, \theta_t, P^x)$ is a Markov process with state space $(E, \mathscr{E}^*)$.

(5.8) PROPOSITION. If $Y \in b\mathscr{F}$, then $x \to E^x(Y)$ is $\mathscr{E}_\Delta^*$ measurable.

Proof. Given μ there exist Y_1 and Y_2 in $b\mathscr{F}^0$ such that $Y_1 \le Y \le Y_2$ and $E^\mu(Y_2 - Y_1) = 0$. Clearly $E^x(Y_1) \le E^x(Y) \le E^x(Y_2)$ for all x and $x \to E^x(Y_j)$ is $\mathscr{E}_\Delta$ measurable, $j = 1, 2$. But

$$\int [E^x(Y_2) - E^x(Y_1)]\, \mu(dx) = E^\mu(Y_2 - Y_1) = 0,$$

and hence $x \to E^x(Y)$ is in $(\mathscr{E}_\Delta)^\mu$. Since μ is arbitrary this yields Proposition 5.8.

(5.9) REMARK. It follows from the proof of Proposition 5.8 that if $Y \in b\mathscr{F}$ then $E^\mu(Y) = \int E^x(Y)\, \mu(dx)$ for each μ. This relation then holds for any nonnegative Y in $\mathscr{F}$.

(5.10) PROPOSITION. For each t, $X_t \in \mathscr{F}_t/\mathscr{E}_\Delta^*$.

Proof. This is an immediate consequence of Lemma 5.6. Actually $X_t \in \bigcap_\mu (\mathscr{F}_t^0)^{P^\mu}/\mathscr{E}_\Delta^*$. See the remark following (5.6).

Combining (5.9) and (5.10) we see that $x \to E^x f(X_t)$ is $\mathscr{E}_\Delta^*$ measurable for all t and $f \in b\mathscr{E}_\Delta^*$. In particular $x \to N_t(x, A) = P^x(X_t \in A)$ is $\mathscr{E}_\Delta^*$ measurable if $A \in \mathscr{E}_\Delta^*$.

(5.11) PROPOSITION. For all $t, h \in \mathbf{T}$, $\theta_h^{-1}\mathscr{F}_t \subset \mathscr{F}_{t+h}$, and so $\theta_h^{-1}\mathscr{F} \subset \mathscr{F}$.

Proof. Since θ_h is in $\mathscr{F}_t^0/\mathscr{F}_{t+h}^0$ and in $\mathscr{F}^0/\mathscr{F}^0$ this will follow from Lemma 5.6 once we show that for each finite μ on $\mathscr{E}_\Delta$ there exists a finite ν on $\mathscr{E}_\Delta$ such that $P^\mu\theta_h^{-1} = P^\nu$. We define a finite measure ν on $\mathscr{E}_\Delta$ by

$$\nu(A) = P^\mu(X_h \in A) = \int P^x(X_h \in A)\, \mu(dx),$$

and check that for $\Lambda \in \mathscr{F}^0$

$$P^{\mu}(\theta_h^{-1}\Lambda) = E^{\mu}\{P^{X(h)}(\Lambda)\}$$
$$= \int P^{\mu}[X_h \in dy]\, P^{y}(\Lambda)$$
$$= \int \nu(dy)\, P^{y}(\Lambda) = P^{\nu}(\Lambda).$$

(5.12) PROPOSITION. If $Y \in b\mathscr{F}$ then for each x and t

$$E^{x}(Y \circ \theta_t \mid \bar{\mathscr{M}}_t) = E^{X(t)}(Y).$$

Proof. From (5.8) and (5.10) it follows that the real-valued function $\omega \to E^{X_t(\omega)}(Y)$ is in $\mathscr{F}_t$, which is contained in $\bar{\mathscr{M}}_t$, and so we need only check that for each $\Lambda \in \bar{\mathscr{M}}_t$

(5.13) $$E^{x}(Y \circ \theta_t; \Lambda) = E^{x}[E^{X(t)}(Y); \Lambda].$$

Clearly, from the definition of $\bar{\mathscr{M}}_t$ it suffices to prove this for $\Lambda \in \mathscr{M}_t$. Given x and t define a measure μ on $\mathscr{E}_{\Delta}^{*}$ by $\mu(A) = P^{x}(X_t \in A)$. The computation made in the proof of (5.11) shows that for $B \in \mathscr{F}^0$ we have $P^{\mu}(B) = P^{x}(\theta_t^{-1}B)$, and of course for any $Y \in b\mathscr{F}$ we have

(5.14) $$E^{\mu}Y = \int E^{y}(Y)\, \mu(dy) = E^{x}[E^{X(t)}(Y)].$$

If $Y \in b\mathscr{F}$ let $Z \in b\mathscr{F}^0$ be such that $\{Y \neq Z\} \subset \Gamma$ where $\Gamma \in \mathscr{F}^0$ and $P^{\mu}(\Gamma) = 0$. Then $\{Y \circ \theta_t \neq Z \circ \theta_t\} \subset \theta_t^{-1}\Gamma$ and $P^{x}(\theta_t^{-1}\Gamma) = P^{\mu}(\Gamma) = 0$. Also by (5.14)

$$E^{x}(E^{X(t)}\,|Y - Z|) = E^{\mu}\,|Y - Z| = 0$$

and so $E^{X(t)}Y = E^{X(t)}Z$ almost surely relative to P^{x}. Thus we may replace Y by Z on each side of (5.13) without altering its validity. But with this replacement the validity of (5.13) is a consequence of Theorem 3.6(b), so the proof is complete.

It is evident now that $(\Omega, \bar{\mathscr{M}}, \bar{\mathscr{M}}_t, X_t, \theta_t, P^{x})$ is a Markov process with state space $(E, \mathscr{E}^{*})$, and hence, in this case, with state space $(E, \mathscr{E})$. Thus nothing is lost by enlarging the σ-algebras $\mathscr{M}$ and $\mathscr{M}_t$ to $\bar{\mathscr{M}}$ and $\bar{\mathscr{M}}_t$. Consequently we will assume *unless explicitly stated otherwise* that

(5.15) $$\mathscr{M} = \bar{\mathscr{M}}; \qquad \mathscr{M}_t = \bar{\mathscr{M}}_t.$$

In particular then $\mathscr{F} \subset \mathscr{M} = \bar{\mathscr{M}}$ and $\mathscr{F}_t \subset \mathscr{M}_t = \bar{\mathscr{M}}_t$.

So far we have made no assumption to insure that P^{x} is actually the probability measure of the process starting from x, although we have stated that P^{x} should be thought of in this manner. We now make this precise.

(5.16) DEFINITION. The Markov process X is called *normal* provided that $\{x\} \in \mathscr{E}$ and that $P^x[X_0 = x] = 1$ for all x in E. The assumption of normality will always be explicitly stated in the hypotheses of those results for which it is needed.

We close this section with following simple, but very useful, result.

(5.17) PROPOSITION. (Zero-One Law). Let X be normal. If $\Lambda \in \mathscr{F}_0$ then $P^x(\Lambda)$ is either zero or one.

Proof. We first observe that if $B \in \mathscr{F}^0$ then $\theta_0^{-1}B = B$, and consequently if $A \in \mathscr{F}$ then $P^\mu(A - \theta_0^{-1}A) = P^\mu(\theta_0^{-1}A - A) = 0$ for all μ. Therefore

$$\begin{aligned} P^x(\Lambda) &= P^x(\Lambda \cap \theta_0^{-1}\Lambda) \\ &= E^x[P^{X(0)}(\Lambda); \Lambda] = [P^x(\Lambda)]^2. \end{aligned}$$

Exercises

(5.18) In the notation of Definition 5.3 show that $f \in \bar{\mathscr{G}}$ if and only if for each $\mu \in \mathbf{U}$ there exist $g \in \mathscr{G}$ and $\Lambda \in \mathscr{E}$ such that $\{f \neq g\} \subset \Lambda$ and $P^\mu(\Lambda) = 0$.

(5.19) Consider the process of uniform motion to the right discussed in Exercise 3.7. Find $\bar{\mathscr{M}}$ and $\mathscr{F}$ for this example.

(5.20) Let X be a Markov process with state space $(E, \mathscr{E})$ where E is a metric space and $\mathscr{E}$ is the Borel sets of E. Suppose X satisfies the conclusion of Proposition 5.17. Let $B = \{x \in E_\Delta \colon P^x(X_0 = x) = 1\}$. Show that $B \in \mathscr{E}_\Delta$ but that B need not equal E_Δ (that is, X need not be normal). On the other hand show that, for each x and t, $P^x(X_t \in B) = 1$.

(5.21) Let $E = \mathbf{R}$, $\mathscr{E} = \mathscr{B}(\mathbf{R})$, $\mathscr{G} = \{\varnothing, \mathbf{R}\}$, and $\mathbf{U} = \{\mu\}$ where μ is a fixed nonzero finite measure on $(E, \mathscr{E})$. Using the notation of (5.2) and (5.3) show that $\mathscr{G}^{\mathbf{U}} \neq \bar{\mathscr{G}}$.

6. Stopping Times

In this section we will introduce a class of random variables that will play a fundamental role in the remainder of this book. Let $(\Omega, \mathscr{F})$ be a measurable space and let $\{\mathscr{F}_t \colon t \in \mathbf{T}\}$ be a fixed increasing family of sub-σ-algebras of $\mathscr{F}$.

Again in this section $\mathbf{T} = [0, \infty]$ and it will be convenient to assume that $\mathscr{F}_\infty = \mathscr{F}$. We will leave to the reader the task of developing the analogous (but simpler) situation in which $\mathbf{T} = \{0, 1, \ldots, \infty\}$.

(6.1) DEFINITION. A map $T: \Omega \to \mathbf{T}$ is called a *stopping time* (with respect to $\{\mathscr{F}_t\}$) provided $\{T \le t\} \in \mathscr{F}_t$ for all t in $[0, \infty)$.

Note that $\{T = \infty\} = \{T < \infty\}^c \in \mathscr{F} = \mathscr{F}_\infty$ and so $\{T \le t\} \in \mathscr{F}_t$ for all $t \in \mathbf{T}$. Hence $T \in \mathscr{F}$. Of course $\{T \le t\} \in \mathscr{F}_t$ if and only if $\{T > t\} \in \mathscr{F}_t$. Clearly any nonnegative constant is a stopping time.

Given the family $\{\mathscr{F}_t\}$ we define for each $t \in [0, \infty)$ a new σ-algebra $\mathscr{F}_{t+} = \bigcap_{s>t} \mathscr{F}_s$. We have used the notation $\mathscr{F}_{t+}$ for this σ-algebra because it is the standard notation. Unfortunately in our notation this same symbol could be used to denote the nonnegative $\mathscr{F}_t$ measurable functions. We will *never* use $\mathscr{F}_{t+}$ in this latter sense. For convenience we set $\mathscr{F}_{\infty+} = \mathscr{F}$. Similarly we define for $0 < t < \infty$, $\mathscr{F}_{t-} = \sigma(\bigcup_{s<t} \mathscr{F}_s)$ and set $\mathscr{F}_{0-} = \mathscr{F}_0$, $\mathscr{F}_{\infty-} = \mathscr{F}$. It is immediate that $\{\mathscr{F}_{t+}\}$ and $\{\mathscr{F}_{t-}\}$ are increasing families of sub-σ-algebras of $\mathscr{F}$, and $\mathscr{F}_{t-} \subset \mathscr{F}_t \subset \mathscr{F}_{t+}$ for all t.

(6.2) DEFINITION. The family $\{\mathscr{F}_t\}$ is right continuous if $\mathscr{F}_t = \mathscr{F}_{t+}$ for each $t < \infty$.

Note that $\{\mathscr{F}_{t+}\}$ is always right continuous.

(6.3) PROPOSITION. *T is a stopping time with respect to $\{\mathscr{F}_{t+}\}$ if and only if $\{T < t\} \in \mathscr{F}_t$ for all $t < \infty$.*

Proof. This follows from the identities

$$\{T \le t\} = \bigcap_n \left\{T < t + \frac{1}{n}\right\} \quad \text{and} \quad \{T < t\} = \bigcup_n \left\{T \le t - \frac{1}{n}\right\}.$$

Note that this actually shows that T is a stopping with respect to $\{\mathscr{F}_{t+}\}$ if and only if $\{T < t\} \in \mathscr{F}_{t-}$ for all t. If T is a stopping time with respect to $\{\mathscr{F}_t\}$, then from the above identity $\{T < t\} \in \mathscr{F}_{t-}$ and so T is a stopping time with respect to $\{\mathscr{F}_{t+}\}$. The converse is not true in general.

Example. If X is a Markov process then according to (3.4) ζ is a stopping time with respect to $\{\mathscr{F}^0_{t+}\}$.

(6.4) PROPOSITION. *If T and S are stopping times (with respect to $\{\mathscr{F}_t\}$), then so are $\sup(T, S)$, $\inf(T, S)$, and $T + S$.*

Proof. The following identities establish the first two statements:

$$\{\sup(T, S) \le t\} = \{T \le t\} \cap \{S \le t\},$$
$$\{\inf(T, S) \le t\} = \{T \le t\} \cup \{S \le t\}.$$

For the third statement we write

$$\begin{aligned}\{T + S > t\} &= \{0 < T < t, T + S > t\} \cup \{T = 0, T + S > t\} \\ &\qquad \cup \{T > t, S = 0\} \cup \{T \ge t, S > 0\} \\ &= \Lambda_1 \cup \Lambda_2 \cup \Lambda_3 \cup \Lambda_4.\end{aligned}$$

We observe that ($\mathbf{Q}$ = rationals)

$$\Lambda_1 = \bigcup_{r \in (0,t) \cap \mathbf{Q}} (\{r < T < t\} \cap \{S > t - r\}) \in \mathscr{F}_t,$$
$$\Lambda_2 = \{T = 0\} \cap \{S > t\} \in \mathscr{F}_t.$$

Similarly $\Lambda_3 \in \mathscr{F}_t$, and finally $\Lambda_4 = \{T \ge t\} \cap \{S = 0\}^c \in \mathscr{F}_t$. Consequently $T + S$ is a stopping time.

(6.5) Proposition. If $\{T_n\}$ is a sequence of stopping times with respect to $\{\mathscr{F}_t\}$, then $\sup T_n$ is such a stopping time. If $\{\mathscr{F}_t\}$ is right continuous then $\inf T_n$, $\limsup T_n$, and $\liminf T_n$ are stopping times with respect to $\{\mathscr{F}_t\}$.

Proof. The proposition results from the following identities:

$$\{\sup T_n \le t\} = \bigcap_n \{T_n \le t\},$$
$$\{\inf T_n < t\} = \bigcup_n \{T_n < t\},$$
$$\limsup T_n = \inf_k \sup_{n \ge k} T_n,$$
$$\liminf T_n = \sup_k \inf_{n \ge k} T_n.$$

(6.6) Definition. If T is a stopping time with respect to $\{\mathscr{F}_t\}$ we define $\mathscr{F}_T$ to consist of all sets Λ in $\mathscr{F}$ such that $\Lambda \cap \{T \le t\} \in \mathscr{F}_t$ for all $t < \infty$.

The reader should have no difficulty in verifying that $\mathscr{F}_T$ is a σ-algebra and that if $T(\omega) = a \ge 0$ for all ω then $\mathscr{F}_T = \mathscr{F}_a$. Intuitively one should think of $\mathscr{F}_t$ as containing all the information in some physical process up to time t; then $\mathscr{F}_T$ contains all the information up to the random time T. The defining property of a stopping time means that we can tell whether or not T is greater than t knowing only the information up to time t.

If T is a stopping time with respect to $\{\mathscr{F}_{t+}\}$ we write $\mathscr{F}_{T+}$ for the corresponding object, that is, the collection of all Λ in $\mathscr{F}$ such that $\Lambda \cap \{T \leq t\} \in \mathscr{F}_{t+}$ for all $t < \infty$. It is easy to see that $\Lambda \in \mathscr{F}_{T+}$ if and only if $\Lambda \in \mathscr{F}$ and $\Lambda \cap \{T < t\} \in \mathscr{F}_t$ for all $t < \infty$.

(6.7) PROPOSITION. Let $\{\mathscr{F}_t\}$ be given.

(i) If T is a stopping time, then T is $\mathscr{F}_T$ measurable.

(ii) If T and S are stopping times and $T \leq S$, then $\mathscr{F}_T \subset \mathscr{F}_S$.

(iii) If $\{\mathscr{F}_t\}$ is right continuous and $\{T_n\}$ is a sequence of stopping times, then if $T = \inf T_n$ one has $\mathscr{F}_T = \bigcap_n \mathscr{F}_{T_n}$.

Proof.

(i) We must show that $\{T \leq a\} \in \mathscr{F}_T$ for all $a \geq 0$. But

$$\{T \leq a\} \cap \{T \leq t\} = \{T \leq a \wedge t\} \in \mathscr{F}_{a \wedge t} \subset \mathscr{F}_t,$$

and so $T \in \mathscr{F}_T$.

(ii) If $\Lambda \in \mathscr{F}_T$ then

$$\Lambda \cap \{S \leq t\} = (\Lambda \cap \{T \leq t\}) \cap \{S \leq t\} \in \mathscr{F}_t,$$

and so $\Lambda \in \mathscr{F}_S$.

(iii) According to (6.5) T is a stopping time and from (ii) we have $\mathscr{F}_T \subset \bigcap_n \mathscr{F}_{T_n}$. On the other hand, if $\Lambda \in \bigcap_n \mathscr{F}_{T_n}$ then

$$\begin{aligned}\Lambda \cap \{T < t\} &= \Lambda \cap \left(\bigcup_n \{T_n < t\}\right) \\ &= \bigcup_n (\Lambda \cap \{T_n < t\}) \in \mathscr{F}_t,\end{aligned}$$

and so $\Lambda \in \mathscr{F}_T$.

(6.8) PROPOSITION. Let $\{\mathscr{F}_t\}$ be given and suppose that T and S are stopping times; then each of the sets $\{T < S\}$, $\{S < T\}$, $\{T \leq S\}$, $\{S \leq T\}$, and $\{S = T\}$ is in both $\mathscr{F}_T$ and $\mathscr{F}_S$.

Proof. We have the equality

$$\{T < S\} \cap \{S \leq t\} = \bigcup_{r \in \mathbf{Q}, r < t} (\{T < r\} \cap \{r < S \leq t\}) \in \mathscr{F}_t,$$

and so $\{T < S\} \in \mathscr{F}_S$. If $D_t = (\mathbf{Q} \cap [0, t]) \cup \{t\}$, then

$$\{T < S\} \cap \{T \leq t\} = \bigcup_{r \in D_t} \{T \leq r\} \cap \{r < S\} \in \mathscr{F}_t,$$

and so $\{T < S\} \in \mathscr{F}_T$. By symmetry $\{S < T\}$ is in both $\mathscr{F}_T$ and $\mathscr{F}_S$, and Proposition 6.8 now follows by taking complements and differences.

The following approximation procedure will be useful in some proofs. Given $H : \Omega \to [0, \infty]$ we define ($n = 1, 2, \ldots; k = 0, 1, \ldots$)

$$H^{(n)}(\omega) = \frac{k+1}{2^n} \quad \text{on} \quad \left\{\frac{k}{2^n} \le H < \frac{k+1}{2^n}\right\}, \tag{6.9}$$

$$= \infty \quad \text{on} \quad \{H = \infty\}.$$

It is obvious that $H^{(n)} \downarrow H$ for all ω and that $H^{(n)} > H$ on $\{H < \infty\}$. Moreover if $H \in \mathscr{F}$, then so is each $H^{(n)}$. Finally it is easy to see that if H is a stopping time so is each $H^{(n)}$. Thus any stopping time T can be approached pointwise from above by stopping times taking on at most countably many values.

Let $(\Omega, \mathscr{F})$ and $\{\mathscr{F}_t;\ t \in \mathbf{T}\}$ be given and suppose that for each t we are given a map X_t from Ω into a measurable space $(E, \mathscr{E})$. Let us call $X = (\Omega, \mathscr{F}, \mathscr{F}_t, X_t)$ a *random function* taking values in $(E, \mathscr{E})$ if $X_t \in \mathscr{F}_t/\mathscr{E}$ for each $t \in \mathbf{T}$. Given $H : \Omega \to [0, \infty]$ we denote the map $\omega \to X_{H(\omega)}(\omega)$ by X_H. Finally we write $\mathscr{R}_t$ for the ordinary Borel sets of the interval $[0, t]$ and set $\mathscr{R} = \mathscr{R}_\infty$.

(6.10) DEFINITION. Let $X = (\Omega, \mathscr{F}, \mathscr{F}_t, X_t)$ be a random function taking values in $(E, \mathscr{E})$. For each $t \in \mathbf{T}$ let Φ_t be the map $(u, \omega) \to X_u(\omega)$ from $[0, t] \times \Omega$ to E and let $\Phi = \Phi_\infty$. Then (i) X is *measurable* if $\Phi \in (\mathscr{R} \times \mathscr{F})/\mathscr{E}$ and (ii) X is *progressively measurable* (with respect to $\{\mathscr{F}_t\}$) if $\Phi_t \in (\mathscr{R}_t \times \mathscr{F}_t)/\mathscr{E}$ for each t.

If E is a metric space and $\mathscr{E}$ the σ-algebra of Borel sets of E and if $t \to X_t(\omega)$ is right continuous for all ω, then X is progressively measurable. The proof of this useful fact is left to the reader as Exercise 6.13.

(6.11) THEOREM. Let X be progressively measurable and let T be an $\{\mathscr{F}_t\}$ stopping time; then X_T is in $\mathscr{F}_T/\mathscr{E}$.

Proof. If $B \in \mathscr{E}$ we must show that $\{X_T \in B\} \in \mathscr{F}_T$, or equivalently $\{X_T \in B\} \cap \{T \le t\} \in \mathscr{F}_t$ for all t. Let $\Psi_t : \{T \le t\} \to [0, t] \times \Omega$ be defined by $\omega \to (T(\omega), \omega)$, and let Φ_t be the map defined in (6.10). If Ψ is the restriction of X_T to $\{T \le t\}$, then $\Psi = \Phi_t \circ \Psi_t$. Now by hypothesis $\Phi_t \in (\mathscr{R}_t \times \mathscr{F}_t)/\mathscr{E}$. Moreover $\Psi_t \in \mathscr{F}_t/(\mathscr{R}_t \times \mathscr{F}_t)$ since if $0 \le u_1 < u_2 \le t$ and $\Lambda \in \mathscr{F}_t$ then $\Psi_t^{-1}((u_1, u_2] \times \Lambda) = \{u_1 < T \le u_2\} \cap \Lambda \in \mathscr{F}_t$. Thus $\Psi \in \mathscr{F}_t/\mathscr{E}$. But $\{X_T \in B\} \cap \{T \le t\} = \Psi^{-1}(B) \in \mathscr{F}_t$, and this establishes (6.11).

(6.12) COROLLARY. If X is measurable and $H : \Omega \to [0, \infty]$ is in $\mathscr{F}$, then $X_H \in \mathscr{F}/\mathscr{E}$.

Proof. Let $\mathscr{G}_t = \mathscr{F}$ for all t and apply (6.11) to the random function $(\Omega, \mathscr{F}, \mathscr{G}_t, X_t)$.

Exercises

(6.13) Prove the assertion following Definition 6.10.

(6.14) Let $(E, \mathscr{E})$ be a measurable space and $\mathbf{U}$ a family of finite measures on $\mathscr{E}$. Let $\{\mathscr{E}_t: t \in \mathbf{R}_+\}$ be an increasing family of sub-σ-algebras of $\mathscr{E}^{\mathbf{U}}$ and let a superscript tilde denote completion in $\mathscr{E}^{\mathbf{U}}$ with respect to $\mathbf{U}$ (Definition 5.3). Prove that $(\mathscr{E}_{t+})^{\sim} = (\tilde{\mathscr{E}}_t)_+$ for all t.

(6.15) Let $(E, \mathscr{E})$ be a measurable space and suppose that $\mathscr{E} \neq \{\varnothing, E\}$. Let $\Omega = E^{\mathbf{T}}$ and let $X_t(\omega) = \omega(t)$ be the coordinate mappings. Here, as usual, $\mathbf{T} = [0, \infty]$. Let $\mathscr{F}_t^0 = \sigma(X_s; s \leq t)$. Show that for each $t < \infty$, $\mathscr{F}_t^0 \neq \mathscr{F}_{t+}^0$.

(6.16) Let $(\Omega, \mathscr{F})$ be a measurable space, E a metric space, and $\mathscr{E}$ the σ-algebra of Borel sets of E. Let $X_t: \Omega \to E$ be in $\mathscr{F}/\mathscr{E}$ for each $t \in \mathbf{R}_+$ and suppose that $t \to X_t(\omega)$ is continuous for each ω. If A is a subset of E define $D_A(\omega) = \inf\{t \geq 0: X_t(\omega) \in A\}$ where the infimum of the empty set is $+\infty$ by convention. Finally let $\mathscr{F}_t^0 = \sigma(X_s: s \leq t)$. (a) If A is open show that D_A is an $\{\mathscr{F}_{t+}^0\}$ stopping time. (b) Give an example to show that D_A need not be an $\{\mathscr{F}_t^0\}$ stopping time when A is open. (c) If A is closed show that D_A is an $\{\mathscr{F}_t^0\}$ stopping time. Note that (a) remains valid if one only assumes that $t \to X_t(\omega)$ is right continuous for each ω.

(6.17) Let Ω be a space and $(E, \mathscr{E})$ be a measurable space. Suppose that for each $t \in \mathbf{R}_+$ we are given a map $X_t: \Omega \to E$. As usual define $\mathscr{F}_\infty^0 = \sigma(X_s: s \in \mathbf{R}_+)$ and $\mathscr{F}_t^0 = \sigma(X_s: s \leq t)$. Suppose further that for each t and ω there exists ω' such that $X_s(\omega') = X_{s \wedge t}(\omega)$ for all $s \in \mathbf{R}_+$. (a) Show that $\Lambda \in \mathscr{F}_t^0$ if and only if (i) $\Lambda \in \mathscr{F}_\infty^0$ and (ii) $\omega_0 \in \Lambda$ and $X_s(\omega) = X_s(\omega_0)$ for all $s \leq t$ together imply that $\omega \in \Lambda$. (b) $T \geq 0$ is an $\{\mathscr{F}_t^0\}$ stopping time if and only if $T \in \mathscr{F}_\infty^0$ and $T(\omega_0) \leq t$, $X_s(\omega) = X_s(\omega_0)$ for all $s \leq t$ imply $T(\omega) \leq t$. (c) Let T be an $\{\mathscr{F}_t^0\}$ stopping time. Show that $\Lambda \in \mathscr{F}_T^0$ if and only if (i) $\Lambda \in \mathscr{F}_\infty^0$ and (ii) $\omega_0 \in \Lambda$ and $X_s(\omega_0) = X_s(\omega)$ for all $s \leq T(\omega_0)$ imply that $\omega \in \Lambda$. [Hint: use Corollary 2.9 of Chapter 0.]

(6.18) Let $(\Omega, \mathscr{F})$ and $\{\mathscr{F}_t; t \in \mathbf{T}\}$ be as in the first paragraph of this section. Show that any $\{\mathscr{F}_t\}$ stopping time is the infimum of a countable family of stopping times T each of which has the following special form: $T = a$ on $\Lambda \in \mathscr{F}_a$ and $T = \infty$ on Λ^c.

7. Stopping Times for Markov Processes

In this section $X = (\Omega, \mathcal{M}, \mathcal{M}_t, X_t, \theta_t, P^x)$ will be a fixed Markov process with state space $(E, \mathcal{E})$. We assume without loss of generality that $\bar{\mathcal{M}} = \mathcal{M}$ and $\bar{\mathcal{M}}_t = \mathcal{M}_t$. We will consider stopping times relative to $\{\mathcal{M}_t\}$ and $\{\mathcal{F}_t\}$ for the most part. In particular we will say that T is a *stopping time for X* provided it is an $\{\mathcal{F}_t\}$ stopping time. If T is an $\{\mathcal{M}_t\}(\{\mathcal{F}_t\})$ stopping time then of course $\mathcal{M}_T(\mathcal{F}_T)$ consists of those sets $\Lambda \in \mathcal{M}(\mathcal{F})$ such that $\Lambda \cap \{T \leq t\} \in \mathcal{M}_t(\mathcal{F}_t)$ for all t. The reader should have no difficulty in verifying that

$$\text{(7.1)} \qquad \bar{\mathcal{M}}_T = \mathcal{M}_T \qquad (\bar{\mathcal{F}}_T = \mathcal{F}_T)$$

where $\bar{\mathcal{M}}_T(\bar{\mathcal{F}}_T)$ is the completion of $\mathcal{M}_T(\mathcal{F}_T)$ in $\mathcal{M}(\mathcal{F})$ with respect to the family $\{P^x;\ x \in E_\Delta\}(\{P^\mu;\ \mu$ a finite measure on $\mathcal{E}_\Delta\})$. We say that X is *progressively measurable* provided that the random function $(\Omega, \mathcal{M}, \mathcal{M}_t, X_t)$ taking values in $(E_\Delta, \mathcal{E}_\Delta)$ is progressively measurable. It then follows from (6.11) that $X_T \in \mathcal{M}_T/\mathcal{E}_\Delta$ whenever T is an $\{\mathcal{M}_t\}$ stopping time. Moreover making use of (7.1) and (5.6) it is easy to see that for such a stopping time

$$\text{(7.2)} \qquad X_T \in \mathcal{M}_T/\mathcal{E}_\Delta^*.$$

The following situation will often arise. Suppose $(E, \mathcal{E})$ is a given measurable space and $X = (\Omega, \mathcal{M}, \mathcal{M}_t, X_t, \theta_t, P^x)$ is a Markov process with state space $(E, \mathcal{E}^*)$. Clearly $\mathcal{M}_t/\mathcal{E}_\Delta^* \subset \mathcal{M}_t/\mathcal{E}_\Delta$ and so we may consider $(\Omega, \mathcal{M}, \mathcal{M}_t, X_t)$ as a random function taking values in $(E_\Delta, \mathcal{E}_\Delta)$ as well as in $(E_\Delta, \mathcal{E}_\Delta^*)$. If this random function (with values in $(E_\Delta, \mathcal{E}_\Delta)$) is progressively measurable, then for any $\{\mathcal{M}_t\}$ stopping time T, $X_T \in \mathcal{M}_T/\mathcal{E}_\Delta$ and consequently (7.2) is again valid. For example, let E_Δ be a metric space and $\mathcal{E}_\Delta = \mathcal{B}(E_\Delta)$ (the Borel sets). If X is a Markov process with state space $(E, \mathcal{E}^*)$ such that $t \to X_t(\omega)$ is right continuous for each ω, then $(\Omega, \mathcal{M}, \mathcal{M}_t, X_t)$ is progressively measurable if we regard it as taking values in $(E_\Delta, \mathcal{E}_\Delta)$ and so (7.2) holds. However if we regard $(\Omega, \mathcal{M}, \mathcal{M}_t, X_t)$ as taking values in $(E_\Delta, \mathcal{E}_\Delta^*)$ it will not, in general, be progressively measurable.

When considering $\{\mathcal{F}_t\}$ stopping times the following theorem allows us to restrict our attention to $\{\mathcal{F}_{t+}^0\}$ stopping times in many situations.

(7.3) THEOREM. Let T be an $\{\mathcal{F}_{t+}\}$ stopping time. Then for each μ there exists a stopping time T_μ relative to $\{\mathcal{F}_{t+}^0\}$ such that $P^\mu(T \neq T_\mu) = 0$.

Proof. For each n define $T^{(n)}$ as in (6.9) Then each $T^{(n)}$ is an $\{\mathcal{F}_t\}$ stopping time taking on the discrete set of values $\{k/2^n;\ k = 1, 2, \ldots, \infty\}$. Suppose the measure μ is given. Fix n for the moment, let $\Lambda_k = \{T^{(n)} = k/2^n\}$, and let $A_k \in \mathcal{F}_{k/2^n}^0$ be such that $P^\mu(\Lambda_k \Delta A_k) = 0$; this is possible because $\Lambda_k \in \mathcal{F}_{k/2^n}$.

Let $B_1 = A_1$, $B_k = A_k - \bigcup_{j<k} A_j$ for $k = 1, \ldots, \infty$, and define $R^{(n)}$ by

$$R^{(n)}(\omega) = \frac{k}{2^n}, \qquad \omega \in B_k$$
$$= \infty, \qquad \omega \in \Omega - \bigcup_k B_k.$$

One checks immediately that $R^{(n)}$ is an $\{\mathscr{F}_t^0\}$ stopping time and that $P^\mu(R^{(n)} \neq T^{(n)}) = 0$. If we set $S_n = \inf_{k \leq n} R^{(k)}$, then according to Propositions (6.4) and (6.5) $\{S_n\}$ is a sequence of $\{\mathscr{F}_t^0\}$ stopping times decreasing to a limit, T_μ, which is an $\{\mathscr{F}_{t+}^0\}$ stopping time. Since $T^{(n)}$ decreases to T it is clear that $P^\mu(S_n \neq T^{(n)}) = 0$ for all n, and consequently $P^\mu(T \neq T_\mu) = 0$.

If $H : \Omega \to [0, \infty]$ we define the translation operator θ_H as follows:

(7.4) $$\theta_H \omega = \theta_{H(\omega)} \omega.$$

Obviously $X_t \circ \theta_H = X_{t+H}$ for each fixed t.

(7.5) PROPOSITION. If X is a Markov process and for all $\{\mathscr{M}_t\}$ stopping times T, $X_T \in \mathscr{M}_T/\mathscr{E}_\Delta$, then for any such T, $\theta_T^{-1}\mathscr{F}_t^0 \subset \mathscr{M}_{t+T}$ and hence $\theta_T^{-1}\mathscr{F}^0 \subset \mathscr{M}$.

Proof. If $\Lambda = \bigcap_{j=1}^n \{X_{t_j} \in B_j\}$ with $B_j \in \mathscr{E}_\Delta$ and $t_1 < t_2 < \ldots < t_n \leq t$, then $\theta_T^{-1}\Lambda = \bigcap_{j=1}^n \{X_{t_j+T} \in B_j\}$ and so $\theta_T^{-1}\Lambda \in \mathscr{M}_{T+t}$. The desired conclusion is now obvious.

8. The Strong Markov Property

In many arguments it is necessary to use the Markov property (3.2) for certain stopping times T as well as for fixed times t. In early work it was more or less tacitly assumed that this was possible. However, this requires proof and in fact is not true for all Markov processes (see Exercise 8.20). In this section we will develop a sufficient condition (Theorem 8.11) for this "strong" or "extended" Markov property.

In this section all stopping times will be $\{\mathscr{M}_t\}$ stopping times unless explicitly mentioned otherwise.

(8.1) DEFINITION. Let $X = (\Omega, \mathscr{M}, \mathscr{M}_t, X_t, \theta_t, P^x)$ be a Markov process with state space $(E, \mathscr{E})$. Then X is said to have the *strong* Markov *property* or simply to be strong Markov, provided that for each stopping time T and $f \in b\mathscr{E}_\Delta$ one has

(S.R.) $$X_T \in \mathscr{M}_T/\mathscr{E}^*_\Delta$$

and

(S.M.) $$E^x\{f \circ X_{t+T} \mid \mathscr{M}_T\} = E^{X(T)}[f(X_t)]$$

for all t and x.

REMARK. $x \to E^x f(X_t)$ is $\mathscr{E}_\Delta$ measurable and so the function $E^{X(T)}[f(X_t)]$ on Ω is $\mathscr{M}_T$ measurable. Hence under (S.R.) the right side of (S.M.) has at least the appropriate measurability property. Note that (S.R.) is satisfied whenever X is progressively measurable.

We now introduce a convention which will prove extremely useful in the remainder of this book: any numerical function f on E will *automatically be extended to E_Δ by setting $f(\Delta) = 0$* unless explicitly mentioned otherwise. From this point of view $b\mathscr{E}$ is identified with the subspace of $b\mathscr{E}_\Delta$ that consists of all elements (in $b\mathscr{E}_\Delta$) vanishing at Δ. Note also that we may now write

$$P_t f(x) = E^x\{f(X_t); X_t \in E\} = E^x f(X_t)$$

if f is in $b\mathscr{E}$ or in $\mathscr{E}_+$. Clearly $P_t f(\Delta) = 0$ and so our convention is consistent.

(8.2) PROPOSITION. If X satisfies (S.R.), then X is strong Markov if and only if for each stopping time T and f in $b\mathscr{E}$ one has

(S.M.)′ $$E^x f(X_{t+T}) = E^x\{E^{X(T)}[f(X_t)]\}$$

for all t and x.

Proof. First of all let us suppose that (S.M.)′ holds for all $f \in b\mathscr{E}_\Delta$. Let T be a stopping time and let $\Lambda \in \mathscr{M}_T$. Define T_Λ as follows: $T_\Lambda = T$ on Λ, $T_\Lambda = \infty$ on Λ^c. Since $\{T_\Lambda \leq t\} = \{T \leq t\} \cap \Lambda \in \mathscr{M}_t$, T_Λ is a stopping time. If $f \in b\mathscr{E}_\Delta$ then

$$E^x f(X_{t+T_\Lambda}) = E^x\{f(X_{t+T}); \Lambda\} + f(\Delta)\, P^x(\Lambda^c)$$

and

$$E^x\{E^{X(T_\Lambda)}[f(X_t)]\} = E^x\{E^{X(T)}[f(X_t)]; \Lambda\} + f(\Delta)\, P^x(\Lambda^c).$$

By assumption the left sides of the above equations are equal and so (S.M.) holds. Thus it remains to show that (S.M.)′ holds for all $f = b\mathscr{E}_\Delta$. Now $\{f \in b\mathscr{E}_\Delta$: (S.M.)′ holds$\}$ is a vector space which contains I_{E_Δ} and, by hypothesis, any $f \in b\mathscr{E}_\Delta$ which vanishes at Δ. But any $f \in b\mathscr{E}_\Delta$ is a linear combination of two such functions, and so (S.M.)′ holds for all $f \in b\mathscr{E}_\Delta$. Thus Proposition 8.2 is established since the converse is obvious.

(8.3) Remark. If $\mathcal{M}_t = \mathcal{F}_t$, then because of Theorem 7.3 it suffices that (S.R.) and (S.M.)′ hold for all $\{\mathcal{F}^0_{t+}\}$ stopping times, and when this is the case then $(\Omega, \mathcal{F}, \mathcal{F}_{t+}, X_t, \theta_t, P^x)$ is strong Markov. The reader should recall that $\tilde{\mathcal{F}}_{t+} = \mathcal{F}_{t+}$ (Exercise 6.14).

(8.4) Proposition. Let X be strong Markov and $Y \in b\mathcal{F}^0$. Then $E^x[Y \circ \theta_T \mid \mathcal{M}_T] = E^{X(T)}(Y)$ for all x and all stopping times T.

Proof. By Proposition 7.5, $\theta_T^{-1}\mathcal{F}^0 \subset \mathcal{M}$, and so $Y \circ \theta_T \in b\mathcal{M}$. Also $\omega \to E^{X_{T(\omega)}}(Y)$ is in $\mathcal{M}_T$ and so it suffices to prove that $E^x[Y \circ \theta_T; \Lambda] = E^x[E^{X(T)}(Y); \Lambda]$ for all $\Lambda \in \mathcal{M}_T$. As usual it is enough to consider only Y of the form $Y = \prod_{j=1}^n f_j \circ X_{t_j}$ with $f_j \in b\mathcal{E}_\Delta$ and $0 \le t_1 < \ldots < t_n$. We argue by induction on n. When $n = 1$ the required equality is just (S.M.). Consider the stopping time $R = T + t_{n-1}$. Since $\Lambda \in \mathcal{M}_T \subset \mathcal{M}_R$ and $\prod_{j=1}^{n-1} f_j[X_{t_j+T}] \in b\mathcal{M}_R$ we have

$$E^x\left[\prod_{j=1}^n f_j(X_{t_j+T}); \Lambda\right] = E^x\left[\prod_{j=1}^{n-1} f_j(X_{t_j+T}) f(X_{R+t_n-t_{n-1}}); \Lambda\right]$$

$$= E^x\left\{\prod_{j=1}^{n-1} f_j(X_{t_j+T})\, E^{X(R)}[f(X_{t_n-t_{n-1}})]; \Lambda\right\}.$$

If we set $g_j = f_j$, $1 \le j \le n-2$, and $g_{n-1}(x) = f_{n-1}(x)\, E^x[f(X_{t_n-t_{n-1}})]$, then each g_j is in $b\mathcal{E}_\Delta$ and the last displayed expression may be written $E^x[\prod_{j=1}^{n-1} g_j(X_{t_j+T}); \Lambda]$. Thus using the induction hypothesis we obtain

$$E^x\left[\prod_{j=1}^n f_j(X_{t_j+T}); \Lambda\right]$$

$$= E^x\left\{E^{X(T)}\left[\prod_{j=1}^{n-1} f_j(X_{t_j})\, E^{X(t_{n-1})}(f \circ X_{t_n-t_{n-1}})\right]; \Lambda\right\}$$

$$= E^x\left\{E^{X(T)}\left[\prod_{j=1}^{n} f_j(X_{t_j})\right]; \Lambda\right\},$$

completing the proof of (8.4).

(8.5) Corollary. If X is strong Markov then for each stopping time T and $t \in \mathbf{T}$ one has $\theta_T^{-1}\mathcal{F}_t \subset \mathcal{M}_{T+t}$, and consequently $\theta_T^{-1}\mathcal{F} \subset \mathcal{M}$.

Proof. Making use of (7.1) this will follow from Lemma 5.6 once we show that for each $x \in E_\Delta$ there exists a finite measure ν on $\mathcal{E}_\Delta$ such that $P^x\theta_T^{-1} = P^\nu$. But using the strong Markov property one checks exactly as in the proof of (5.11) that $\nu(A) = P^x[X_T \in A]$ is such a measure.

(8.6) COROLLARY. Let X be strong Markov. Then for each $Y \in b\mathscr{F}$ one has $E^x[Y \circ \theta_T \mid \mathscr{M}_T] = E^{X(T)}(Y)$ for all stopping times T and all x.

Proof. Corollary 8.5 implies that $Y \circ \theta_T \in b\mathscr{M}$ and as before $E^{X(T)}(Y)$ is $\mathscr{M}_T$ measurable as a function of ω. The proof of (8.6) then goes exactly as the proof of Proposition 5.12.

Suppose R and T are stopping times; then one can form $R + T \circ \theta_R$. Intuitively if we think of R and T as the times at which certain physical events, α and β, occur, then $R + T \circ \theta_R$ would be the first time β occurs *after* α has occurred and hence should be a stopping time itself. The following theorem makes this precise.

(8.7) THEOREM. Suppose X is strong Markov and let R be an $\{\mathscr{M}_{t+}\}$ stopping time and T be an $\{\mathscr{F}_{t+}\}$ stopping time. Then $R + T \circ \theta_R$ is an $\{\mathscr{M}_{t+}\}$ stopping time.

Proof. If $S = R + T \circ \theta_R$ then

$$\{S < t\} = \bigcup_{r \in \mathbf{Q}} \{R < t - r;\ T \circ \theta_R < r\}.$$

But $\{T \circ \theta_R < r\} = \theta_R^{-1}\{T < r\} \in \mathscr{M}_{R+r}$ by (8.5), and if $\Lambda \in \mathscr{M}_{R+r}$ then $\Lambda \cap \{R < t - r\} = \Lambda \cap \{R + r < t\} \in \mathscr{M}_t$. Thus $\{S < t\} \in \mathscr{M}_t$ and so S is an $\{\mathscr{M}_{t+}\}$ stopping time.

Of course, one can replace $\{\mathscr{M}_{t+}\}$ and $\{\mathscr{F}_{t+}\}$ by $\{\mathscr{M}_t\}$ and $\{\mathscr{F}_t\}$, respectively, in the *hypothesis* of (8.7) without altering the conclusion. However it is not clear (and perhaps not true) that with this replacement one can also replace $\{\mathscr{M}_{t+}\}$ by $\{\mathscr{M}_t\}$ in the conclusion.

We are now going to give some sufficient conditions for the strong Markov property. We begin by introducing the "potential operators" which will play a central role in the remainder of this book. In Section 3 we defined the transition operators P_t for $f \in b\mathscr{E}$ by

(8.8)
$$P_t f(x) = \int P_t(x, dy) f(y)$$
$$= E^x[f(X_t);\ X_t \in E] \qquad \text{for} \quad t \geq 0.$$

It follows from the Chapman–Kolmogorov equations that $\{P_t;\ t \geq 0\}$ is a semigroup of (linear) operators on either $b\mathscr{E}$ or $b\mathscr{E}^*$ to itself. Both $b\mathscr{E}$ and $b\mathscr{E}^*$ are Banach spaces under the supremum norm, $\|\cdot\|$, and $\|P_t\| \leq 1$. If X is normal, then $P_0 = I$. Recall that any numerical function f on E is extended to E_Δ by setting $f(\Delta) = 0$.

Let us now assume that X is measurable relative to $\mathscr{F}^0$; that is, the map $\Phi : (t, \omega) \to X_t(\omega)$ is in $(\mathscr{R} \times \mathscr{F}^0)/\mathscr{E}_\Delta$, $\mathscr{R}$ denoting as usual the Borel sets of $[0, \infty)$. It now follows from Lemma 5.6 that if λ is a finite measure on $\mathscr{R}$ and μ is a finite measure on $\mathscr{E}_\Delta$, then $\Phi \in (\mathscr{R} \times \mathscr{F}^0)^{\lambda,\mu}/\mathscr{E}_\Delta^*$ where $(\mathscr{R} \times \mathscr{F}^0)^{\lambda,\mu}$ is the completion of $\mathscr{R} \times \mathscr{F}^0$ with respect to the product measure $\lambda \times P^\mu$. In particular one can take λ to be equivalent to Lebesgue measure. Therefore if f is in $\mathscr{E}_\Delta$ $(\mathscr{E}_\Delta^*)$, then $(t, \omega) \to f[X_t(\omega)]$ is in $\mathscr{R} \times \mathscr{F}^0$ $((\mathscr{R} \times \mathscr{F}^0)^{\lambda,\mu}$ for all $\lambda, \mu)$, respectively. It is now evident that if $f \in b\mathscr{E}$ then $(t, x) \to P_t f(x) = E^x f(X_t)$ is in $b(\mathscr{R} \times \mathscr{E})$, while if $f \in b\mathscr{E}^*$ then the above map is $(\mathscr{R} \times \mathscr{E})^{\lambda \times \mu}$ measurable for all λ and μ where the σ-algebra in question is the completion of $\mathscr{R} \times \mathscr{E}$ with respect to the product measure $\lambda \times \mu$. Therefore if $\alpha > 0$ and $f \in b\mathscr{E}^*$ we can define

$$U^\alpha f(x) = \int_0^\infty e^{-\alpha t} P_t f(x)\, dt = E^x \int_0^\infty e^{-\alpha t} f(X_t)\, dt. \tag{8.9}$$

One verifies easily that U^α maps $b\mathscr{E}(b\mathscr{E}^*)$ into $b\mathscr{E}(b\mathscr{E}^*)$ and that $\|U^\alpha\| \le \alpha^{-1}$. The operator U^α is called the *α-potential operator of* X and the family $\{U^\alpha; \alpha > 0\}$ is called the *resolvent of the semi-group* $\{P_t; t \ge 0\}$. It is easy to check, using the semigroup property of $\{P_t; t \ge 0\}$, that the resolvent equation

$$U^\alpha - U^\beta = (\beta - \alpha) U^\alpha U^\beta, \qquad \alpha, \beta > 0, \tag{8.10}$$

holds, and consequently $U^\alpha U^\beta = U^\beta U^\alpha$.

We now assume that E_Δ is a metric space and that $\mathscr{E}_\Delta$ is the σ-algebra of Borel sets in E_Δ. A Markov process X is said to be *right continuous* if almost surely the mapping $t \to X_t(\omega)$ is a right continuous function from $[0, \infty]$ to E_Δ. If X is a right continuous Markov process with state space $(E, \mathscr{E}^*)$, then the random function $(\Omega, \mathscr{M}, \mathscr{M}_t, X_t)$ with values in $(E_\Delta, \mathscr{E}_\Delta)$ is progressively measurable and so the above discussion applies.

We come now to the main theorem of this section.

(8.11) THEOREM. Let $(E_\Delta, \mathscr{E}_\Delta)$ be as above and let $X = (\Omega, \mathscr{M}, \mathscr{M}_t, X_t, \theta_t, P^x)$ be a right continuous Markov process with state space $(E, \mathscr{E}^*)$. Suppose there exists a linear space $\mathbf{L}$ of bounded continuous functions on E such that (i) for every $f \in \mathbf{L}$ and $\alpha > 0$ the map $t \to U^\alpha f(X_t)$ is right continuous on $[0, \zeta)$ almost surely and (ii) whenever G is an open subset of E there exists an increasing sequence $\{f_n\}$ in $\mathbf{L}$ with $f_n \uparrow I_G$. Then $(\Omega, \mathscr{M}, \mathscr{M}_{t+}, X_t, \theta_t, P^x)$ is strong Markov.

REMARKS. In particular X is then Markov with respect to $\{\mathscr{M}_{t+}\}$. Note that if $U^\alpha : \mathbf{C}_K(E) \to \mathbf{C}(E)$ for each $\alpha > 0$, then $\mathbf{L} = \mathbf{C}_K(E)$ certainly satisfies the conditions in (8.11) provided E is σ-compact.

Proof. The discussion following (7.2) implies that (S.R.) holds. Let T be an $\{\mathscr{M}_{t+}\}$ stopping time and define $T^{(n)}$ as in (6.9) for each n. If $f \in \mathbf{L}$ and $\alpha > 0$, then

$$E^x \int_0^\infty e^{-\alpha t} f(X_{T+t})\, dt = \lim_n E^x \int_0^\infty e^{-\alpha t} f[X(T^{(n)} + t)]\, dt$$

$$= \lim_n \sum_{k=1}^\infty E^x\left\{\int_0^\infty e^{-\alpha t} f[X(t + k2^{-n})]\, dt;\, T^{(n)} = k2^{-n}\right\}.$$

But $\{T^{(n)} = k2^{-n}\} = \{(k-1)2^{-n} \leq T < k2^{-n}\} \in \mathscr{M}_{k2^{-n}}$, and so using the Markov property (relative to $\{\mathscr{M}_t\}$) the last displayed expression becomes

$$\lim_n \sum_{k=1}^\infty E^x\left\{E^{X(k2^{-n})}\left[\int_0^\infty e^{-\alpha t} f(X_t)\, dt\right];\, T^{(n)} = k2^{-n}\right\}$$

$$= \lim_n E^x U^\alpha f[X(T^{(n)})].$$

But $t \to U^\alpha f(X_t)$ is right continuous on $[0, \zeta)$ and equal to zero on $[\zeta, \infty]$ and hence is right continuous on $[0, \infty]$ (a.s.). Since $T^{(n)} \downarrow T$ we obtain

$$\int_0^\infty e^{-\alpha t}\, E^x[f(X_{t+T})]\, dt = E^x[U^\alpha f(X_T)]$$

$$= \int_0^\infty e^{-\alpha t}\, E^x\{E^{X(T)}[f(X_t)]\}\, dt$$

for all $\alpha > 0$. The functions $t \to E^x[f(X_{t+T})]$ and $t \to E^x\{E^{X(T)}[f(X_t)]\}$ are both right continuous and the above calculation shows that they have the same Laplace transform. Therefore by the uniqueness theorem for Laplace transforms, they are equal. Now using the second property of $\mathbf{L}$ we obtain

$$P^x[X_{t+T} \in G] = E^x\{P^{X(T)}(X_t \in G)\}$$

for all open sets G in E and hence by MCT for all G in $\mathscr{E}$. An application of Proposition 8.2 now completes the proof of Theorem 8.11.

Since $\mathscr{F}_{t+} \subset M_{t+}$ we see that X is Markovian with respect to $\{\mathscr{F}_{t+}\}$ under the hypotheses of Theorem 8.11. In this situation things are very nice indeed as the following proposition shows.

(8.12) PROPOSITION. If X is Markov relative to $\{\mathscr{F}^0_{t+}\}$, then $\mathscr{F}_t = \mathscr{F}_{t+}$ for each t.

Proof. By assumption for any $Y \in b\mathscr{F}^0$, t, and μ we have

$$E^\mu[Y \circ \theta_t \mid \mathscr{F}^0_{t+}] = E^{X(t)}(Y)$$

$$= E^\mu[Y \circ \theta_t \mid \mathscr{F}^0_t].$$

Suppose we have established the fact that for all μ and $Y \in b\mathscr{F}^0$

(8.13) $$E^\mu(Y \mid \mathscr{F}_{t+}^0) = E^\mu(Y \mid \mathscr{F}_t^0).$$

If we set $Y = I_\Lambda$ with $\Lambda \in \mathscr{F}_{t+}^0$, then (8.13) implies that, for each μ, I_Λ differs from an $\mathscr{F}_t^0$ measurable function on a P^μ null set. (See the discussion above Theorem 1.3.) Since $\Lambda \in \mathscr{F}_{t+}^0 \subset \mathscr{F}$, this implies that $\Lambda \in \tilde{\mathscr{F}}_t^0 = \mathscr{F}_t$. Consequently $\mathscr{F}_{t+}^0 \subset \mathscr{F}_t$. But $(\mathscr{F}_{t+}^0)^\sim = \mathscr{F}_{t+}$ by (6.14), and so $\mathscr{F}_{t+} \subset \mathscr{F}_t$. Thus Proposition 8.12 follows from (8.13). It evidently suffices to prove (8.13) for $Y = \prod_{j=1}^n f_j \circ X_{t_j}$, with $f_j \in b\mathscr{E}_\Delta$ and $0 \leq t_1 < \ldots < t_i \leq t < t_{i+1} < \ldots < t_n$. Such a Y can be written as $(\prod_{j=1}^i f_j \circ X_{t_j})(G \circ \theta_t)$ where $G = \prod_{j=i+1}^n f_j \circ X_{t_j - t}$. Consequently

$$E^\mu(Y \mid \mathscr{F}_{t+}^0) = \prod_{j=1}^i f_j \circ X_{t_j} E^{X(t)}(G)$$
$$= E^\mu(Y \mid \mathscr{F}_t^0).$$

This completes the proof of (8.12).

It is clear that whenever the hypotheses of Theorem 8.11 hold we can assume without loss of generality that

(8.14) $$\mathscr{M}_{t+} = \mathscr{M}_t \quad \text{and} \quad \bar{\mathscr{M}}_t = \mathscr{M}_t.$$

Whenever (8.14) is in force, (8.12) also obtains and in this situation Theorem 8.7 appears more natural. However we will explicitly mention it when we assume (8.14).

(8.15) REMARK. If $X = (\Omega, \mathscr{M}, \mathscr{M}_t, X_t, \theta_t, P^x)$ is strong Markov and if $X_T \in \mathscr{F}_T/\mathscr{E}^*$ whenever T is an $\{\mathscr{F}_t\}$ stopping time, then $(\Omega, \mathscr{F}, \mathscr{F}_t, X_t, \theta_t, P^x)$ is also strong Markov. In particular this will be the case whenever E_Δ is a metric space ($\mathscr{E}_\Delta = \mathscr{B}(E_\Delta)$ or $\mathscr{B}(E_\Delta)^*$) and X is right continuous.

Exercises

(8.16) Let $X = (\Omega, \mathscr{M}, \mathscr{M}_t, X_t, \theta_t, P^x)$ be a strong Markov process with state space $(E, \mathscr{E})$. Let T be a stopping time and let $G : \Omega \times \Omega \to \mathbf{R}$ be in $b(\mathscr{M}_T \times \mathscr{F})$. Let $H(\omega) = G(\omega, \theta_T \omega)$. Show that

$$E^x\{H \mid \mathscr{M}_T\}(\omega) = \int G(\omega, \omega') P^{X_T(\omega)}(d\omega')$$

for each x.

(8.17) Let X be as in (8.16) and assume further that X is $\mathscr{F}$ measurable.

Let T be a stopping time and $S \geq 0$ be $\mathscr{M}_T$ measurable. If $f \in b\mathscr{E}_\Delta^*$, then

$$E^x\{f(X_{T+S}) \mid \mathscr{M}_T\}(\omega) = N_{S(\omega)}(X_T(\omega), f)$$

for all x where $N_t(x,f) = E^x f(X_t)$. [Hint: suppose first that $f \in b\mathscr{E}_\Delta$ and use (8.16) with $G(\omega, \omega') = f[X_{S(\omega)}(\omega')]$.]

(8.18) Let E be a metric space and $\mathscr{E}_\Delta = \mathscr{B}(E_\Delta)$. Let $X = (\Omega, \mathscr{M}, \mathscr{M}_t, X_t, \theta_t, P^x)$ be a right continuous Markov process with state space $(E, \mathscr{E})$. For each $x \in E$ let $T_x = \inf\{t\colon X_t \neq x\}$. Show that for each x there exists $\lambda(x)$, $0 \leq \lambda(x) \leq \infty$, such that $P^x(T_x > t) = e^{-\lambda(x)t}$. [Hint: consider instead $h(t) = P^x(T_x \geq t)$ and use the fact that $h(t) = P^x(T_x > t)$ except for countably many values of t.]

(8.19) Let $E = \mathbf{R}^n$ and $\mathscr{E} = \mathscr{B}(\mathbf{R}^n)$. Let $X = (\Omega, \mathscr{M}, \mathscr{M}_t, X_t, \theta_t, P^x)$ be a Markov process with state space $(E, \mathscr{E})$ whose transition function is given by (2.12) and which has right continuous paths. (The existence of such a process for given $\{\mu_t; t > 0\}$ will be established in Section 9.) (a) Show that X is strong Markov with respect to $\{\mathscr{M}_{t+}\}$. (b) Show that $P^x(X_t \in B) = P^0(X_t + x \in B)$ for all x, t, and $B \in \mathscr{E}$. (c) Let T be a finite stopping time and let $Y(t) = X(t + T) - X(T)$ for $t \geq 0$. Let $\mathscr{G} = \sigma(Y_t;\ t \in \mathbf{R}_+)$. Show that $\mathscr{G}$ and $\mathscr{F}_T$ are independent with respect to each P^μ. [Hint: begin by showing that for each μ

$$E^\mu\left\{\prod_{k=1}^{N} e^{i(y_k, Y(t_k))} \mid \mathscr{F}_T\right\} = E^0\left\{\prod_{k=1}^{N} e^{\,i(y_k, X(t_k))}\right\}$$

where $y_k \in E$, $1 \leq k \leq N$, and $0 \leq t_1 < \ldots < t_N$.]

(8.20) Consider the process defined in (2.22). For each $t \geq 0$ define $Z_t(\omega) = \lim_{s\uparrow t} X_s(\omega)$. Show that $P(Z_t \neq X_t) = 0$ for each t and hence that $\{Z_t\}$ is a Markov process (in the sense of Definition 1.1) having as a transition function the one defined in (2.21) (which we denote by $N_t(x, A)$). Let $\mathscr{F}_t^0 = \sigma(Z_s;\ s \leq t)$. Find an $\{\mathscr{F}_{t+}^0\}$ stopping time T and a bounded continuous f such that $Ef(Z_{t+T}) \neq E\{N_t(Z_T, f)\}$.

9. Standard Processes

In this section we introduce the class of Markov processes with which we will be mainly concerned in the remainder of this book—the "standard" processes. In particular we will give conditions for the existence of a standard process with a prescribed transition function.

We begin with some definitions of fundamental importance.

(9.1) DEFINITION. Let $X = (\Omega, \mathcal{M}, \mathcal{M}_t, X_t, \theta_t, P^x)$ be a Markov process with state space $(E, \mathcal{E})$. We assume that E_Δ is a metric space and that $\mathcal{E}_\Delta \supset \mathcal{B}(E_\Delta)$, the class of Borel sets in E_Δ. Suppose that $X_T \in \mathcal{M}_T/\mathcal{E}_\Delta^*$ for all $\{\mathcal{M}_t\}$ stopping times T—this is just condition (S.R.) of (8.1). Under these conditions we say that X is *quasi-left-continuous* provided that whenever $\{T_n\}$ is an increasing sequence of $\{\mathcal{M}_t\}$ stopping times with limit T, then almost surely $X(T_n) \to X(T)$ on $\{T < \zeta\}$.

It is useful to introduce a property somewhat stronger than (9.1): namely, we will say that X is quasi-left-continuous on $[0, \infty)$ provided the convergence $X(T_n) \to X(T)$ holds almost surely on $\{T < \infty\}$ rather than just on $\{T < \zeta\}$. In general quasi-left-continuity will refer to Definition 9.1; when confusion is possible we will refer to the property of (9.1) as quasi-left-continuity on $[0, \zeta)$ as opposed to quasi-left-continuity on $[0, \infty)$. We are, of course, using the topology of E_Δ in discussing the convergence of $\{X(T_n)\}$.

We come now to the definition of a standard process.

(9.2) DEFINITION. A normal Markov process $X = (\Omega, \mathcal{M}, \mathcal{M}_t, X_t, \theta_t, P^x)$ with state space $(E, \mathcal{E})$ is called a *standard process* provided:

(i) E is a locally compact space with a countable base and Δ is adjoined to E as the point at infinity if E is noncompact and as an isolated point if E is compact. Furthermore $\mathcal{E}_\Delta$ is the σ-algebra of Borel sets of E_Δ (or, equivalently, $\mathcal{E}$ is the σ-algebra of Borel sets of E).

(ii) $\mathcal{M}_{t+} = \mathcal{M}_t = \bar{\mathcal{M}}_t$ for all t.

(iii) The paths functions $t \to X_t(\omega)$ are right continuous on $[0, \infty)$ and have left-hand limits on $[0, \zeta)$ almost surely.

(iv) X is strong Markov.

(v) X is quasi-left-continuous.

REMARKS. Proposition 8.12 implies that, for a standard process, $\mathcal{F}_{t+} = \mathcal{F}_t$ for all t. Also the assumption in (iii) that $t \to X_t(\omega)$ has left-hand limits on $[0, \zeta)$ almost surely is in fact a consequence of the other hypotheses (see Exercise 9.15). *A standard process which is quasi-left-continuous on* $[0, \infty)$ *is called a* Hunt *process.*

Recall that a subset A of E is *bounded* provided that there exists a compact subset of K of E such that $A \subset K$. The next proposition implies that almost surely the left-hand limits of $t \to X_t$ on $[0, \zeta)$ must lie in E.

(9.3) PROPOSITION. If X is a standard process, then for each t the set $A(\omega) = \{X_s(\omega): 0 \le s \le t, t < \zeta(\omega)\}$ is almost surely bounded.

Proof. Let $\{K_n\}$ be an increasing sequence of compacts in E such that $K_n \subset K_{n+1}^0$ for all n and $\bigcup K_n = E$. (Here K^0 denotes the interior of K.) Let $T_n = \inf\{t: X_t \notin K_n\}$. Then T_n is an $\{\mathscr{F}_t\}$ stopping time because $E_\Delta - K_n$ is open. The sequence $\{T_n\}$ is increasing and if $T = \lim T_n$, then, because of the quasi-left-continuity, $X_{T_n} \to X_T$ almost surely on $\{T < \zeta\}$. However, by the right continuity of the paths $X_{T_{n+1}}$ is not in K_n, and so $X_T = \Delta$ almost surely on $\{T < \zeta\}$ and, hence, also on Ω. Thus $T = \zeta$ almost surely. The conclusion of (9.3) is now immediate because $\{X_s(\omega): 0 \leq s < T_n(\omega)\}$ is contained in K_n.

We come now to the basic existence theorem for standard processes. This theorem will be used in the sequel only in the discussion of examples and so a reader might skip the details of the proof on a first reading. However, the techniques of the proof are of general interest.

(9.4) THEOREM. Let E be locally compact with a countable base and let $\mathscr{E}$ be the Borel sets of E. Let $P_t(x, A)$ be a sub-Markov transition function on $(E, \mathscr{E})$ with $P_0(x, \cdot) = \varepsilon_x$. Let $\mathbf{C}_0$ denote the space of continuous functions on E which vanish at ∞ and suppose that (1) $P_t \mathbf{C}_0 \subset \mathbf{C}_0$ for each $t \geq 0$ and (2) $P_t f \to f$ uniformly on E as $t \to 0$ for each $f \in \mathbf{C}_0$. Then there exists a standard process (in fact, a Hunt process) with state space $(E, \mathscr{E})$ and transition function the given $P_t(x, A)$.

REMARK. It follows from standard semigroup considerations (see Exercise 9.13) that, in the presence of Condition (1), Condition (2) is equivalent to the apparently weaker requirement that $P_t f \to f$ pointwise as $t \to 0$ for each $f \in \mathbf{C}_0$.

Proof. Define E_Δ as in (9.2i) and extend $P_t(x, A)$ to a Markov transition function, $N_t(x, A)$ on $(E_\Delta, \mathscr{E}_\Delta)$, by setting for $A \in \mathscr{E}_\Delta$

$$\begin{aligned} N_t(x, A) &= P_t(x, A \cap E) + I_A(\Delta)(1 - P_t(x, E)), \qquad & x \in E, \\ &= I_A(\Delta), & x = \Delta. \end{aligned}$$

One checks immediately that N is indeed a transition function on $(E_\Delta, \mathscr{E}_\Delta)$ with $N_t(x, E_\Delta) = 1$ for all $t \geq 0$, $x \in E_\Delta$. Let $\mathbf{C} = \mathbf{C}(E_\Delta)$ and observe that if $f \in \mathbf{C}$ and if g is the restriction of $f - f(\Delta)$ to E, then $N_t f(x) = P_t g(x) + f(\Delta)$ for $x \in E$ and $N_t f(\Delta) = f(\Delta)$. It follows from this that, for every $f \in \mathbf{C}$, $N_t f \in \mathbf{C}$ and that $N_t f \to f$ uniformly on E_Δ as $t \to 0$. Also if $f \in \mathbf{C}$ then $N_{t+s} f - N_t f = N_t(N_s f - f) \to 0$ as $s \downarrow 0$ and so $t \to N_t f(x)$ is right continuous. Therefore given $\alpha > 0$ we may form $U^\alpha f(x) = \int_0^\infty e^{-\alpha t} N_t f(x)\, dt$ and $U^\alpha \mathbf{C} \subset \mathbf{C}$. Since $\alpha U^\alpha f(x) = \int_0^\infty e^{-t} N_{t/\alpha} f(x)\, dt$, it follows that $\alpha U^\alpha f \to f$ uniformly as $\alpha \to \infty$ for each $f \in \mathbf{C}$. We may regard $\mathbf{C}_0$ as the subspace of $\mathbf{C}$

consisting of all $f \in \mathbf{C}$ satisfying $f(\Delta) = 0$. With this interpretation $U^\alpha \mathbf{C}_0 \subset \mathbf{C}_0$ for each $\alpha > 0$. In particular if $f \in \mathbf{C}_0$ is strictly positive on E and if $\alpha > 0$, then $U^\alpha f$ is strictly positive on E. These remarks will be used in the course of the proof.

Next let $\mathbf{T} = [0, \infty)$, $W = E_\Delta^{\mathbf{T}}$, and $\mathscr{G} = \mathscr{E}_\Delta^{\mathbf{T}}$ so that $(W, \mathscr{G})$ is the usual product measurable space. By Kolmogorov's theorem, (2.11), and the ensuing discussion there exists for each $x \in E_\Delta$ a probability measure P^x on $(W, \mathscr{G})$ such that over $(W, \mathscr{G}, P^x)$ the coordinate mappings $X_t(w) = w(t)$ form a temporally homogeneous Markov process in the sense of Section 2 with N as transition function and initial distribution ε_x. Let $\mathscr{G}_t = \sigma(X_s;\ s \leq t)$, and let $\mathbf{Q}$ denote the rationals in $[0, \infty)$. Let Λ consist of all those w in W which have the following two properties:

(a) $\lim_{s \uparrow t, s \in \mathbf{Q}} w(s)$ exists (in E_Δ) for each $t > 0$,

$\lim_{s \downarrow t, s \in \mathbf{Q}} w(s)$ exists (in E_Δ) for each $t \geq 0$;

(b) $w(\mathbf{Q} \cap [0, t])$ is bounded in E for every $t \in \mathbf{Q}$ such that $w(t) \in E$.

Denote by Λ_a and Λ_b the set of functions defined by conditions (a) and (b), respectively, so that $\Lambda = \Lambda_a \cap \Lambda_b$. We now assert that $\Lambda \in \mathscr{G}$ and that $P^x(\Lambda) = 1$ for all $x \in E_\Delta$.

To get started on this let $f \in \mathbf{C}_+$ and $g = U^\alpha f$, $\alpha > 0$. Then

$$e^{-\alpha t} N_t g(x) = \int_t^\infty e^{-\alpha u} N_u f(x)\, du \leq g(x). \tag{9.5}$$

Thus given $0 \leq t < s$, $\Gamma \in \mathscr{G}_t$, and $x \in E_\Delta$ we have, using (2.6),

$$\begin{aligned} E^x\{e^{-\alpha s} g(X_s); \Gamma\} &= e^{-\alpha t} E^x\{e^{-\alpha(s-t)} N_{s-t}\, g(X_t); \Gamma\} \\ &\leq E^x\{e^{-\alpha t} g(X_t); \Gamma\}. \end{aligned}$$

Therefore for each $x \in E_\Delta$ the family $\{e^{-\alpha t} g(X_t), \mathscr{G}_t;\ t \geq 0\}$ is a nonnegative supermartingale over $(W, \mathscr{G}, P^x)$. Choose $f \in \mathbf{C}_0$ so that f is strictly positive on E, and let $g = U^\alpha f$, $\alpha > 0$. Then, as noted above, $g \in \mathbf{C}_0$ and g is strictly positive on E. Observe that $(\Lambda_b)^c$ is precisely the union over all $t \in \mathbf{Q}$ of the sets

$$\Gamma_t = \left\{e^{-\alpha t} g(X_t) > 0,\ \inf_{s \in \mathbf{Q} \cap [0,t]} e^{-\alpha s} g(X_s) = 0\right\}.$$

But each $\Gamma_t \in \mathscr{G}_t$ and so $\Lambda_b \in \mathscr{G}$. Moreover according to (1.6) of Chapter 0, $P^x(\Gamma_t) = 0$ for each $x \in E_\Delta$ and $t \geq 0$. Therefore $P^x(\Lambda_b) = 1$ for all $x \in E_\Delta$.

Next let d be a metric for E_Δ and define for $\varepsilon > 0$, $h_\varepsilon(x, y) = 1$ if $d(x, y) \geq \varepsilon$

and $h_\varepsilon(x, y) = 0$ if $d(x, y) < \varepsilon$. If U is a finite subset of $[0, \infty)$ with an even number of elements, say $u_1 < \ldots < u_{2n}$, define

$$H_\varepsilon(U) = H_\varepsilon(U)(w) = \sum_{i=1}^{n} h[X_{u_{2i-1}}(w), X_{u_{2i}}(w)].$$

Clearly $H_\varepsilon(U) \in \mathscr{G}$. For each $D \subset [0, \infty)$ define $H_\varepsilon(D) = \sup H_\varepsilon(U)$ where the supremum is taken over all finite subsets U of D with an even number of elements. If D is countable, then $H_\varepsilon(D)$ is again in $\mathscr{G}$. Next observe that

$$\Lambda_a = \bigcap_{n=1}^{\infty} \bigcap_{m=1}^{\infty} \{H_{1/n}(\mathbf{Q} \cap [0, m]) < \infty\},$$

and so $\Lambda_a \in \mathscr{G}$. According to Theorem 1.5 of Chapter 0, for each $f \in \mathbf{C}_+$ the restriction to $\mathbf{Q}$ of $t \to e^{-\alpha t} g(X_t)$ has right- and left-hand limits at all points of $[0, \infty)$ almost surely P^x for all x. This assertion is then also valid for any $f \in \mathbf{C}$. (Of course $\alpha > 0$ and $g = U^\alpha f$.) Let $\Lambda(\alpha, f)$ denote the set of w's such that the restriction of $t \to g[X_t(w)]$ to $\mathbf{Q}$ has right- and left-hand limits at each point of $[0, \infty)$. By the above argument $\Lambda(\alpha, f) \in \mathscr{G}$ and $P^x[\Lambda(\alpha, f)] = 1$ for all $x \in E_\Delta$, $\alpha > 0$, and $f \in \mathbf{C}$. Let $\{\alpha_n\}$ be a sequence approaching ∞ and let $\{f_k\}$ be a countable dense subset of $\mathbf{C}$. Then $\{\alpha_n U^{\alpha_n} f_k;\ n \geq 1, k \geq 1\}$ is uniformly dense in $\mathbf{C}$ since $\alpha U^\alpha f \to f$ uniformly as $\alpha \to \infty$ whenever $f \in \mathbf{C}$. Now let $\Lambda' = \{w\colon$ for every $f \in \mathbf{C}$ the restriction of $t \to f[X_t(w)]$ to $\mathbf{Q}$ has right- and left-hand limits at all points of $[0, \infty)\}$. Clearly $\Lambda' = \bigcap_{n,k} \Lambda(\alpha_n, f_k)$ and so $P^x(\Lambda') = 1$ for all x. But it is also evident that $\Lambda' = \Lambda_a$ and so the assertion that $P^x(\Lambda) = 1$ for all $x \in E_\Delta$ is established.

We now simply delete the set Λ^c from W; that is, we define $W' = W \cap \Lambda$, $\mathscr{G}_t' = \{A \cap \Lambda\colon A \in \mathscr{G}_t\}$, $\mathscr{G}' = \{A \cap \Lambda\colon A \in \mathscr{G}\}$, and we take $(P^x)'$ to be the restriction of P^x to $\mathscr{G}'$ and X_t' to be the restriction of X_t to W'. Clearly for each x, X_t' is a Markov process with respect to $\{\mathscr{G}_t'\}$ over $(W', \mathscr{G}', (P^x)')$ having the original N as transition function. At this point we will drop the primes from our notation; that is, we will assume that Λ^c is deleted from W so that $W = \Lambda$.

Now for each $t \geq 0$ and w in W set

$$Z_t(w) = \lim_{s \downarrow t,\, s \in \mathbf{Q}} X_s(w). \tag{9.6}$$

Of course this limit exists since $W = \Lambda$. Furthermore it is easy to see that for each w the mapping $t \to Z_t(w)$ is everywhere right continuous and has left-hand limits, and is such that $Z_s(w) = \Delta$ whenever $Z_t(w) = \Delta$ and $s \geq t$. Clearly each Z_t is in $(\mathscr{G}_{t+})/\mathscr{E}_\Delta$. We next assert that for each x, $\{Z_t;\ t \geq 0\}$ is a Markov process with respect to $\{\mathscr{G}_{t+}\}$ over $(W, \mathscr{G}, P^x)$ having N as transition function, and that $P^x\{X_t = Z_t\} = 1$ for all t. Indeed suppose we are given x, t, $\Gamma \in \mathscr{G}_{t+}$, $s > t$ with $s - t$ rational, and $f \in \mathbf{C}$. Let $\{t_n\}$ be a sequence of rationals decreasing to t; then $s_n = s - t + t_n$ is a sequence of rationals decreasing to s. Now $\Gamma \in \mathscr{G}_{t_n}$ for each n, and so using the Markov property for X

$$(9.7) \qquad E^x\{f(X_{s_n}); \Gamma\} = E^x\{N_{s-t}f(X_{t_n}); \Gamma\}$$

for each n. But $X_{s_n} \to Z_s$ and $X_{t_n} \to Z_t$ as $n \to \infty$ while $x \to N_{s-t}f(x)$ is continuous, and so letting $n \to \infty$ in (9.7) we obtain

$$(9.8) \qquad E^x\{f(Z_s); \Gamma\} = E^x\{N_{s-t}f(Z_t); \Gamma\}.$$

The restriction that $s - t$ be rational is removed by observing that each side of (9.8) is right continuous in s. Finally, the validity of (9.8) for all $f \in \mathbf{C}$ implies its validity for all $f \in b\mathscr{E}_\Delta$. Thus we have established that $\{Z_t, \mathscr{G}_{t+}; t \geq 0\}$ is a Markov process over $(W, \mathscr{G}, P^x)$ with transition function N for each x. Now suppose that $g \in b\mathscr{E}_\Delta$, $f \in \mathbf{C}$, $t \in [0, \infty)$, and $\{s_n\}$ is a sequence of rationals decreasing to t. Then

$$E^x\{f(X_{s_n})\, g(X_t)\} = E^x\{N_{s_n - t}f(X_t)\, g(X_t)\},$$

and letting $n \to \infty$ we obtain

$$E^x\{f(Z_t)\, g(X_t)\} = E^x\{f(X_t)\, g(X_t)\}.$$

It now follows from MCT that $E^x\{h(Z_t, X_t)\} = E^x\{h(X_t, X_t)\}$ for all $h \in b(\mathscr{E}_\Delta \times \mathscr{E}_\Delta)$, and this implies that $P^x\{X_t = Z_t\} = 1$ (take, for example, $h(x, y)$ to be a bounded metric for E_Δ). Of course we need this conclusion only for $t = 0$ since we already know that X and Z have the same transition function N. We remark at this point that $P^x(Z_t \in A) = P_t(x, A)$ for x in E and A in $\mathscr{E}$.

For the next step in the construction let Ω denote the set of all functions ω from $[0, \infty]$ to E_Δ such that $t \to \omega(t)$ is right continuous and has left-hand limits throughout $[0, \infty)$, $\omega(\infty) = \Delta$, and $\omega(s) = \Delta$ if $s \geq t$ and $\omega(t) = \Delta$. By construction for each w the functions $t \to Z_t(w)$ (with $Z_\infty(w)$ set equal to Δ) is in Ω. Let $Y_t(\omega) = \omega(t)$ and $\mathscr{F}_t^0 = \sigma(Y_s;\ s \leq t)$ for all $t \in [0, \infty]$. Define a mapping $\pi : W \to \Omega$ by $\pi\, w(t) = Z_t(w)$ and note that $\pi^{-1}\mathscr{F}_\infty^0 \subset \mathscr{G}$ and that $\pi^{-1}\mathscr{F}_t^0 \subset \mathscr{G}_{t+}$ for each $t \in [0, \infty)$. Let $\hat{P}^x = P^x\pi^{-1}$. Define translation operators θ_t on Ω by $\theta_t\, \omega(s) = \omega(t + s)$. It is then easy to see that $(\Omega, \mathscr{F}_\infty^0, \mathscr{F}_t^0, Y_t, \theta_t, \hat{P}^x)$ is a Markov process with state space $(E, \mathscr{E})$ in the sense of Definition 3.1 and having the given $P_t(x, A)$ as transition function. Moreover if we form the "completions" $\mathscr{F}$ and $\mathscr{F}_t$ as in Section 5, then $Y = (\Omega, \mathscr{F}, \mathscr{F}_{t+}, Y_t, \theta_t, P^x)$ is a strong Markov process because the hypotheses of Theorem 8.11 are certainly satisfied. In particular $\mathscr{F}_t = \mathscr{F}_{t+}$ by (8.12). We will drop the "ˆ" from $\hat{P}^x$ at this point. We now assert that Y is quasi-left-continuous on $[0, \infty)$. Once this is established the proof of Theorem 9.4 will be complete.

To this end let $\{T_n\}$ be an increasing sequence of stopping times with limit T. In trying to prove that $Y_{T_n} \to Y_T$ almost surely on $\{T < \infty\}$ there is no loss of generality in assuming that T is bounded, as we now do. Let $L = \lim Y(T_n)$ and for $t > 0$, $L_t = \lim Y(T_n + t)$, these limits existing by the definition of Y. Since $T_n + t$ is in $[T, T + t]$ for all large n and $t \to Y_t$ is right continuous, it follows that $\lim_{t \downarrow 0} L_t = Y_T$. Let f and g be elements of $\mathbf{C}$. Then for each x

$$\begin{aligned} E^x\{f(L)\, g(Y_T)\} &= \lim_{t\to 0} \lim_{n\to\infty} E^x\{f(Y_{T_n})\, g(Y_{T_n+t})\} \\ &= \lim_{t\to 0} \lim_{n\to\infty} E^x\{f(Y_{T_n})\, N_t\, g(Y_{T_n})\} \\ &= \lim_{t\to 0} E^x\{f(L)\, N_t\, g(L)\} = E^x\{f(L)\, g(L)\}. \end{aligned}$$

An application of MCT then shows that

$$E^x\{h(L, Y_T)\} = E^x\{h(L, L)\} \tag{9.9}$$

for all $h \in b(\mathscr{E}_\Delta \times \mathscr{E}_\Delta)$, and consequently $L = Y_T$ almost surely. The proof of Theorem 9.4 is now complete.

We will conclude this section by giving a condition under which the paths of the process constructed above may be assumed to be continuous on $[0, \zeta)$. In order to obtain a simple statement we will retain the assumptions of Theorem 9.4 and work with the process Y constructed in the proof of (9.4).

(9.10) Proposition. Assume the hypothesis of (9.4), and in addition suppose that for each compact subset K of E and neighborhood G of K, $t^{-1} P_t(x, E - G) \to 0$ as $t \to 0$ uniformly on K. Then almost surely the sample functions $t \to Y_t$ are continuous on $[0, \zeta)$.

Proof. Let d be a metric for E. By (9.3) the set $\{Y_s(\omega): 0 \le s \le t, t < \zeta(\omega)\}$ is almost surely bounded. Consequently it will suffice to show that, for each x, each $t > 0$, each compact $K \subset E$, and each $\varepsilon > 0$,

$$P^x\left[\bigcup_{k=0}^{n-1} \{d[Y(tk/n), Y(t(k+1)/n)] > \varepsilon\},\ Y([0, t]) \subset K\right] \tag{9.11}$$

approaches zero as $n \to \infty$. Define

$$\alpha_\varepsilon(t, K) = \sup_{x \in K} P_t(x, E - B_\varepsilon(x)) \tag{9.12}$$

where $B_\varepsilon(x) = \{y \in E: d(x, y) < \varepsilon\}$. The reader will verify easily that the additional hypothesis in (9.10) (in addition to the assumptions of (9.4)) is equivalent to the assertion that $t^{-1} \alpha_\varepsilon(t, K) \to 0$ as $t \to 0$ for each $\varepsilon > 0$ and compact $K \subset E$. Now the expression in (9.11) is dominated by

$$\sum_{k=0}^{n-1} E^x\{P^{Y(tk/n)}[d(Y_0, Y_{t/n}) > \varepsilon]; Y(tk/n) \in K\} \le n\, \alpha_\varepsilon(t/n, K) \to 0$$

as $n \to \infty$, and so (9.10) is established.

Exercises

(9.13) Verify the Remark following the statement of Theorem 9.4. [Hint: let $P_t(x, A)$ satisfy the hypotheses of Theorem 9.4. First show that $U^\alpha \mathbf{C}_0 \subset \mathbf{C}_0$ and that $\mathbf{L} = U^\alpha \mathbf{C}_0$ is independent of $\alpha > 0$. (See (8.9) and the remarks following it.) Next check that $\|P_t f - f\| \to 0$ as $t \to 0$ for all $f \in \mathbf{L}$ and hence for all $f \in \mathbf{L}'$, the uniform closure of $\mathbf{L}$. Finally use the Hahn–Banach theorem to show that $\mathbf{L}' = \mathbf{C}_0$.]

(9.14) Show that any transition function of the form (2.12) satisfies the hypotheses of Theorem 9.4, and that if X is the resulting Hunt process one may assume that $\zeta(\omega) = \infty$ whenever $X_0(\omega) \in \mathbf{R}^N$.

(9.15) Show that the condition that $t \to X_t$ has left-hand limits on $[0, \zeta)$ almost surely in the definition of a standard process is a consequence of the other hypotheses. [Hint: let d be a metric for E_Δ. Show that for each $\varepsilon > 0$, $T = \inf\{t\colon d(X_0, X_t) > \varepsilon\}$ is an $\{\mathscr{F}_t\}$ stopping time, and then consider the sequence of stopping times $T_0 = 0$, $T_{n+1} = T_n + T \circ \theta_{T_n}$.] Note that the same argument shows that paths of a Hunt process have left-hand limits (in E_Δ) on $[0, \infty)$ almost surely.

(9.16) Let $E = \mathbf{R}$, $\mathscr{E} = \mathscr{B}(\mathbf{R})$, and define

$$\begin{aligned} P_t(x, \cdot) &= \varepsilon_{x+t} && \text{if} \quad x \ge 0 \quad \text{or} \quad x + t < 0, \\ &= \tfrac{1}{2}\varepsilon_{x+t} && \text{if} \quad x < 0 \quad \text{and} \quad x + t \ge 0. \end{aligned}$$

Verify that $P_t(x, A)$ is a transition function, but note that $U^\alpha \mathbf{C}_0$ is not contained in $\mathbf{C}_0$. Prove that any standard process with state space $(E, \mathscr{E})$ and this transition function will fail to be quasi-left-continuous on $[0, \infty)$. (The existence of a standard process with this transition function will follow from the results of Chapter III.)

(9.17) Show that for the Brownian motion transition function in R^N (see (2.17)) the hypotheses of (9.4) and of (9.10) are satisfied. The resulting Hunt process is called the *Brownian motion process in* $\mathbf{R}^N$. In view of (9.10) we may assume that all paths of this process are continuous on $[0, \infty)$.

(9.18) Let X be a Hunt process with state space $(\mathbf{R}^N, \mathscr{B}(\mathbf{R}^N))$ and a transition function of the form (2.12). Prove that if almost surely $t \to X_t(\omega)$ is continuous, then the transition function must satisfy the condition of (9.10). [Hint: in this situation we must show that $t^{-1}\,\mu_t(G^c) \to 0$ as $t \to 0$ where $\{\mu_t\}$ is the semigroup appearing in (2.12) and G is a neighborhood of the origin. Let $\{t_n\}$ be a sequence of numbers decreasing to 0 and let r_n be the greatest integer in t_n^{-1}. Use the equality

$$P^0\{|X(kt_n) - X((k+1)t_n)| \le \varepsilon \text{ for all } k \le r_n\} = [\mu_{t_n}(\{x: |x| \le \varepsilon\})]^{r_n}$$

to deduce the desired conclusion.]

(9.19) Let X be a standard process and define $\beta(t) = \sup_{s \le t, x \in E} N_s(x, E_\Delta - \{x\})$. Prove that if $\beta(t) \to 0$ as $t \to 0$, then almost surely the sample functions $t \to X_t$ are step functions with only finitely many jumps in any bounded t interval. [Hint: recall from (8.18) the equality $P^x(T_x > t) = e^{-\lambda(x)t}$ where $T_x = \inf\{t: X_t \ne x\}$. Obtain the estimate $P^x\{T_x \le t\} \le 2\,\beta(t)$ by considering separately $P^x\{T_x \le t;\ X_t \ne x\}$ and $P^x\{T_x \le t;\ X_t = x\}$. Conclude from this estimate that $\lambda(x)$ is bounded. Define $T_1 = \inf\{t: X_t \ne X_0\}$ and $T_{n+1} = T_n + T_1 \circ \theta_{T_n}$. Prove that T_1 is a stopping time and that almost surely $T_n \to \infty$.]

10. Measurability of Hitting Times

In this section we will introduce certain stopping times which will be of central importance in the following chapters. Suppose $X = (\Omega, \mathscr{M}, \mathscr{M}_t, X_t, \theta_t, P^x)$ is a Markov process with state space $(E, \mathscr{E})$. If A is a subset of E_Δ we define two functions

(10.1)
$$D_A(\omega) = \inf\{t \ge 0: X_t(\omega) \in A\}$$
$$T_A(\omega) = \inf\{t > 0: X_t(\omega) \in A\}$$

where in both cases the infimum of the empty set is understood to be $+\infty$. We call D_A the *first entry time* of A and T_A the *first hitting time* of A. It is obvious that these definitions are valid for any stochastic process X. We often omit the adjective "first."

The following properties are immediate consequences of the definitions and we leave their verification to the reader.

(10.2) PROPOSITION
(a) $s + D_A(\theta_s\omega) = \inf\{t \ge s: X_t(\omega) \in A\}$,
(b) $s + T_A(\theta_s\omega) = \inf\{t > s: X_t(\omega) \in A\}$,
(c) $t + D_A \circ \theta_t = D_A$ on the set $\{D_A \ge t\}$,

(d) $t + T_A \circ \theta_t = T_A$ on the set $\{T_A > t\}$,
(e) $D_A \leq T_A$ and $D_A(\omega) = T_A(\omega)$ if $X_0(\omega) \notin A$.

It is sometimes convenient to refer to $s + D_A \circ \theta_s (s + T_A \circ \theta_s)$ as the first entry (hitting) time of A after s. It is immediate from (10.2) that these are increasing functions of s and that

(10.3) $$\lim_{s \downarrow 0}(s + D_A \circ \theta_s) = \lim_{s \downarrow 0}(s + T_A \circ \theta_s) = T_A.$$

The next proposition contains additional elementary properties of D_A and T_A.

(10.4) PROPOSITION. Let A and B be subsets of E_Δ. Then
(a) $A \subset B$ implies $D_A \geq D_B$; $T_A \geq T_B$;
(b) $D_{A \cup B} = \inf(D_A, D_B)$; $T_{A \cup B} = \inf(T_A, T_B)$;
(c) $D_{A \cap B} \geq \sup(D_A, D_B)$; $T_{A \cap B} \geq \sup(T_A, T_B)$;
(d) if $\{A_n\}$ is a sequence of subsets of E_Δ and $A = \bigcup A_n$, then $D_A = \inf D_{A_n}$; $T_A = \inf T_{A_n}$.

Proof. Since the first three statements are obvious we restrict our attention to the fourth, and in particular to T_A. Clearly $\inf T_{A_n} \geq T_A$. On the other hand if $T_A(\omega) < \infty$ then for each $\varepsilon > 0$ there exists a t such that $T_A(\omega) \leq t < T_A(\omega) + \varepsilon$ and $X_t(\omega) \in A$. But then $X_t(\omega)$ is in some A_n and so $\inf T_{A_n} \leq T_A + \varepsilon$ which completes the proof. (Choose $t > 0$ if $T_A(\omega) = 0$.)

We are interested in formulating conditions under which D_A and T_A are stopping times for a large class of sets A. To this end we introduce the theory of capacity. See Brelot [1], Bourbaki [1], Dynkin [1], or Meyer [1] for a more detailed discussion.

(10.5) DEFINITION. Let E be a locally compact separable metric space and let $\mathscr{K}$ be the class of all compact subsets of E. A function $\varphi : \mathscr{K} \to \mathbf{R}$ is called a Choquet *capacity* provided:
(i) if $A, B \in \mathscr{K}$ and $A \subset B$, then $\varphi(A) \leq \varphi(B)$;
(ii) given $A \in \mathscr{K}$ and $\varepsilon > 0$ there exists an open set $G \supset A$ such that for every $B \in \mathscr{K}$ with $A \subset B \subset G$ one has $\varphi(B) - \varphi(A) < \varepsilon$;
(iii) $\varphi(A \cup B) + \varphi(A \cap B) \leq \varphi(A) + \varphi(B)$ for all $A, B \in \mathscr{K}$.

Given such a capacity φ one defines the *inner capacity*, $\varphi_*(A)$, of an *arbitrary* set $A \subset E$ by $\varphi_*(A) = \sup_{K \subset A} \varphi(K)$ where the supremum is taken over all *compact* subsets of A. One next defines the *outer capacity*, $\varphi^*(A)$, of an *arbitrary* set $A \subset E$ by $\varphi^*(A) = \inf_{G \supset A} \varphi_*(G)$ where the infimum is taken over all *open* sets containing A. An arbitrary set A is said to be *capacitable* provided $\varphi^*(A) = \varphi_*(A)$ and we define the capacity of A to be the common value, which we denote by $\varphi(A)$. In view of (10.5ii) the notation is consistent; that is, if $K \in \mathscr{K}$ then K is capacitable and $\varphi_*(K) = \varphi^*(K) = \varphi(K)$. We are now ready to state Choquet's fundamental theorem. We refer the reader to Brelot [1], Bourbaki [1], or Dynkin [1] for a proof. Also a very general discussion in an abstract setting will be found in Meyer [1].

(10.6) THEOREM (*Choquet*). Every Borel (more generally, analytic) subset of E is capacitable.

We are now going to apply (10.6) to the study of hitting and entry times. We assume in the remainder of this section that E is a locally compact separable metric space, that Δ is adjoined to E as the point at infinity if E is noncompact or as an isolated point if E is compact, and that $\mathscr{E}(\mathscr{E}_\Delta)$ is the σ-algebra of Borel subsets of $E(E_\Delta)$. We further assume that the given Markov process X has state space $(E, \mathscr{E}^*)$ and is right continuous, quasi-left-continuous, and that $\mathscr{M}_t \supset \mathscr{F}_{t+}$. Consequently $\mathscr{F}_t = \mathscr{F}_{t+}$. For convenience of exposition we will actually assume that $t \to X_t(\omega)$ is right continuous for all ω. This causes no loss of generality for we can always accomplish this by removing from Ω a fixed set in $\mathscr{F}$ having P^μ measure zero for all μ. The next theorem and Theorem 10.19 are the main results of this section.

(10.7) THEOREM. Under the above assumptions, D_A and T_A are $\{\mathscr{F}_t\}$ stopping times for all Borel (more generally, analytic) subsets A of E_Δ.

We will break up the proof into several steps. First note that if D_A is an $\{\mathscr{F}_t\}$ stopping time then so is $s + D_A \circ \theta_s$ for all $s \in \mathbf{R}_+$ (see the proof of (8.7)), and, consequently, in light of (10.3), we may restrict our attention to D_A.

(10.8) LEMMA. If G is open (in E_Δ) then $D_G = T_G$ and D_G is an $\{\mathscr{F}^0_{t+}\}$ stopping time.

Proof. This results from the right continuity of the paths, which implies that

$$\{D_G < t\} = \bigcup_{r \in \mathbf{Q} \cap [0,t)} \{X_r \in G\} \in \mathscr{F}^0_t.$$

Of course, if $t \to X_t(\omega)$ is only almost surely right continuous one can only assert that $D_G = T_G$ almost surely and that D_G is an $\{\mathscr{F}_t\}$ stopping time.

Until the proof of Theorem 10.7 is completed A, B, K, G, etc., will always denote subsets of E. If $A \subset E$ we define

$$d_A = \min(D_A, \zeta) = D_{A \cup \{\Delta\}}.$$

(10.9) LEMMA. D_A is an $\{\mathcal{F}_t\}$ stopping time if and only if d_A is.

Proof. Since ζ is an $\{\mathcal{F}^0_{t+}\}$ stopping time the "only if" statement is immediate. On the other hand $D_A = d_A$ on $\{d_A < \zeta\}$ and $D_A = \infty$ on $\{d_A = \zeta\}$ and so

$$\{D_A < t\} = \{d_A < t\} \cap \{d_A < \zeta\}$$

which is in $\mathcal{F}_t$ if d_A is an $\{\mathcal{F}_t\}$ stopping time.

For any $A \subset E$ and $t \geq 0$ let

$$R_t(A) = \{\omega\colon X_s(\omega) \in A \cup \{\Delta\}, \text{ for some } s, 0 \leq s \leq t\},$$

$$R_t^*(A) = \{\omega\colon X_s(\omega) \in A \text{ for some } s, 0 \leq s \leq t\}.$$

Obviously $R_t(G)$ and $R_t^*(G)$ are in $\mathcal{F}^0_t$ if G is open. Moreover for any set A, $\{d_A < t\} = \bigcup_n R_{t-1/n}(A)$, and so to show that d_A is an $\{\mathcal{F}_t\}$ stopping time it suffices to show that $R_t(A) \in \mathcal{F}_t$ for all t. Finally note that one always has $R_t(A) \subset \{d_A \leq t\}$.

(10.10) LEMMA. Let $K \subset E$ be compact; then

(i) $R_t(K) \in \mathcal{F}_t$ for all t;

(ii) if $\{G_n\}$ is a decreasing sequence of open subsets of E with $G_n \supset \bar{G}_{n+1} \supset K$ for all n and such that $K = \bigcap G_n = \bigcap \bar{G}_n$, then $d_{G_n} \uparrow d_K$ a.s. and $P^\mu(R_t(G_n)) \downarrow P^\mu(R_t(K))$ for all μ and t.

Proof. Let $\{G_n\}$ be as in (ii); then $\{d_{G_n}\}$ is an increasing sequence of $\{\mathcal{F}^0_{t+}\}$ stopping times, and so its limit, which we denote by T, is an $\{\mathcal{F}^0_{t+}\}$ stopping time. Clearly $T \leq d_K \leq \zeta$. If $T = \zeta$ then $T = d_K$, while on $\{T < \zeta\}$, $d_{G_n} = D_{G_n}$ for all n and $X(D_{G_n}) \to X(T)$ a.s. on $\{T < \zeta\}$ by the quasi-left-continuity. But $X(D_{G_n}) \in \bar{G}_n$ on $\{D_{G_n} < \infty\}$, and consequently $X(T) \in \bigcap \bar{G}_n = K$ a.s. on $\{T < \zeta\}$. Hence $d_K \leq T$ a.s. on $\{T < \zeta\}$. Therefore $d_{G_n} \uparrow d_K$ a.s. and so d_K is an $\{\mathcal{F}_t\}$ stopping time.

If $\omega \in \{d_K \leq t\}$, then either $X_s(\omega) \in K \cup \{\Delta\}$ for some $s \leq t$ or there exists a sequence $\{s_n\}$ decreasing to t with $X_{s_n}(\omega) \in K \cup \{\Delta\}$ for all n. But $K \cup \{\Delta\}$ is closed and so the right continuity of the paths implies that in this case $X_t(\omega) \in K \cup \{\Delta\}$. In other words $\{d_K \leq t\} \subset R_t(K)$ and hence these two sets are equal. Therefore $R_t(K) \in \mathcal{F}_t$. Finally $\{d_{G_n} \leq t\}$ decreases to $\{d_K \leq t\}$ a.s. as $n \to \infty$, and since

$$\{d_{G_n} \leq t\} \supset R_t(G_n) \supset R_t(K) = \{d_K \leq t\}$$

it follows that $P^\mu[R_t(G_n)] \downarrow P^\mu[R_t(K)]$. Thus Lemma 10.10 is established.

(10.11) REMARK. If X is assumed quasi-left-continuous on $[0, \infty)$ rather than on $[0, \zeta)$, then one sees by exactly the same argument that (10.10) remains valid if we replace R by R^* and d by D throughout.

(10.12) LEMMA. Let μ and t be fixed; then $\varphi(K) = P^\mu[R_t(K)]$ is a Choquet capacity on the compact subsets of E.

Proof. We must check that the three properties of Definition 10.5 are valid. Property (i) is obvious. Property (ii) is an immediate consequence of Lemma (10.10ii). As to Property (iii), if A and B are compact, then

$R_t(A \cup B) - R_t(B)$

$= \{\omega: X_s(\omega) \in A \cup B \cup \{\Delta\}$ for some $s \leq t$ but $X_s(\omega) \notin B \cup \{\Delta\}$ for any $s \leq t\}$

$\subset \{\omega: X_s(\omega) \in A$ for some $s \leq t$ but $X_s(\omega) \notin (A \cap B) \cup \{\Delta\}$

for any $s \leq t\}$

$\subset R_t(A) - R_t(A \cap B)$,

and therefore $\varphi(A \cup B) - \varphi(B) \leq \varphi(A) - \varphi(A \cap B)$. Hence φ is a Choquet capacity.

(10.13) REMARK. Under the assumption of (10.11) the set function $K \to P^\mu[R_t^*(K)]$ is a Choquet capacity for all μ and t.

(10.14) LEMMA. For any Borel set B, $R_t(B) \in \mathscr{F}_t$.

Proof. Let φ be the capacity $\varphi(K) = P^\mu[R_t(K)]$. If G is open and K is any compact contained in G, then $\varphi(K) = P^\mu[R_t(K)] \leq P^\mu[R_t(G)]$ and hence $\varphi_*(G) = \sup_{K \subset G} \varphi(K) \leq P^\mu[R_t(G)]$. Let $\{K_n\}$ be an increasing sequence of compact subsets of G whose union is G. Clearly $R_t(K_n) \uparrow R_t(G)$ and so $\varphi(K_n) \uparrow P^\mu[R_t(G)]$. Thus $\varphi_*(G) = P^\mu[R_t(G)]$ for any open set G. If B is a Borel set, then by Choquet's theorem (10.6), $\varphi_*(B) = \varphi^*(B)$. Hence for each n there exist a compact set $K_n \subset B$ and an open set $G_n \supset B$ such that

$$\varphi_*(G_n) - \varphi(K_n) = P^\mu[R_t(G_n)] - P^\mu[R_t(K_n)] < 1/n.$$

Let $\Lambda_1 = \bigcup_n R_t(K_n)$ and $\Lambda_2 = \bigcap_n R_t(G_n)$; then Λ_1 and Λ_2 are in $\mathscr{F}_t$ and

$\Lambda_1 \subset R_t(B) \subset \Lambda_2$. But $P^\mu(\Lambda_2 - \Lambda_1) \leq P^\mu[R_t(G_n)] - P^\mu[R_t(K_n)] < 1/n$ for each n, and hence $P^\mu(\Lambda_2 - \Lambda_1) = 0$. Since μ is arbitrary it is now immediate that $R_t(B) \in \mathscr{F}_t$.

(10.15) REMARK. The proof of (10.14) actually shows that $\varphi(B) = P^\mu[R_t(B)]$ for any Borel set B where $\varphi(B)$ is the capacity of B. Finally note that under the assumptions of (10.11) if $\psi(K)$ is defined to be $P^\mu[R_t^*(K)]$ then again ψ is a Choquet capacity and $\psi(B) = P^\mu[R_t^*(B)]$ for any Borel set B.

Theorem 10.7 is now an immediate corollary of the above lemmas. Thus we now know that T_A, D_A, and d_A are $\{\mathscr{F}_t\}$ stopping times for any Borel set A. However useful this information may be, the following approximation theorems are even more useful.

(10.16) THEOREM. If B is a Borel subset of E, then for each μ there exist an increasing sequence $\{K_n\}$ of compact subsets of B and a decreasing sequence $\{G_n\}$ of open sets containing B such that $d_{K_n} \downarrow d_B$ and $d_{G_n} \uparrow d_B$ almost surely P^μ.

Proof. Let μ be given. Since $\varphi(B) = P^\mu[R_t(B)]$, there exists for each $t \geq 0$ an increasing sequence $\{K_n^t\}$ of compact subsets of B such that $P^\mu[R_t(K_n^t)] \uparrow P^\mu[R_t(B)]$. Let $\{q_j\}$ be an enumeration of the nonnegative rationals. Define $K_n = K_n^{q_1} \cup \ldots \cup K_n^{q_n}$. Clearly $\{K_n\}$ is an increasing sequence of compact subsets of B, and so $\{d_{K_n}\}$ decreases to a limit $T \geq d_B$. Now $\{d_B < T\} = \bigcup_n \{d_B < q_n < T\}$ and for each m

$$\begin{aligned}\{d_B < q_n < T\} &\subset \{d_B < q_n < d_{K_m}\}\\ &\subset R_{q_n}(B) - R_{q_n}(K_m).\end{aligned}$$

But if $m > n$ then $K_m \supset K_m^{q_n}$ and so

$$\{d_B < q_n < T\} \subset R_{q_n}(B) - R_{q_n}(K_m^{q_n}).$$

Consequently $P^\mu[d_B < q_n < T] = 0$, and this implies that $T = \lim d_{K_n} = d_B$ almost surely P^μ.

The required sequence $\{G_n\}$ of open sets is constructed in a similar manner. See Exercise 10.22.

(10.17) COROLLARY. If B is a Borel subset of E, then for each μ there exist an increasing sequence $\{K_n\}$ of compact subsets of B and a decreasing sequence $\{G_n\}$ of open sets containing B such that $D_{K_n} \downarrow D_B$ almost surely P^μ on Ω and $D_{G_n} \uparrow D_B$ almost surely P^μ on $\{D_B < \infty\}$.

Proof. By (10.16) there exists such a sequence $\{K_n\}$ with $d_{K_n} \downarrow d_B$ almost surely

P^μ. Now on $\{D_B < \infty\} = \{D_B < \zeta\}$ one has $d_B = D_B$ and hence $d_{K_n} < \zeta$ for n sufficiently large. But then $d_{K_n} = D_{K_n}$ and consequently $D_{K_n} \downarrow D_B$ on $\{D_B < \infty\}$ a.s. P^μ. But on $\{D_B = \infty\}$ each $D_{K_n} = \infty$ since $B \supset K_n$. Again from (10.16) there exists an appropriate sequence $\{G_n\}$ with $d_{G_n} \uparrow d_B = D_B \wedge \zeta$ a.s. P^μ, and on $\{D_B < \infty\} = \{D_B < \zeta\}$ one has $d_{G_n} = D_{G_n} < \zeta$. Therefore $D_{G_n} \uparrow D_B$ a.s. P^μ on $\{D_B < \infty\}$.

(10.18) REMARK. If X is quasi-left-continuous on $[0, \infty)$ then making use of (10.11), (10.13), and (10.15) one can repeat the proof of (10.16) to obtain for a given μ a decreasing sequence $\{G_n\}$ of open sets containing B such that $D_{G_n} \uparrow D_B$ almost surely P^μ.

(10.19) THEOREM. *If B is a Borel subset of E then for each μ there exists an increasing sequence $\{K_n\}$ of compact subsets of B such that $T_{K_n} \downarrow T_B$ a.s. P^μ.*

Proof. Let $\{t_k\}$ be a strictly decreasing sequence of positive numbers with limit 0. We know that $t_k + D_A \circ \theta_{t_k} \downarrow T_A$ as $k \to \infty$ for any set A. Suppose that we could find an increasing sequence $\{K_n\}$ of compact subsets of B such that $D_{K_n} \circ \theta_{t_k} \downarrow D_B \circ \theta_{t_k}$ a.s. P^μ as $n \to \infty$ for each k. Then

$$\begin{aligned} T_B &= \lim_k \lim_n (t_k + D_{K_n} \circ \theta_{t_k}) \\ &= \lim_n \lim_k (t_k + D_{K_n} \circ \theta_{t_k}) \\ &= \lim_n T_{K_n} \quad \text{a.s.} \quad P^\mu, \end{aligned}$$

where the interchange of limits is justified since the double sequence in question is decreasing in both n and k. Thus the proof of (10.19) is reduced to constructing a sequence $\{K_n\}$ with the above property. For the given μ define for $A \in \mathscr{E}$

$$\nu_k(A) = \int \mu(dx)\, P^x[X(t_k) \in A] = P^\mu[X(t_k) \in A].$$

According to Corollary 10.17 there exists for each k an increasing sequence $\{K_n^k\}$ of compact subsets of B such that $D_{K_n^k} \downarrow D_B$ a.s. P^{ν_k}. Define $K_n = K_n^1 \cup K_n^2 \cup \ldots K_n^n$. Clearly $\{K_n\}$ is an increasing sequence of compact subsets of B. Moreover $K_n^k \subset K_n \subset B$ if $n \geq k$ and so $D_B \leq D_{K_n} \leq D_{K_n^k}$ for $n \geq k$. Thus $D_{K_n} \downarrow D_B$ a.s. P^{ν_k} for all k. If $T = \lim_n D_{K_n}$ (which exists since $\{K_n\}$ is increasing), then we have shown so far that $P^{\nu_k}[T \neq D_B] = 0$ for all k. Clearly $D_{K_n} \circ \theta_{t_k} \downarrow T \circ \theta_{t_k}$ as $n \to \infty$ and

$$\begin{aligned} P^\mu[T \circ \theta_{t_k} \neq D_B \circ \theta_{t_k}] &= E^\mu\, P^{X(t_k)}(T \neq D_B) \\ &= P^{\nu_k}(T \neq D_B) = 0. \end{aligned}$$

Therefore $D_{K_n} \circ \theta_{t_k} \downarrow D_B \circ \theta_{t_k}$ as $n \to \infty$ a.s. P^μ for all k, and thus the construction is complete.

Let us emphasize that the approximating sequences constructed in (10.16)–(10.19) depend in general on the measure μ. See Exercise 10.24 for a sharpening of (10.19) under an additional hypothesis. The only place that we have used the Markov property in the present section is in the proof of (10.19). We should also point out that the analogous approximation theorem by open sets $G_n \supset B$ is not valid in general. A simple example is given by uniform motion to the right along the real axis (see (3.7)). If $B = \{0\}$ then $P^0(T_B < \infty) = 0$, while if G is any open set containing 0 then $P^0(T_G = 0) = 1$.

In the remainder of this section we will assume that X is a standard process although somewhat weaker hypotheses would suffice. The above discussion of course applies to any standard process.

(10.20) Proposition. Let X be a standard process and suppose that B is a Borel set. Define

$$R_t'(B) = R_t(B) \cup \{\omega : X_{s-}(\omega) \text{ exists and is in } B \cup \{\Delta\} \text{ for some } s,\ 0 \le s \le t\}.$$

Then $P^\mu[R_t'(B) - R_t(B)] = 0$ for all μ. In particular $R_t'(B) \in \mathscr{F}_t$.

Proof. Clearly $R_t(A) \subset R_t'(A)$ for any set A. For the given B and μ there exists a decreasing sequence $\{G_n\}$ of open sets containing B such that $P^\mu[R_t(G_n) - R_t(B)] \le 1/n$ for each n. Also for any open set G, $R_t(G) = R_t'(G)$. Therefore

$$R_t(B) \subset R_t'(B) \subset \bigcap_n R_t(G_n)$$

and Proposition 10.26 now follows since the extreme members in this string of inclusions have the same P^μ measure.

Let $\zeta'(\omega) = \inf\{t\colon X_t(\omega) = \Delta \text{ or } X_{t-}(\omega) \text{ exists and equals } \Delta\}$. Clearly $\zeta' \le \zeta$ and

$$\{\zeta' < r < \zeta\} \subset \{s \to X_s \text{ is unbounded in } E \text{ on } [0, r],\ r < \zeta\}.$$

Consequently according to (9.3), $\zeta = \zeta'$ almost surely. Combining this with (10.20) one easily sees that almost surely $D_B = \inf\{t\colon X_t \in B \text{ or } X_{t-} \text{ exists and is in } B\}$ for any Borel set B.

In many situations it is necessary to deal with a class of sets that lies between the Borel sets and the universally measurable sets.

(10.21) Definition. A set $A \subset E_\Delta$ is a *nearly* Borel *set* (relative to the given process X) if for each μ there exist Borel subsets B and B' of E_Δ such that

$B \subset A \subset B'$ and

$$P^{\mu}[X_t \in B' - B \text{ for some } t \geq 0] = 0.$$

This is equivalent to $P^{\mu}[D_{B'-B} < \infty] = 0$. Roughly speaking a set is nearly Borel if the process can not distinguish it from a Borel set. Clearly the above definition makes sense for any process and not just for standard processes. The reader should verify for himself that the class of nearly Borel sets is a σ-algebra in E_{Δ}. We denote it by $\mathscr{E}^n_{\Delta}$. The σ-algebra of nearly Borel subsets of E will be denoted by $\mathscr{E}^n$. Obviously $\mathscr{E} \subset \mathscr{E}^n \subset \mathscr{E}^*$ and a similar relationship holds for E_{Δ}. A function $f \in \mathscr{E}^n$ is called nearly Borel measurable. Such functions are characterized by the property that for each μ there exist $f_1, f_2 \in \mathscr{E}$ such that $f_1 \leq f \leq f_2$ and

$$P^{\mu}[f_1(X_t) \neq f_2(X_t) \text{ for some } t \geq 0] = 0.$$

It is easy to check that all of the results of this section which were proved for Borel sets extend to nearly Borel sets, and we will use them for nearly Borel sets without special mention. We leave the details to the reader.

Exercises

(10.22) Give a detailed construction of the sequence of open sets $\{G_n\}$ in Theorem 10.16.

(10.23) Let X be a standard process having as transition function the one defined in (9.16). Let $B = \{0\}$. Show that if μ is a measure with $\mu[(-\infty, 0)] > 0$, then there is no sequence $\{G_n\}$ of open sets containing B and such that $D_{G_n} \to D_B$ almost surely P^{μ}. Consequently (10.17) can not in general be strengthened.

(10.24) Let X be a standard process with transition function $P_t(x, A)$ and suppose that there exists a σ-finite measure λ on $\mathscr{E}$ such that for each x in E and $t > 0$ the measure $P_t(x, \cdot)$ is absolutely continuous with respect to λ. Show that if B is a Borel set then there is an increasing sequence $\{K_n\}$ of compact subsets of B such that T_{K_n} decreases to T_B almost surely. [Hint: let μ be a finite measure equivalent to λ and choose the sequence $\{K_n\}$ such that $T_{K_n} \downarrow T_B$ almost surely P^{μ}.]

(10.25) Let X be a standard process and let A be a nearly Borel subset of E. Let $T = T_A$ and let $\alpha(t) = \sup_{x \in E} P^x[T > t]$. Show that for each $x \in E$ and $t > 0$ one has $E^x(T) \leq t[1 - \alpha(t)]^{-1}$. [Hint: if $\alpha(t) = 1$, there is nothing to prove. If $\alpha(t) < 1$, show that $P^x[T > nt] \leq [\alpha(t)]^n$, and then write $E^x(T) = \int_0^\infty P^x(T \geq s)\, ds$.]

(10.26) Prove that the class of nearly Borel sets in E_Δ forms a σ-algebra contained in $\mathscr{E}_\Delta^*$.

11. Further Properties of Hitting Times

Throughout this section X will denote a given standard process with state space $(E, \mathscr{E})$. If T is any $\{\mathscr{M}_t\}$ stopping time and $\alpha \geq 0$ we define the kernel P_T^α as follows: If $f \in b\mathscr{E}_\Delta^*$ then

$$P_T^\alpha f(x) = E^x\{e^{-\alpha T} f(X_T);\ T < \infty\}.$$

In particular if $f \in b\mathscr{E}^*$, then $P_T^\alpha f(x) = E^x\{e^{-\alpha T} f(X_T);\ T < \zeta\}$. If T is an $\{\mathscr{M}_t\}$ stopping time this need not be a measurable function of x. However if T is an $\{\mathscr{F}_t\}$ stopping time then P_T^α is a bounded linear operator from $b\mathscr{E}_\Delta^*$ to $b\mathscr{E}_\Delta^*$ (or from $b\mathscr{E}^*$ to $b\mathscr{E}^*$). As usual the definition extends to any nonnegative function in $\mathscr{E}_\Delta^*$. If $\alpha = 0$ we write P_T in place of P_T^0. If $T = T_A$ where $A \in \mathscr{E}_\Delta^n$ we write P_A^α in place of $P_{T_A}^\alpha$, and P_A when $\alpha = 0$. The measure $P_A^\alpha(x, \cdot)$ is called the *α-order hitting distribution of A starting from x* or the *α-order harmonic measure of A relative to x*. When $\alpha = 0$ we drop it from our terminology as usual.

If A is a nearly Borel subset of E, then the right continuity of the paths implies that all of the measures $P_A^\alpha(x, \cdot)$ are concentrated on $\bar{A}$ (the closure of A in E). In actual fact one can make a good deal more precise statement than this, and we next develop the necessary machinery for this more precise statement.

If T is an $\{\mathscr{F}_t\}$ stopping time then $\{T = 0\} \in \mathscr{F}_0$, and so, for each $x \in E_\Delta$, $P^x(T = 0)$ is either zero or one according to the zero-one law (5.17). The point x is said to be *irregular for T* if this probability is zero and *regular for T* if this probability is one.

(11.1) DEFINITION. If $A \in \mathscr{E}_\Delta^n$ then a point x is *regular for A* provided it is regular for T_A, and x is *irregular for A* provided it is irregular for T_A.

Thus x is regular for A if $P^x(T_A = 0) = 1$ and irregular for A if $P^x(T_A > 0) = 1$. Intuitively x is regular for A if the process starting from x is in A at arbitrarily small strictly positive times with probability one. If A^r is the set of all points which are regular for the nearly Borel set A, then $A^r = \{x\colon P^x(T_A = 0) = 1\}$ and consequently A^r is universally measurable. In fact A^r is itself nearly Borel but we cannot prove this until later. If A^0 denotes the interior of A, then plainly $A^0 \subset A^r \subset \bar{A}$.

We can now complete the approximation theorems given in Section 10.

(11.2) THEOREM. Let $A \in \mathscr{E}^n$ and let μ be a finite measure such that $\mu(A - A^r) = 0$. Then there exists a decreasing sequence $\{G_n\}$ of open sets containing A such that $T_{G_n} \uparrow T_A$ almost surely P^μ on $\{T_A < \infty\}$ and $T_{G_n} \wedge \zeta \uparrow T_A \wedge \zeta$ almost surely P^μ (on Ω).

Proof. According to (10.16) and (10.17) there exists such a sequence of open sets for which a.s. P^μ one has $D_{G_n} \uparrow D_A$ on $\{D_A < \infty\}$ and $D_{G_n} \wedge \zeta \uparrow D_A \wedge \zeta$. But for any open set G, $D_G = T_G$, and so to complete the proof we need only show that $P^\mu(D_A \neq T_A) = 0$. But

$$P^\mu(D_A < T_A) = \int \mu(dx)\, P^x(D_A < T_A).$$

If $x \in A^c$ then $P^x(D_A = T_A) = 1$, while if $x \in A^r$ then $0 \leq D_A \leq T_A = 0$ almost surely P^x. Since μ attributes all its mass to $A^c \cup A^r$ it follows that $P^\mu(D_A < T_A) = 0$ and so the proof of (11.2) is complete.

(11.3) REMARK. If X is quasi-left-continuous on $[0, \infty)$, then using (10.18) and the above argument one can strengthen the conclusion of (11.2) to $T_{G_n} \uparrow T_A$ almost surely P^μ.*

The next result gives us the precise information alluded to earlier regarding $P_A^\alpha(x, \cdot)$.

(11.4) THEOREM. Let $A \in \mathscr{E}_\Delta^n$; then

(i) $X(T_A) \in A \cup A^r$ a.s. on $\{T_A < \infty\}$;

(ii) for each x and α, $P_A^\alpha(x, \cdot)$ is concentrated on $A \cup A^r$.

Proof. We will prove only (i), for (ii) is an obvious consequence of (i). Since $X_t \notin A$ for $0 < t < T_A$ it is the case that if $X_{T_A} \notin A$ then $T_A + T_A \circ \theta_{T_A} = T_A$. Consequently for any μ

$$\begin{aligned} P^\mu(T_A < \infty, X_{T_A} \notin A) &= P^\mu(T_A \circ \theta_{T_A} = 0, T_A < \infty, X_{T_A} \notin A) \\ &= E^\mu\{P^{X(T_A)}(T_A = 0);\ T_A < \infty, X_{T_A} \notin A\}. \end{aligned}$$

Therefore almost surely P^μ on $\{T_A < \infty, X_{T_A} \notin A\}$ we must have

* Theorem 6.1 of Chapter III contains another important extension of (11.2).

$$P^{X(T_A)}(T_A = 0) = 1;$$

that is, $X_{T_A} \in A^r$. This clearly yields Statement (i) of Theorem 11.4.

Exercises

(11.5) If T and R are stopping times show that $X_T \circ \theta_R = X_{R+T\circ\theta_R}$.

(11.6) Let $A \in \mathscr{E}^n_\Delta$ and $\alpha > 0$. Prove that $x \in A^r$ if and only if $P^\alpha_A(x, E_\Delta) = 1$.

(11.7) Let $A \in \mathscr{E}^n_\Delta$. Prove that if $x \notin A$ then $x \in A^r$ if and only if $P_A(x, \cdot) = \varepsilon_x$.

(11.8) Show by an example that the condition "$x \notin A$" in (11.7) cannot be eliminated. [Hint: consider the Hunt process "uniform motion around a circle." The definition of this process should be clear by analogy with Exercise 3.7.]

(11.9) Let A and B be in $\mathscr{E}^n_\Delta$ with $A \subset B$. Show that if $X_{T_B} \notin A - A^r$ almost surely on $\{T_B > 0\}$ then $P^\alpha_B P^\alpha_A = P^\alpha_A$ for all $\alpha \geq 0$, and conversely that if $P^\alpha_B P^\alpha_A = P^\alpha_A$ for some $\alpha > 0$ then $X_{T_B} \notin A - A^r$ almost surely on $\{T_B > 0\}$. Show that $P^\alpha_A P^\alpha_B = P^\alpha_A$ for all $\alpha \geq 0$ provided that $X_{T_A} \in B^r$ almost surely on $\{T_A < \infty\}$, and that if $P^\alpha_A P^\alpha_B = P^\alpha_A$ for some $\alpha > 0$, then $X_{T_A} \in B^r$ almost surely on $\{T_A < \infty\}$.

12. Regular Step Processes

In this section we will introduce an important class of Markov processes. However, the material in this section will not be needed in the sequel except as a source of examples, and so the reader may omit a detailed reading of the proofs. The processes to be constructed have a very simple intuitive description; however, we will go into some detail regarding the construction.

By a *Markov kernel* Q on a measurable space $(E, \mathscr{E})$ we mean a function $Q(x, A)$ defined for $x \in E$ and $A \in \mathscr{E}$ such that (a) for each $A \in \mathscr{E}$, $x \to Q(x, A)$ is in $\mathscr{E}$, and (b) for each $x \in E$, $A \to Q(x, A)$ is a probability measure on $\mathscr{E}$. If in (b) above we only assume that, for each $x \in E$, $A \to Q(x, A)$ is a measure on $\mathscr{E}$ with $Q(x, E) \leq 1$, then the resulting object Q is called a *sub-Markov kernel* on $(E, \mathscr{E})$. We now fix a measurable space $(E, \mathscr{E})$, and in this section we will assume that $\{x\} \in \mathscr{E}$ for each $x \in E$. The basic data from which we will construct the process are (i) a function $\lambda \in \mathscr{E}$ satisfying $0 < \lambda(x) < \infty$ for all $x \in E$, and (ii) a Markov kernel $Q(x, A)$ on $(E, \mathscr{E})$ satisfying $Q(x, \{x\}) = 0$

for all x. We can now give an intuitive description of the process to be constructed. A particle starting from a point $x_0 \in E$ remains there for an exponentially distributed holding time T_1 with parameter $\lambda(x_0)$ at which time it "jumps" to a new position x_1 according to the probability distribution $Q(x_0, \cdot)$. It then remains at x_1 for a length of time T_2 which is exponentially distributed with parameter $\lambda(x_1)$, but which is conditionally independent, given x_1, of T_1. Then it jumps to x_2 according to $Q(x_1, \cdot)$, and so on. Most of this section is devoted to a rigorous construction of such a process.

Let $F = E \times \mathbf{R}_+$ and $\mathscr{F} = \mathscr{E} \times \mathscr{B}(\mathbf{R}_+)$, where $\mathbf{R}_+ = [0, \infty)$ as usual. Let $\mathrm{N} = \{0, 1, 2, \ldots\}$, and set $\Omega = F^{\mathrm{N}}$ and $\mathscr{G} = \mathscr{F}^{\mathrm{N}}$. Thus $(F, \mathscr{F})$ is a measurable space and $(\Omega, \mathscr{G})$ is the usual infinite product measurable space over $(F, \mathscr{F})$. The points $\omega \in \Omega$ are sequences $\{(x_n, t_n); n \geq 0\}$ with $x_n \in E$ and $t_n \in \mathbf{R}_+$ for all n. If $\omega = \{(x_n, t_n); n \geq 0\}$ let $Y_n(\omega) = (x_n, t_n)$, $Z_n(\omega) = x_n$, and $\tau_n(\omega) = t_n$. Thus Y_n is the nth coordinate map and $Y_n = (Z_n, \tau_n)$. Obviously $Y_n \in \mathscr{G}/\mathscr{F}$, $Z_n \in \mathscr{G}/\mathscr{E}$, and $\tau_n \in \mathscr{G}$ for each n.

We next define a Markov kernel π on the measurable space $(F, \mathscr{F})$ as follows:

$$\text{(12.1)} \qquad \pi(x, t; dy, ds) = \begin{cases} 0, & s < t, \\ Q(x, dy)\, \lambda(x)\, e^{-\lambda(x)(s-t)}\, ds, & s \geq t. \end{cases}$$

Note that π is translation invariant in the $\mathbf{R}_+$ variables. We now invoke a theorem of Ionescu Tulcea (see Doob [1, pp. 613–615] or Neveu [1]) which states the following in the present situation: if γ is a probability measure on $(F, \mathscr{F})$, then there exists a probability measure P^γ on $(\Omega, \mathscr{G})$ such that the coordinate mappings $\{Y_n; n \in \mathrm{N}\}$ form a (temporally homogeneous) Markov process over $(\Omega, \mathscr{G}, P^\gamma)$ with γ as initial measure and π as transition function; that is,

$$\text{(12.2)} \qquad P^\gamma(Y_{n+1} \in A \mid Y_0, \ldots, Y_n) = \int_A \pi(Z_n, \tau_n, dy, ds)$$

for all $A \in \mathscr{F}$ and $n \geq 0$. If γ is unit mass at $(x, 0) \in F$ we will write P^x for P^γ.

One easily calculates that for each γ

$$E^\gamma\{\exp[-\alpha(\tau_{n+1} - \tau_n)]\} = E^\gamma\{\lambda(Z_n)/[\alpha + \lambda(Z_n)]\},$$

and hence letting $\alpha \to \infty$ we see that $P^\gamma(\tau_{n+1} \leq \tau_n) = 0$. Moreover from the assumption that $Q(x, \{x\}) = 0$ for all x it easily follows that $P^\gamma(Z_{n+1} = Z_n) = 0$ for all n and γ. Finally $P^x(\tau_0 = 0) = 1$ for all x. Consequently if $\Omega' = \{Z_{n+1} \neq Z_n \text{ and } \tau_{n+1} > \tau_n \text{ for all } n \text{ and } \tau_0 = 0\}$, then $\Omega' \in \mathscr{G}$ and $P^x(\Omega') = 1$ for all x. If $\mathscr{G}'$ denotes the trace of $\mathscr{G}$ on Ω', then, for each x, $\{Y_n; n \in \mathrm{N}\}$ is a Markov process over $(\Omega', \mathscr{G}', P^x)$ with $\varepsilon_{x,0}$ as initial measure and π as transition function. We now drop the prime from our notation; that is, we now let Ω denote the set of all sequences $\{(x_n, t_n); n \geq 0\}$ with $0 = t_0 < t_1 < \ldots$ and

$x_{n+1} \neq x_n$ for all $n \geq 0$, and let $\mathscr{G}$ be the σ-algebra in Ω generated by the coordinate mappings $\{Y_n; n \in \mathbf{N}\}$. The measures P^x are now regarded as probability measures on this $(\Omega, \mathscr{G})$.

Finally we are ready to define the process which interests us. First let $\zeta(\omega) = \lim_n \tau_n(\omega)$ and then define for $t \geq 0$

$$X_t(\omega) = Z_n(\omega) \quad \text{if} \quad \tau_n(\omega) \leq t < \tau_{n+1}(\omega),$$

(12.3)

$$= \Delta \quad \text{if} \quad \zeta(\omega) \leq t.$$

where Δ is a point adjoined to E in the usual manner. Of course we set $X_\infty(\omega) = \Delta$ for all ω. Clearly $t \to X_t(\omega)$ is a step function on $[0, \zeta(\omega))$ for each ω. Consequently if E is given the discrete topology then $t \to X_t(\omega)$ is right continuous on $[0, \infty)$ and has left-hand limits on $[0, \zeta(\omega))$ for each ω. Moreover $X_t \in \mathscr{G}/\mathscr{E}_\Delta$ for each t. We adjoin a point ω_Δ to Ω and set $X_t(\omega_\Delta) = \Delta$ for all t. We put $\{\omega_\Delta\} \in \mathscr{G}$ and *set* $P^x(\{\omega_\Delta\}) = 0$ for all $x \in E$. Also we set $Z_n(\omega_\Delta) = \Delta$ and $\tau_n(\omega_\Delta) = \infty$ for all n. Next we define translation operators as follows: $\theta_t \omega_\Delta = \omega_\Delta$ for all t; if $\omega = \{(x_n, t_n); n \geq 0\}$ and $t \geq \zeta(\omega) = \lim_n t_n$ then $\theta_t \omega = \omega_\Delta$, while if $t_k \leq t < t_{k+1}$, $k \geq 0$, then $\theta_t \omega = \{(x_{n+k}, [t_{n+k} - t] \vee 0); n \geq 0\}$. One checks immediately that $X_s \circ \theta_t = X_{s+t}$ for all s and t. Finally we let P^Δ be unit mass at ω_Δ.

We now come to the main result of this section.

(12.4) THEOREM. $X = (\Omega, \mathscr{G}, X_t, \theta_t, P^x)$ is a Markov process with state space $(E, \mathscr{E})$.

Proof. We will carry out the proof in a series of observations and lemmas, leaving detailed verification to the reader. First of all let $\mathscr{G}_n = \sigma(Y_i; 0 \leq i \leq n)$ and let $\mathscr{F}_t^0 = \sigma(X_s; s \leq t)$. We assert that if $\Lambda \in \mathscr{F}_t^0$, then for each n there is a set $\Lambda_n \in \mathscr{G}_n$ such that

$$\Lambda \cap \{\tau_n \leq t < \tau_{n+1}\} = \Lambda_n \cap \{t < \tau_{n+1}\}.$$

Indeed, the class of sets Λ with this property is a σ-algebra and obviously it contains the sets $\{X_s \in A\}$ with $A \in \mathscr{E}$ and $s \leq t$.

We pause to introduce some additional notation. If $\alpha \geq 0$, we define for $x \in E$ and $A \in \mathscr{E}$

$$\textbf{(12.5)} \qquad Q_\alpha(x, A) = \frac{\lambda(x)}{\alpha + \lambda(x)} Q(x, A).$$

If $K(r, B)$, $r \in C$, $B \in \mathscr{C}$, is any sub-Markov kernel on a measurable space $(C, \mathscr{C})$ we define the iterates, K^n, of K by $K^0(r, B) = \varepsilon_r(B)$ and $K^{n+1}(r, B) = \int K(s, B) K^n(r, ds)$ for $n \geq 0$. In the present case, and with this notation, a simple induction argument shows that

(12.6) $$\int_0^\infty e^{-\alpha s}\,\pi^n(x, 0; A, ds) = Q_\alpha^n(x, A)$$

for all $\alpha \geq 0$, $n \geq 0$, $x \in E$, and $A \in \mathscr{E}$. It also follows by induction that π^n is translation invariant in its $\mathbf{R}_+$ variables.

The following lemma contains the basic calculations which we will need. The proof is a straightforward calculation which we will leave to the reader. It makes use of (12.6) and the Markov property for $\{Y_n\}$ (that is, the fact that the finite-dimensional distributions of $\{Y_n\}$ can be written down in terms of the iterates of π).

(12.7) LEMMA. Let $g \in \mathscr{E}_+$ and $\alpha \geq 0$. Then for each x

(i) $E^x\{g(Z_k)\,[e^{-\alpha\tau_k} - e^{-\alpha\tau_{k+1}}] \mid \mathscr{G}_n\}$

$$= \alpha\, e^{-\alpha\tau_n}\, Q_\alpha^{k-n}\left(Z_n, \frac{g}{\alpha + \lambda}\right), \qquad k \geq n;$$

(ii) $E^x\{e^{-\alpha\tau_{n+1}}\, g(Z_{n+1});\ \tau_{n+1} > t \mid \mathscr{G}_n\}$

$$= Q_\alpha(Z_n, g)\exp[\lambda(Z_n)\,\tau_n - (\alpha + \lambda(Z_n))(t \vee \tau_n)],$$

for each $n \geq 0$ and $t \geq 0$.

Next given $\alpha \geq 0$ and $f \in \mathscr{E}_+$ (with f extended to Δ by $f(\Delta) = 0$) we define

$$U^\alpha f(x) = E^x \int_0^\infty e^{-\alpha t} f(X_t)\,dt.$$

Since by definition

$$\int_0^\infty e^{-\alpha t} f(X_t)\,dt = \int_0^\zeta e^{-\alpha t} f(X_t)\,dt$$

$$= \sum_{k=0}^\infty \int_{\tau_k}^{\tau_{k+1}} e^{-\alpha t} f(X_t)\,dt = \sum_{k=0}^\infty f(Z_k)\left[\frac{e^{-\alpha\tau_k} - e^{-\alpha\tau_{k+1}}}{\alpha}\right]$$

an application of (12.7i) yields

(12.8) $$U^\alpha f(x) = \sum_{k=0}^\infty Q_\alpha^k\left(x, \frac{f}{\alpha + \lambda}\right)$$

$$= \frac{f(x)}{\alpha + \lambda(x)} + Q_\alpha(x, U^\alpha f).$$

(12.9) LEMMA. If $f \in \mathscr{E}_+$, $t \geq 0$, $\alpha \geq 0$, and $x \in E$, then

$$E^x\left\{\int_t^\infty e^{-\alpha u} f(X_u)\,du \mid \mathscr{F}_t^0\right\} = e^{-\alpha t}\, U^\alpha f(X_t).$$

Proof. We must show that for $\Lambda \in \mathscr{F}_t^0$ we have

$$\text{(12.10)} \qquad E^x\left(\int_t^\infty e^{-\alpha u} f(X_u)\, du; \Lambda\right) = E^x\left(e^{-\alpha t}\, U^\alpha f(X_t); \Lambda\right).$$

We need not concern ourselves with $\{t \geq \zeta\}$ and so we may write $\Lambda \cap \{t < \zeta\} = \bigcup_n (\Lambda \cap \{\tau_n \leq t < \tau_{n+1}\})$. But $\Lambda \cap \{\tau_n \leq t < \tau_{n+1}\} = \Lambda_n \cap \{t < \tau_{n+1}\}$ with $\Lambda_n \in \mathscr{G}_n$ and $\Lambda_n \subset \{\tau_n \leq t\}$. Thus to establish (12.9) we need show only that (12.10) is valid when Λ is replaced by $\Lambda_n \cap \{t < \tau_{n+1}\}$, with $\Lambda_n \in \mathscr{G}_n$ and $\Lambda_n \subset \{\tau_n \leq t\}$. With this replacement we may calculate the right-hand side of (12.10) using (12.7ii) with $\alpha = 0$ and $g = 1$. The result is

$$\text{(12.11)} \qquad e^{-\alpha t}\, E^x(e^{-\lambda(Z_n)(t-\tau_n)}\, U^\alpha f(Z_n); \Lambda_n).$$

We calculate the left side of (12.10) by writing the integral as $\int_t^{\tau_{n+1}} + \sum_{k=n+1}^\infty \int_{\tau_k}^{\tau_{k+1}}$. Using (12.7ii) twice, the first term becomes

$$e^{-\alpha t}\, E^x\{f(Z_n)\,[\alpha + \lambda(Z_n)]^{-1}\, e^{-\lambda(Z_n)(t-\tau_n)}; \Lambda_n\}.$$

For the second term we use (12.7i) and (12.7ii) in that order to obtain

$$e^{-\alpha t}\, E^x\left\{e^{-\lambda(Z_n)(t-\tau_n)} \sum_{k=1}^\infty Q_\alpha^k\left(Z_n, \frac{f}{\alpha+\lambda}\right); \Lambda_n\right\}.$$

Combining these calculations with (12.8) we obtain (12.11) for the left side of (12.10) also. Hence Lemma 12.9 is established.

The conclusion of Theorem 12.4 is now immediate. Indeed if $f \in b\mathscr{E}_+$, then Lemma 12.9 implies that

$$\text{(12.12)} \quad E^x\left\{\int_t^\infty \varphi(u) f(X_u)\, du \mid \mathscr{F}_t^0\right\} = E^{X(t)}\left\{\int_0^\infty \varphi(u+t) f(X_u)\, du\right\}$$

whenever φ is a linear combination of exponentials and hence, by uniform approximation, whenever φ is continuous and vanishes at ∞. Since $f(X_u)$ is right continuous and bounded, given $s \geq 0$ one can choose a sequence $\{\varphi_n\}$ such that upon replacing φ by φ_n the left side of (12.12) approaches $E^x\{f(X_{t+s}) \mid \mathscr{F}_t^0\}$ while the right side approaches $E^{X(t)}\{f(X_s)\}$. Thus (3.2) holds and so the proof of Theorem 12.4 is complete.

Observe that $\tau_1 = \inf\{t\colon X_t \neq X_0\}$ and hence τ_1 is an $\{\mathscr{F}_{t+}^0\}$ stopping time. Moreover $\tau_{n+1} = \tau_n + \tau_1 \circ \theta_{\tau_n}$ and each τ_n is an $\{\mathscr{F}_{t+}^0\}$ stopping time. If T is an $\{\mathscr{F}_t\}$ stopping time, then $X_T = Z_n$ on $\{\tau_n \leq T < \tau_{n+1}\}$ and so $X_T \in \mathscr{F}/\mathscr{E}_\Delta$. If f is any bounded function on E_Δ, then $t \to f(X_t)$ is right continuous since $t \to X_t$ is right continuous when E_Δ is given the discrete topology. Therefore the proof of Theorem 8.11 can be repeated essentially word for word to show that X is a strong Markov process relative to $\{\mathscr{F}_{t+}\}$.

Finally let us show that X is quasi-left-continuous when E_Δ is given the discrete topology. Let $\{T_n\}$ be an increasing sequence of $\{\mathscr{F}_t\}$ stopping times with limit T. We may assume that $T \le \zeta$. In view of the form of the path functions it is clear that $X_{T_n} \to X_T$ for all ω in $\{T < \zeta\} - \bigcup_k \{T = \tau_k\}$. Consequently for a given path $t \to X_t(\omega)$ with $T(\omega) < \zeta$ the convergence of X_{T_n} to X_T can fail only if for some k we have $T = \tau_{k+1}$ and $\tau_k \le T_n < \tau_{k+1}$ for all large n. Let k and an initial measure μ be given. On the set $\{\tau_k \le T_n < \tau_{k+1}\}$ we have $\tau_{k+1} = T_n + \tau_1 \circ \theta_{T_n}$ and $X(T_n) = Z_k$. Thus if $\Lambda = \{\tau_k \le T_n < \tau_{k+1}$ for all large $n\}$, then

$$
\begin{aligned}
P^\mu(T = \tau_{k+1}; \Lambda) &= \lim_{\alpha\to\infty} E^\mu(e^{-\alpha(\tau_{k+1} - T)}; \Lambda) \\
&\le \lim_{\alpha\to\infty} \liminf_n E^\mu(e^{-\alpha(\tau_{k+1} - T_n)}; \tau_k \le T_n < \tau_{k+1}) \\
&= \lim_{\alpha\to\infty} \liminf_n E^\mu(e^{-\alpha\tau_1 \circ \theta_{T_n}}; \tau_k \le T_n < \tau_{k+1}) \\
&\le \lim_{\alpha\to\infty} E^\mu\{E^{Z_k}(e^{-\alpha\tau_1})\} \\
&= \lim_{\alpha\to\infty} E^\mu\left\{\frac{\lambda(Z_k)}{\alpha + \lambda(Z_k)}\right\} = 0.
\end{aligned}
$$

This establishes the quasi-left-continuity of X.

If E is a locally compact space with a countable base and $\mathscr{E}$ is the Borel sets of E then it is an immediate consequence of the above results that X is a standard process.

Exercises

(12.13) Given $\alpha > 0$ and $x \in E$ show that a necessary and sufficient condition that $P^x(\zeta = \infty) = 1$ is that $Q_\alpha^n(x, E) \to 0$ as $n \to \infty$. Conclude from this that if λ is bounded then, for all x, $P^x(\zeta = \infty) = 1$. [Hint: use (12.6).]

(12.14) Show that relative to each P^y if $h \in b\mathscr{G}_n$ then $\tau_{n+1} - \tau_n$, Z_{n+1}, and h are conditionally independent given Z_n.

(12.15) Let E be locally compact with a countable base and let $\mathscr{E} = \mathscr{B}(E)$. Show that if λ is a bounded continuous function and Q maps $\mathbf{C}_0$ into $\mathbf{C}_0$, then, for each $\alpha > 0$, U^α maps $\mathbf{C}_0$ into $\mathbf{C}_0$. One may replace $\mathbf{C}_0$ by $\mathbf{C}$ in the hypothesis and conclusion. [Hint: show that if Q maps $\mathbf{C}_0$ into $\mathbf{C}_0$, then so does each Q_α^n. Then use (12.8).]

II

EXCESSIVE FUNCTIONS

1. Introduction

Most of the paragraphs which follow have as fixed data a given standard process $X = (\Omega, \mathcal{M}, \mathcal{M}_t, X_t, \theta_t, P^x)$ with state space $(E, \mathcal{E})$. Often the definitions and theorems involve only the action of the corresponding semigroup $\{P_t\}$ of transition operators and nothing more about X. However our viewpoint is that the process X as well as the semigroup $\{P_t\}$ is of fundamental interest, and so we will not attempt to separate statements involving only the semigroup from those which also involve the process. When starting with a standard process X it is to be understood that objects such as nearly Borel sets, potential operators U^α, hitting operators P_T^α, and so forth are defined relative to X.

Frequently we will make calculations of the following sort: Let T be an $\{\mathcal{M}_t\}$ stopping time, $f \in \mathcal{E}_+^*$, and α a nonnegative number. Recall that

$$U^\alpha f(x) = E^x \int_0^\infty e^{-\alpha t} f(X_t)\, dt.$$

We will refer to $U^\alpha f$ as the *α-potential of f*. As usual when $\alpha = 0$ we will drop it from our notation and terminology. Observe that $U(x, B) = U^0(x, B)$ is the expected amount of time that the process starting from x spends in B. The integral on t in the above expression for $U^\alpha f(x)$ can be written $\int_0^T + \int_T^\infty$. If $T(\omega) < \infty$, then using the change of variable $t - T = u$ the second integral becomes

$$\int_T^\infty e^{-\alpha t} f(X_t)\, dt = e^{-\alpha T} \int_0^\infty e^{-\alpha u} f(X_u \circ \theta_T)\, du$$
$$= e^{-\alpha T} \left\{ \int_0^\infty e^{-\alpha u} f(X_u)\, du \right\} \circ \theta_T,$$

and this is valid also if $T = \infty$ since $f(\Delta) = 0$. The function in braces is an element of $\mathscr{F}$ and so by the strong Markov property

(1.1)
$$E^x \int_T^\infty e^{-\alpha t} f(X_t)\, dt = E^x\left\{e^{-\alpha T}\, E^{X(T)} \int_0^\infty e^{-\alpha t} f(X_t)\, dt\right\}$$
$$= E^x\{e^{-\alpha T}\, U^\alpha f(X_T)\}$$
$$= P_T^\alpha U^\alpha f(x).$$

Consequently

(1.2)
$$U^\alpha f(x) = E^x \int_0^T e^{-\alpha t} f(X_t)\, dt + P_T^\alpha U^\alpha f(x).$$

This shows, for example, that $P_T^\alpha U^\alpha f \leq U^\alpha f$ and that $P_T^\alpha U^\alpha f$ decreases as T increases. Generally we will simply write down the results of such calculations, leaving the intermediate steps to the reader.

(1.3) THEOREM.
(a) If $f \in \mathscr{E}_+^*$ and A is nearly Borel, then $P_A^\alpha U^\alpha f \leq U^\alpha f$.
(b) If in addition f vanishes off A, then $P_A^\alpha U^\alpha f = U^\alpha f$.
(c) If f and g are in $\mathscr{E}_+^*$ with f vanishing off the nearly Borel set A and $U^\alpha f \leq U^\alpha g$ on A, then $U^\alpha f \leq U^\alpha g$ everywhere.

Proof. The first two statements are immediate consequences of (1.2) applied to the stopping time T_A. We will prove (c) as an application of some of the results from Chapter I. If K is a compact subset of A then $P_K^\alpha U^\alpha f \leq P_K^\alpha U^\alpha g$, for the measures P_K^α are carried by K. Fix x and let $\{K_n\}$ be an increasing sequence of compact subsets of A such that $T_{K_n} \downarrow T_A$ a.s. P^x. If B is nearly Borel, then

$$E^x \int_{T_B}^\infty e^{-\alpha t} f(X_t)\, dt = P_B^\alpha U^\alpha f(x),$$

and so Lebesgue's monotone convergence theorem shows that $P_{K_n}^\alpha U^\alpha f(x) \uparrow P_A^\alpha U^\alpha f(x)$. Then from (a) and (b) we have

$$U^\alpha f(x) = P_A^\alpha U^\alpha f(x) = \lim_n P_{K_n}^\alpha U^\alpha f(x) \leq \lim_n P_{K_n}^\alpha U^\alpha g(x)$$
$$\leq U^\alpha g(x),$$

which establishes (c).

For $\alpha > 0$ we write $P_t^\alpha = e^{-\alpha t} P_t$. Note that this agrees with our definition of P_T^α if $T = t$. The reader should note that, for each $\alpha \geq 0$, $\{P_t^\alpha;\, t \geq 0\}$ is a semigroup of bounded operators on $b\mathscr{E}$ (or $b\mathscr{E}^*$) whose resolvent $\{V^\beta;\, \beta > 0\}$

is given by $V^{\beta} = U^{\alpha+\beta}$. In particular $V^0 = U^{\alpha}$ is a bounded operator if $\alpha > 0$. Note that we now have $U^{\alpha} = \int_0^{\infty} P_t^{\alpha}\, dt$. This notation will be used in the sequel without special mention.

Exercises

(1.4) Prove that if $f \in b\mathscr{E}^*$ and T is an $\{\mathscr{M}_t\}$ stopping time then

$$P_t^{\alpha} P_T^{\alpha} U^{\alpha} f(x) = E^x \int_{t+T\circ\theta_t}^{\infty} e^{-\alpha u} f(X_u)\, du$$

for all x, t, and $\alpha \geq 0$.

(1.5) Let X be the Brownian motion process in $\mathbf{R}^n$ (see (2.17) and (9.17) of Chapter I). (a) Prove that if $n \geq 3$ the potential operator for this process is given by

$$U f(x) = \{\Gamma(n/2 - 1)/4\pi^{n/2}\} \int |x - y|^{2-n} f(y)\, dy$$

for $f \geq 0$, where the integral is over $\mathbf{R}^n$ and dy denotes integration with respect to n-dimensional Lebesgue measure. (The precise value of the constant appearing before the integral sign is, of course, relatively unimportant.) Specifically, Uf is the ordinary Newtonian potential of f in $\mathbf{R}^n$. (b) Prove that if $n = 1$ or 2 then, for $f \geq 0$, $Uf = 0$ if $f = 0$ a.e. (Lebesgue measure) and otherwise $Uf = \infty$.

(1.6) Let X be the one-sided stable process of index β, $0 < \beta < 1$, in $\mathbf{R}$ (see (2.19) of Chapter I). Show that

$$U f(x) = \frac{1}{\Gamma(\beta)} \int_x^{\infty} (y - x)^{\beta-1} f(y)\, dy.$$

[Hint: let $h_{\beta}(t, x)$ be the continuous density function for the measure η_t^{β} (Chapter I, (2.19)). Then evaluate the integral $k_{\beta}(x) = \int_0^{\infty} h_{\beta}(t, x)\, dt$ by taking its Laplace transform.]

(1.7) Let X be the symmetric stable process of index α, $0 < \alpha < 2$, in $\mathbf{R}^n$ (see (2.20) of Chapter I). If $\alpha < n$ prove that for $f \geq 0$

$$U f(x) = \Gamma\left(\frac{n-\alpha}{2}\right)\left[2^{\alpha}\pi^{n/2}\,\Gamma\left(\frac{\alpha}{2}\right)\right]^{-1} \int |x - y|^{\alpha-n} f(y)\, dy.$$

Specifically Uf is the Riesz potential of order α. (b) Prove that if $\alpha \geq n$ then Uf has the same description as in Part (b) of Exercise 1.5. [Hint: because of

Exercise 2.20 of Chapter I it suffices to calculate

$$\int_0^\infty dt \int_0^\infty h_{\alpha/2}(t, u)\, g_u(x)\, du,$$

where $h_{\alpha/2}$ is defined in (1.6) above and g_u is defined in (2.17) of Chapter I. Make this calculation by interchanging the order of integration and then using the results of (1.6).]

2. Excessive Functions

In this section $X = (\Omega, \mathcal{M}, \mathcal{M}_t, X_t, \theta_t, P^x)$ will be a fixed standard process with state space $(E, \mathcal{E})$.

(2.1) Definition. Let $\alpha \geq 0$. A function $f \in \mathcal{E}_+^*$ is called *α-excessive* (*relative to X*) if (a) $P_t^\alpha f \leq f$ for every $t \geq 0$ and also (b) $P_t^\alpha f \to f$ pointwise as $t \to 0$.

Of course Condition (a) implies that $P_t^\alpha f$ increases as t decreases. If $\alpha = 0$ we simply say "excessive" rather than "0-excessive." Let $\mathcal{S}^\alpha$ denote the class of functions which are α-excessive. Since the process X is fixed in each discussion it need not be referred to in the notation. The next four propositions collect some elementary but important properties of α-excessive functions. Statements involving α are understood to hold for all $\alpha \geq 0$ unless explicitly stated otherwise.

(2.2) Proposition. (a) Constant nonnegative functions are in $\mathcal{S}^\alpha$. (b) If $f, g \in \mathcal{S}^\alpha$ then $f + g \in \mathcal{S}^\alpha$ and $cf \in \mathcal{S}^\alpha$ (c a nonnegative constant). (c) If $\{f_n\} \in \mathcal{S}^\alpha$ and $f_1 \leq f_2 \leq \ldots$, then $\lim f_n \in \mathcal{S}^\alpha$. (d) $\mathcal{S}^\alpha = \bigcap \mathcal{S}^\beta$, the intersection being over all $\beta > \alpha$. (e) If $f \in \mathcal{E}_+^*$ then $U^\alpha f \in \mathcal{S}^\alpha$.

Proof. Properties (a)–(d) are immediate consequences of the definition. For Property (e) we write

$$P_t^\alpha U^\alpha f = P_t^\alpha \int_0^\infty P_u^\alpha f\, du = \int_0^\infty P_{t+u}^\alpha f\, du = \int_t^\infty P_u^\alpha f\, du.$$

The term on the right is less than $U^\alpha f = \int_0^\infty P_u^\alpha f\, du$ and approaches it as t decreases to 0.

(2.3) Proposition. If $f \in \mathcal{E}_+^*$ then $f \in \mathcal{S}^\alpha$ if and only if (a) $\beta U^{\beta+\alpha} f \leq f$ for all $\beta \geq 0$ and (b) $\beta U^{\beta+\alpha} f \to f$ as $\beta \to \infty$.

Proof. If $f \in \mathscr{S}^\alpha$ then clearly $\beta U^{\alpha+\beta} f \leq f$, while

$$\beta U^{\beta+\alpha} f = \beta \int_0^\infty e^{-\beta t} P_t^\alpha f \, dt = \int_0^\infty e^{-t} P_{t/\beta}^\alpha f \, dt \uparrow f$$

as $\beta \to \infty$ by the monotone convergence theorem. For the sufficiency we first note that if f satisfies (a) and (b) for a particular value of α, then it follows readily from the resolvent equation (8.10) of Chapter I that f satisfies (a) and (b) for every larger value of α. Thus in view of Property (d) of (2.2) we may (and do) assume that $\alpha > 0$ without loss of generality. Let us first suppose that $f \in \mathscr{E}_+^*$ satisfies (a) only (with $\alpha > 0$). Then $f_n = \min(f, n)$ satisfies (a) also, and the resolvent equation implies that for $\eta > \beta$ we have

$$\text{(2.4)} \qquad \begin{aligned} \beta U^{\beta+\alpha} f_n &= \beta U^{\eta+\alpha} f_n + (\eta - \beta) U^{\eta+\alpha} (\beta U^{\beta+\alpha} f_n) \\ &\leq \beta U^{\eta+\alpha} f_n + (\eta - \beta) U^{\eta+\alpha} f_n = \eta U^{\eta+\alpha} f_n . \end{aligned}$$

Letting $n \to \infty$ we obtain $\beta U^{\beta+\alpha} f \leq \eta U^{\eta+\alpha} f$ and consequently $\beta \to \beta U^{\beta+\alpha} f$ is an increasing function of β whenever $f \in \mathscr{E}_+^*$ satisfies (a). Again from the resolvent equation we obtain

$$\text{(2.5)} \qquad U^{\beta+\alpha} f_n = U^\alpha (f_n - \beta U^{\beta+\alpha} f_n).$$

Consequently from Property (e) of (2.2) we see that $\beta U^{\beta+\alpha} f_n \in \mathscr{S}^\alpha$ because $f_n - \beta U^{\beta+\alpha} f_n \in b\mathscr{E}_+^*$. As $\beta \to \infty$, $\beta U^{\beta+\alpha} f_n$ increases to g_n which is in $\mathscr{S}^\alpha$ by (c) of (2.2), and, as $n \to \infty$, g_n increases to $g \in \mathscr{S}^\alpha$. Clearly $g_n \leq f_n$ and so $g \leq f$. But for each $\beta > 0$, $g \geq \lim_n \beta U^{\beta+\alpha} f_n = \beta U^{\beta+\alpha} f$. Thus if f also satisfies (b) we obtain, on letting $\beta \to \infty$, $g \geq f$. Therefore $f = g \in \mathscr{S}^\alpha$.

(2.6) PROPOSITION. If $f \in \mathscr{S}^\alpha$ and $\alpha > 0$ there is a sequence $\{h_n\} \in b\mathscr{E}_+^*$ such that $U^\alpha h_n$ increases to f as $n \to \infty$.

Proof. Taking $\beta = n$ in (2.5) we have

$$n U^{n+\alpha} f_n = U^\alpha (n[f_n - n U^{n+\alpha} f_n]).$$

Since $\beta U^{\beta+\alpha} f_n$ is increasing in both β and n it follows from the proof of (2.3) that $n U^{n+\alpha} f_n$ increases to f as $n \to \infty$. Hence we obtain (2.6) by setting $h_n = n(f_n - n U^{n+\alpha} f_n)$.

(2.7) REMARK. The condition in (2.6) that α be strictly positive cannot generally be removed. For example, if X is Brownian motion in $\mathbf{R}$ the results of Exercise 1.5 state that, for $f \geq 0$, Uf is either identically infinite or identically 0. For conditions under which (2.6) is valid when $\alpha = 0$ see Exercise 2.19.

(2.8) PROPOSITION. Let $f \in \mathscr{S}^\alpha$. If T is an $\{\mathscr{M}_t\}$ stopping time, then $P_T^\alpha f \leq f$. If A is a nearly Borel subset of E_Δ, then $P_A^\alpha f \in \mathscr{S}^\alpha$.

Proof. Since $P_T^\alpha f$ increases to $P_T^\beta f$ as α decreases to β it follows from (2.2) that we may assume $\alpha > 0$. Then by Proposition 2.6 it will suffice to consider only f of the form $f = U^\alpha g$ with $g \in b\mathscr{E}_+^*$. The first statement in (2.8) is now an immediate consequence of Eq. (1.2). Furthermore we have (Exercise 1.4)

$$\text{(2.9)} \qquad P_t^\alpha P_T^\alpha U^\alpha g(x) = E^x \int_{t+T\circ\theta_t}^\infty e^{-\alpha s} g(X_s)\, ds.$$

Recall that if A is in $\mathscr{E}_\Delta^n$ then T_A is an $\{\mathscr{F}_t\}$ stopping time for which $t + T_A \circ \theta_t \geq T_A$ and $t + T_A \circ \theta_t \downarrow T_A$ as $t \downarrow 0$. Combining this with (2.9) we see that $P_A^\alpha U^\alpha g \in \mathscr{S}^\alpha$, completing the proof of Proposition 2.8.

It follows from Proposition 2.8 taking $f = 1$ on E that if A is a nearly Borel subset of E_Δ, then $\Phi_A^\alpha(x) = E^x(e^{-\alpha T_A}; T_A < \zeta)$ is α-excessive.

The main theorem in this section, Theorem 2.12, concerns the behavior of an excessive function composed with the process X. Before stating it we need some preliminary remarks. Recall from Section 11 of Chapter I, that if $A \in \mathscr{E}^n$, then A^r denotes the set of points regular for A and that $A^r \in \mathscr{E}^*$.

(2.10) PROPOSITION. Suppose $f \in \mathscr{S}^\alpha$, $A \in \mathscr{E}^n$, and $x \in A^r$. Then

$$\inf\{f(y)\colon y \in A\} \leq f(x) \leq \sup\{f(y)\colon y \in A\}.$$

Proof. To prove the left inequality let I be the infimum in question and K be a compact subset of A. By (2.8)

$$f(x) \geq P_K^\alpha f(x) \geq I\, E^x(e^{-\alpha T_K}; T_K < \zeta),$$

because $X(T_K) \in A$ if $T_K < \infty$. By (10.19) of Chapter I and the fact that $x \in A^r$ we may choose K so that $E^x(e^{-\alpha T_K}; T_K < \zeta)$ is close to 1. Hence $f(x) \geq I$. In proving the second inequality we may assume that $\alpha > 0$ and then by (2.6) that $f = U^\alpha g$, $g \in b\mathscr{E}_+^*$. By a now familiar calculation, if K is a compact subset of A

$$U^\alpha g(x) \leq E^x \int_0^{T_K} e^{-\alpha t} g(X_t)\, dt + \sup\{U^\alpha g(y)\colon y \in A\}.$$

Since there is a sequence $\{K_n\}$ of compacts in A such that $P^x(T_{K_n} \to 0) = 1$ and since $E^x \int_0^\infty e^{-\alpha t} g(X_t)\, dt < \infty$ it follows that the first term on the right may be made arbitrarily close to 0. Hence the desired inequality follows.

Let $f \in \mathscr{E}^n$ take on only finite values. For $\varepsilon > 0$ define

$$T(\omega) = \inf\{t\colon |f(X_0(\omega)) - f(X_t(\omega))| > \varepsilon\}.$$

If we set $A_{k,n} = \{f < k/n - \varepsilon\} \cup \{f > (k+1)/n + \varepsilon\}$, for $n = 1, 2, \ldots,$ and

$k = 0, \pm 1, \ldots$, then for any $t > 0$ we have

$$\{T < t\} = \bigcup_{n=1}^{\infty} \bigcup_{k=-\infty}^{\infty} \left\{\frac{k}{n} \leq f(X_0) \leq \frac{k+1}{n},\ T_{A_{k,n}} < t\right\}.$$

The sets $A_{k,n}$ are nearly Borel and so T is an $\{\mathscr{F}_t\}$ stopping time.

(2.11) PROPOSITION. Let $f \in \mathscr{S}^\alpha$ be nearly Borel measurable and with only finite values, and let T be defined as above. Then no point is regular for T. Also $|f(X_0) - f(X_T)| \geq \varepsilon$ on $\{T < \infty\}$ almost surely.

Proof. Given x let $B_1 = \{y : f(y) > f(x) + \varepsilon\}$, $B_2 = \{y : f(y) < f(x) - \varepsilon\}$, and $B = B_1 \cup B_2$. Clearly $P^x(T = T_B) = 1$. By (2.10), x is not in $B_1^r \cup B_2^r = B^r$. Consequently x is not regular for T, establishing the first assertion in (2.11). By (11.4) of Chapter I, $X(T_B) \in B \cup B^r$ almost surely on $\{T_B < \infty\}$. But (2.10) implies that $B \cup B^r$ is contained in $\{y : f(y) \geq f(x) + \varepsilon$ or $f(y) \leq f(x) - \varepsilon\}$, and this yields the second assertion in (2.11).

(2.12) THEOREM. Let $f \in \mathscr{S}^\alpha$. Then (a) f is nearly Borel measurable, (b) almost surely the mapping $t \to f(X_t)$ is right continuous and has left-hand limits on $[0, \infty)$, (c) if $A = \{f < \infty\}$ then almost surely $f(X_t) < \infty$ for every $t > T_A$.

Proof. The regularity assertions in (b) are of course relative to the topology of $[0, \infty]$. To get started assume $f \in \mathscr{S}^\alpha$ and that f is nearly Borel measurable. Let us first show that as far as (b) is concerned there is no loss of generality in assuming that f is bounded. Indeed define $q : [0, \infty] \to [0, 1]$ by

$$q(x) = 1 - e^{-x}, \qquad x < \infty,$$
$$q(\infty) = 1.$$

Then q is continuous, concave, strictly increasing, and satisfies $q(xy) \geq x\,q(y)$ for $y \geq 0$ and $0 \leq x \leq 1$. From this and Jensen's inequality we have

$$\begin{aligned} E^x\{e^{-\alpha t} q[f(X_t)]\} &\leq E^x\{q[e^{-\alpha t} f(X_t)]\} \\ &\leq q[E^x\{e^{-\alpha t} f(X_t)\}] \\ &\leq q[f(x)]; \end{aligned}$$

that is, $P_t^\alpha(q \circ f) \leq q \circ f$ for all t. Given x and $\varepsilon > 0$ let $A = \{y\colon q[f(y)] > q[f(x)] + \varepsilon\}$. Then A is empty if $f(x) = \infty$ while if $f(x) < \infty$

$$\inf\{f(y)\colon y \in A\} \geq q^{-1}[q[f(x)] + \varepsilon] > f(x),$$

and so x is not regular for A according to Proposition 2.10. Similarly x is not regular for $\{y: q[f(y)] < q[f(x)] - \varepsilon\}$. Therefore $q[f(X_t)] \to q[f(x)]$ as $t \to 0$ almost surely P^x. Hence $q \circ f \in \mathscr{S}^\alpha$ whenever $f \in \mathscr{S}^\alpha$. We may analyze the truth of Statement (b) using $q \circ f$ as readily as f. So we will assume f is bounded.

Now given $\varepsilon > 0$ define T as in (2.11) and then define

$$T_0 = 0,$$

$$T_{n+1} = T_n + T \circ \theta_{T_n}, \qquad n = 0, 1, \ldots.$$

Given $x \in E$ and $\Lambda \in \mathscr{F}_{T_n}$ we have

$$\begin{aligned} E^x\{e^{-\alpha T_{n+1}} f(X_{T_{n+1}}); \Lambda\} &= E^x\{e^{-\alpha T_n} P_T^\alpha f(X_{T_n}); \Lambda\} \\ &\leq E^x\{e^{-\alpha T_n} f(X_{T_n}); \Lambda\}. \end{aligned}$$

That is, relative to P^x the family of random variables $\{e^{-\alpha T_n} f(X_{T_n}); \mathscr{F}_{T_n}\}$ is a supermartingale. It is bounded since f is bounded and hence has a limit almost surely P^x as $n \to \infty$; see Theorem 1.4 of Chapter 0. On the other hand $|f(X_{T_n}) - f(X_{T_{n+1}})| \geq \varepsilon$ a.s. on $\{T_{n+1} < \infty\}$ according to (2.11), and so $P^x(\lim T_n = \infty) = 1$. Since x is arbitrary and $\{\lim T_n = \infty\} \in \mathscr{F}$ it follows that $\lim T_n = \infty$ almost surely. This argument can be repeated for $\varepsilon = 1/k$, $k = 1, 2, \ldots$, with the corresponding iterates being denoted by T_n^k. Almost surely $\lim_n T_n^k = \infty$ for all k. On the other hand it is clear that if ω is such that $T_n^k(\omega) \to \infty$ as $n \to \infty$ for all k, then $t \to f(X_t(\omega))$ is right continuous and has left-hand limits on $[0, \infty)$. This proves Assertion (b) if f is known to be nearly Borel measurable.

In proving (a) we may assume that $\alpha > 0$. Then in view of Proposition 2.6 it is enough to prove (a) for $f = U^\alpha g$, $g \in b\mathscr{E}_+^*$. Given a measure μ on $\mathscr{E}$ define a measure ν on $\mathscr{E}$ by $\nu(B) = \int \mu(dx)\, U^\alpha(x, B)$. Then there exist g_1, g_2 in $b\mathscr{E}_+$ such that $g_1 \leq g \leq g_2$ and $\nu(g_2 - g_1) = 0$. Now $U^\alpha g_1 \leq U^\alpha g \leq U^\alpha g_2$ and the extreme members of this inequality are in $b\mathscr{E}_+$. For each fixed $t \geq 0$

$$\begin{aligned} E^\mu\{U^\alpha(g_2 - g_1)(X_t)\} &= \int \mu(dy)\, P_t U^\alpha(g_2 - g_1)(y) \\ &\leq e^{\alpha t} \int \mu(dy)\, U^\alpha(g_2 - g_1)(y) \\ &= e^{\alpha t} \nu(g_2 - g_1) = 0. \end{aligned}$$

It now follows that $U^\alpha g_1(X_t) = U^\alpha g_2(X_t)$ for all rational $t \geq 0$ almost surely P^μ. But we have already shown that $t \to U^\alpha g_i(X_t)$ is right continuous in t almost surely since $U^\alpha g_i$ is Borel measurable. Consequently $U^\alpha g_1(X_t) = U^\alpha g_2(X_t)$ for all $t \geq 0$ almost surely P^μ, and since μ was arbitrary this implies that $U^\alpha g$ is nearly Borel measurable. Thus the proofs of (a) and (b) are complete.

Let $A = \{f < \infty\}$ and $B = \{f = \infty\}$. We now know that A and B are nearly Borel sets. If $x \in B^r$, then $x \in B$ by (2.10). From (2.8), $f(x) \geq P_B^\alpha f(x)$ and this implies that $P^x(T_B < \infty) = 0$ if x is in A. Consequently if K is a compact subset of A, then $P^x[T_B \circ \theta_{T_K} < \infty] = E^x\{P^{X(T_K)}(T_B < \infty)\} = 0$. For a given x we can find an increasing sequence $\{K_n\}$ of compact subsets of A such that $T_{K_n} \downarrow T_A$ almost surely P^x. But then $T_B \circ \theta_{T_{K_n}} \to T_B \circ \theta_{T_A}$ almost surely P^x, and hence $P^x[T_B \circ \theta_{T_A} = \infty] = 1$. Since x is arbitrary this yields Statement (c) of Theorem 2.12.

The proof of (2.12) yields a little more when $\alpha = 0$. Specifically, suppose $f \in \mathscr{S}$. Then the exponential factor in the supermartingale $e^{-\alpha T_n} f(X_{T_n})$ is absent. It follows that almost surely $T_n = \infty$ for some n and so $\lim f(X_t)$ exists also as $t \to \infty$. It is easy to see that this limit is finite almost surely on $\{T_A < \infty\}$ where $A = \{f < \infty\}$.

(2.13) COROLLARY. If $A \in \mathscr{E}^n$, then $A^r \in \mathscr{E}^n$.

Proof. Let $T = T_A$. If $\alpha > 0$, then $\Phi_A^\alpha(x) = E^x(e^{-\alpha T})$ is α-excessive according to the remark following Proposition 2.8, and hence is nearly Borel measurable. Since $A^r = \{\Phi_A^\alpha = 1\}$ we obtain Corollary 2.13.

(2.14) COROLLARY. If $f, g \in \mathscr{S}^\alpha$ then $\min(f, g) \in \mathscr{S}^\alpha$.

Proof. Let $h = \min(f, g)$. Obviously $h \geq P_t^\alpha h$. The fact that $P_t^\alpha h \to h$ as $t \to 0$ is an immediate consequence of Theorem 2.12(b).

Exercises

(2.15) Let f be in $\mathscr{S}^\alpha$. Let R and T be $\{\mathscr{M}_t\}$ stopping times with $R \leq T$. If $\Lambda \in \mathscr{M}_R$ show that

$$E^x\{e^{-\alpha R} f(X_R); \Lambda\} \geq E^x\{e^{-\alpha T} f(X_T); \Lambda\}$$

for all x.

(2.16) A standard process X is called a *strong* Feller *process* if $U^\alpha f$ is continuous for all $f \in b\mathscr{E}_+$ having compact support and all $\alpha > 0$. Show that in this case any α-excessive function ($\alpha \geq 0$) is lower semicontinuous.

(2.17) A function $f \in \mathscr{E}_+^*$ satisfying (a) of Definition 2.1 is called *α-super-mean-valued.* If f is α-super-mean-valued show that $\bar{f} = \lim_{t \to 0} P_t^\alpha f$ exists (pointwise), that $\bar{f} \leq f$, and that $\bar{f}$ is α-excessive. Also show that, for all $\beta \geq 0$, $U^\beta f = U^\beta \bar{f}$. [Hint: for each x the mapping $t \to P_t^\alpha f(x)$ has only countably many discontinuities; if t is a continuity point then $P_t f(x) = P_t \bar{f}(x)$.] Furthermore characterize $\bar{f}$ as the largest α-excessive function dominated by f. The function $\bar{f}$ is called the (α-*excessive*) *regularization* of f.

(2.18) An $\{\mathscr{F}_t\}$ stopping time T is called a *terminal time* provided $t + T \circ \theta_t = T$ almost surely on $\{T > t\}$ for all $t \geq 0$. Note that if A is a nearly Borel set, then both T_A and D_A are terminal times. If $f \in \mathscr{S}^\alpha$ and T is a terminal time show that $P_T^\alpha f$ is α-super-mean-valued. If in addition T has the property that whenever $\{t_n\}$ is a sequence decreasing to zero one has $t_n + T \circ \theta_{t_n} \to T$ almost surely, then $P_T^\alpha f \in \mathscr{S}^\alpha$. Note that if A is a nearly Borel set, then T_A always has this additional property while D_A may fail to have it.

(2.19) Let X be a standard process such that $U(x, K) = \int_0^\infty P_t(x, K)\, dt$ is bounded in x for each compact subset K of E. Show that in this situation if f is excessive, then there exists a sequence $\{g_n\}$ of bounded functions in $\mathscr{E}_+^*$ such that $\{Ug_n\}$ increases to f and each Ug_n is bounded. [Hint: (i) let f be a bounded excessive function such that $\lim_{t\to\infty} P_t f = 0$. If $g_t = t^{-1}(f - P_t f)$ show that Ug_t increases to f as $t \to 0$. (ii) Let $\{K_n\}$ be an increasing sequence of compact subsets of E whose union is E. If $h_n = nI_{K_n}$ show that $Uh_n \uparrow \infty$ and each Uh_n is bounded. (iii) If f is excessive show that $f_n = \min(f, Uh_n)$ satisfies the hypotheses of (i). Use this to construct the desired sequence $\{g_n\}$.]

(2.20) (i) Let $(E, \mathscr{E})$ and $(F, \mathscr{F})$ be measurable spaces, μ a measure on $\mathscr{F}$ which is a countable sum of finite measures, and $u: E \to F$ be such that (a) $u \in \mathscr{E}/\mathscr{F}$, (b) there exists a mapping $\tau: F \to E$ such that $\tau \in \mathscr{F}/\mathscr{E}$ and $u[\tau(x)] = x$ for all x in F. Let $\nu = \mu\tau^{-1}$. Then show that $\int_E (f \circ u)\, d\nu = \int_F f\, d\mu$ for each $f \in \mathscr{F}_+$. In particular derive from this the fact that if u is a continuous nondecreasing function on the interval $[a, b]$ and if $F(s)$ is a right continuous nondecreasing numerical function defined on $[u(a), u(b)]$, then

$$\int_{[a,b]} f[u(t)]\, dF[u(t)] = \int_{[u(a),u(b)]} f(t)\, dF(t)$$

for all nonnegative Borel functions f on $[u(a), u(b)]$.

(ii) Let $\alpha > 0$ and let h and φ be nonnegative functions in $b\mathscr{E}^*$. Define

$$\psi(x) = E^x\left\{\int_0^\infty e^{-\alpha t}\, h(X_t)\, \varphi(X_t) \exp\left[-\int_0^t h(X_s)\, ds\right] dt\right\}.$$

Show that $\psi = U^\alpha[h(\varphi - \psi)]$. [Hint: use (i) to show that

$$1 - \exp\left[-\int_0^t h(X_s)\,ds\right] = \int_0^t h(X_s)\exp\left[-\int_s^t h(X_u)\,du\right] ds,$$

and substitute this into the expression defining ψ.]

(iii) If $\varphi = U^\alpha f$ with f a nonnegative function in $b\mathscr{E}^*$, then show that (notation as in (ii))

$$\psi(x) = E^x \int_0^\infty e^{-\alpha t} f(X_t)\left[1 - \exp\left(-\int_0^t h(X_s)\,ds\right)\right] dt.$$

Consequently $\psi \leq \varphi$ and ψ increases as h increases in this case. Finally show that this last statement continues to hold if φ is only assumed to be a bounded α-excessive function.

3. Exceptional Sets

In this section we are going to introduce various classes of sets that will play the role of exceptional sets in the theory to be developed. The importance and the appropriateness of these notions will only become apparent later. In the present section we will merely give the definitions and develop some elementary consequences of them. Again in this section $X = (\Omega, \mathscr{M}, \mathscr{M}_t, X_t, \theta_t, P^x)$ will be a fixed standard process with state space $(E, \mathscr{E})$. Our definitions will involve the process X, or at least its semigroup, but since we are regarding X as fixed the qualifying phrase "relative to X" will generally be omitted.

(3.1) DEFINITION. (a) A set $A \in \mathscr{E}^*$ is of *potential zero* (or simply "*null*") if $U(x, A) = 0$ for all $x \in E$. (b) A set $A \subset E$ is *polar* if there exists a set $D \in \mathscr{E}^n$ such that $A \subset D$ and $P^x(T_D < \infty) = P^x(T_D < \zeta) = 0$ for all x. (c) A set $A \subset E$ is *thin* if there exists a set $D \in \mathscr{E}^n$ such that $A \subset D$ and D^r is empty, that is $P^x(T_D = 0) = 0$ for all x. (d) A set $A \subset E$ is *semipolar* if A is contained in a countable union of thin sets. (e) A set $A \subset E$ is *thin at* x if there exists a set $D \in \mathscr{E}^n$ such that $D \supset A$ and x is irregular for D; that is, $P^x(T_D = 0) = 0$.

Since $U^\alpha(x, A) = \int_0^\infty e^{-\alpha t} P_t(x, A)\,dt$, it is clear that A is of potential zero if and only if $U^\alpha(x, A) = 0$ for all x for some, and hence for all, $\alpha \geq 0$. Intuitively a polar set is one which the process never enters at a strictly positive time almost surely. If $A \in \mathscr{E}^n$, then A is polar if and only if $\Phi_A^\alpha(x) = E^x(e^{-\alpha T_A}; T_A < \zeta) = E^x(e^{-\alpha T_A}; T_A < \infty) = 0$ for all x for some, and hence for all, $\alpha \geq 0$. On the other hand a thin set is one which the process avoids during some initial open time interval almost surely. If $A \in \mathscr{E}^n$, then A is thin if and only if $\Phi_A^\alpha < 1$ for some, and hence for all, $\alpha > 0$. Clearly a polar set is thin and a thin set is semipolar. In addition, it will follow from Proposition 3.4 that a semipolar set A in $\mathscr{E}^*$ is of potential zero. It is evident that a countable union of polar (semipolar) sets is again polar (semipolar). A set may be thin at every point without being a thin set. (See Exercise 3.14.) Often instead of saying A is thin at x we will say that x is irregular for A.

(3.2) PROPOSITION. If f and g are in $\mathscr{S}^\alpha$ and $f \geq g$ except on a null set, then $f \geq g$ everywhere. In particular if $f = g$ except on a null set then $f = g$ everywhere.

Proof. It is immediate from the definition of a null set that $\beta U^{\alpha+\beta} f \geq \beta U^{\alpha+\beta} g$ for all $\beta > 0$. Letting $\beta \to \infty$ and using Proposition 2.3 yields the desired conclusion.

The next proposition is a very useful result.

(3.3) PROPOSITION. If $A \in \mathscr{E}^n$, then $A - A^r$ is semipolar.

Proof. Let $\alpha > 0$, and let $A_n = A \cap \{\Phi_A^\alpha \leq 1 - 1/n\}$. Clearly $A - A^r$ is the union of the A_n. If $\Phi_A^\alpha(x) < 1$, then x is not regular for A and hence is not regular for A_n. If $\Phi_A^\alpha(x) = 1$, then by (2.10) x is not regular for $\{\Phi_A^\alpha \leq 1 - 1/n\}$. Thus no point is regular for A_n; that is, A_n is thin. This yields Proposition 3.3.

An immediate consequence of our next result is the fact, mentioned earlier, that a semipolar set in $\mathscr{E}^*$ is null.

(3.4) PROPOSITION. Let A be semipolar. Then almost surely $X_t \in A$ for only countably many values of t.

Proof. We may suppose that A is thin and in $\mathscr{E}^n$. Pick $\alpha > 0$, $\beta < 1$, and let $B = A \cap \{\Phi_A^\alpha \leq \beta\}$. The set A is a countable union of sets such as B and so it will suffice to prove $X_t \in B$ only countably often. Let $T_1 = T_B$ and $T_{n+1} = T_n + T_B \circ \theta_{T_n}$. Since $B \subset A$ and A is thin, B^r is empty. Therefore almost surely $X_{T_n} \in B$ on $\{T_n < \infty\}$. Now

$$E^x(e^{-\alpha T_{n+1}}) = E^x[\Phi_B^\alpha(X_{T_n})\, e^{-\alpha T_n}]$$
$$\leq \beta\, E^x(e^{-\alpha T_n}).$$

Consequently $E^x(e^{-\alpha T_n}) \to 0$ as $n \to \infty$ and hence $T_n \to \infty$ almost surely. Since $X_t \notin B$ if $T_n < t < T_{n+1}$ the proof of Proposition 3.4 is complete.

(3.5) PROPOSITION. Let $f \in \mathscr{S}^\alpha$ and $A = \{f = \infty\}$. Then A is polar if and only if A is null.

Proof. Assume A is null and let $x \in E$. Since $U(x, A) = 0$ there is a sequence $\{t_n\}$ approaching 0 such that $P^x(X_{t_n} \in A) = 0$. By Theorem 2.12(c), $P^x(X_t \in A$ for some $t > t_n) = 0$. Thus $P^x(X_t \in A$ for some $t > 0) = 0$. Since x is arbitrary and $A \in \mathscr{E}^n$, this implies that A is polar.

Recall (see Exercise 2.17) that a function f in $\mathscr{E}_+^*$ is α-super-mean-valued provided $P_t^\alpha f \leq f$ for all t, and that for such an f, $\bar{f} = \lim_{t \downarrow 0} P_t^\alpha f$ is the largest α-excessive function which is dominated by f. Moreover for each $\beta \geq 0$, $U^\beta f = U^\beta \bar{f}$. Consequently $\{\bar{f} \neq f\} = \{\bar{f} < f\}$ is of potential zero. However the following theorem of Doob which generalizes an important theorem of Cartan in classical potential theory allows us to say much more in an important special case.

(3.6) THEOREM. Let $\{f_n\}$ be a decreasing sequence of α-excessive functions and let $f = \lim_n f_n$. Then f is α-super-mean-valued and $\{\bar{f} < f\}$ is semipolar.

Proof. It is evident that f is α-super-mean-valued and so only the second assertion requires proof. In order to simplify the notation we will carry out the proof when $\alpha = 0$. The reader should check for himself that the argument extends to other values of α with only notational changes. Obviously f and $\bar{f}$ are nearly Borel measurable. We first claim that we may restrict our attention to bounded f. Indeed $\{\bar{f} < f\}$ is a countable union of sets $\{\bar{f} < r < f\}$ with r a positive rational, so it suffices to show that each of these is semipolar. Now if $g_r = \min(f, r)$ then the sequence of bounded excessive functions $\min(f_n, r)$ decreases to g_r. But $\bar{g}_r \leq \bar{f}$ and so $\{\bar{f} < r < f\} \subset \{\bar{f} < g_r\} \subset \{\bar{g}_r < g_r\}$. Hence we may assume f bounded in the remainder of the proof.

Let $A = \{\bar{f} = f\}$. Then $E - A$ is of potential zero and hence we must have $A^r = E$. The following two statements are valid for f because they are valid for any exessive function and their validity is preserved under the taking of decreasing limits. Namely: (i) if T is any $\{\mathscr{M}_t\}$ stopping time, then $f \geq P_T f$; and (ii) if $B \in \mathscr{E}^n$ and $x \in B^r$, then $f(x) \geq \inf\{f(y) : y \in B\}$. Let $A_\varepsilon = \{f \geq \bar{f} + \varepsilon\}$. Suppose $x \notin A_\varepsilon$ and pick numbers α and β such that $f(x) < \beta$, $\bar{f}(x) > \alpha$, and $\beta - \alpha < \varepsilon$. According to (2.10), x is not regular for $\{\bar{f} \leq \alpha\}$ and by what was said above x is not regular for $\{f \geq \beta\}$. Consequently $A_\varepsilon^r \subset A_\varepsilon$. To complete the proof of Theorem 3.6 it clearly suffices to show that each A_ε is thin. This is accomplished by the following lemma.

(3.7) LEMMA. Suppose that x is regular for A_ε, that $\beta \geq 0$, and that there is a sequence $\{T_n\}$ of $\{\mathscr{F}_t\}$ stopping times such that (a) $P^x(T_n \to 0) = 1$ and (b) $f(x) \geq P_{T_n} f(x) + \beta$ for all n. Then there is a sequence $\{Q_n\}$ of $\{\mathscr{F}_t\}$ stopping times such that (a') $P^x(Q_n \to 0) = 1$ and (b') $f(x) \geq P_{Q_n} f(x) + \beta + \varepsilon/2$ for all n.

Indeed repeated application of (3.7) with β first equal to 0, then equal to $\varepsilon/2$, and so on, shows that if $x \in A_\varepsilon^r$ then $f(x) = \infty$ which contradicts the fact that f is bounded. Thus (3.7) implies that A_ε is thin and hence the proof of (3.6) will be complete as soon as (3.7) is established.

Proof of (3.7). Let $\delta > 0$ and $0 < \eta < 1$, and let $R = T_{A_\varepsilon}$. Since almost surely P^x, $T_n \to 0$ and $X_t \in A_\varepsilon$ at arbitrarily small strictly positive values of t $(x \in A_\varepsilon^r)$, we may choose n so large that $S = T_n + R \circ \theta_{T_n}$ satisfies $P^x(S \geq \delta/2) < \eta/2$. Moreover $X(S) \in A_\varepsilon$ almost surely on $\{S < \infty\}$ because $A_\varepsilon^r \subset A_\varepsilon$. Let $\mu(dy) = P^x[X(S) \in dy;\ S < \infty]$. Since $A^r = E$, $P^\mu(T_A > 0) = \int \mu(dy)\, P^y(T_A > 0) = 0$. Consequently we can choose a compact subset K of A such that

$$P^x\left[T_K \circ \theta_S \geq \frac{\delta}{2};\ S < \infty\right] = P^\mu\left(T_K \geq \frac{\delta}{2}\right) < \frac{\eta}{2}.$$

If $Q = S + T_K \circ \theta_S$, then

$$P^x(Q \geq \delta) \leq P^x\left(S \geq \frac{\delta}{2}\right) + P^x\left[T_K \circ \theta_S \geq \frac{\delta}{2};\ S < \infty\right] < \eta.$$

But $f(X_{T_K}) = \bar{f}(X_{T_K})$ almost surely on $\{T_K < \infty\}$ because $X(T_K) \in K \subset A$. Therefore the following inequalities obtain:

$$\begin{aligned} E^x f(X_Q) &= E^x\{E^{X(S)}[\bar{f}(X_{T_K})]\} \\ &\leq E^x\, \bar{f}(X_S) \\ &\leq E^x f(X_S) - \varepsilon\, P^x(S < \infty) \\ &\leq E^x f(X_{T_n}) - \varepsilon\, P^x(S < \infty) \\ &\leq f(x) - \beta - \varepsilon(1 - \eta/2) \\ &\leq f(x) - \beta - \varepsilon/2. \end{aligned}$$

In view of the fact that δ and η are arbitrary, these estimates yield Lemma 3.7.

Exercises

(3.8) (a) Give an example of a thin set which is not polar. (b) Give an example of a semipolar set which is not thin. (c) Give an example of a null set which is not semipolar. [Hint: consider the process of uniform motion to the right.]

(3.9) Let $A \subset E$ be null. Show that each point of E is regular for A^c.

(3.10) Let X be a standard process and assume that there exist a $\beta > 0$ and a Radon measure λ on $\mathscr{E}$ such that $f \in \mathscr{S}^\beta$ and $f = 0$ almost everywhere λ implies that $f = 0$. Show that there exists a finite measure ξ on $\mathscr{E}$ such that $A \in \mathscr{E}^*$ is null if and only if $\xi(A) = 0$. Show that under this assumption any α-excessive function is Borel measurable. (See Section 1 of Chapter V.)

(3.11) Let $A \in \mathscr{E}^n$ and let $f \in \mathscr{S}^\alpha$. Show that if every point of A is regular for A, then $P_A^\alpha f$ is the smallest α-excessive function dominating f on A. Give an example to show that $P_A^\alpha f$ need not dominate f on A if $A - A^r$ is not empty.

(3.12) Let $A \in \mathscr{E}^n$ and let $x \in E$ be fixed. Let $\{T_n\}$ be an increasing sequence of $\{\mathscr{F}_t\}$ stopping times with $\lim T_n = T_A$ almost surely P^x. If $\Phi = \Phi_A^\alpha$ show that $\Phi(X_{T_n}) \to 1$ almost surely P^x on $\{T_n < T_A \text{ for all } n,\ T_A < \infty\}$ and hence that $\lim_{t \uparrow T_A} \Phi(X_t) = 1$ almost surely P^x on this set. [Hint: since Φ_A^α increases as α decreases it suffices to consider $\alpha > 0$. Use (2.15) to show that $\{e^{-\alpha T_n} \Phi(X_{T_n});\ \mathscr{F}_{T_n}\}$ is a supermartingale relative to P^x and hence that $\lim_n e^{-\alpha T_n} \Phi(X_{T_n}) = L$ exists and $L \leq e^{-\alpha T_A}$ almost surely P^x. Next use the strong Markov property to show that if $\Lambda = \{T_n < T_A \text{ for all } n\}$, then $E^x(L; \Lambda) = E^x(e^{-\alpha T_A}; \Lambda)$. Conclude from this that $\lim_n \Phi(X_{T_n}) = 1$ almost surely P^x on $\Lambda \cap \{T_A < \infty\}$.]

(3.13) Let $A \in \mathscr{E}^n$ and suppose that $\limsup_{t \downarrow 0} P_t(x, A) > 0$. Then show that x is regular for A.

(3.14) Construct an example of a set A which is thin at every point without being a thin set. [Hint: let X be uniform motion to the right in $\mathbf{R}^2$; that is, starting from any point $(x_0, y_0) \in \mathbf{R}^2$ the process moves to the right along the line $y = y_0$ with speed one. The reader should have no difficulty writing down a rigorous definition of this process. Let A be a nonmeasurable set (with respect to two-dimensional Lebesgue measure) which intersects each horizontal line in at most one point. See Hewitt and Stromberg [1, 21.27] for the construction of such a set.]

(3.15) Let X be the symmetric stable process of index α in $\mathbf{R}$. If $\alpha \leq 1$ show that for each x the singleton $\{x\}$ is polar. [Hint: using the notation of (1.7) let $f_\alpha(t, x) = \int_0^\infty h_{\alpha/2}(t, u)\, g_u(x)\, du$. If $\alpha < 1$ show $y \to |x - y|^{\alpha - 1} = c \int_0^\infty f_\alpha(t, y - x)\, dt$ is excessive for each x, and then use Proposition 3.5. If $\alpha = 1$ and $\beta > 0$ let $U_\beta(x) = \int_0^\infty e^{-\beta t} f_1(t, x)\, dt$; then show that $U_\beta(0) = \infty$, $U_\beta(x) < \infty$ for $x \neq 0$, and that $y \to U_\beta(x - y)$ is β-excessive for each x.]

(3.16) Let X be the Brownian motion process in $\mathbf{R}$. Show that for all points x, x is regular for $\{x\}$. Consequently the only thin set is the empty set. [Hint:

use (3.13) and the continuity of the paths.] This result is also valid for stable processes in **R** of index α with $1 < \alpha < 2$ as will be proved later.

(3.17) Show that if f is α-super-mean-valued then $\{\bar{f} < f\}$ need not be semi-polar. [Hint: use the result of (3.16).]

(3.18) Let X be Brownian motion in **R**. If $f = I_{[0,\infty)}$, $T = T_{\{0\}}$, and $x < 0$, then compute

$$E^x \int_0^\infty e^{-\alpha t} f(X_t)\,dt = E^x \int_T^\infty e^{-\alpha t} f(X_t)\,dt$$

to obtain $P^x(T \leq t) = 2P^x(X_t \geq 0)$. Now use the explicit form of the Gauss kernel $g_t(x)$ (see (2.17) of Chapter I) to show that $E^x\{\exp(-\alpha T_y)\} = e^{-\sqrt{\alpha}|x-y|}$ for all $x, y \in \mathbf{R}$ and $\alpha > 0$, where $T_y = T_{\{y\}}$. In particular show that $P^x(T_y < \infty) = 1$.

(3.19) The notation is that of (3.18). Let $a < x < b$ and let $\alpha > 0$. Let $T = \min(T_a, T_b) = \inf\{t : X_t \notin (a, b)\}$ almost surely P^x and let $J_a = E^x(e^{-\alpha T}; X(T) = a)$ and $J_b = E^x(e^{-\alpha T}; X(T) = b)$. Express $E^x(e^{-\alpha T_a})$ and $E^x(e^{-\alpha T_b})$ in terms of J_a and J_b and solve the resulting equations to obtain

$$J_a = \frac{\sinh\sqrt{\alpha}(b - x)}{\sinh\sqrt{\alpha}(b - a)}; \qquad J_b = \frac{\sinh\sqrt{\alpha}(x - a)}{\sinh\sqrt{\alpha}(b - a)}.$$

Conclude from this that $P^x(T_a < T_b) = (b - x)(b - a)^{-1}$ and that $E^x(T) = \frac{1}{2}(x - a)(b - x)$.

(3.20) Let $f \in \mathcal{S}^\alpha$ and let $\{B_n\}$ be a decreasing sequence of nearly Borel subsets of E_Δ. Then $\{P^\alpha_{B_n} f\}$ decreases to an α-super-mean-valued function g. Show that $g = \bar{g}$ except possibly at those points x in E such that either $g(x) = \infty$ or $x \in B_n^r$ for all n. [Hint: first show that if S and T are stopping timer with $S \leqslant T$, then for all $t > 0$,

$$E^x\{e^{-\alpha S} f(X_S); S \leqslant t\} \geqslant E^x\{e^{-\alpha T} f(X_T); T \leqslant t\}.]$$

4. The Fine Topology

Once again, in this section the basic datum is a given standard process $X = (\Omega, \mathcal{M}, \mathcal{M}_t, X_t, \theta_t, P^x)$ with state space $(E, \mathcal{E})$, and all definitions are relative to this given process.

(4.1) Definition. A set $A \subset E_\Delta$ is called *finely open* if $A^c = E_\Delta - A$ is thin at each x in A. In other words for each $x \in A$ there is a set $D \in \mathscr{E}_\Delta^n$ such that $A^c \subset D$ and $P^x(T_D > 0) = 1$.

Intuitively a set A is finely open provided the process remains in A for an initial interval of time almost surely P^x for each $x \in A$. By the right continuity of the paths any open set is finely open. Let $\mathcal{O}(\mathcal{O}_\Delta)$ be the collection of all finely open subsets of $E(E_\Delta)$. One checks without difficulty that $\mathcal{O}(\mathcal{O}_\Delta)$ is a topology on $E(E_\Delta)$. It is called the *fine topology* on $E(E_\Delta)$. We already observed that the fine topology on $E(E_\Delta)$ is finer than the original topology on $E(E_\Delta)$. Note that $\mathcal{O}$ is just the topology that E inherits when we regard E as a subspace of $(E_\Delta, \mathcal{O}_\Delta)$, and that Δ is isolated in $(E_\Delta, \mathcal{O}_\Delta)$. Consequently we will restrict our attention to the fine topology on E.

In this section we will give an alternate description of $\mathcal{O}$ which will show that in reality $\mathcal{O}$ depends only on the transition function of X. We will also discuss the behavior of $t \to f(X_t)$ when f is nearly Borel measurable and finely continuous, that is, continuous relative to $\mathcal{O}$. As usual the continuity of numerical functions is defined relative to the topology of $\overline{\mathbf{R}}$, the extended real line. Let us also recall that an element f of $\mathscr{S}^\alpha$ is regarded as a function on E, and hence sets of the form $f^{-1}(D)$ where $D \subset \overline{\mathbf{R}}$ are subsets of E. Of course in certain formulas we must set $f(\Delta) = 0$; but $\{f < 1\}$, for example, is a subset of E.

(4.2) Proposition. If $f \in \mathscr{S}^\alpha$, then f is finely continuous.

Proof. Let I be an open interval in $\overline{\mathbf{R}}$ and let $B = f^{-1}(I)$. By Proposition 2.10 no point in B is regular for B^c. Consequently B is finely open.

(4.3) Proposition. Let $A \subset E$ and suppose that x is in A and A^c is thin at x. Then there exists a compact set K such that $x \in K \subset A$ and K^c is thin at x.

Proof. By definition there exists a $B \in \mathscr{E}^n$ such that $B \subset A$ and $P^x(T_{B^c} > 0) = 1$. Since $x \in A$ we may assume that $x \in B$, and also that B has compact closure $\overline{B}$ in E. Of course $B^c = E_\Delta - B$. By Theorem 11.2 of Chapter I there exists a decreasing sequence $\{G_n\}$ of open subsets of E containing $B^c - \{\Delta\} = E - B \supset E - \overline{B}$ such that $\min(T_{G_n}, \zeta) \uparrow \min(T_{E-B}, \zeta)$ almost surely P^x. But $\min(T_{E-B}, \zeta) = T_{B^c}$, $\min(T_{G_n}, \zeta) = T_{G_n \cup \{\Delta\}}$, and $G_n \cup \{\Delta\}$ is open in E_Δ since $G_n \supset E - \overline{B}$. Consequently, by the zero-one law, there exists an open set G in E_Δ such that $\Delta \in G$, $G^c \subset B$, and $P^x(T_G > 0) = 1$. If we let $K = G^c$ the proof is complete.

(4.4) PROPOSITION. Let $\mathscr{U} = \{B : B \in \mathcal{O} \cap \mathscr{E}^n$ and $\bar{B}$ is compact in $E\}$. Then $\mathscr{U}$ is a base for the fine topology.

Proof. Let A be a fine neighborhood of the point x. By (4.3) there is a compact subset K of E such that $x \in K \subset A$ and $P^x(T_{K^c} > 0) = 1$. Let $\alpha > 0$ and $\Phi^\alpha(x) = E^x(e^{-\alpha T_{K^c}})$, and let $B = \{\Phi^\alpha < 1\}$. Since $\Phi^\alpha \in \mathscr{S}^\alpha$, the set B is in $\mathcal{O} \cap \mathscr{E}^n$, and of course $x \in B$. If $y \in K^c$ then $\Phi^\alpha(y) = 1$ because K^c is open, so $B \subset K$. Finally $\bar{B}$ is compact in E because K is compact in E. This completes the proof.

(4.5) THEOREM. Let $\alpha > 0$. Then the fine topology is the coarsest topology relative to which the elements of $\mathscr{S}^\alpha$ are continuous.

Proof. In the proof of (4.4) we produced a base for the fine topology consisting of sets of the form $\{\Phi_G^\alpha < 1\}$ where G ranges over the open sets of E_Δ containing the point Δ. Hence we obtain Theorem 4.5.

REMARK. We have actually shown that the fine topology is the coarsest topology rendering the functions Φ_G^α, G as above, upper semicontinuous.

(4.6) THEOREM. Suppose that $U(x, K)$ is finite for all x whenever K is a compact subset of E. Then the fine topology is the coarsest topology relative to which the elements of $\mathscr{S} = \mathscr{S}^0$ are continuous.

Proof. Let $\mathscr{T}$ be the coarsest topology relative to which the elements of $\mathscr{S}$ are continuous. By (4.2), $\mathscr{T} \subset \mathcal{O}$, and, by the preceding remark, it will suffice to show that $\{\Phi_G^1 < 1\} \in \mathscr{T}$ for all open sets G in E_Δ. Let $\{G_n\}$ be a sequence of open sets with compact closures such that $\bar{G}_n \subset G_{n+1}$ for $n \geq 1$ and $\bigcup G_n = E$. If

(4.7)
$$\varphi_n(x) = U\, I_{G_n}(x) - P_G U\, I_{G_n}(x)$$
$$= E^x \int_0^{T_G} I_{G_n}(X_t)\, dt,$$

then φ_n is $\mathscr{T}$-continuous since both UI_{G_n} and $P_G\, UI_{G_n}$ are excessive and finite. Clearly φ_n increases with n. If $\varphi = \lim_n \varphi_n$, then φ is $\mathscr{T}$-lower-semicontinuous. Thus to complete the proof it suffices to show that $\{\Phi_G^1 < 1\} = \{\varphi > 0\}$. If $\varphi_n(x) > 0$ then $P^x(T_G > 0) = 1$, and so $\Phi_G^1(x) < 1$. Hence $\{\varphi > 0\} \subset \{\Phi_G^1 < 1\}$. On the other hand if $\Phi_G^1(x) < 1$ and $x \in G_n$, then $\varphi_n(x) > 0$. Consequently $\{\Phi_G^1 < 1\} \subset \{\varphi > 0\}$, completing the proof of Theorem 4.6.

(4.8) THEOREM. Suppose $f \in \mathscr{E}^n$. Then f is finely continuous if and only if the mapping $t \to f(X_t)$ is right continuous almost surely.

Proof. The implication from the right continuity of the composition to the fine continuity of f is immediate; in fact only the right continuity at $t = 0$ is needed. Coming to the converse implication, suppose that $t \to f[X_t(\omega)]$ is not right continuous and that $t \to X_t(\omega)$ is right continuous. Then there is a point $t_0 < \zeta(\omega)$, a sequence $\{t_n\}$ decreasing to t_0 with $t_n < \zeta(\omega)$ for all n, and a pair of intervals I_1, I_2 of the form $[-\infty, r)$, $(s, \infty]$ (perhaps not in that order) with r and s rational and $r < s$, such that $f[X_{t_0}(\omega)] \in I_1$ and $f[X_{t_n}(\omega)] \in I_2$ for all $n \geq 1$. Let $A = f^{-1}(I_1)$ and $B = f^{-1}(I_2)$ so that A and B are disjoint finely open sets in $\mathscr{E}^n$. If $\varphi = \Phi_B^1$, then $\varphi[X_{t_0}(\omega)] < 1$ and $\varphi[X_{t_n}(\omega)] = 1$ for all n. But $\varphi \in \mathscr{S}^1$ and hence almost surely $t \to \varphi(X_t)$ is right continuous. Since there are only countably many pairs (I_1, I_2) of the above form and $t \to X_t(\omega)$ is right continuous almost surely, the proof of Theorem 4.8 is complete.

Exercises

(4.9) Let $A \in \mathscr{E}^n$. Show that A is finely closed if and only if $A^r \subset A$, and that the fine closure of A is $A \cup A^r$. [Hint: first show that almost surely $T_A \leq T_{A^r}$.] Consequently if $x \notin A$, then A is thin at x if and only if x is not in the fine closure of A. Show that in this last statement the condition "$x \notin A$" may not be eliminated. See (1.8) of Chapter V for an extension of this result to arbitrary subsets A of E.

(4.10) Assume that $\{x\}$ is polar for each $x \in E$. Under this assumption show that a sequence $\{x_n\}$ approaches x in the fine topology if and only if $x_n = x$ for all large n.

(4.11) Let B be a null set. Show that $E - B$ is finely dense in E.

(4.12) Show that $(E, \mathscr{O})$ is a completely regular Hausdorff space.

(4.13) Let $A \in \mathscr{E}^n$, $x \in \bar{A}$ (the closure of A in E), and consider the following statements: (i) x is not in A and A is thin at x; (ii) there exists an $f \in \mathscr{S}^\alpha$ such that $f(x) < \liminf_{y \to x,\, y \in A} f(y)$. Show that (ii) implies (i) and that (i) implies (ii) if $\alpha > 0$. [Hint: to show that (i) implies (ii) for $\alpha \wedge 0$, use Proposition 4.3. Make use of Proposition 2.10 for the other assertion.]

(4.14) Let $Y \in b\mathscr{F}$. Suppose that for each μ, $Y \circ \theta_{T_n} \to Y$ almost surely P^μ whenever $\{T_n\}$ is a sequence of $\{\mathscr{F}_t\}$ stopping times decreasing to zero almost surely P^μ. Show that $f(x) = E^x(Y)$ is nearly Borel measurable and finely continuous. [Hint: one may assume $Y \geq 0$. Show that $\alpha U^\alpha f \to f$ and so

f is nearly Borel measurable. If $\varepsilon > 0$ show that x is not regular for $A = \{y : f(y) \geq f(x) + \varepsilon\}$ and $B = \{y : f(y) \leq f(x) - \varepsilon\}$, and conclude from this that f is finely continuous.]

(4.15) Let $\alpha > 0$ and $\varphi \in \mathscr{S}^\alpha$. If $A \in \mathcal{O} \cap \mathscr{E}^n$, show that there is a sequence $\{g_n\}$ in $\mathscr{E}_+^n$ such that each g_n vanishes off A and $U^\alpha g_n \uparrow P_A^\alpha \varphi$. Each g_n may be taken to have compact support in A, if A is a countable union of closed sets. [Hint: assume first that φ is bounded. Let $\{h_n\}$ be an increasing sequence of nonnegative elements of $b\mathscr{E}^n$ such that $\lim h_n$ is infinite on A and each h_n vanishes off A—or off some variable compact subset of A if A is a countable union of closed sets. Let ψ_n be defined as in Exercise 2.20 using φ and h_n. Show that

$$\psi_n(x) = E^x\left\{\int_0^\infty e^{-\alpha t}\, \phi(X_t)\, d_t\left[1 - \exp\left(-\int_0^t h_n(X_s)\, ds\right)\right]\right\},$$

and use this to show that $\psi_n \to P_A^\alpha \varphi$. Use the results of (2.20) to complete the proof when φ is bounded. If φ is unbounded apply the above argument to $\min(\varphi, a)$ and let $a \to \infty$.]

(4.16) Let X be Brownian motion in $\mathbf{R}$. Prove that $\Phi_{\{y\}}^\alpha(x) \to 1$ as $y \to x$. [Hint: see (3.18).] Prove that the fine topology coincides with the usual topology of $\mathbf{R}$. [Hint: by the remark following (4.5) it suffices to show that Φ_G^α is upper semicontinuous when $\alpha > 0$ and G is open. For this use the result of the first part of (4.16).]

(4.17) Let X be a standard process such that for all $B \in \mathscr{E}^n$ either $UI_B = 0$ or $UI_B = \infty$; that is, $UI_B = \infty$ whenever B is not of potential zero. Show that if $B \in \mathscr{E}^n$ and $UI_B \neq 0$, then $\Phi_B = 1$. Of course $\Phi_B(x) = P^x(T_B < \zeta) = P^x(T_B < \infty)$ is regarded as a function on E and $\Phi_B \in \mathscr{S}$. [Hint: let $\Psi_B = \lim_{t\to\infty} P_t \Phi_B$ and $\Theta_B = \Phi_B - \Psi_B$. Then $P_t \Psi_B = \Psi_B$. Use (2.19i) and the hypothesis to show that $\Theta_B = 0$, and hence that $P_t \Phi_B = \Phi_B$. Show that there exists an x_0 such that $\Phi_B(x_0) = 1$. Finally show that if $D = \{\Phi_B < 1\}$, then $U(x_0, D) = 0$ and conclude from this that D must be empty.]

(4.18) Let X be as in (4.17). Show that if $B \in \mathscr{E}^n$ and $\Phi_B \neq 0$, then $\Phi_B = 1$. [Hint: assume B compact and let $\eta = \sup \Phi_B > 0$. If $0 < \delta < \eta$ and $A = \{\Phi_B > \delta\}$ use (4.17) to show $\Phi_A = 1$. Conclude from this that $\Phi_B = \eta$. Finally show that $\Phi_B = 1$.]

(4.19) Let X be a standard process such that $\Phi_B = 1$ whenever $B \in \mathscr{E}^n$ is nonvoid and finely open. Show that the only excessive functions are constants. Combine this with (4.17) (or (4.18)), (1.5b), and (1.7b) to conclude that the only excessive functions for Brownian motion in $\mathbf{R}$ or $\mathbf{R}^2$, or the symmetric stable process in $\mathbf{R}$ with $\alpha \geq 1$, are constants.

(4.20) Let X be as in (4.19) and assume that E contains at least two distinct points. Show that $P^x(\zeta < \infty) = 0$ for all $x \in E$. [Hint: show that $\psi(x) = E^x(1 - e^{-\zeta})$ is excessive and hence constant. Use this to show that $E^x(e^{-\zeta}) = 0$ for all x in E.]

(4.21) Let X be as in (4.20). If $B \in \mathscr{E}^n$ show that $UI_B = 0$ or $UI_B = \infty$. [Hint: suppose that $\eta = \sup UI_B > 0$. If $0 < \delta < \eta$ and $A = \{UI_B > \delta\}$, then $\Phi_A = 1$. Use this and (4.20) to show that $UI_B = \infty$.] Note that if excessive functions are lower semicontinuous, then one need only assume $\Phi_B = 1$ whenever B is a nonvoid open subset of E.

(4.22) Combine the results of (4.17)–(4.21) to show the equivalence of the following statements. (Assume that E contains at least two distinct points.) (i) $UI_B = 0$ or $UI_B = \infty$ for $B \in \mathscr{E}^n$. (ii) $\Phi_B = 1$ whenever $B \in \mathscr{E}^n$ is nonvoid and finely open (or whenever B is a nonvoid open subset of E if excessive functions are lower semicontinuous). (iii) $\Phi_B = 1$ whenever B is in $\mathscr{E}^n$ and is not polar. (iv) The only excessive functions are constants. A process satisfying any, and hence all, of these conditions is called *recurrent*.

(4.23) Let X be a standard process such that $UI_B = 0$ or $\Phi_B > 0$ for all $B \in \mathscr{E}^n$. Then if B in $\mathscr{E}^n$ is not polar, $\Phi_B > 0$. [Hint: suppose $\Phi_B(x_0) = \eta > 0$ for some x_0. Show that if $D = \{\Phi_B > \eta/2\}$, then $\Phi_D > 0$. Conclude from this that $\Phi_B > 0$.] Note that if there exists a measure ξ on $\mathscr{E}$ such that $P_t(x, dy) = f(t, x, y)\, \xi(dy)$ with f strictly positive, then X satisfies the above condition. In particular it is satisfied for Brownian motion or the symmetric stable process in $\mathbf{R}^n$.

(4.24) Let X be a standard process and assume (i) for some α, $U^\alpha : \mathbf{C}_K \to \mathbf{C}$ and (ii) for each compact subset K of E, $U(x, K)$ is everywhere finite. Show that almost surely $X_t \to \Delta$ as $t \to \infty$. [Hint: let K be a compact subset of E, let G be an open set containing K and having compact closure in E, and let α be such that $U^\alpha : \mathbf{C}_K \to \mathbf{C}$. Then the assertion is a consequence of the following three observations: (i) $U(x, G) \geq U^\alpha(x, G) \geq \delta > 0$ for x in K, (ii) $P_t U I_G(x) \geq \delta\, P^x(T_K \circ \theta_t < \infty)$, and (iii) $P_t UI_G \to 0$ as $t \to \infty$.]

5. Alternative Characterization of Excessive Functions

According to (2.8) if $f \in \mathscr{S}^\alpha$, then $P_K^\alpha f \leq f$ for every compact K. In this section we will derive various converses to this assertion. These are useful for finding the functions excessive relative to specific semigroups and for proving that in various situations excessiveness is a local property. The fundamental theorem, which we will now prove, is due to Dynkin [2] and [4]. As usual $X = (\Omega, \mathcal{M}, \mathcal{M}_t, X_t, \theta_t, P^x)$ is a fixed standard process with state space $(E, \mathcal{E})$.

(5.1) THEOREM. Suppose $f \in \mathcal{E}_+^n$ and $f \geq P_K^\alpha f$ for every compact subset K of E. Then $\eta U^{\eta+\alpha} f \leq f$ for all $\eta > 0$.

Proof. We will take $\alpha = 0$ to simplify notation. Clearly it is enough to prove the theorem for bounded f. Let $\beta > 0, \eta > 0$, and define $h = U^{\eta+\beta} f, F = \eta h - f$. The resolvent equation implies that for $\gamma > 0$

$$h = U^\gamma[(\gamma - \beta)h - F] = U^\gamma g,$$

where we have put $g = (\gamma - \beta)h - F$. Since h is the γ potential of the bounded function g we have for any stopping time T

$$h(x) = E^x \int_0^T e^{-\gamma t} g(X_t)\, dt + P_T^\gamma h(x),$$

from which we obtain

$$P_T^\gamma F(x) - F(x) = -\eta E^x \int_0^T e^{-\gamma t} g(X_t)\, dt + f(x) - P_T^\gamma f(x). \tag{5.2}$$

Let $A = \{F < 0\}$. Choose $\gamma < \beta$ and let $T = T_K$, K any compact subset of A. By hypothesis, $f - P_T f$ is positive and so $f - P_T^\gamma f$ is also positive. Of course $(\beta - \gamma)h$ is positive and $P_T^\gamma F$ is negative because $X_T \in K \subset A$ almost surely on $\{T < \infty\}$. Consequently, using the definition of g,

$$F(x) \leq -\eta E^x \int_0^T e^{-\gamma t} F(X_t)\, dt.$$

But $E^x \int_0^{T_A} e^{-\gamma t} F(X_t)\, dt \geq 0$ and we can find compacts $K_n \subset A$ such that $P^x(T_{K_n} \to T_A) = 1$. Hence $F(x) \leq 0$; that is, $\eta U^{\eta+\beta} f \leq f$. Letting β decrease to 0 completes the proof of Theorem 5.1.

(5.3) COROLLARY. If f satisfies the hypotheses of Theorem 5.1 and also $\liminf_{t \downarrow 0} P_t^\alpha f \geq f$, then $f \in \mathcal{S}^\alpha$.

Proof. By (5.1) and the argument in (2.3), $\eta U^{\eta+\alpha} f$ increases as $\eta \to \infty$. Let $g = \lim_\eta \eta U^{\eta+\alpha} f$. By (2.3), $f \in \mathcal{S}^\alpha$ if and only if $g = f$. It follows from the condition in (5.3) and Fatou's lemma that $\liminf_{\eta \to \infty} \eta U^{\eta+\alpha} f \geq f$, and since $g \leq f$ this implies that $g = f$.

Before coming to the applications of (5.3) we need some definitions.

(5.4) DEFINITION. Let $T = \inf\{t: X_t \neq X_0\}$. A point $x \in E_\Delta$ is called a *holding point* (for X) provided $P^x(T > 0) = 1$. In the alternative case, $P^x(T = 0) = 1$, the point x is called an *instantaneous point*.

Clearly Δ is a holding point. Define a sequence of stopping times $\{S_n\}$ by $S_0 = 0$ and $S_{n+1} = S_n + T \circ \theta_{S_n}$.

(5.5) DEFINITION. The process X is called a *regular step process* if $S_n \to \zeta$ as $n \to \infty$ almost surely P^x for all x in E.

The definition implies of course that for a regular step process every point is a holding point, and that the path functions $t \to X_t$ are step functions taking the value $X(S_n)$ on $[S_n, S_{n+1})$ for all n almost surely. These processes are just the ones constructed in Section 12 of Chapter I. Most of the familiar processes which one encounters are either regular step processes or else all points are instantaneous. The next two theorems treat these two cases.

(5.6) THEOREM. Let X be a regular step process. Then $f \in \mathscr{E}_+^*$ is in $\mathscr{S}^\alpha$ if and only if $f \geq P_T^\alpha f$ (where T is defined in (5.4)).

Proof. First note that for each $x \in E$ the set $\{x\}$ is finely open so that f is finely continuous. Next we observe that $f \in \mathscr{E}^n$; indeed given an initial measure μ, let μ_n be the measure

$$\mu_n(A) = P^\mu(X(S_n) \in A),$$

and let $v_n, u_n \in \mathscr{E}$ be such that $v_n \leq f \leq u_n$ and $\mu_n(\{v_n < u_n\}) = 0$. Of course, we set $v_n(\Delta) = u_n(\Delta) = 0$. If $v = \sup v_n$ and $u = \inf u_n$ then $v, u \in \mathscr{E}$, $v \leq f \leq u$, and since $X_t = X(S_n)$ on $[S_n, S_{n+1})$ we have

$$P^\mu[v(X_t) < u(X_t) \text{ for some } t]$$
$$\leq \sum_n P^\mu[v_n(X(S_n)) < u_n(X(S_n))] = 0.$$

Consequently $f \in \mathscr{E}^n$.

Let K be a compact subset of E and let $Q = T_K$. We next claim that for each $n \geq 0$

$$f(x) \geq E^x(e^{-\alpha Q} f(X_Q); Q \leq S_n) + E^x(e^{-\alpha S_n} f(X_{S_n}); Q > S_n). \tag{5.7}$$

We will prove this by an induction argument. Clearly (5.7) is valid when $n = 0$.

Now assume that (5.7) holds for some fixed value of n. Denoting the second summand on the right side of (5.7) by J and using the inequality $f(X_{S_n}) \geq P_T^\alpha f(X_{S_n})$ we have

$$J \geq E^x\{e^{-\alpha S_n} E^{X(S_n)}[e^{-\alpha T} f(X_T)]; Q > S_n\}$$
$$= E^x\{e^{-\alpha S_{n+1}} f(X_{S_{n+1}}); Q > S_n\}.$$

But $\{Q > S_n\}$ differs from

$$\{Q > S_n; Q \circ \theta_{S_n} = T \circ \theta_{S_n}\} \cup \{Q > S_n; Q \circ \theta_{S_n} > T \circ \theta_{S_n}\}$$

by a set having P^x measure 0, and $Q = S_n + Q \circ \theta_{S_n}$ on $\{Q > S_n\}$. Consequently

$$J \geq E^x\{e^{-\alpha Q} f(X_Q); Q > S_n, Q = S_{n+1}\}$$
$$+ E^x\{e^{-\alpha S_{n+1}} f(X_{S_{n+1}}); Q > S_{n+1}\},$$

and since $\{S_n < Q \leq S_{n+1}\} = \{S_n < Q; Q = S_{n+1}\}$ almost surely, we obtain (5.7) with n replaced by $n + 1$. Thus (5.7) is valid for all n. Now letting $n \to \infty$ in (5.7) we obtain $f \geq P_K^\alpha f$ for all compact subsets K of E. Finally the fine continuity of f and Fatou's lemma imply that $\liminf_{t \downarrow 0} P_t^\alpha f \geq f$, and so Theorem 5.6 follows from Corollary 5.3.

Before stating the next theorem we let d be a fixed metric on E compatible with the topology and define

(5.8) $$T_n = \inf\{t: d(X_t, X_0) > 1/n\}.$$

(5.9) THEOREM. Suppose all points of E are instantaneous. Let $f \in \mathscr{E}_+^n$ be finely continuous and suppose that for every compact $K \subset E$ there is an integer N such that $f(x) \geq P_{T_n}^\alpha f(x)$ for all $x \in K$ and $n \geq N$. Then $f \in \mathscr{S}^\alpha$.

Proof. Once again we assume $\alpha = 0$ to simplify the writing. Let K be a compact subset of E and pick a positive integer N such that $f(x) \geq P_{T_n} f(x)$ for all $x \in K$ and $n \geq N$. Fix $n \geq N$ and define

$$R = T_n, \quad \text{if } X_0 \in K,$$
$$= 0, \quad \text{if } X_0 \notin K,$$

and $R_0 = 0$, $R_{k+1} = R_k + R \circ \theta_{R_k}$. Obviously $f \geq P_R f$, and because $d(X(R_k), X(R_{k+1})) \geq 1/n$ if $R_{k+1} < T_{K^c} \leq \zeta$ it follows from the quasi-left-continuity of X that $\lim_{k \to \infty} R_k \geq T_{K^c}$ almost surely. Given an integer k and $\Gamma \in \mathscr{F}_{R_k}$ we have from the strong Markov property

$$E^x\{f[X(R_k)]; \Gamma\} \geq E^x\{P_R f(X(R_k)); \Gamma\}$$
$$= E^x\{f(X(R_{k+1})); \Gamma\}.$$

From this it follows easily that for any given t

$$f(x) \geq \sum_{k=0}^{\infty} E^x\{f(X(R_{k+1})); R_k \leq t < R_{k+1}\}.$$

Recalling the dependence of R_k on the integer n, define

$$\begin{aligned} S_n &= R_{k+1}, && \text{if} \quad R_k \leq t < R_{k+1}, \\ &= \infty, && \text{if} \quad R_k \leq t \text{ for all } k. \end{aligned}$$

Then for all $n \geq N$

(5.10) $$f(x) \geq E^x\{f(X(S_n)); t < T_{K^c}\}.$$

It is clear from the definitions that if $t < T_{K^c}$ then S_{2n} is in the interval $(t, t + T_n \circ \theta_t]$. As there are no holding points $T_n \circ \theta_t$ approaches 0 as $n \to \infty$ almost surely on $\{\zeta > t\}$. By hypothesis and Theorem 4.8 the mapping $s \to f(X_s)$ is continuous on the right almost surely, and so letting n approach ∞ in (5.10) we obtain $f(x) \geq E^x(f(X_t); t < T_{K^c})$. By choosing K large enough we may make T_{K^c} arbitrarily close to ζ and so finally $f \geq P_t f$ for all t. This inequality, along with the fact that f is finely continuous, implies that $f \in \mathscr{S}$. Thus the proof of Theorem 5.9 is complete.

We have not attempted to find the best way of combining (5.6) and (5.9). Rather we will content ourselves with the following result in the general situation.

(5.11) THEOREM. Let f be defined, nonnegative, and lower semicontinuous on E, and suppose that, for each x in E, $f(x) \geq P^{\alpha}_{T_n} f(x)$ for a sequence of values of n approaching infinity (the sequence may depend on x; T_n is defined in (5.8)). Then $f \in \mathscr{S}^{\alpha}$.

Proof. Once again we assume $\alpha = 0$ to simplify the notation. Clearly f is Borel measurable in the present case. Let K be a compact subset of E. We are going to show that $f \geq P_K f$. To this end define

$$A_n = \{y \in E: d(y, K) > 1/n \text{ and } f(y) \geq P_{T_n} f(y)\}.$$

It then follows from our hypothesis that $E - K = \bigcup_n A_n$. Let

$$\begin{aligned} R_1 &= T_n, && \text{if} \quad X_0 \in A_n - \bigcup_{k<n} A_k, \\ &= 0, && \text{if} \quad X_0 \notin E - K. \end{aligned}$$

Clearly $f \geq P_{R_1} f$. Fix $x \in E - K$. If Q is an $\{\mathscr{F}_t\}$ stopping time, $\Lambda \in \mathscr{F}_Q$, and $R = Q + R_1 \circ \theta_Q$, then

(5.12)
$$E^x\{f(X_Q); \Lambda\} \geq E^x\{P_{R_1} f(X_Q); \Lambda\} = E^x\{f(X_R); \Lambda\}.$$

Let β be any countable ordinal and suppose that R_γ has been defined for all $\gamma < \beta$. If β has a predecessor, $\beta - 1$, define $R_\beta = R_{\beta-1} + R_1 \circ \theta_{R_{\beta-1}}$. If β is a limit ordinal define $R_\beta = \sup_{\gamma<\beta} R_\gamma$. In either case R_β is an $\{\mathscr{F}_t\}$ stopping time and $R_\beta \geq R_\gamma$ whenever $\gamma \leq \beta$. Since $R_1 > 0$ almost surely on $\{X_0 \in E - K\}$, we have for $\gamma < \beta$

$$P^x[R_\gamma = R_\beta; R_\gamma < \min(\zeta, T_K)] = 0.$$

For each countable ordinal β, let $q(\beta) = E^x[R_\beta/(1 + R_\beta)]$. Since $x \in E - K$, $0 < q(\beta) \leq 1$ for all β and $q(\gamma) \leq q(\beta)$ whenever $\gamma \leq \beta$. Moreover if $\gamma < \beta$ and $P^x[R_\gamma < \min(\zeta, T_K)] > 0$, then $q(\gamma) < q(\beta)$. Consequently there exists a countable ordinal β_0 such that $P^x[R_{\beta_0} = \min(\zeta, T_K)] = 1$. We know $f(x) \geq P_{R_1} f(x)$. Suppose that $f(x) \geq P_{R_\gamma} f(x)$ for all $\gamma < \beta$. If β has a predecessor, $\beta - 1$, then (5.12) implies that $f(x) \geq P_{R_\beta} f(x)$. If β is a limit ordinal, then we can find an increasing sequence $\{\beta_n\}$ such that $R_{\beta_n} \uparrow R_\beta$ almost surely P^x, and consequently $X(R_{\beta_n}) \to X(R_\beta)$ on $\{R_\beta < \zeta\}$ almost surely P^x. Therefore using the lower semicontinuity of f and Fatou's lemma we obtain $f(x) \geq P_{R_\beta} f(x)$. In particular $f(x) \geq P_{R_{\beta_0}} f(x) = P_K f(x)$.

So far we have shown that $f(x) \geq P_K f(x)$ for any compact subset K of E and $x \in E - K$. If x is regular for K, then $P_K f(x) = f(x)$. Finally suppose that $x \in K - K^r$. Then x must be regular for $E - K$ and so there exists an increasing sequence $\{K_n\}$ of compact subsets of $E - K$ such that $P^x(T_{K_n} \downarrow 0) = 1$. Since $X(T_{K_n}) \in E - K$ almost surely on $\{T_{K_n} < \infty\}$ and $x \in E - K_n$ we obtain with two applications of what has already been established

$$E^x\{f[X(T_{K_n} + T_K \circ \theta_{T_{K_n}})]\} = E^x\{P_K f[X(T_{K_n})]\} \leq P_{K_n} f(x) \leq f(x).$$

On the other hand $X(T_{K_n} + T_K \circ \theta_{T_{K_n}}) \to X(T_K)$ almost surely P^x; in fact there is equality as soon as $T_{K_n} < T_K$. Thus Fatou's lemma implies that $f(x) \geq P_K f(x)$ for $x \in K - K^r$ as well. One final appeal to the lower semicontinuity of f and Fatou's lemma yields $\liminf_{t \downarrow 0} P_t f \geq f$, and so Theorem 5.11 follows from Corollary 5.3.

We are now going to identify some of the objects defined in this chapter with objects from classical potential theory when the process X in question is Brownian motion. Therefore let $X = (\Omega, \mathscr{M}, \mathscr{M}_t, X_t, \theta_t, P^x)$ be the Brownian motion process in $\mathbf{R}^n$. We assume that X is given in its natural function space representation; that is, Ω consists all maps $\omega : [0, \infty] \to (\mathbf{R}^n \cup \{\Delta\})$ which are *continuous* on $[0, \infty)$, $\omega(\infty) = \Delta$, and $\omega(t) \in \mathbf{R}^n$ for all $t < \infty$;

$X_t(\omega) = \omega(t)$, $\theta_t \omega(s) = \omega(s + t)$; and $\mathscr{M}$ and $\mathscr{M}_t$ are equal to $\mathscr{F}$ and $\mathscr{F}_t$, respectively, for all t. If $r > 0$ let $T_r = \inf\{t: |X_t - X_0| > r\}$ where $|x - y|$ is the usual Euclidean metric on $\mathbf{R}^n$. The key fact which we need is that $P_{T_r}(x, \cdot)$ is the uniform distribution of unit mass on $S_r(x) = \{y: |x - y| = r\}$. Clearly the continuity of the paths implies that $P_{T_r}(x, \cdot)$ is carried by $S_r(x)$.

First of all let us show that $P^x(T_r < \infty) = 1$, which implies that $P_{T_r}(x, S_r(x)) = 1$. Fix x and $r > 0$ and let T be the first hitting time of $\{y: |y - x| > r\}$. Let $g_t(x)$ be the Gauss kernel defined in (2.17) of Chapter I. Then

$$\begin{aligned} P^x(T_r \leq t) = P^x(T \leq t) &\geq P^x[|X_t - x| > r] \\ &= \int_{|y|>r} g_t(y)\, dy, \end{aligned}$$

and using the form of the Gauss kernel it is easy to see that this last expression approaches one as t approaches infinity. Hence $P^x(T_r < \infty) = 1$. Next let O be a distance-preserving transformation of $\mathbf{R}^n$. Then O induces a transformation $\mathbf{O}$ on Ω by $(\mathbf{O}\omega)(t) = \mathrm{O}\omega(t)$ if $t < \infty$ and $(\mathbf{O}\omega)(\infty) = \Delta$. We now claim that $P^x\mathbf{O}^{-1} = P^{\mathrm{O}x}$ for any such O and all x. Indeed first observe that if $P_t(x, A)$ is the transition function of X, then $P_t(x, \mathrm{O}^{-1}A) = P_t(\mathrm{O}x, A)$ for any such O. Consequently if $0 \leq t_1 < \ldots < t_k$ and $A_1, \ldots, A_k$ are in $\mathscr{B}(\mathbf{R}^n)$, then

$$\begin{aligned} P^x\left[\mathbf{O}^{-1}\bigcap_{j=1}^{k}\{X_{t_j} \in A_j\}\right] &= P^x\left[\bigcap_{j=1}^{k}\{X_{t_j} \in \mathrm{O}^{-1}A_j\}\right] \\ &= P^{\mathrm{O}x}\left[\bigcap_{j=1}^{k}\{X_{t_j} \in A_j\}\right]. \end{aligned}$$

Thus $P^x\mathbf{O}^{-1} = P^{\mathrm{O}x}$ on $\mathscr{F}^0$, and hence on $\mathscr{F}$. Finally note that

$$\begin{aligned} T_r(\mathbf{O}\omega) &= \inf\{t: |X_t(\mathbf{O}\omega) - X_0(\mathbf{O}\omega)| > r\} \\ &= \inf\{t: |\mathrm{O}X_t(\omega) - \mathrm{O}X_0(\omega)| > r\} \\ &= T_r(\omega), \end{aligned}$$

and so $X_{T_r} \circ \mathbf{O} = \mathrm{O}X_{T_r}$. Let x and $r > 0$ be fixed and suppose $\mathrm{O}x = x$. Then if Γ is a Borel set

$$\begin{aligned} P_{T_r}(x, \mathrm{O}^{-1}\Gamma) &= P^x[X(T_r) \in \mathrm{O}^{-1}\Gamma] \\ &= P^x[\mathrm{O}X(T_r) \in \Gamma] \\ &= P^{\mathrm{O}x}[X(T_r) \in \Gamma] \\ &= P_{T_r}(x, \Gamma); \end{aligned}$$

that is, the measure $P_{T_r}(x, \cdot)$ is invariant under all distance-preserving transformations of $\mathbf{R}^n$ which leave x fixed. Combining this with our previous observations it follows that $P_{T_r}(x, \cdot)$ is indeed the uniform distribution of unit mass on $S_r(x)$.

In this paragraph we assume that the reader is familiar with classical potential theory as expounded, for example, in Brelot's monograph [1]. Clearly the Brownian motion process X is a strong Feller process, (2.16), and hence excessive functions are lower semicontinuous. It now follows from Theorem 5.11 and the above evaluation of $P_{S_r(x)}(x, \cdot)$ that $f \in \mathcal{S}$ if and only if $f = \infty$ or f is a nonnegative superharmonic function. In particular, since the only nonnegative superharmonic functions in $\mathbf{R}$ or $\mathbf{R}^2$ are the constants, we see that the only excessive functions for Brownian motion in one or two dimensions are the nonnegative constants. On the other hand if $n \geq 3$, then X satisfies the condition of (4.6) and so the fine topology is the coarsest topology relative to which the nonnegative superharmonic functions are continuous; that is, it coincides with the Cartan fine topology of classical potential theory; see Brelot [1, p. 90]. The reader familiar with classical potential theory should now have no difficulty in checking the following facts ($n \geq 3$): (i) A is thin at x as defined in this chapter if and only if A is thin at x in the potential theoretic sense; and (ii) the concept of polar set in this theory and in potential theory agree. Moreover it follows from results of potential theory that a thin set, and hence a semipolar set, is polar. Finally we remark that a set is polar if and only if it has capacity zero. These last two statements will be proved in much greater generality by probabilistic methods in Chapter VI.

III

MULTIPLICATIVE FUNCTIONALS AND SUBPROCESSES

In this chapter we develop the fundamental properties of multiplicative functionals of a Markov process. Multiplicative functionals arise naturally when one "kills" a given process according to some rule (see Sections 2 and 3). The reader could skip Section 3 and the second half of Section 5 on a first reading. Also Section 6 stands somewhat apart from the remainder of the chapter. This section contains Hunt's characterization of the hitting operators P_B^α. This result is one of the most important in probabilistic potential theory and could have been included in Chapter II. However we deferred it to the present chapter in order to discuss it in somewhat greater generality than would have been possible in Chapter II. The reader particularly interested in this result might well go immediately from Section 6 of this chapter to Section 1 of Chapter V where related questions are discussed.

1. Multiplicative Functionals

Let $X = (\Omega, \mathcal{M}, \mathcal{M}_t, X_t, \theta_t, P^x)$ be a Markov process with state space $(E, \mathcal{E})$. We will shortly impose more conditions on X but the following definition makes sense for any Markov process.

(1.1) DEFINITION. A family $M = \{M_t;\ 0 \leq t < \infty\}$ of real-valued random variables on $(\Omega, \mathcal{F})$ is called a *multiplicative functional* of X provided:

(i) $M_t \in \mathcal{F}_t$ for each $t \geq 0$;
(ii) $M_{t+s} = M_t(M_s \circ \theta_t)$ a.s. for each $t, s \geq 0$;
(iii) $0 \leq M_t(\omega) \leq 1$ for all t and ω.

Most authors do not assume that multiplicative functionals satisfy Condition (iii). Thus we should perhaps call a multiplicative functional satisfying (iii) a multiplicative functional with values in [0, 1]; however, this is the only type of multiplicative functional we will consider in this book. From now on we will write "MF" in place of "multiplicative functional." We emphasize that the exceptional set in (ii) depends on t and s in general. Also it follows from (ii) and (iii) that, for each t and s, $M_{t+s} \leq M_s$ almost surely. Sometimes it will be convenient to write $M(t, \omega)$ for $M_t(\omega)$ and $M(t)$ for M_t.

We say that a multiplicative functional M is *measurable* provided the family $\{M_t\}$ is progressively measurable relative to $\{\mathscr{F}_t\}$. We say that M is *right continuous* (or *continuous*) provided $t \to M_t(\omega)$ is right continuous (or continuous) almost surely. In particular a right continuous multiplicative functional is measurable. It will be convenient to let $M_\infty = 0$ for any MF. The relationship $M_0 = M_0(M_0 \circ \theta_0) = M_0^2$ a.s. implies that almost surely M_0 is either zero or one. We call a point x in E *permanent* for M if $P^x(M_0 = 1) = 1$, and we let E_M denote the set of permanent points. Clearly $E_M \in \mathscr{E}^*$. In case X is normal, the zero-one law implies that $x \in E - E_M$ if and only if $P^x(M_0 = 0) = 1$.

We next give a few examples. The reader should verify for himself that these are indeed MF's.

(1.2) For each $\alpha \geq 0$, $M_t = e^{-\alpha t}$ is a MF and $E_M = E$.

Recall from (2.18) of Chapter II that a *terminal time* T is an $\{\mathscr{F}_t\}$ stopping time satisfying, for each $t \geq 0$,

(1.3) $T = t + T \circ \theta_t$ almost surely on $\{T > t\}$.

(1.4) Let T be a terminal time and define $M_t(\omega) = 1$ if $t < T(\omega)$ and $M_t(\omega) = 0$ if $t \geq T(\omega)$; that is, $M_t = I_{[0,T)}(t)$. In this case E_M consists of those points x in E which are irregular for T, that is, such that $P^x(T > 0) = 1$.

Terminal times will play an important role in this and later chapters. Intuitively one should think of a terminal time as the *first* time some physical event occurs. Since $t + T \circ \theta_t$ is then the first time the event occurs after t, the relationship (1.3) becomes intuitively clear.

(1.5) Let X be progressively measurable with respect to $\{\mathscr{F}_t\}$ and let $f \in \mathscr{E}_+$. Define

$$M_t = \exp\left(-\int_0^t f(X_s)\, ds\right).$$

Obviously Example (1.2) is continuous and (1.4) is right continuous. Example (1.5) is continuous if f is bounded, but need not be even right continuous if f is unbounded. However if $\{\mathscr{F}_t\}$ is right continuous and we define

$$T = \inf\left\{t: \int_0^t f(X_s)\, ds = \infty\right\},$$

then T is a terminal time and

$$N_t = I_{[0,T)}(t) \exp(-\int_0^t f(X_s)\, ds)$$

is a right continuous multiplicative functional. In addition for each ω the functions $t \to N_t(\omega)$ and $t \to M_t(\omega)$ agree for all values of t except possibly $t = T(\omega)$.

In general we will only be interested in a MF during the time the trajectory $t \to X_t$ is in E. With this in mind we make the following definition.

(1.6) DEFINITION. Two MF's of X, say M and N, are *equivalent* provided $P^x[M_t \neq N_t;\ X_t \in E] = 0$ for all t and x.

If E_Δ is a metric space and X is right continuous, and if, in addition, both M and N are right continuous, then this is equivalent to the statement that $t \to M_t(\omega)$ and $t \to N_t(\omega)$ are identical functions on $[0, \zeta(\omega))$ almost surely. Observe that if M is a MF, then $N_t = M_t I_E(X_t)$ is a MF and that M and N are equivalent. We will ordinarily be interested in MF's only up to equivalence and consequently when this is the case we may assume without loss of generality that $M_t(\omega) = 0$ whenever $X_t(\omega) = \Delta$. Finally note that if E_Δ is a metric space and X is right continuous then $I_E(X_t)$ and $I_{[0,\zeta)}(t)$ are equivalent right continuous MF's.

The following proposition is of some interest. We leave its proof to the reader as Exercise 1.10.

(1.7) PROPOSITION. Let M be a MF of X and let $\{\mathscr{F}_t\}$ be right continuous. Then a necessary and sufficient condition that M be equivalent to a right continuous MF of X is that $t \to E^x(M_t)$ be right continuous at $t = 0$ for all x.

If M is a MF of X we define for each $t \geq 0$ an operator Q_t on $b\mathscr{E}^*$ by

$$\textbf{(1.8)} \qquad Q_t f(x) = E^x\{f(X_t)\, M_t\} = E^x\{f(X_t)\, M_t;\ X_t \in E\}$$

(recall that numerical functions on E are extended to E_Δ by setting $f(\Delta) = 0$). Clearly Q_t is a positive linear map from $b\mathscr{E}^*$ to $b\mathscr{E}^*$ such that $Q_t \leq P_t$ where

P_t is the transition operator of X. Moreover

$$\begin{aligned} Q_{t+s}f(x) &= E^x[f(X_{t+s})\, M_{t+s}] \\ &= E^x[(f \circ X_t \circ \theta_s)\, M_s(M_t \circ \theta_s)] \\ &= E^x\{M_s E^{X(s)}[f(X_t)\, M_t]\} \\ &= Q_s Q_t f(x), \end{aligned}$$

and so $\{Q_t;\, t \geq 0\}$ is a semigroup—the *semigroup generated by M*. Obviously equivalent multiplicative functionals generate the same semigroup. The following proposition contains the converse of this statement.

(1.9) PROPOSITION. Two multiplicative functionals are equivalent if and only if they generate the same semigroup $\{Q_t;\, t \geq 0\}$.

Proof. Let M and N be MF's. We must show that if $E^x\{f(X_t)\, M_t\} = E^x\{f(X_t)\, N_t\}$ for all $f \in b\mathscr{E}^*$, then M and N are equivalent. For this it suffices to show that if t and x are fixed, then for each $H \in b\mathscr{F}_t$ one has $E^x\{HM_t;\, X_t \in E\} = E^x\{HN_t;\, X_t \in E\}$. But the set of $H \in b\mathscr{F}_t$ for which this last relation holds is a vector space containing the constants and closed under monotone convergence. Thus by MCT we need only consider H of the form $\prod_{i=1}^n f_i(X_{t_i})$ where $0 \leq t_1 < \ldots < t_n = t$ and each $f_i \in b\mathscr{E}^*$. If $n = 1$ the desired equality is just our hypothesis. Suppose $0 \leq t_1 < \ldots < t_n < t_{n+1} = t$; then

$$\begin{aligned} &E^x\left\{\prod_{j=1}^n f_j(X_{t_j})\, f_{n+1}(X_t)\, M_t\right\} \\ &\qquad = E^x\left\{\prod_{j=1}^n f_j(X_{t_j})\, M_{t_n} E^{X(t_n)}[f_{n+1}(X_{t-t_n})\, M_{t-t_n}]\right\}. \end{aligned}$$

If we let $g_n(x) = f_n(x)\, Q_{t-t_n} f_{n+1}(x)$, the induction step is clear. Thus (1.9) is proved.

Let us point out that Q_0 need not be the identity operator; in fact $Q_0 f(x) = E^x\{f(X_0)\, M_0\}$. In particular if X is normal $Q_0 f = I_{E_M} f$. Finally, if E_Δ is a metric space and X is right continuous and if M is right continuous, then $Q_t 1(x) = E^x\{M_t;\, X_t \in E\}$ approaches $Q_0 1(x) = E^x\{M_0;\, X_0 \in E\}$ as $t \to 0$.

In Section 3 we will give conditions under which there exists a Markov process whose transition semigroup is $\{Q_t\}$. However, before that, we are going to show in the next section that under mild restrictions any semigroup that is dominated by $\{P_t\}$ is generated by a multiplicative functional.

Exercises

(1.10) Prove Proposition 1.7. [Hint: show that, under the hypotheses of (1.7), $N_t = \sup_{r>t, r\in\mathbf{Q}} M_r$ defines a right continuous MF equivalent to M.]

(1.11) Let M be a MF such that almost surely the mapping $t \to M_t$ is nonincreasing. Let $T = \inf\{t\colon M_t = 0\}$. Show that T is an $\{\mathcal{F}_{t+}\}$ stopping time and that $T = t + T \circ \theta_t$ almost surely on $\{T > t\}$ for all t. In particular if $\{\mathcal{F}_t\}$ is right continuous, then T is a terminal time.

2. Subordinate Semigroups

Let $X = (\Omega, \mathcal{M}, \mathcal{M}_t, X_t, \theta_t, P^x)$ be a Markov process with state space $(E, \mathcal{E})$. Let $\{P_t\}$ denote the semigroup of transition operators of X and $\mathbf{B} = b\mathcal{E}^*$. We have seen in Chapter I that $P_t\mathbf{B} \subset \mathbf{B}$ for each $t \geq 0$.

(2.1) Definition. A semigroup $\{Q_t\colon t \geq 0\}$ of nonnegative linear operators on $\mathbf{B}$ is *subordinate* to $\{P_t\}$ if $Q_t f \leq P_t f$ for each $t \geq 0$ and $f \in \mathbf{B}_+$.

It is an immediate consequence of the definition that $\|Q_t\| \leq \|P_t\| \leq 1$. Since $P_t f(x) = \int P_t(x, dy) f(y)$ for $f \in \mathbf{B}$, it follows that for each t and x there exists a measure $Q_t(x, \cdot)$ on $\mathcal{E}^*$ such that $Q_t(x, \cdot) \leq P_t(x, \cdot)$ and $Q_t f(x) = \int Q_t(x, dy) f(y)$ for all $f \in \mathbf{B}$. In particular $Q_t(x, A)$ is a transition function on $(E, \mathcal{E}^*)$. We showed in Section 1 that if M is a MF of X, then the semigroup generated by M is subordinate to $\{P_t\}$. The next theorem, which is due to Meyer [2], states that under mild regularity assumptions on X any semigroup subordinate to $\{P_t\}$ is generated by a MF. Before stating this theorem we introduce the following condition on X:

(2.2) (a) There exists a countable family $\mathcal{B} \subset \mathcal{E}$ such that $\sigma(\mathcal{B})^* = \mathcal{E}^*$ and $\{x\} \in \mathcal{E}$ for each $x \in E$. (b) Given $t > 0$ and J, a countable dense subset of $[0, t]$ containing t, then $\sigma(X_s\colon s \in J)^\sim = \mathcal{F}_t$. Recall that if $\mathcal{G} \subset \mathcal{F}$ then $\tilde{\mathcal{G}}$ is the completion of $\mathcal{G}$ in $\mathcal{F}$ relative to the family $\{P^\mu\}$. (See (5.3) of Chapter I.)

Observe that if E_Δ is a separable metric space with $\mathcal{B}(E_\Delta) \subset \mathcal{E}_\Delta \subset \mathcal{B}(E_\Delta)^*$ and if X is right continuous, then (2.2) is certainly satisfied. In particular (2.2) is satisfied if X is a standard process.

(2.3) Theorem. Let X be normal and satisfy (2.2). If $\{Q_t\}$ is a semigroup subordinate to $\{P_t\}$, then there exists a multiplicative functional M of X which generates $\{Q_t\}$. If $\{\mathcal{F}_t\}$ is right continuous and $t \to Q_t 1(x)$ is right continuous at $t = 0$ for all x, then one may take M to be right continuous.

Proof. Since X is normal $P_0(x, \cdot) = \varepsilon_x$. But $Q_0(x, \cdot) \le P_0(x, \cdot)$ and therefore $Q_0(x, \cdot) = c(x)\, \varepsilon_x$. Plainly the fact that $Q_0 = Q_0^2$ implies that $c(x)$ is either zero or one. Let $E_0 = \{x: Q_0 1(x) = 1\}$; then $E_0 \in \mathscr{E}^*$ and $Q_0(x, \cdot) = I_{E_0}(x)\, \varepsilon_x$. The points in E_0 will be called *permanent* for $\{Q_t\}$. Observe that for any t and x

$$Q_t(x, E - E_0) = \int Q_t(x, dy)\, Q_0(y, E - E_0) = 0,$$

so that all of the measures $Q_t(x, \cdot)$ are concentrated on E_0. Also $Q_t(x, E) = \int Q_0(x, dy)\, Q_t(y, E) = 0$ if $x \in E - E_0$; that is, $Q_t(x, \cdot) = 0$ for all t if $x \notin E_0$.

Since $Q_t(x, \cdot) \le P_t(x, \cdot)$ for each fixed x and t, it follows from the Radon–Nikodym theorem that there exists a function $q_t(x, y)$ defined on $\mathbf{R}_+ \times E \times E$ such that $Q_t(x, dy) = q_t(x, y)\, P_t(x, dy)$. We may assume that $0 \le q_t(x, y) \le 1$ for all (x, y). Since $x \to P_t(x, A)$ and $x \to Q_t(x, A)$ are $\mathscr{E}^*$ measurable for each $t > 0$ and $A \in \mathscr{E}^*$ it follows from (2.2a) and a theorem of Doob [1, p. 344] (see also Dynkin [2, p. 218]) that we may assume that $q_t \in \mathscr{E}^* \times \mathscr{E}^*$ for each t. By the remarks in the preceding paragraph we may suppose that q_t vanishes off $E_0 \times E_0$. It will be convenient to extend q_t to $E_\Delta \times E_\Delta$ by letting q_t be zero if either of its arguments is Δ. The extended function is again denoted by q_t. Evidently $q_t \in \mathscr{E}_\Delta^* \times \mathscr{E}_\Delta^*$ and vanishes off $E_0 \times E_0$.

Given $t > 0$ let $U = \{0 = t_0 < t_1 < \ldots < t_n = t\}$ be a finite subset of $[0, t]$ containing 0 and t. Let $\mathscr{F}(U) = \sigma(X_s \colon s \in U)^\sim$, and define

$$M_t(U) = q_{t_1}(X_0, X_{t_1})\, q_{t_2 - t_1}(X_{t_1}, X_{t_2}) \ldots q_{t - t_{n-1}}(X_{t_{n-1}}, X_t). \tag{2.4}$$

Clearly $M_t(U)$ is in $\mathscr{F}(U) \subset \mathscr{F}_t$, and $0 \le M_t(U) \le 1$. Also note that $M_t(U)(\omega) = 0$ if $X_t(\omega) = \Delta$. As in Eq. (2.8) of Chapter I the formula

$$\int \varepsilon_x(dx_0) \int Q_{t_1}(x_0, dx_1) \ldots \int Q_{t - t_{n-1}}(x_{n-1}, dx_n)\, h(x_0, \ldots, x_n),$$

for $h \in b\mathscr{E}^{n+1}$, defines a measure $Q^U(x, h)$ on $\mathscr{E}^{n+1}$. In view of (2.4) and the normality of X we have

$$Q^U(x, h) = E^x\{M_t(U)\, h(X_{t_0}, \ldots, X_{t_n})\}.$$

In particular, for any $f \in \mathbf{B}$ and $x \in E$

$$Q_t f(x) = E^x\{f(X_t)\, M_t(U)\}. \tag{2.5}$$

As we observed just following the statement of Theorem 2.11 of Chapter I, the family $\{Q^U(x, \cdot)\}$, as U ranges over the finite subsets of $[0, t]$ containing 0 and t, is a projective system. In particular, if $U = \{0 = t_0 < t_1 < \ldots < t_n = t\}$, V is a finite subset of $[0, t]$ containing U, and Φ is an ω function of the form

$\Phi = h(X_{t_0}, \ldots, X_{t_n})$ with $h \in b\mathscr{E}^{n+1}$, then

$$E^x\{\Phi\, M_t(U)\} = E^x\{\Phi\, M_t(V)\}.$$

Consequently we have

$$E^x\{M_t(V) \mid \mathscr{F}(U)\} = M_t(U) \tag{2.6}$$

whenever $U \subset V$.

Let $\{U_n\}$ be an increasing sequence of finite subsets of $[0, t]$, each containing 0 and t, and such that $D = \bigcup U_n$ is dense in $[0, t]$. Let $M_t^n = M_t(U_n)$ and $\mathscr{F}^n = \mathscr{F}(U_n)$. It is immediate from (2.6) that $\{M_t^n; \mathscr{F}^n\}$ is a martingale relative to the measure P^x for each x. Since each M_t^n is in $\mathscr{F}_t$ and $0 \leq M_t^n \leq 1$, it follows from the martingale convergence theorem ((1.4) of Chapter 0) that there exists an $\mathscr{F}_t$ measurable random variable M_t such that $0 \leq M_t \leq 1$ and $M_t^n \rightarrow M_t$ as $n \rightarrow \infty$ almost surely. Moreover we may assume that $M_t = 0$ if $X_t = \Delta$ since each M_t^n has this property. Suppose $\{\bar{U}_n\}$ is another increasing sequence of finite subsets of $[0, t]$, each containing 0 and t, and having $\bar{D} = \bigcup \bar{U}_n$ dense in $[0, t]$. If $\bar{M}_t$ is defined relative to the sequence $\{\bar{U}_n\}$ in the same manner as M_t is defined relative to $\{U_n\}$, then we claim that $M_t = \bar{M}_t$ almost surely. To prove this it clearly suffices to suppose that $U_n \subset \bar{U}_n$ for each n. But then (2.6) implies that for $p > n$ we have

$$E^x(\bar{M}_t^p \mid \mathscr{F}^n) = M_t^n = E^x(M_t^p \mid \mathscr{F}^n)$$

for all x. Letting $p \rightarrow \infty$ we obtain $E^x(\bar{M}_t - M_t; \Lambda) = 0$ for all $\Lambda \in \bigcup_n \mathscr{F}^n$ and hence, from (2.2b), for any $\Lambda \in \mathscr{F}_t$. Therefore $M_t = \bar{M}_t$ almost surely since both M_t and $\bar{M}_t$ are in $\mathscr{F}_t$.

For each $t > 0$ we define M_t as above using an increasing sequence $\{U_n\}$. If $t = 0$ we define $M_0 = I_{E_0}(X_0)$. Observe that $I_{E_0}(x) = Q_0(x, \{x\}) = q_0(x, x)\, P_0(x, \{x\}) = q_0(x, x)$. We will now show that $\{M_t\}$ is a MF and that

$$Q_t f(x) = E^x[f(X_t)\, M_t]; \qquad f \in \mathbf{B}. \tag{2.7}$$

Obviously $0 \leq M_t \leq 1$ and $M_t \in \mathscr{F}_t$ for all t. It is also clear from (2.5) and the definition of M_0 that (2.7) holds. Moreover $M_t(\omega) = 0$ if $X_t(\omega) = \Delta$. Thus it remains to verify that

$$M_{t+s} = M_s(M_t \circ \theta_s) \quad \text{almost surely.} \tag{2.8}$$

Assume first that $t > 0$ and $s > 0$. Let D consist of the rationals in $[0, t + s]$, $t + s$, and all numbers of the form $s + r$ with r a rational in $[0, t]$. Let $\{U_n\}$ be an increasing sequence of finite subsets of D, each of which contains 0, s, and $t + s$ and such that $D = \bigcup U_n$. Let $V_n = U_n \cap [0, s]$ and let $W_n = \{r \geq 0: r + s \in U_n\}$. Clearly $\{W_n\}$ is an increasing sequence of finite subsets of $[0, t]$, each of which contains 0 and t, whose union is dense in

$[0, t]$. If $U_n = \{0 = t_0 < t_1 < \ldots < t_j < s < t_{j+1} < \ldots < t_p = t + s\}$, then it is easy to verify that

$$M_{t+s}(U_n) = M_s(V_n)\,[M_t(W_n) \circ \theta_s],$$

and so letting $n \to \infty$ we obtain (2.8) in this case. If $t = s = 0$ then (2.8) is clear. Finally if $t > 0$ and $s = 0$ or $t = 0$ and $s > 0$, then using the fact that q_u vanishes off $E_0 \times E_0$ one can easily verify that (2.8) holds.

We have now proved the first assertion of Theorem 2.3. But $E^x\{M_t\} = Q_t 1(x)$ since $M_t = 0$ if $X_t = \Delta$, and so if $t \to Q_t 1(x)$ is continuous at $t = 0$ for all x so is $t \to E^x\{M_t\}$. Therefore we can apply Proposition 1.7 to obtain a right continuous MF, N, equivalent to M provided $\{\mathcal{F}_t\}$ is right continuous. This establishes the last sentence of Theorem 2.3.

REMARK. Note that E_0 is the set of permanent points of M, that is $E_0 = E_M$.

Exercises

(2.9) Let X be normal and satisfy (2.2). Let $Q_t(x, A)$ be a transition function on $(E, \mathcal{E}^*)$ such that for each $x \in E$ and $t \geq 0$ the measure $Q_t(x, \cdot)$ is absolutely continuous with respect to $P_t(x, \cdot)$. Let $q_t(x, y)$ be a density for $Q_t(x, \cdot)$ with respect to $P_t(x, \cdot)$ that is jointly measurable in x and y, and assume that $q_t(x, y) = 0$ if either x or y equals Δ. If U is a finite subset of $[0, t]$ containing 0 and t, then define $M_t(U)$ as in (2.4). Show that for each $t \geq 0$ there exists a nonnegative $M_t \in \mathcal{F}_t$ such that $M_t(U_n) \to M_t$ almost surely whenever $\{U_n\}$ is an increasing sequence of finite subsets of $[0, t]$, each containing 0 and t, such that $\bigcup U_n$ is dense in $[0, t]$. Show that $\{M_t;\, t \geq 0\}$ satisfies (i) and (ii) of Definition 1.1. Let us call such a family a *generalized nonnegative multiplicative functional* (GMF for short). Show that $Q_t f(x) \geq E^x\{f(X_t)\, M_t\}$ if $f \in \mathbf{B}_+$. Finally show that $Q_t f(x) = E^x\{f(X_t)\, M_t\}$ for all $f \in \mathbf{B}_+$ if and only if $\{M_t(U)$; U a finite subset of $[0, t]$ containing 0 and $t\}$ is uniformly integrable relative to P^x.

(2.10) Let $X = (\Omega, \mathcal{F}^0, \mathcal{F}_t^0, X_t, \theta_t, P^x)$ be a Markov process of function space type having $(E, \mathcal{E})$ as state space (see (4.2) of Chapter I). Assume X is normal and satisfies (2.2). Let $\hat{X} = (\Omega, \mathcal{F}^0, \mathcal{F}_t^0, X_t, \theta_t, \hat{P}^x)$ be another Markov process of function space type with the same state space $(E, \mathcal{E})$. Show that the following two conditions are necessary and sufficient that the restriction of $\hat{P}^x$ to $\mathcal{F}_t^0$ be absolutely continuous with respect to the restriction of P^x to $\mathcal{F}_t^0$ for all x and t: (i) $\hat{P}_t(x, \cdot) \ll P_t(x, \cdot)$ for all t and x, and (ii) using the notation of (2.9) with $Q_t(x, A) = \hat{P}_t(x, A)$, for each t and x, $\{M_t(U)$; U a finite subset of $[0, t]$ containing 0 and $t\}$ is uniformly integrable relative to

P^x. Moreover there exists a GMF, $\{M_t\}$, of X such that, for each x, M_t is a version of the Radon–Nikodym derivative of $\hat{P}^x \mid \mathscr{F}_t^0$ with respect to $P^x \mid \mathscr{F}_t^0$.

3. Subprocesses

Let $X = (\Omega, \mathscr{M}, \mathscr{M}_t, X_t, \theta_t, P^x)$ be a Markov process with state space $(E, \mathscr{E})$. Let $E_0 \in \mathscr{E}^*$ and let $\mathscr{E}_0^*$ be the trace of $\mathscr{E}^*$ on E_0. A Markov process Y with state space $(E_0, \mathscr{E}_0^*)$ is called a *subprocess* of X provided its transition semigroup $\{Q_t\}$ is dominated by $\{P_t\}$ in the following sense: $Q_t f(x) \leq P_t f(x)$ for all $x \in E_0$ and $f \geq 0$ in $b\mathscr{E}_0^*$, where f is taken to vanish on $E - E_0$ in forming $P_t f$. If we define $\bar{Q}_t$ on $\mathbf{B} = b\mathscr{E}^*$ by setting $\bar{Q}_t f(x) = Q_t f_{E_0}(x)$ for $x \in E_0$ where f_{E_0} is the restriction of f to E_0, and $\bar{Q}_t f(x) = 0$ for $x \in E - E_0$, then $\{\bar{Q}_t\}$ is a semigroup of nonnegative linear operators on $\mathbf{B}$ which agrees with Q_t on $b\mathscr{E}_0^*$. The above definition is then equivalent to the statement that the semigroup $\{\bar{Q}_t\}$ is subordinate to $\{P_t\}$ as defined in Section 2. We will now drop the notation $\bar{Q}_t$ and write Q_t for the extended operators defined on $\mathbf{B}$. In the remainder of this section *we will assume that X is normal* although this is not necessary for the validity of all that follows.

If X satisfies the conditions of Theorem 2.3 and Y is a subprocess of X, then there exists a multiplicative functional M of X such that for each $x \in E_0$, $t \geq 0$, and $f \in b\mathscr{E}^*$ vanishing off E_0 one has $\hat{E}^x\{f(Y_t)\} = E^x\{f(X_t) M_t\}$ where $\hat{E}$ is the expectation operator corresponding to Y. In this section we are going to establish a partial converse; namely, given a *right continuous* MF, M, of X we are going to construct a subprocess Y satisfying the above condition. Roughly speaking Y_t is obtained by "killing" X_t at a rate $-dM_t/M_t$; that is, $-dM_t/M_t$ is the "conditional probability" that X_t is killed in the interval $(t, t + dt)$ given that it is "still alive" at time t. We will now make this discussion precise.

We begin with some definitions. Let $\hat{\Omega} = \Omega \times [0, \infty]$ and write $\hat{\omega} = (\omega, \lambda)$ for the generic point in $\hat{\Omega}$. Let $\mathscr{R}$ be the σ-algebra of Borel subsets of $[0, \infty]$ and set $\hat{\mathscr{M}} = \mathscr{M} \times \mathscr{R}$. Let $\pi : \hat{\Omega} \to \Omega$ and $\gamma : \hat{\Omega} \to [0, \infty]$ be the natural projections; that is, $\pi(\omega, \lambda) = \omega$ and $\gamma(\omega, \lambda) = \lambda$. Let $\hat{\omega}_\Delta = (\omega_\Delta, 0)$ and set $\hat{\zeta} = (\zeta \circ \pi) \wedge \gamma$; that is, $\hat{\zeta}(\omega, \lambda) = \zeta(\omega) \wedge \lambda$. Define $\hat{X}_t(\hat{\omega}) = X_t(\omega)$ if $t < \lambda$ and $\hat{X}_t(\hat{\omega}) = \Delta$ if $t \geq \lambda$; here, of course, $\hat{\omega} = (\omega, \lambda)$. Note that $\hat{\zeta} = \inf\{t : \hat{X}_t = \Delta\}$. We will often write simply ω and λ for $\pi(\hat{\omega})$ and $\gamma(\hat{\omega})$. Define $\hat{\theta}_t \hat{\omega} = (\theta_t \omega, (\lambda - t) \vee 0)$ where $\infty - \infty$ is taken to be zero. Note that $\hat{\theta}_\infty \hat{\omega} = \hat{\omega}_\Delta$ and $\hat{X}_t \circ \hat{\theta}_h = \hat{X}_{t+h}$ for all t and h. Let $\hat{\Omega}_t = \Omega \times (t, \infty] \in \hat{\mathscr{M}}$, and define $\hat{\mathscr{M}}_t$ to consist of all sets $\hat{\Lambda} \in \hat{\mathscr{M}}$ for which there exists a $\Lambda \in \mathscr{M}_t$ such that $\hat{\Lambda} \cap \hat{\Omega}_t = \Lambda \times (t, \infty]$.

(3.1) Proposition. $\{\hat{\mathscr{M}}_t\}$ is an increasing sequence of sub-σ-algebras of

$\hat{\mathscr{M}}$ and $\hat{X}_t \in \hat{\mathscr{M}}_t/\mathscr{E}_\Delta$ for all t. Moreover if $\{\mathscr{M}_t\}$ is right continuous so is $\{\hat{\mathscr{M}}_t\}$.

Proof. The reader will easily verify that $\{\hat{\mathscr{M}}_t\}$ is an increasing sequence of sub-σ-algebras of $\hat{\mathscr{M}}$. If $A \in \mathscr{E}$, then $\{\hat{X}_t \in A\} = \{X_t \in A\} \times (t, \infty] \in \hat{\mathscr{M}}_t$. Thus the first sentence of (3.1) is established. Regarding the second, fix $t < \infty$ and suppose $\hat{\Lambda} \in \hat{\mathscr{M}}_{t+}$. Then for each $n \geq 1$, $\hat{\Lambda} \in \hat{\mathscr{M}}_{t+1/n}$ and so there is a set $\Lambda_n \in \mathscr{M}_{t+1/n}$ such that $\hat{\Lambda} \cap \hat{\Omega}_{t+1/n} = \Lambda_n \times (t + 1/n, \infty]$. It is easy to check that all the sets Λ_n are the same. If we set $\Lambda = \Lambda_1$ then $\Lambda \in \mathscr{M}_{t+1/n}$ for all n, and since $\{\mathscr{M}_t\}$ is right continuous $\Lambda \in \mathscr{M}_t$. Clearly then $\hat{\Lambda} \cap \hat{\Omega}_t = \Lambda \times (t, \infty]$ and so $\hat{\Lambda} \in \hat{\mathscr{M}}_t$. Thus the proof of (3.1) is complete.

We now suppose that we are given a *right continuous* multiplicative functional M of X. Let Ω_0 be the set of ω's such that $t \to M_t(\omega)$ is right continuous, nonincreasing, and $M_0(\omega)$ is either 1 or 0. By definition $P^x(\Omega_0) = 1$ for all x. For $\omega \in \Omega_0$ we define a measure α_ω on $[0, \infty]$ by $\alpha_\omega((\lambda, \infty]) = M_\lambda(\omega)$ for $\lambda \in [0, \infty]$ and $\alpha_\omega(\{0\}) = 0$. This defines a unique measure, α_ω, on the Borel sets of $[0, \infty]$. Since $M_\infty = 0$ by convention, α_ω is a probability measure if $M_0(\omega) = 1$ and $\alpha_\omega = 0$ if $M_0(\omega) = 0$. For convenience we define $\alpha_\omega = \varepsilon_0$ if $\omega \notin \Omega_0$. Since $M_\lambda \in \mathscr{F}_\lambda$ and $\Omega_0 \in \mathscr{F}$ it is immediate that $\omega \to \alpha_\omega(\Gamma)$ is $\mathscr{F}$ measurable for any $\Gamma \in \mathscr{R}$.

We next define measures $\hat{P}^x$ on $\hat{\mathscr{M}}$ as follows: if $\hat{\Lambda} \in \hat{\mathscr{M}} = \mathscr{M} \times \mathscr{R}$ and if $\hat{\Lambda}^\omega = \{\lambda: (\omega, \lambda) \in \hat{\Lambda}\}$, then $\hat{\Lambda}^\omega \in \mathscr{R}$ and $\alpha_\omega(\hat{\Lambda}^\omega)$ exists for each $\omega \in \Omega$. Moreover $\omega \to \alpha_\omega(\hat{\Lambda}^\omega)$ is in $\mathscr{F}$ since this is true for rectangles in $\mathscr{M} \times \mathscr{R}$. (Use MCT.) Recall that $E_M = \{x \in E: P^x(M_0 = 1) = 1\}$ is the set of permanent points of M. We now define

$$\hat{P}^x(\hat{\Lambda}) = E^x[\alpha_\omega(\hat{\Lambda}^\omega)] \qquad \text{if} \quad x \in E_M, \tag{3.2}$$

$$\hat{P}^x = \text{unit mass at } \hat{\omega}_\Delta = (\omega_\Delta, 0) \qquad \text{if} \quad x \in E_\Delta - E_M.$$

Clearly each $\hat{P}^x$ is a probability measure on $(\hat{\Omega}, \hat{\mathscr{M}})$. In order to avoid trivial notational difficulties we will now assume, as we may without loss of generality, that $\Omega_0 = \Omega$. We come now to the basic result of this section. The notation is that established above.

(3.3) THEOREM. $\hat{X} = (\hat{\Omega}, \hat{\mathscr{M}}, \hat{\mathscr{M}}_t, \hat{X}_t, \hat{\theta}_t, \hat{P}^x)$ is a Markov process with state space $(E, \mathscr{E}^*)$ such that $\hat{E}^x\{f(\hat{X}_t)\} = E^x\{f(X_t)\, M_t\}$ for all $f \in b\mathscr{E}^*$; that is, $\hat{X}$ is a subprocess of X whose transition semigroup is generated by M.

Proof. Since $\Delta \notin E_M$ it is clear that $\hat{P}^\Delta(\hat{X}_0 = \Delta) = 1$. If $A \in \mathscr{E}^*$, then $\{\hat{X}_t \in A\} = \{X_t \in A\} \times (t, \infty]$ and so

$$\hat{P}^x(\hat{X} \in A) = E^x\{M_t;\, X_t \in A\}. \tag{3.4}$$

Therefore $x \to \hat{P}^x(\hat{X}_t \in A)$ is $\mathscr{E}^*$ measurable for each $A \in \mathscr{E}^*$. Since (3.4) remains true if one replaces I_A by any $f \in b\mathscr{E}^*$, the only thing that remains to be checked is the Markov property of $\hat{X}$; that is, for all $B \in \mathscr{E}^*$, t, s, and x we must show that

$$\text{(3.5)} \qquad \hat{E}^x\{I_B(\hat{X}_{t+s}); \hat{\Lambda}\} = \hat{E}^x\{\hat{E}^{\hat{X}(t)}[I_B(\hat{X}_s)]; \hat{\Lambda}\}$$

for all $\hat{\Lambda} \in \hat{\mathscr{M}}_t$. If $x \notin E_M$ both sides of (3.5) are zero, and so we may suppose that $x \in E_M$. Observe first of all that MCT implies that if $Y \in b\hat{\mathscr{M}}$ and $x \in E_M$ then

$$\hat{E}^x(Y) = E^x\left\{\int Y(\omega, \lambda)\, \alpha_\omega(d\lambda)\right\}.$$

By definition of $\hat{\mathscr{M}}_t$, $\hat{\Lambda} \cap \hat{\Omega}_t = \Lambda \times (t, \infty]$ where $\Lambda \in \mathscr{M}_t$. Since $B \subset E$, $\{\hat{X}_{t+s} \in B\} \subset \hat{\Omega}_{t+s}$, and so $\{\hat{X}_{t+s} \in B\} \cap \hat{\Lambda} = (\{X_{t+s} \in B\} \cap \Lambda) \times (t+s, \infty]$. Therefore the left side of (3.5) may be written

$$E^x\{I_B(X_{t+s})\, M_{t+s}; \Lambda\} = E^x\{E^{X(t)}[I_B(X_s)\, M_s] M_t; \Lambda\}.$$

On the other hand if $F(y) = \hat{E}^y[I_B(\hat{X}_s)] = E^y[I_B(X_s)\, M_s]$, then since $F(\Delta) = 0$ the right side of (3.5) becomes

$$\hat{E}^x[F(\hat{X}_t); \hat{\Lambda} \cap \hat{\Omega}_t] = E^x\{F(X_t)\, M_t; \Lambda\},$$

and hence (3.5) is verified. Thus Theorem 3.3 is established.

In the sequel we will refer to the process $\hat{X}$ constructed above as the *canonical subprocess* corresponding to M. Note that we do *not* complete the σ-algebras $\hat{\mathscr{M}}$ and $\hat{\mathscr{M}}_t$ in the appropriate manner relative to $\{\hat{P}^x; x \in E_\Delta\}$. Thus for canonical subprocesses we do *not* assume condition (5.15) of Chapter I. Of course, $\hat{\zeta}$ is the lifetime of $\hat{X}$. It is immediate that if M and N are equivalent right continuous multiplicative functionals, then the corresponding canonical subprocesses are equivalent processes. Also note that if $x \in E_M$, then $\hat{P}^x[\hat{X}_0 = x] = P^x[X_0 = x] = 1$ while $\hat{P}^x[\hat{X}_0 \in E] = 0$ if $x \in E - E_M$. Consequently it is natural to ask if X can be considered as a Markov process with state space $(E_M, \mathscr{E}_M^*)$ where $\mathscr{E}_M^*$ is the trace of $\mathscr{E}^*$ on E_M. Unfortunately this is not always possible. See Exercise 3.19.

(3.6) Remark. Let $\mathscr{M}^* = \mathscr{M} \times \{\emptyset, [0, \infty]\}$. One should think of $\mathscr{M}^*$ as the σ-algebra in $\hat{\Omega}$ containing the information in the *original* process X. From the definition of the measures $\hat{P}^x$ it follows immediately that

$$\hat{P}^x\{\gamma > t \mid \mathscr{M}^*\} = M_t \circ \pi$$

for all $x \in E_\Delta$. In particular the conditional probability depends only on the behavior of the original process in the time interval $[0, t]$. Also it is this equality that is taken as justification for the intuitive statement that $\hat{X}$ is obtained by "killing" X at a rate $-dM_t/M_t$.

(3.7) REMARK. Suppose that T is an $\{\mathscr{F}_t\}$ stopping time such that $t + T \circ \theta_t = T$ on $\{T > t\}$ for each $t \geq 0$. This is slightly stronger than assuming T is a terminal time for X, which requires that the above identity should hold only almost surely for each $t \geq 0$. For simplicity assume that $T \leq \zeta$. This can always be accomplished by replacing T by $\min(T, \zeta)$. If $M_t = I_{[0,T)}(t)$, then M_t is a right continuous MF. In this case E_M is the set of points in E which are *not* regular for T, and α_ω is easily seen to be unit mass at $T(\omega)$ if $T(\omega) > 0$. If $\Phi : \Omega \to \hat{\Omega}$ is the map $\omega \to (\omega, T(\omega))$, then $\Phi \in \mathscr{M}/\hat{\mathscr{M}}$. Finally define $\tilde{X}_t(\omega) = X_t(\omega)$ if $t < T(\omega)$ and $\tilde{X}_t(\omega) = \Delta$ if $t \geq T(\omega)$; $\tilde{\theta}_t \omega = \theta_t \omega$ if $t < T(\omega)$ and $\tilde{\theta}_t \omega = \omega_\Delta$ if $t \geq T(\omega)$. It follows from the definitions that $\hat{X}_t \circ \Phi = \tilde{X}_t$ and $\tilde{X}_{t+s} = \tilde{X}_t \circ \tilde{\theta}_s$. Also if $\Lambda \in \mathscr{M}$ and $\hat{\Lambda} = \Lambda \times (\lambda, \infty]$, then $\Phi^{-1}\hat{\Lambda} = \Lambda \cap \{T > \lambda\}$ and consequently $P^x\Phi^{-1}\hat{\Lambda} = P^x(\Lambda \cap \{T > \lambda\}) = E^x(M_\lambda; \Lambda) = \hat{P}^x(\hat{\Lambda})$. Therefore $P^x\Phi^{-1} = \hat{P}^x$ on $\hat{\mathscr{M}}$ and this makes it clear that $\tilde{X} = (\Omega, \mathscr{F}, \tilde{X}_t, \tilde{\theta}_t, P^x)$ is a Markov process equivalent to $\hat{X}$. It is usually simpler to work with $\tilde{X}$ than with $\hat{X}$ when M has this special form.

(3.8) REMARK. If T is merely assumed to be a terminal time for X and $M_t = I_{[0,T)}(t)$ one can still carry out the above construction of $\tilde{X}$, and it is evident that, for each x, $\tilde{X}$ is a Markov process over $(\Omega, \mathscr{M}, P^x)$ in the sense of Definition 1.1 of Chapter I and that it is equivalent to $\hat{X}$. The only drawback is that $\tilde{X}_{t+h} = \tilde{X}_t \circ \tilde{\theta}_h$ only almost surely in this case. This is an unimportant difficulty, but as a result $\tilde{X}$ does not satisfy the assumptions of Definition 3.1 of Chapter I as it stands.

We are now going to investigate the regularity properties of $\hat{X}$. For example if E_Δ is a metric space and X is right continuous (or has left-hand limits), then it is obvious from the construction that $\hat{X}$ has the same properties. In order to discuss the strong Markov property for $\hat{X}$ or the quasi-left-continuity of $\hat{X}$ we prepare the following lemma. We use the notation already developed in this section.

(3.9) LEMMA. Let $\hat{T}$ be a stopping time relative to $\{\hat{\mathscr{M}}_t\}$. Then:

(a) There is a unique $\{\mathscr{M}_t\}$ stopping time T (defined on Ω) such that $\hat{T} \wedge \gamma = (T \circ \pi) \wedge \gamma$.

(b) Let $\hat{T}$ and T be as above. Then given $\hat{\Lambda} \in \hat{\mathscr{M}}_{\hat{T}}$ there is a set $\Lambda \in \mathscr{M}_T$ such that $\hat{\Lambda} \cap \{\hat{T} < \gamma\} = (\Lambda \times [0, \infty]) \cap \{(\omega, \lambda) : T(\omega) < \lambda\}$.

REMARKS. γ is an $\{\hat{\mathscr{M}}_t\}$ stopping time since $\{\gamma \leq t\} \cap \hat{\Omega}_t$ is empty for each t. Observe that $\gamma \circ \hat{\theta}_t = \gamma - t$ on $\{\gamma > t\}$. The relationship whose existence is asserted in (a) may, of course, be written

(3.10) $$\hat{T}(\omega, \lambda) \wedge \lambda = T(\omega) \wedge \lambda$$

for all ω and λ.

Proof of (a). First of all suppose that $\hat{T}$ is some numerical function on $\hat{\Omega}$ and that T is a numerical function on Ω which is related to $\hat{T}$ by (3.10); then it is evident that T is uniquely determined. Suppose in addition that $\hat{T}$ is an $\{\hat{\mathscr{M}}_t\}$ stopping time. If $\Lambda = \{\omega: T(\omega) \le a\}$, then $\Lambda \times (a, \infty] = \{\hat{T} \le a\} \cap \hat{\Omega}_a$. But by the definition of $\{\hat{\mathscr{M}}_t\}$ and the fact that $\hat{T}$ is an $\{\hat{\mathscr{M}}_t\}$ stopping time this last intersection is of the form $\Lambda_a \times (a, \infty]$ with $\Lambda_a \in \mathscr{M}_a$. Consequently T is an $\{\mathscr{M}_t\}$ stopping time. Let $\hat{\mathbf{T}}$ denote the class of all $\{\hat{\mathscr{M}}_t\}$ stopping times $\hat{T}$ for which there exists a numerical function T on Ω related to $\hat{T}$ by (3.10). Write $T = \Phi(\hat{T})$. We have already seen that T is uniquely determined by $\hat{T}$ and that T is an $\{\mathscr{M}_t\}$ stopping time. Next let $\{\hat{T}_\alpha\}$ be any family of elements of $\hat{\mathbf{T}}$ and $T_\alpha = \Phi(\hat{T}_\alpha)$. If $\hat{T} = \inf \hat{T}_\alpha$ is known to be an $\{\hat{\mathscr{M}}_t\}$ stopping time, then obviously $\hat{T} \in \hat{\mathbf{T}}$ and $\Phi(\hat{T}) = \inf T_\alpha$. Finally suppose that $\hat{T}$ is an $\{\hat{\mathscr{M}}_t\}$ stopping time of the form

$$\begin{aligned} \hat{T}(\omega, \lambda) &= a \qquad \text{if} \quad (\omega, \lambda) \in \hat{\Gamma}_a, \\ &= \infty \qquad \text{if} \quad (\omega, \lambda) \notin \hat{\Gamma}_a, \end{aligned}$$

where $\hat{\Gamma}_a \in \hat{\mathscr{M}}_a$. Now $\hat{\Gamma}_a \cap \hat{\Omega}_a = \Gamma_a \times (a, \infty]$ with $\Gamma_a \in \mathscr{M}_a$. If we define $T(\omega) = a$ for $\omega \in \Gamma_a$ and $T(\omega) = \infty$ for $\omega \notin \Gamma_a$, then T is an $\{\mathscr{M}_t\}$ stopping time and one checks easily that $\hat{T}$ and T are related by (3.10). But any $\{\hat{\mathscr{M}}_t\}$ stopping time $\hat{T}$ is the infimum of a countable family of stopping times of this special form and so from our previous remarks it follows that $\hat{T} \in \hat{\mathbf{T}}$. Thus Assertion (a) is established.

Proof of (b). Recall that saying $\hat{\Lambda} \in \hat{\mathscr{M}}_T$ means that $\hat{\Lambda} \in \hat{\mathscr{M}}$ and that $\hat{\Lambda} \cap \{\hat{T} \le a\} \in \hat{\mathscr{M}}_a$ for all $a \in [0, \infty)$. Given $\hat{\Lambda}$ let

$$\hat{\Lambda}_a = \hat{\Lambda} \cap \{(\omega, \lambda): T(\omega) \le a, \lambda > a\} = \hat{\Lambda} \cap \{\hat{T} \le a\} \cap \hat{\Omega}_a$$

for each $a \ge 0$, so that $\hat{\Lambda} \cap \{\hat{T} < \gamma\} = \bigcup_a \hat{\Lambda}_a$, the union being over the nonnegative rationals, for example. By the definition of $\hat{\mathscr{M}}_a$ each $\hat{\Lambda}_a$ is of the form $\Lambda_a \times (a, \infty]$ with $\Lambda_a \in \mathscr{M}_a$. Plainly $\Lambda_a \subset \{T \le a\}$ for all a, $\Lambda_b \subset \Lambda_a$ if $b \le a$, and $\Lambda_a \cap \{T \le b\} = \Lambda_b \cap \{T \le b\}$ if $b \le a$. Consequently if $b \ge a$ then $\Lambda_a \cap \{T \le b\} = \Lambda_a \in \mathscr{M}_a \subset \mathscr{M}_b$, while if $b \le a$, $\Lambda_a \cap \{T \le b\} = \Lambda_b \cap \{T \le b\} \in \mathscr{M}_b$. Therefore each Λ_a is in $\mathscr{M}_T$. If $\Lambda = \bigcup_a \Lambda_a$, the union being over the nonnegative rationals, then $\Lambda \in \mathscr{M}_T$ and $\Lambda \cap \{T \le a\} = \Lambda_a$ for any nonnegative rational a. Therefore

$$(\Lambda \times [0, \infty]) \cap \{(\omega, \lambda): T(\omega) < \lambda\} = \bigcup_a (\Lambda \cap \{T \leq a\}) \times (a, \infty]$$
$$= \bigcup_a (\Lambda_a \times (a, \infty])$$
$$= \bigcup_a \hat{\Lambda}_a = \hat{\Lambda} \cap \{\hat{T} < \gamma\},$$

completing the proof of (b).

Let us assume now that X is strong Markov (as well as normal) and that M is a (right continuous) MF of X, and let $\hat{X}$ be the canonical subprocess corresponding to M. We are now interested in conditions that insure that $\hat{X}$ is strong Markov. Simple examples show that this is not always the case. (see Exercise 3.19). First note that since M is right continuous $\{M_t\}$ is progressively measurable with respect to $\{\mathscr{F}_t\}$, and hence with respect to $\{\mathscr{M}_t\}$ since we are, of course, assuming that $\mathscr{M}_t \supset \mathscr{F}_t$. See (5.15) of Chapter I. If T is an $\{\mathscr{M}_t\}$ stopping time then Theorem 6.11 of Chapter I implies that $M_T \in \mathscr{M}_T$.

(3.11) Definition. A MF, M, of X is said to be a *strong multiplicative functional* (SMF) provided that M is right continuous and satisfies

$$E^x[f(X_{t+T})\, M_{t+T}] = E^x\{E^{X(T)}[f(X_t)\, M_t]\, M_T\}$$

for all x, t, $\{\mathscr{M}_t\}$ stopping times T, and $f \in b\mathscr{E}$.

(3.12) Proposition. Let X be a strong Markov process and let M be a SMF of X. Then the canonical subprocess $\hat{X}$ corresponding to M is strong Markov.

Proof. Let $\hat{T}$ be an $\{\hat{\mathscr{M}}_t\}$ stopping time. By (3.9) there exists an $\{\mathscr{M}_t\}$ stopping time T such that $\hat{T}(\omega, \lambda) = T(\omega)$ on $\{\hat{T}(\omega, \lambda) < \lambda\}$. Therefore if $A \in \mathscr{E}^*$, then $\{\hat{X}_{\hat{T}} \in A\} \cap \{\hat{T} \leq t\} \cap \hat{\Omega}_t = (\{X_T \in A\} \cap \{T \leq t\}) \times (t, \infty]$ and so $\hat{X}_{\hat{T}} \in \hat{\mathscr{M}}_{\hat{T}}/\mathscr{E}^*_\Delta$. Hence according to (8.2) of Chapter I it suffices to show that

$$\hat{E}^x\{f(\hat{X}_{t+\hat{T}})\} = \hat{E}^x\{\hat{E}^{\hat{X}(\hat{T})} f(\hat{X}_t)\}$$

for all $f \in b\mathscr{E}^*$. Since $f(\Delta) = 0$ we may assume that $\hat{T} \leq \gamma$. Now

$$\hat{E}^x\{f(\hat{X}_{t+\hat{T}})\} = \hat{E}^x\{f(X_{t+T});\, t + T < \gamma\}$$
$$= E^x\{f(X_{t+T})\, M_{t+T}\}$$
$$= E^x\{E^{X(T)}[f(X_t)\, M_t] M_T\}$$
$$= E^x\{\hat{E}^{X(T)}[f(\hat{X}_t)] M_T\}$$
$$= \hat{E}^x\{\hat{E}^{\hat{X}(\hat{T})}[f(\hat{X}_t)]\},$$

and Theorem 3.12 is established.

(3.13) PROPOSITION. Suppose E_Δ is a metric space, $\mathscr{E}_\Delta \supset \mathscr{B}(E_\Delta)$, and X is quasi-left-continuous; then so is $\hat{X}$.

Proof. Let $\{\hat{T}_n\}$ be an increasing sequence of $\{\hat{\mathscr{M}}_t\}$ stopping times with limit $\hat{T}$; then we must show that $\hat{X}(\hat{T}_n) \to \hat{X}(\hat{T})$ almost surely on $\{\hat{T} < \hat{\zeta}\}$. Since $\hat{\zeta} = \zeta \wedge \gamma$ we may assume $\hat{T} \le \gamma$. Thus there exist $\{\mathscr{M}_t\}$ stopping times T and T_n such that $\hat{T} = T \wedge \gamma$ and $\hat{T}_n = T_n \wedge \gamma$. Consequently $T_n \uparrow T$, and therefore if $x \in E_M$

$$\begin{aligned}\hat{P}^x[\hat{X}(\hat{T}_n) \not\to \hat{X}(\hat{T});\ \hat{T} < \hat{\zeta}] \\ = \hat{P}^x[X(T_n) \not\to X(T);\ T < \zeta \wedge \gamma] \\ = E^x[M_T;\ X(T_n) \not\to X(T);\ T < \zeta] = 0.\end{aligned}$$

If $x \notin E_M$, then $\hat{X}(\hat{T}_n) = \Delta = \hat{X}(\hat{T})$ almost surely $\hat{P}^x$, and so the proof of Proposition 3.13 is complete.

In order to study the quasi-left-continuity of $\hat{X}$ on $[0, \infty)$ we introduce the following condition.

(3.14) Let $\{R_n\}$ be an increasing sequence of $\{\mathscr{M}_t\}$ stopping times with limit R; then $M_{R_n} \to M_R$ almost surely on $\{R < \zeta\}$.

(3.15) PROPOSITION. Suppose E_Δ is a metric space, $\mathscr{E}_\Delta \supset \mathscr{B}(E_\Delta)$, and M satisfies condition (3.14). Then if X is quasi-left-continuous on $[0, \infty)$, so is $\hat{X}$.

Proof. Let $\{\hat{T}_n\}$ be an increasing sequence of $\{\hat{\mathscr{M}}_t\}$ stopping times with limit $\hat{T}$. Let T_n and T be $\{\mathscr{M}_t\}$ stopping times such that $\hat{T}_n \wedge \gamma = T_n \wedge \gamma$ and $\hat{T} \wedge \gamma = T \wedge \gamma$. A simple inspection of the situation, using the definition of $\hat{X}$ and the fact that X is quasi-left-continuous on $[0, \infty)$, shows that it suffices to verify that

$$\Gamma = \{(\omega, \lambda): \zeta(\omega) > \lambda,\ T_n(\omega) < \lambda \text{ for all } n,\ T(\omega) = \lambda\}$$

has $\hat{P}^x$ measure zero for all $x \in E_M$. But for each n

$$\Gamma \subset \{(\omega, \lambda): T_n(\omega) < \lambda \le T(\omega),\ T(\omega) < \zeta(\omega)\},$$

and so for each $x \in E_M$

$$\hat{P}^x(\Gamma) \le E^x\{M_{T_n} - M_T;\ T < \zeta\} \to 0$$

as $n \to \infty$ by (3.14). Thus Proposition 3.15 is established.

(3.16) COROLLARY. Suppose X is a standard process and M is a SMF of X satisfying:

(i) $E_M = E$, i.e., all points are permanent for M;

(ii) $M_t \in \mathscr{F}_t^0$ for each t.

Then $\hat{X} = (\hat{\Omega}, \hat{\mathscr{F}}, \hat{\mathscr{F}}_t, \hat{X}_t, \hat{\theta}_t, \hat{P}^x)$, where $\hat{\mathscr{F}}$ and $\hat{\mathscr{F}}_t$ have their usual meanings relative to the process $\hat{X}$, is a standard process (with the same state space as X). If in addition X is a Hunt process and M satisfies (3.14), then $\hat{X}$ is a Hunt process.

REMARK. Theorem 4.12 of the next section implies that if M is a right continuous MF with $E_M = E$, then M is a SMF. Consequently in the hypothesis of Corollary 3.16 one need not assume that M is a strong MF; this actually follows from Condition (i) of (3.16).

Proof. Since X is a standard process, $\{\mathscr{M}_t\}$ is right continuous and hence so is $\{\hat{\mathscr{M}}_t\}$. But $\hat{\mathscr{F}}_t^0 \subset \hat{\mathscr{M}}_t$ and so $\hat{X}$ is Markov relative to $\{\hat{\mathscr{F}}_{t+}^0\}$. Consequently it follows from (8.12) of Chapter I that $\{\hat{\mathscr{F}}_t\}$ is right continuous. Clearly $\hat{X}$ is right continuous and has left-hand limits on $[0, \zeta)$ almost surely since X enjoys these properties. Next, in view of Theorem 7.3 of Chapter I, in checking the strong Markov property and the quasi-left-continuity of $\hat{X}$ it suffices to consider $\{\hat{\mathscr{F}}_{t+}^0\}$ stopping times; since $\hat{\mathscr{F}}_{t+}^0 \subset \hat{\mathscr{M}}_t$ these properties follow from Propositions 3.12 and 3.13. So far we have not made any use of the two assumptions on M in our theorem. However (i) implies that $\hat{X}$ is normal. (See the discussion following the proof of Theorem 3.3.) Finally if $A \in \mathscr{E}$ then $\hat{P}^x(\hat{X}_t \in A) = E^x(M_t; X_t \in A)$ and so (ii) implies that $x \to \hat{P}^x(\hat{X}_t \in A)$ is $\mathscr{E}$ measurable. Therefore we may regard $(E, \mathscr{E})$ as the state space of $\hat{X}$, and the first sentence of (3.16) is proved. The second sentence follows immediately from (3.15) of this chapter and (7.3) of Chapter I.

(3.17) EXAMPLE. Let X be a standard process and let $M_t = e^{-\beta t}$, $\beta > 0$. Let X^β denote the process in (3.16) corresponding to this MF. Then X^β is called the β-subprocess of X. It follows from Corollary 3.16 that X^β is a standard process, and if X is a Hunt process then so is X^β. If P_t^β denotes the transition semigroup of X^β, then obviously $P_t^\beta = e^{-\beta t} P_t$. In particular a function is α-excessive for X^β if and only if it is $\alpha + \beta$ excessive for X. The use of the β-subprocess will turn out to be an important tool in studying the original process X.

Let us close this section by giving conditions under which $(E_M, \mathscr{E}_M^*)$, where $\mathscr{E}_M^*$ is the trace of $\mathscr{E}^*$ on E_M, may be taken as the state space of $\hat{X}$. An obvious necessary condition is that $\hat{P}^x(\hat{X}_t \in E - E_M$ for some $t < \hat{\zeta}) = 0$ for all x in E_M. If we are willing to delete from $\hat{\Omega}$ a set which is almost surely null, this is also sufficient. With this in mind let $R = T_{E-E_M}$ be the hitting time (for X) of $E - E_M$ and let us assume that R is an $\{\mathscr{F}_t\}$ stopping time. Since $\hat{X}_t = X_t$ if $t < \hat{\zeta}$ the probability in question is just

$$\hat{P}^x[R < \zeta] = \hat{P}^x[R < \gamma; R < \zeta]$$
$$= E^x\{M_R; R < \zeta\}.$$

Thus if $R = T_{E-E_M}$ is a stopping time, then a necessary and sufficient condition that we may regard $(E_M, \mathscr{E}_M^*)$ as the state space for $\hat{X}$ (after a trivial modification of $\hat{\Omega}$) is that $M_R = 0$ almost surely on $\{R < \zeta\}$. See Proposition 4.21 in this connection.

Exercises

(3.18) Show that there exists a standard process having the transition function defined in (9.16) of Chapter I. [Hint: let X be uniform motion to the right in $\mathbf{R}$ and obtain the desired standard process as a subprocess of X.] Show directly that condition (3.14) is not satisfied by the MF involved in the construction.

(3.19) Let X be the Brownian motion process in $\mathbf{R}$. Let $M_t(\omega) = 1$ for all t if $X_0(\omega) \neq 0$ and let $M_t(\omega) = 0$ for all t if $X_0(\omega) = 0$. Show that $M = \{M_t\}$ is a continuous MF of X. Show that M is not a SMF and that the corresponding canonical subprocess $\hat{X}$ is not strong Markov. [Hint: consider $T = \inf\{t: X_t = 0\}$ and use (3.18) of Chapter II.] Describe $\hat{X}$ in this case. Note that $E - E_M = \{0\}$ and that $\hat{P}^x[\hat{X}_t \in E_M$ for all $t < \hat{\zeta}] = 0$ if $x \neq 0$. Consequently E_M can not be taken as the state space of $\hat{X}$.

(3.20) The notation is that of Proposition 3.1. Show that if $\hat{f}(\omega, \lambda)$ is $\hat{\mathscr{M}}_t$ measurable, then $\hat{f}(\omega, \lambda) = \hat{f}(\omega, \mu)$ whenever $\lambda, \mu > t$. [Hint: show that if $\hat{f} \in b\hat{\mathscr{M}}_t$, then there exists an $f \in b\mathscr{M}_t$ such that $f(\omega, \lambda)\, I_{\hat{\Omega}_t}(\omega, \lambda) = f(\omega) I_{(t,\infty]}(\lambda)$.]

(3.21) Let X be a standard process and let G be an open subset of E such that each x in $E_\Delta - G$ is regular for $E_\Delta - \bar{G}$. Show that $T_{E_\Delta - \bar{G}} = T_{E_\Delta - G}$ almost surely. Let $T = T_{E_\Delta - \bar{G}}$ and let $\tilde{X}$ be constructed as in (3.7) using this T. Show that we may regard $(G, \mathscr{G})$ as the state space of $\tilde{X}$ where $\mathscr{G} = \mathscr{B}(G)$ is the σ-algebra of Borel subsets of G. If Δ is redefined as the point at infinity of G, then show that $\tilde{X}$ is a standard process. $\tilde{X}$ is called the *restriction of* X *to* G.

4. Resolvents and Strong Multiplicative Functionals

In this section we are going to develop some criteria to insure that a right continuous MF is a SMF. Our main tool is a result about resolvents (Theorem

4.9) which is due to Meyer. Therefore we will first develop some general facts about resolvents and prove Theorem 4.9, after which we will return to the main problem of this section—the characterization of strong multiplicative functionals. We will assume throughout this section that $X = (\Omega, \mathcal{M}, \mathcal{M}_t, X_t, \theta_t, P^x)$ is a given standard process with state space $(E, \mathcal{E})$, although slightly less stringent assumptions would suffice. We will let $\mathbf{B}$ denote the Banach space $b\mathcal{E}^*$.

Let $M = \{M_t\}$ be a right continuous MF of X and let $Q_t f(x) = E^x\{f(X_t) M_t\}$ denote the corresponding semigroup on $\mathbf{B}$. If $f \in \mathbf{C} = \mathbf{C}(E)$, then $t \to Q_t f(x)$ is right continuous for each x and hence $(s, x) \to Q_s f(x)$ from $[0, t) \times E \to \mathbf{R}$ is in $b(\mathcal{R}_t \times \mathcal{E}^*)$ for each t when $f \in \mathbf{C}$, where, as usual, $\mathcal{R}_t$ denotes the Borel sets of $[0, t)$. This last statement then extends to any $f \in b\mathcal{E}$ by use of MCT. Thus for any $\alpha > 0$ we can define $V^\alpha f(x) = \int_0^\infty e^{-\alpha t} Q_t f(x)\, dt$, and V^α maps $b\mathcal{E}$ into $\mathbf{B}$. If $f \in b\mathcal{E}$ and $f \geq 0$, then $V^\alpha f \leq U^\alpha f$ and so V^α is given by a measure on $(E, \mathcal{E})$ which we denote by $V^\alpha(x, dy)$. Evidently $V^\alpha f(x) = \int V^\alpha(x, dy) f(y)$ and $V^\alpha 1(x) \leq \alpha^{-1}$. Each of the measures $V^\alpha(x, \cdot)$ extends uniquely to $\mathcal{E}^*$ and hence we can define $V^\alpha f(x) = \int V^\alpha(x, dy) f(y)$ for $f \in \mathbf{B}$. We next show that $V^\alpha \mathbf{B} \subset \mathbf{B}$. To this end let μ be a finite measure on $\mathcal{E}^*$. Then $\mu V^\alpha(B) = \int \mu(dx)\, V^\alpha(x, B)$ defines a finite measure on $\mathcal{E}$. Thus given $f \in \mathbf{B}$ there exist $g, h \in b\mathcal{E}$ with $g \leq f \leq h$ such that $\mu V^\alpha g = \mu V^\alpha h$. But this implies that $V^\alpha g \leq V^\alpha f \leq V^\alpha h$ and $V^\alpha g = V^\alpha h$, a.e. μ. Consequently $V^\alpha f \in \mathbf{B}$.

Let us next show that $V^\alpha f(x) = \int_0^\infty e^{-\alpha t}\, Q_t f(x)\, dt$ for $f \in \mathbf{B}$. We know that $(t, x) \to Q_t f(x)$ is in $b(\mathcal{R} \times \mathcal{E}^*)$ and that this formula holds for $f \in b\mathcal{E}$. Given a finite measure ν on $\mathcal{E}^*$ and a finite measure λ on $\mathcal{R}$ we can define for $B \in \mathcal{E}$

$$\mu(B) = \iint Q_t(x, B)\, \lambda(dt)\, \nu(dx),$$

and obviously μ is a finite measure on $\mathcal{E}$. Now if $f \in \mathbf{B}$ there exist $g, h \in b\mathcal{E}$ such that $g \leq f \leq h$ and $\mu(g) = \mu(h)$. Thus $Q_t g(x) \leq Q_t f(x) \leq Q_t h(x)$ for all (t, x) and

$$\iint Q_t\, g(x)\, \lambda(dt)\, \nu(dx) = \mu(g) = \mu(h) = \iint Q_t\, h(x)\, \lambda(dt)\, \nu(dx).$$

Therefore $(t, x) \to Q_t f(x)$ is in $(\mathcal{R} \times \mathcal{E}^*)^{\lambda, \nu}$ (this is the completion of $\mathcal{R} \times \mathcal{E}^*$ with respect to $\lambda \times \nu$) for all λ and ν. Thus for each x and $f \in \mathbf{B}$ we can form $\int_0^\infty e^{-\alpha t}\, Q_t f(x)\, dt$, and since this is a measure in f and agrees with $V^\alpha f(x)$ for $f \in b\mathcal{E}$ we must have $V^\alpha f(x) = \int_0^\infty e^{-\alpha t}\, Q_t f(x)\, dt$ for $f \in \mathbf{B}$. In particular the above argument shows that $t \to Q_t f(x)$ is Lebesgue measurable for each x. Finally note that we have enough joint measurability in $Q_t f(x)$ to use Fubini's theorem.

There remains one more measurability detail to discuss: namely, to show that

(4.1) $$V^\alpha f(x) = E^x \int_0^\infty e^{-\alpha t} f(X_t)\, M_t\, dt$$

for all $f \in \mathbf{B}$. If $f \in \mathbf{C}$ it follows from the right continuity of the paths that $(t, \omega) \to f(X_t(\omega))$ is in $\mathscr{R} \times \mathscr{F}$, and this continues to hold for $f \in b\mathscr{E}$. One now shows by an argument similar to that used above that $(t, \omega) \to f(X_t(\omega))$ is in $(\mathscr{R} \times \mathscr{F})^{\lambda,\mu}$ for all λ and μ when f is in $\mathbf{B}$, where the σ-algebra in question is the completion of $\mathscr{R} \times \mathscr{F}$ with respect to $\lambda \times P^\mu$, λ a finite measure on $\mathscr{R}$ and μ a finite measure on $\mathscr{E}_\Delta$. Thus we can form the expression on the right side of (4.1) for any $f \in \mathbf{B}$ and since it is a measure which agrees with the left side of (4.1) on $\mathbf{C}$ the equality in (4.1) must obtain for $f \in \mathbf{B}$. Again we have enough joint measurability in $f[X_t(\omega)]$ to use the Fubini theorem.

The family $\{V^\alpha; \alpha > 0\}$ has the following properties:

(4.2)
$$\text{(i)}\quad \|V^\alpha\| \le \alpha^{-1},$$
$$\text{(ii)}\quad V^\alpha f \le U^\alpha f \text{ if } f \in \mathbf{B}_+;$$
$$\text{(iii)}\quad V^\alpha - V^\beta = (\beta - \alpha) V^\alpha V^\beta,\ \alpha, \beta > 0.$$

This last relation is the resolvent equation and is an easy consequence of the fact that $\{Q_t; t \ge 0\}$ is a semigroup (or (4.1) and the Markov property). We now make the following definition.

(4.3) Definition. A family $\{V^\alpha; \alpha > 0\}$ of positive linear operators on $\mathbf{B}$ is called a *resolvent subordinate to* $\{U^\alpha\}$ provided it satisfies (4.2).

In particular it follows from the above discussion that the resolvent of a semigroup generated by a *right continuous* MF is subordinate to $\{U^\alpha\}$.

It follows from (4.2iii) that $V^\alpha V^\beta = V^\beta V^\alpha$ for $\alpha, \beta > 0$. Since $f \to U^\alpha f(x)$ is a finite measure, Condition (4.2ii) implies that $f \to V^\alpha f(x)$ is a finite measure also. As usual, we denote it by $V^\alpha(x, \cdot)$. Hence if $f \ge 0$ is in $\mathscr{E}^*$, $V^\alpha f = V^\alpha(\cdot, f)$ exists and is in $\mathscr{E}^*_+$. Also if $f \in \mathbf{B}$ and $\alpha > 0$, $\beta \ge 0$, then (4.2) implies

(4.4) $$\eta V^{\eta+\beta} V^\alpha f = V^\alpha f - V^{\eta+\beta} f + (\alpha - \beta) V^{\eta+\beta} V^\alpha f$$
$$\to V^\alpha f \text{ (in norm) as } \eta \to \infty.$$

(4.5) Definition. A function $f \in \mathscr{E}^*_+$ is called $\alpha - V$ *supermedian* if $\beta V^{\alpha+\beta} f \le f$ for all $\beta > 0$, and is called $\alpha - V$ *excessive* if, in addition, $\beta V^{\alpha+\beta} f \to f$ pointwise as $\beta \to \infty$.

Obviously the minimum of two $\alpha - V$ supermedian functions is again

$\alpha - V$ supermedian. Also $\beta V^{\alpha+\beta}1 \leq \beta U^{\alpha+\beta}1 \leq \beta/(\alpha+\beta)$, implies that nonnegative constants are $\alpha - V$ supermedian for any α.

(4.6) PROPOSITION:

(i) If $f \in \mathscr{E}_+^*$ then $V^\alpha f$ is $\alpha - V$ excessive.

(ii) If f is $\alpha - V$ supermedian, then $\beta \to \beta V^{\alpha+\beta} f$ is increasing and $\bar{f} = \lim_{\beta\to\infty} \beta V^{\alpha+\beta} f$ is the largest $\alpha - V$ excessive function dominated by f. We call $\bar{f}$ the $\alpha - V$ *excessive regularization* of f, and we have $V^\beta f = V^\beta \bar{f}$ for any $\beta > 0$.

Proof. For bounded f the first statement is an immediate consequence of (4.4). This statement for unbounded f then follows from (4.6ii) if we approximate f by $f_n = f \wedge n$. Thus we need only prove (4.6ii). If $\beta > \eta$, then

$$\beta V^{\alpha+\beta} - \eta V^{\alpha+\eta} = \beta V^{\alpha+\eta} + \beta(\eta - \beta) V^{\alpha+\beta} V^{\alpha+\eta} - \eta V^{\alpha+\eta}$$
$$= (\beta - \eta) V^{\alpha+\eta}[I - \beta V^{\alpha+\beta}].$$

It is immediate from this identity that if f is a bounded $\alpha - V$ supermedian function, then $\beta \to V^{\alpha+\beta} f(x)$ is increasing. For general $\alpha - V$ supermedian f let $f_n = f \wedge n$; then $f_n \uparrow f$ and each f_n is $\alpha - V$ supermedian. Hence for each n if $\beta > \eta$, then $\beta V^{\beta+\alpha} f_n \geq \eta V^{\eta+\alpha} f_n$ and letting $n \to \infty$ we obtain the first conclusion of (4.6ii).

Thus $\bar{f}(x) = \lim_{\beta\to\infty} \beta V^{\alpha+\beta} f(x)$ exists when f is $\alpha - V$ supermedian. Clearly $\bar{f} \in \mathscr{E}_+^*$ and $\bar{f} \leq f$. Suppose g is $\alpha - V$ excessive and $g \leq f$; then $\beta V^{\alpha+\beta} g \leq \beta V^{\alpha+\beta} f$ and letting $\beta \to \infty$ we find that $g \leq \bar{f}$. Thus to complete the proof of Proposition 4.6 we must show that $\bar{f}$ is $\alpha - V$ excessive. Suppose first that f is bounded; then from the monotone convergence theorem and (4.4) we have

$$V^\beta \bar{f} = V^\beta \left[\lim_{\eta\to\infty} \eta V^{\alpha+\eta} f \right]$$
$$= \lim_{\eta\to\infty} \eta V^\beta V^{\alpha+\eta} f = V^\beta f,$$

for any $\beta > 0$. Consequently $\beta V^{\alpha+\beta} \bar{f} = \beta V^{\alpha+\beta} f \uparrow \bar{f}$ as $\beta \to \infty$, and so $\bar{f}$ is $\alpha - V$ excessive. If f is unbounded let $f_n = f \wedge n$ again; then $\beta V^{\alpha+\beta} f_n$ is increasing in both β and n. Therefore $\bar{f} = \lim_\beta \lim_n \beta V^{\alpha+\beta} f_n = \lim_n \bar{f}_n$. But it is easy to see from what has already been proved that the limit of an increasing sequence of $\alpha - V$ excessive functions is $\alpha - V$ excessive, and so $\bar{f}$ is $\alpha - V$ excessive. We already know that $V^\beta f_n = V^\beta \bar{f}_n$ for each n and $\beta > 0$, and hence letting $n \to \infty$ we obtain the last conclusion of (4.6).

(4.7) PROPOSITION. If $f \in \mathbf{B}_+$, then $U^\alpha f - V^\alpha f$ is $\alpha - U$ supermedian for any $\alpha > 0$.

Proof. If $f \in \mathbf{B}_+$, then $(U^\alpha f - V^\alpha f) \in \mathbf{B}_+$ for any $\alpha > 0$. Moreover

$$U^\alpha f - V^\alpha f - \beta U^{\beta+\alpha}(U^\alpha f - V^\alpha f) \geq U^\alpha f - V^\alpha f - \beta U^{\beta+\alpha} U^\alpha f + \beta V^{\beta+\alpha} V^\alpha f$$
$$= U^{\alpha+\beta} f - V^{\alpha+\beta} f \geq 0,$$

and so Proposition 4.7 is proved.

Recall from (2.3) of Chapter II that f is α-excessive (for X) if and only if f is $\alpha - U$ excessive in the sense of Definition 4.5.

(4.8) DEFINITION. A resolvent $\{V^\alpha\}$ is *exactly subordinate* to $\{U^\alpha\}$ provided it is subordinate and, in addition, $U^\alpha f - V^\alpha f$ is α-excessive (relative to X) for all $f \in \mathbf{B}_+$ and $\alpha > 0$.

Example. Let B be a nearly Borel subset of E and $T = T_B$. If $V^\alpha f(x) = E^x \int_0^T e^{-\alpha t} f(X_t)\, dt$ for $f \in \mathbf{B}$, then $\{V^\alpha\}$ is a resolvent subordinate to $\{U^\alpha\}$; in fact it is just the resolvent corresponding to the right continuous MF, $M_t = I_{[0,T)}(t)$. If $f \in \mathbf{B}_+$, then one easily computes that $U^\alpha f - V^\alpha f = P_B^\alpha U^\alpha f$, and so $\{V^\alpha\}$ is exactly subordinate to $\{U^\alpha\}$ according to Proposition 2.8 of Chapter II.

(4.9) THEOREM. Let $\{V^\alpha\}$ be a resolvent subordinate to $\{U^\alpha\}$. Then $W^\alpha f(x) = \lim_{\beta \to \infty} \beta U^\beta V^\alpha f(x)$ exists for all $f \in \mathbf{B}$, $\alpha > 0$, and $x \in E$ and the family $\{W^\alpha\}$ is a resolvent exactly subordinate to $\{U^\alpha\}$. Moreover for each $\alpha > 0$, $V^\alpha(x, \cdot) \leq W^\alpha(x, \cdot)$ for all x, with equality precisely for those x for which $\beta U^\beta V^\gamma 1(x) \to V^\gamma 1(x)$ as $\beta \to \infty$ for some $\gamma > 0$. In particular equality holds at any x for which $\beta V^\beta 1(x) \to 1$ as $\beta \to \infty$.

Proof. We first prove the existence of the limit in question. From the resolvent equation

$$U^\alpha = U^{\beta+\alpha} + \beta U^{\beta+\alpha} U^\alpha$$
$$U^\beta = U^{\beta+\alpha} + \alpha U^{\beta+\alpha} U^\beta,$$

and so if $f \in \mathbf{B}_+$ we have

$$U^\alpha f - \beta U^\beta V^\alpha f = U^{\beta+\alpha} f + \beta U^{\beta+\alpha} U^\alpha f - \beta U^{\beta+\alpha} V^\alpha f - \beta\alpha U^{\beta+\alpha} U^\beta V^\alpha f$$
$$= \beta U^{\beta+\alpha}[U^\alpha f - V^\alpha f] + O(1/\beta)$$

since $\|U^\gamma\| \leq \gamma^{-1}$. Because of this equality, Propositions 4.6 and 4.7 imply that $W^\alpha f = \lim_{\beta \to \infty} \beta U^\beta V^\alpha f$ exists for $f \in \mathbf{B}_+$ and that $U^\alpha f - W^\alpha f$ is the $\alpha - U$ excessive regularization of $U^\alpha f - V^\alpha f$. Consequently $W^\alpha f = \lim_{\beta \to \infty} \beta U^\beta V^\alpha f$

exists for all $f \in \mathbf{B}$. Clearly $V^\alpha f \leq W^\alpha f \leq U^\alpha f$ for $f \in \mathbf{B}_+$, and hence, for each $\alpha > 0$, W^α is a bounded positive linear operator on $\mathbf{B}$ which is given by measures $W^\alpha(x, \cdot)$ satisfying $V^\alpha(x, \cdot) \leq W^\alpha(x, \cdot) \leq U^\alpha(x, \cdot)$ for all x in E.

Let us next show that the family $\{W^\alpha\}$ is a resolvent. If $f \in \mathbf{B}_+$ it follows from the above remarks and Proposition 4.6 that, for any $\beta > 0$, $U^\beta(U^\alpha f - V^\alpha f) = U^\beta(U^\alpha f - W^\alpha f)$, and hence that $U^\beta V^\alpha f = U^\beta W^\alpha f$ for any $f \in \mathbf{B}_+$. Therefore $U^\beta V^\alpha = U^\beta W^\alpha$. Furthermore for $f \in \mathbf{B}_+$ one has

$$0 = U^\beta[W^\alpha - V^\alpha]f \geq V^\beta[W^\alpha - V^\alpha]f \geq 0,$$

and consequently $V^\beta W^\alpha = V^\beta V^\alpha$. Now

$$V^\alpha - V^\beta = (\beta - \alpha)V^\beta V^\alpha = (\beta - \alpha)V^\beta W^\alpha,$$

and operating on this relation by ηU^η and letting $\eta \to \infty$ one obtains

$$W^\alpha - W^\beta = (\beta - \alpha)W^\beta W^\alpha.$$

Thus $\{W^\alpha\}$ is a resolvent and since $U^\alpha f - W^\alpha f$ is α-excessive (it is the $\alpha - U$ excessive regularization of $U^\alpha f - V^\alpha f$ and hence is α-excessive) for any $f \in \mathbf{B}_+$, it follows that $\{W^\alpha\}$ is a resolvent exactly subordinate to $\{U^\alpha\}$.

We have already seen that $V^\alpha(x, \cdot) \leq W^\alpha(x, \cdot)$ for all x and α. Suppose $\beta U^\beta V^\gamma 1(y) \to V^\gamma 1(y)$ as $\beta \to \infty$ for some fixed $\gamma > 0$ and $y \in E$. By definition $\beta U^\beta V^\gamma 1 \to W^\gamma 1$ pointwise as $\beta \to \infty$ and hence $V^\gamma(y, \cdot) = W^\gamma(y, \cdot)$. Therefore as $\beta \to \infty$, $\beta U^\beta V^\gamma f(y) \to W^\gamma f(y) = V^\gamma f(y)$ for all $f \in \mathbf{B}$. Now given any $\alpha > 0$, $V^\alpha 1 - V^\gamma 1 = (\gamma - \alpha)V^\gamma V^\alpha 1$ and operating on this by βU^β and letting $\beta \to \infty$ we obtain $W^\alpha 1(y) = V^\gamma 1(y) + (\gamma - \alpha)V^\gamma V^\alpha 1(y) = V^\alpha 1(y)$. In other words if $\beta U^\beta V^\gamma 1(y) \to V^\gamma 1(y)$ for some $\gamma > 0$, then this convergence holds for all $\gamma > 0$. We have now proved all except the last sentence of Theorem 4.9.

If $\beta V^\beta 1(y) \to 1$ as $\beta \to \infty$, then $\beta U^\beta 1(y) - \beta V^\beta 1(y) \to 0$. Let $\alpha > 0$. Then $\alpha V^\alpha 1 \leq 1$ and so $(\beta U^\beta - \beta V^\beta)V^\alpha 1(y) \leq (1/\alpha)(\beta U^\beta - \beta V^\beta) 1(y) \to 0$. But $\beta V^\beta V^\alpha 1 \to V^\alpha 1$ and hence $\beta U^\beta V^\alpha 1(y) \to V^\alpha 1(y)$. Therefore $W^\alpha(y, \cdot) = V^\alpha(y, \cdot)$ and the proof of Theorem 4.9 is now complete.

(4.10) Corollary. Let M be a right continuous MF of X and let $\{Q_t\}$ and $\{V^\alpha\}$ denote the corresponding semigroup and resolvent. Let $\{W^\alpha\}$ be the exactly subordinate resolvent corresponding to $\{V^\alpha\}$; then $V^\alpha(x, \cdot) = W^\alpha(x, \cdot)$ for all $x \in E_M$.

Proof. If $x \in E_M$ then $Q_t 1(x) = E^x\{M_t; t < \zeta\} \to 1$ as $t \to 0$, and so $\beta V^\beta 1(x) \to 1$ as $\beta \to \infty$, proving (4.10).

We turn now to the main topic of this section—the characterization of strong multiplicative functionals, which were defined in (3.11). We remind the reader that for any MF, M, we have $M_\infty = 0$ by convention.

(4.11) DEFINITION. A right continuous MF, $M = \{M_t\}$, is called *regular* (RMF) provided $P^x[X_T \in E - E_M;\ M_T > 0] = 0$ for all x and all $\{\mathscr{M}_t\}$ stopping times T.

Perhaps we should point out explicitly that the properties of being a regular MF or a strong MF depend only on the equivalence class (in the set of all right continuous MF's) to which a MF belongs. In other words if M is a RMF (SMF) and N is a right continuous MF equivalent to M, then N is a RMF (SMF). We will eventually show that these two notions are equivalent; that is, M is a RMF if and only if it is a SMF.

We begin with the implication in one direction.

(4.12) THEOREM. Any RMF is a SMF.

Proof. Let M be a RMF and let $\{Q_t\}$ and $\{V^\alpha\}$ be the corresponding semigroup and resolvent. Let $\{W^\alpha\}$ be the exactly subordinate resolvent associated with $\{V^\alpha\}$ in Theorem 4.9. It follows from (4.10) that for each $f \in \mathbf{C}_+$ and $\alpha > 0$ the function $g_\alpha = W^\alpha f = U^\alpha f - (U^\alpha f - W^\alpha f)$ is bounded, nearly Borel measurable, finely continuous, and agrees with $V^\alpha f$ on E_M.

Given an $\{\mathscr{M}_t\}$ stopping time T recall that $T^{(n)} = (k+1)2^{-n}$ if $k2^{-n} \le T < (k+1)2^{-n}$ and $T^{(n)} = \infty$ if $T = \infty$, $k = 0, 1, \ldots,$ is an $\{\mathscr{M}_t\}$ stopping time and $T^{(n)} \downarrow T$. Consider for $\alpha > 0$ and $f \in \mathbf{C}_+$

$$
\begin{aligned}
q^\alpha(x) &= E^x \int_0^\infty e^{-\alpha t} f(X_{t+T})\, M_{t+T}\, dt \\
&= \lim_n E^x \int_0^\infty e^{-\alpha t} f[X(t + T^{(n)})]\, M(t + T^{(n)})\, dt \\
&= \lim_n \sum_k E^x\left\{\int_0^\infty e^{-\alpha t} f(X_{t+k2^{-n}})\, M_{t+k2^{-n}}\, dt;\ T^{(n)} = k2^{-n}\right\} \\
&= \lim_n \sum_k E^x\left\{E^{X(k2^{-n})}\left[\int_0^\infty e^{-\alpha t} f(X_t)\, M_t\, dt\right] M_{k2^{-n}};\ T^{(n)} = k2^{-n}\right\} \\
&= \lim_n E^x\left\{E^{X(T^{(n)})}\left[\int_0^\infty e^{-\alpha t} f(X_t)\, M_t\, dt\right] M_{T^{(n)}}\right\} \\
&= \lim_n E^x\{V^\alpha f[X(T^{(n)})]\, M(T^{(n)})\}.
\end{aligned}
$$

Since M is regular this last expression becomes

$$
\lim_n E^x\{g_\alpha[X(T^{(n)})]\, M(T^{(n)})\} = E^x\{g_\alpha(X(T))\, M_T\} = E^x\{V^\alpha f(X_T)\, M_T\},
$$

and so

$$q^\alpha(x) = E^x[V^\alpha f(X_T)\, M_T] = E^x\left\{E^{X(T)}\left[\int_0^\infty e^{-\alpha t} f(X_t)\, M_t\, dt\right] M_T\right\}.$$

Consequently the functions $\varphi(t) = E^x\{f(X_{t+T})\, M_{t+T}\}$ and $\psi(t) = E^x\{E^{X(T)}[f(X_t)\, M_t] M_T\}$ have the same Laplace transforms, namely $q^\alpha(x)$, and since φ and ψ are both right continuous the uniqueness theorem for Laplace transforms implies that they must be identical. But this is just the statement that (3.11) holds for any $f \in \mathbf{C}_+$ and consequently for any $f \in \mathbf{B}$. Thus Theorem 4.12 is proved.

(4.13) Definition. A right continuous MF is said to be *exact* (EMF) provided the corresponding resolvent $\{V^\alpha\}$ is exactly subordinate to $\{U^\alpha\}$.

The following proposition is a consequence of the proof of Theorem 4.12.

(4.14) Proposition. Any EMF is a SMF.

(4.15) Proposition. Let M be a SMF. Then for all $\{\mathscr{M}_t\}$ stopping times T and $Y \in b\mathscr{F}$ one has

$$E^x\{(Y \circ \theta_T) M_{t+T}; \Lambda\} = E^x\{E^{X(T)}(Y M_t)\, M_T; \Lambda\}$$

for all $\Lambda \in \mathscr{M}_T$, t, and x.

Proof. As usual it suffices to consider the case $\Lambda = \Omega$ and Y of the form $\prod_{j=1}^n f_j(X_{t_j})$ with $t_1 < \ldots < t_n$ and $f_j \in b\mathscr{E}_\Delta^*$, $j = 1, \ldots, n$. Of course, Proposition 4.15 must essentially express the fact that the canonical subprocess $\hat{X}$ corresponding to M is a strong Markov process, and we will deduce (4.15) from this fact. Suppose first of all that we have proved (4.15) for Y's of the above form with $t_n \le t$. If $t_1 < \ldots < t_j \le t < t_{j+1} < \ldots < t_n$, then the strong Markov property for X yields

$$E^x\left\{\prod_{i=1}^n f_i \circ X_{t_i+T} M_{t+T}\right\}$$
$$= E^x\left\{\left(\prod_{i=1}^j f_i \circ X_{t_i+T}\right) M_{t+T} E^{X(t+T)}\left(\prod_{i=j+1}^n f_i \circ X_{t_i-t}\right)\right\},$$

which, according to the supposition we have just made, becomes

$$E^x\left\{E^{X(T)}\left[\left(\prod_{i=1}^j f_i \circ X_{t_i}\right) E^{X(t)}\left(\prod_{i=j+1}^n f_i \circ X_{t_i-t}\right) M_t\right] M_T\right\}$$
$$= E^x\left\{E^{X(T)}\left[\left(\prod_{i=1}^n f_i \circ X_{t_i}\right) M_t\right] M_T\right\}.$$

Thus the general case is reduced to the case $t_n \le t$.

To handle the case $t_n \le t$ we use the fact that $\hat{X}$, the canonical subprocess corresponding to M, is a strong Markov process. The notation is that of Section 3. If $\hat{T} = T \wedge \gamma$, then $\hat{T}$ is an $\{\hat{\mathscr{M}}_t\}$ stopping time, and so

$$\begin{aligned} E^x\left\{\left(\prod_{j=1}^n f_j \circ X_{t_j+T}\right) M_{t+T}\right\} &= \hat{E}^x\left\{\prod_{j=1}^n f_j \circ X_{t_j+T};\, t + T < \gamma\right\} \\ &= \hat{E}^x\left\{\prod_{j=1}^n f_j \circ \hat{X}_{t_j+\hat{T}};\, t + \hat{T} < \gamma\right\} \\ &= \hat{E}^x\left\{\hat{E}^{\hat{X}(\hat{T})}\left(\prod_{j=1}^n f_j \circ \hat{X}_{t_j};\, t < \gamma\right);\, \hat{T} < \gamma\right\} \\ &= E^x\left\{E^{X(T)}\left[\left(\prod_{j=1}^n f_j \circ X_{t_j}\right) M_t\right] M_T\right\}, \end{aligned}$$

and hence Proposition 4.15 is proved.

(4.16) COROLLARY. Let M be a SMF, T an $\{\mathscr{M}_t\}$ stopping time, $Y \in b\mathscr{F}$, and $R \in \mathscr{F}$, $R \ge 0$. Then

$$E^x\{(Y \circ \theta_T) M(T + R \circ \theta_T);\, \Lambda\} = E^x\{E^{X(T)}(Y M_R) M_T;\, \Lambda\}$$

for all $\Lambda \in \mathscr{M}_T$ and x.

Proof. We leave it to the reader to check that $M_R \in \mathscr{F}$ and $M_{T+R\circ\theta_T} \in \mathscr{M}$ under the stated conditions. Let $R^{(n)}$ have its usual meaning; then $\{R^{(n)} = k2^{-n}\} \in \mathscr{F}$ and $\{R^{(n)} \circ \theta_T = k2^{-n}\} = \theta_T^{-1}\{R^{(n)} = k2^{-n}\}$. Recalling that $M_\infty = 0$ by assumption, the left side of the desired equality equals

$$\begin{aligned} &\lim_n E^x\{(Y \circ \theta_T) M(T + R^{(n)} \circ \theta_T);\, \Lambda\} \\ &\quad = \lim_n \sum_k E^x\{(Y \circ \theta_T) M_{T+k2^{-n}};\, R^{(n)} \circ \theta_T = k2^{-n};\, \Lambda\} \\ &\quad = \lim_n \sum_k E^x\{E^{X(T)}[Y M_{k2^{-n}};\, R^{(n)} = k2^{-n}] M_T;\, \Lambda\} \\ &\quad = E^x\{E^{X(T)}[Y M_R] M_T;\, \Lambda\}, \end{aligned}$$

completing the proof of Corollary (4.16).

(4.17) THEOREM. Let M be a SMF, T an $\{\mathscr{F}_t\}$ stopping time, and $R \in \mathscr{M}$, $R \ge 0$. Then

$$M_{T(\omega)+R(\omega)}(\omega) = M_{T(\omega)}(\omega)\, M_{R(\omega)}(\theta_T \omega) \tag{4.18}$$

almost surely.

Proof. In general we will omit the ω's when writing expressions such as (4.18). For example (4.18) will be written $M_{T+R} = M_T\, M_R(\theta_T)$. The reader should distinguish carefully between $M_R(\theta_T) = M_{R(\omega)}(\theta_T\omega)$ and $M_R \circ \theta_T = M_{R(\theta_T\omega)}(\theta_T\omega)$. First note that (4.18) holds on $\{R = \infty\}$ since $M_\infty = 0$ by convention. Therefore it suffices to consider the case $R(\omega) \equiv t$ since we can approximate R from above in the usual manner by countably valued random variables. Thus we must show that $M_{t+T} = M_T(M_t \circ \theta_T)$ almost surely for each t. The fact that T is an $\{\mathscr{F}_t\}$ stopping time implies that M_{t+T}, M_T, and $M_t \circ \theta_T$ are all in $\mathscr{F}$. Let $\mathscr{H}_T = \sigma\{X_{T+s};\, s \geq 0\}^\sim$ and $\mathscr{G} = \sigma(\mathscr{H}_T \cup \mathscr{F}_T)^\sim$. We will show that

$$\text{(4.19)} \qquad E^x\{M_{T+t};\, \Lambda\} = E^x\{M_T(M_t \circ \theta_T);\, \Lambda\}$$

for all $\Lambda \in \mathscr{G}$, t, and x. It suffices to consider $\Lambda = D \cap \{X_{T+t_1} \in A_1\} \cap \ldots \cap \{X_{T+t_n} \in A_n\}$ where $D \in \mathscr{F}_T$ and $A_j \in \mathscr{E}_\Delta$, $1 \leq j \leq n$. But for such a Λ Proposition 4.15 implies that the left side of (4.19) reduces to

$$E^x\{E^{X(T)}[\textstyle\prod_{j=1}^n I_{A_j}(X_{t_j})\, M_t] M_T;\, D\},$$

while the strong Markov property for X implies that the right side of (4.19) reduces to the same thing. Thus in order to complete the proof it suffices to show that $\mathscr{G} = \mathscr{F}$. Since this is of some independent interest we formulate it as a proposition.

(4.20) PROPOSITION. Let $\mathscr{G}$ be as above. Then $\mathscr{G} = \mathscr{F}$.

Proof. If $f \in \mathbf{C}$ and $\alpha > 0$, then

$$\int_0^\infty e^{-\alpha t} f(X_t)\, dt = \int_0^T e^{-\alpha t} f(X_t)\, dt + e^{-\alpha T} \int_0^\infty e^{-\alpha t} f(X_{t+T})\, dt,$$

and so $\int_0^\infty e^{-\alpha t} f(X_t)\, dt \in b\mathscr{G}$. It now follows from the Stone–Weierstrass approximation theorem that $\int_0^\infty g(t)\, e^{-t} f(X_t)\, dt \in b\mathscr{G}$ for any bounded continuous function g on $[0, \infty)$. Since $t \to f(X_t)$ is right continuous almost surely, for each $u \geq 0$ we can find a sequence $\{g_n\}$ in $\mathbf{C}([0, \infty))$ such that $\int g_n(t)\, e^{-t} f(X_t)\, dt \to f(X_u)$ almost surely. Consequently $\mathscr{F}^0 \subset \mathscr{G} \subset \mathscr{F}$, and since $\tilde{\mathscr{F}}^0 = \mathscr{F}$ we obtain Proposition 4.20.

We next complete the characterization of strong multiplicative functionals.

(4.21) PROPOSITION. *M is a SMF if and only if M is a RMF.*

Proof. In view of Theorem 4.12 we need only show that a SMF is regular. Let M be a SMF and define $R = \inf\{t\colon M_t = 0\}$. Clearly R is an $\{\mathscr{F}_t\}$ stopping time and the right continuity of M implies that $M_R = 0$ almost surely ($M_\infty = 0$). Let T be any $\{\mathscr{M}_t\}$ stopping time; then $H = T + R \circ \theta_T$ is also an $\{\mathscr{M}_t\}$ stopping time according to (8.7) of Chapter I. Using Corollary 4.16 we have for each x

$$E^x\{M_H;\, T < \zeta\} = E^x\{E^{X(T)}[M_R]\, M_T;\, T < \zeta\}$$
$$= 0.$$

Consequently $T + R \circ \theta_T = H \geq R$ almost surely on $\{T < \zeta\}$, and so

$$\begin{aligned} P^x[X_T \in E - E_M;\, M_T > 0] &= P^x[X_T \in E - E_M;\, T < R] \\ &\leq P^x[X_T \in E - E_M;\, R \circ \theta_T > 0] \\ &= E^x\{P^{X(T)}(R > 0);\, X_T \in E - E_M\} \\ &= 0, \end{aligned}$$

since $P^y(R > 0) = 0$ for all $y \in E - E_M$.

(4.22) PROPOSITION. Suppose that M is a right continuous MF and that E_M is nearly Borel measurable. If $T = T_{E-E_M}$, then M is regular if and only if $M_T = 0$ almost surely.

Proof. Let x be fixed. Then there exists an increasing sequence $\{K_n\}$ of compact subsets of $E - E_M$ such that $T_n = T_{K_n} \downarrow T$ almost surely P^x. Since $X(T_n) \in K_n \subset E - E_M$ almost surely on $\{T_n < \infty\}$ one has $M(T_n) = 0$ almost surely on $\{T_n < \infty\}$ if M is regular. But $M(T_n) \to M(T)$ almost surely P^x and hence $M(T) = 0$ almost surely P^x on $\{T < \infty\}$. Since x is arbitrary and $M_\infty = 0$ we obtain $M_T = 0$ almost surely. Conversely suppose $M_T = 0$ almost surely. If R is any $\{\mathscr{M}_t\}$ stopping time and $X_R \in E - E_M$, then either $R \geq T$ or $R = 0$. If $R \geq T$, then $M_R = 0$ almost surely. On the other hand

$$P^x(X_R \in E - E_M,\, M_R > 0;\, R = 0) \leq P^x(X_0 \in E - E_M,\, M_0 > 0)$$
$$= 0,$$

from the definition of E_M and the fact that X is normal. Consequently M is regular.

(4.23) REMARK. It is sometimes convenient to *normalize* a right continuous MF, $M = \{M_t\}$, as follows: Define $N_t(\omega) = 1$ for all t if $X_0(\omega) = \Delta$, $N_t(\omega) = M_t(\omega)$ if $t < \zeta(\omega)$ and $X_0(\omega) \in E$, and finally $N_t(\omega) = \inf_{s < \zeta(\omega)} M_s(\omega)$ if $t \geq \zeta(\omega)$ and $X_0(\omega) \in E$. Then $N_t = N_{\zeta-} = M_{\zeta-}$ almost surely on $\{\zeta \leq t\}$. The reader should check for himself that $N = \{N_t\}$ is a right continuous MF equivalent to M, although N_∞ need not be zero. We will say that a right continuous MF, M, is *normalized* provided $t \to M_t(\omega)$ and $t \to N_t(\omega)$ are identical functions of t almost surely where N is defined from M as above. The definitions of strong, regular, or exact MF do not depend on the convention $M_\infty = 0$, and, in fact, among right continuous MF's they depend only on the equivalence class to which the MF belongs. The only results above which appear possibly to depend on the convention $M_\infty = 0$ are (4.15), (4.16), and (4.17). However the reader can easily check that these results hold for normalized MF's. For example let us check (4.17) for a normalized MF, M. It follows from (4.17) that $M_{T+R} = M_T\, M_R(\theta_T)$ on $\{T + R < \infty\}$. Here and in the rest of this paragraph equality means equality almost surely. But writing this equality for $R_n = R \wedge n$ and letting $n \to \infty$ we obtain the desired equality on $\{T < \infty, R = \infty\}$. Finally since $\theta_\infty \omega = \omega_\Delta$, $M_t \circ \theta_T = 1$ for all t on $\{T = \infty\}$, and so the desired equality holds on this set also.

We close this section with several definitions that will prove useful in the sequel. Let T be a terminal time. Then, as we have observed several times, $M_t = I_{[0,T)}(t)$ is a right continuous MF. We say that T is an *exact* terminal time provided the corresponding MF is exact. The reader should check that a terminal time T is exact if and only if $P_T^\alpha U^\alpha f \in \mathscr{S}^\alpha$ for all $f \in \mathscr{E}_+^*$ and $\alpha > 0$. A terminal time T is a *strong* terminal time provided that $R + T \circ \theta_R = T$ almost surely on $\{R < T\}$ for all $\{\mathscr{F}_t\}$ stopping times R. Again the reader should check that, under the assumption that $\mathscr{M}_t = \mathscr{F}_t$ for all t, T is a strong terminal time if and only if the corresponding MF is a SMF (see Exercise 4.26).

Exercises

(4.24) Let X be a standard process such that there is a set $A \in \mathscr{E}$ which is nonempty and polar. Let $M_t(\omega) = 1$ for all t if $X_0(\omega) \notin A$ and let $M_t(\omega) = 0$ for all t if $X_0(\omega) \in A$. Show that $\{M_t\}$ is a RMF, and hence a SMF, but that it is not exact. What is the corresponding exactly subordinate resolvent $\{W^\alpha\}$?

(4.25) Let X be a standard process and let M be a right continuous MF. Let $\{Q_t\}$ be the semigroup generated by M and let $\{V^\alpha\}$ be the corresponding resolvent. (a) Show that if $r < s < t$, then $M_{t-r} \circ \theta_r \leq M_{t-s} \circ \theta_s$ almost surely. (b) Use (a) to show that if $t > 0$, $x \in E$, and $f \in \mathbf{B}$, then $\bar{Q}_t f(x) = \lim_{r \downarrow 0} P_r\, Q_{t-r}\, f(x)$ exists and that $\bar{Q}_t f \leq P_t f$ if $f \in \mathbf{B}_+$. (c) Show that if $t > 0$ and $s \geq 0$, then $\bar{Q}_{t+s} = \bar{Q}_t\, Q_s$. (d) Show that if f is a bounded nearly Borel measurable finely continuous function, then $t \to \bar{Q}_t f(x)$ is right continuous on $(0, \infty)$. (e) Let $\{W^\alpha\}$ be the exactly subordinate resolvent corresponding to $\{V^\alpha\}$. Show that $P_t V^\alpha f \to W^\alpha f$ as $t \to 0$ for each $f \in \mathbf{B}$. (f) Show $W^\alpha f = \int_0^\infty e^{-\alpha t}\, \bar{Q}_t f\, dt$ for each $f \in \mathbf{B}$. (g) Use (f) to show that $\{\bar{Q}_t;\, t > 0\}$ is a semigroup. (h) Show that there exists a MF, M′, of X such that $\bar{Q}_t f(x) = E^x\{f(X_t)\, M_t'\}$ for $t > 0$, $x \in E$, and $f \in \mathbf{B}$. (i) Show that $\bar{Q}_0 f(x) = \lim_{r \downarrow 0, r \in \mathbf{Q}} \bar{Q}_r f(x)$ for $f \in \mathbf{C}_0(E)$ defines a measure on $\mathscr{E}$. Prove that $h(x) = \lim_{t \downarrow 0} \bar{Q}_t 1(x)$ exists, that $\bar{Q}_0(x, \cdot) = h(x)\, \varepsilon_x$ for each x, and that $h(x)$ is either zero or one. Show that $\{\bar{Q}_t;\, t \geq 0\}$ is a semigroup subordinate to P_t and that $\{\bar{Q}_t;\, t \geq 0\}$ is generated by a right continuous MF, say $\bar{M}$, (j) Finally show that $E_M \subset E_{\bar{M}}$ and that if $x \in E_M$, then P^x almost surely $M_t = \bar{M}_t$ for all $t \geq 0$.

(4.26) Let X be a standard process with $\mathscr{M}_t = \mathscr{F}_t$ for all t and $\mathscr{M} = \mathscr{F}$. Show that T is a strong terminal time if and only if $M_t = I_{[0,T)}(t)$ is a SMF.

5. Excessive Functions

Throughout this section $X = (\Omega, \mathscr{M}, \mathscr{M}_t, X_t, \theta_t, P^x)$ will be a fixed standard process with state space $(E, \mathscr{E})$. The main purpose of this section is to study functions which are excessive for a subordinate semigroup $\{Q_t\}$; in particular to study the regularity properties of such a function composed with X. We will always assume that $\{Q_t\}$ is generated by a *right continuous* multiplicative functional M, which in view of Theorem 2.3 is equivalent to assuming that $t \to Q_t 1(x)$ is right continuous at $t = 0$. As in Section 4, $\{V^\alpha\}$ will denote the corresponding resolvent and we will write V for V^0. We will very shortly assume that M is exact; however, the basic definition and elementary properties do not depend on this assumption and so we will not introduce it for awhile.

(5.1) DEFINITION. Let $\alpha \geq 0$. A nonnegative function f in $\mathscr{E}^*$ is said to be α-excessive for (X, M) (or $\alpha-(X, M)$ excessive) provided that (i) $e^{-\alpha t}\, Q_t f \leq f$ for all $t \geq 0$ and (ii) $Q_t f \to f$ pointwise as $t \to 0$.

As usual when $\alpha = 0$ we will drop it entirely from our notation and terminology. Let $\mathscr{S}^\alpha(M)$ denote the class of all $\alpha - (X, M)$ excessive functions and let $Q_t^\alpha = e^{-\alpha t}\, Q_t$. Since Propositions 2.2, 2.3, and 2.6 of Chapter II depend only on the semigroup in question for their statement and proof they remain valid (both statements and proofs) provided one replaces P_t^α, U^α, and $\mathscr{S}^\alpha$ throughout by Q_t^α, V^α, and $\mathscr{S}^\alpha(M)$, respectively. We will use these results without special mention. Note that $f \in \mathscr{S}^\alpha(M)$ if and only if f is $\alpha - V$ excessive in the sense of Definition 4.5. Also note that if $\beta > 0$, then $M_t^\beta = e^{-\beta t} M_t$ is a MF and that $\mathscr{S}^\alpha(M^\beta) = \mathscr{S}^{\alpha+\beta}(M)$ for any $\alpha \geq 0$. Moreover the resolvent corresponding to M^β is given by $V^{\alpha+\beta}$. Using this device we can often assume $\alpha = 0$ and V bounded when discussing $\alpha-(X, M)$ excessive functions.

If T is an $\{\mathcal{M}_t\}$ stopping time we define $Q_T^\alpha f(x) = E^x\{e^{-\alpha T} f(X_T) M_T\}$ for $\alpha \geq 0$ and $f \in \mathbf{B} = b\mathscr{E}^*$ or $f \in \mathscr{E}_+^*$; in particular if T is an $\{\mathscr{F}_t\}$ stopping time, then Q_T^α is a bounded operator on $\mathbf{B}$ whose norm does not exceed one. If $A \in \mathscr{E}_\Delta^n$ we write Q_A^α for $Q_{T_A}^\alpha$.

We *assume now that M is exact*. There is no loss of generality in assuming that $M_t = 0$ whenever $t \geq \zeta$ since we can always replace M_t by the equivalent right continuous MF, $M_t I_{[0,\zeta)}(t)$. *These assumptions will be in force without special mention in the remainder of this section.* Consequently M is a SMF. By Theorem 4.17 if T is an $\{\mathscr{F}_t\}$ stopping time, then $M_{T+t} = M_T(M_t \circ \theta_T)$ almost surely for each t and hence this relation holds for all t almost surely (consider rational t and use the right continuity of M). It is now easy to check that

$$Q_T^\alpha V^\alpha g(x) = E^x \int_T^\infty e^{-\alpha u} g(X_u) M_u \, du$$

whenever T is an $\{\mathscr{F}_t\}$ stopping time and $g \in \mathscr{E}_+^*$. Hence Proposition 2.8 of Chapter II is valid if one replaces $\mathscr{S}^\alpha$, $\{\mathcal{M}_t\}$, and P_T^α by $\mathscr{S}^\alpha(M)$, $\{\mathscr{F}_t\}$, and Q_T^α respectively. Since $Q_t(x, \cdot) = 0$ if $x \notin E_M$ it is clear that any $\alpha-(X, M)$ excessive function ($\alpha \geq 0$) vanishes off E_M. One can now prove the analog of Proposition 2.10 of Chapter II. We will state this explicitly but omit the proof since it is exactly the same as that given in Chapter II.

(5.2) PROPOSITION. Suppose $f \in \mathscr{S}^\alpha(M)$, $A \in \mathscr{E}^n$, and $x \in (A^r \cap E_M)$. Then

$$\inf\{f(y): y \in A\} \leq f(x) \leq \sup\{f(y): y \in A\}.$$

Note that the right-hand inequality holds for any $x \in A^r$.

By the very definition of exactness $V^\alpha f$ is nearly Borel measurable and finely continuous for $\alpha > 0$ and $f \in \mathbf{B}_+$, and hence any $f \in \mathscr{S}^\alpha(M)$, $\alpha \geq 0$, is nearly Borel measurable and finely lower semicontinuous. Let $\varphi = V^1 1$; that is, $\varphi(x) = E^x \int_0^\infty e^{-t} M_t \, dt$. Since $M_t = 0$ if $t \geq \zeta$, $\varphi(\Delta) = 0$ and so $E_M = \{\varphi > 0\}$. Therefore E_M is nearly Borel measurable and finely open. Define $E_n = \{\varphi > 1/n\}$; then $E_M = \bigcup E_n$. We now introduce three stopping times associated with M which will be of importance in the sequel:

$$\begin{aligned} T_n &= T_{E_\Delta - E_n}, \qquad n \geq 1, \\ T &= \lim_n T_n, \\ R &= T_{E_\Delta - E_M}, \\ S &= \inf\{t: M_t = 0\}. \end{aligned}$$

Note that $P^x(T = 0) = P^x(S = 0) = 1$ if $x \notin E_M$.

(5.3) PROPOSITION:

(i) R, S, and T are $\{\mathscr{F}_t\}$ stopping times;

(ii) R, S, and T are strong terminal times (see the last paragraph of Section 4);

(iii) $S \leq T \leq R$ almost surely.

Proof. Since $\{T_n\}$ is an increasing sequence of hitting times, T is a terminal time which does not exceed R. If H is a hitting time of a nearly Borel set, then $t + H \circ \theta_t = H$ on $\{H > t\}$ without an exceptional set and so $Q + H \circ \theta_Q = H$ on $\{H > Q\}$ for any nonnegative function Q. Consequently R and T are strong terminal times. Clearly S is an $\{\mathscr{F}_t\}$ stopping time. Given an $\{\mathscr{F}_t\}$ stopping time Q one has $M_{u+Q} = M_Q(M_u \circ \theta_Q)$ for all $u \geq 0$ almost surely, and if $Q < S$ then $M_Q > 0$. Thus almost surely on $\{Q < S\}$, $M_u \circ \theta_Q = 0$ if and only if $M_{u+Q} = 0$, and it is now easy to see that S is a strong terminal time. The only thing remaining to be checked is that $S \leq T$ almost surely. First note that $M_{t+T} \leq M_{t+T_n} = M_{T_n}(M_t \circ \theta_{T_n}) \leq M_t \circ \theta_{T_n}$ for all t almost surely. Therefore

$$\frac{1}{n} \geq E^x\{\varphi(X_{T_n});\ T_n < S\}$$

$$= E^x\left\{\int_0^\infty e^{-t}\,(M_t \circ \theta_{T_n})\,dt;\ T_n < S\right\}$$

$$\geq E^x\left\{\int_0^\infty e^{-t}\,M_{t+T}\,dt;\ T < S\right\},$$

and so $\int_0^\infty e^{-t} M_{t+T}\,dt = 0$ almost surely on $\{T < S\}$. But $M_T > 0$ if $T < S$ and consequently, by the right continuity of M, $\int_0^\infty e^{-t} M_{t+T}\,dt > 0$. Therefore $P^x(T < S) = 0$ for all x and this completes the proof of Proposition 5.3.

It is intuitively clear that if $\varphi(x) \geq a > 0$, then $t \to M_t$ must be bounded away from zero on some interval $[0, t_0]$ with positive P^x probability. In fact this is precisely the reason for introducing the sets E_n. The following lemma gives a precise meaning to this statement. The argument is one that is often used, and is generally omitted.

(5.4) LEMMA. There exist positive constants τ and γ independent of x such that $P^x(M_\tau > \gamma) \geq a/2$ whenever $\varphi(x) \geq a$.

Proof. If for a fixed x we let $G_t(\gamma) = P^x(M_t > \gamma)$, then for every positive τ and $\gamma < 1$

$$\varphi(x) = E^x \int_0^\infty e^{-t} M_t \, dt = \int_0^\tau e^{-t} E^x(M_t) \, dt + \int_\tau^\infty e^{-t} E^x(M_t) \, dt$$
$$\leq (1 - e^{-\tau}) + e^{-\tau} E^x(M_\tau),$$

while

$$E^x(M_\tau) = E^x\{M_\tau; M_\tau \leq \gamma\} + E^x\{M_\tau; M_\tau > \gamma\}$$
$$\leq \gamma[1 - G_\tau(\gamma)] + G_\tau(\gamma).$$

Thus whenever $\varphi(x) \geq a$

$$G_\tau(\gamma) \geq 1 - e^\tau(1 - a)/(1 - \gamma).$$

By choosing τ and γ small enough we may make the right side of this inequality exceed $a/2$, so the proof is complete.

(5.5) Remark. Let $\varphi^\beta(x) = E^x \int_0^\infty e^{-\beta t} M_t \, dt$ for $\beta > 0$. If $P^x(M_\tau \geq \gamma) \geq a/2$ then $\varphi^\beta(x) \geq a\{\gamma(1 - e^{-\beta\tau})/2\beta\}$. Consequently if T_n^β is the hitting time of $\{\varphi^\beta \leq 1/n\}$ and $T^\beta = \lim_n T_n^\beta$, then $T^\beta = T$, as defined just before (5.3). In particular if $M_t^\beta = e^{-\beta t} M_t$, then the stopping times S, T, and R are the same regardless of whether they are defined relative to M_t or to M_t^β.

Define $S_n = \inf\{t: M_t \leq 1/n\}$. Clearly $\{S_n\}$ is an increasing sequence of $\{\mathscr{F}_t\}$ stopping times and $S_n \leq S$ for all n. Moreover it is immediate that $S_n \uparrow S$. Note that if $M_{S-} = \lim_{t \uparrow S} M_t$, then $\{M_{S-} = 0\} = \{S_n < S \text{ for all } n\}$ and $\{M_{S-} > 0\} = \{S_n = S \text{ for some } n\}$. The next result shows that $T = S$ almost surely on $\{M_{S-} = 0\}$.

(5.6) Proposition. For all x, $P^x(S_n < S \text{ for all } n, S < T) = 0$.

Proof. Since $T_p \uparrow T$ it suffices to show that, for all $x \in E_M$, $P^x(S_n < S$ for all n, $S < T_p) = 0$ for all p. But if $0 < S_n < T_p$, then $\varphi(X_{S_n}) > 1/p$ and consequently by Lemma 5.4 there exist positive constants τ and γ such that $P^{X(S_n)}[M_\tau > \gamma] \geq 1/2p$. Thus given $x \in E_M$ and $n \geq 2$ one has

$$P^x(S_n < S, S_n < T_p) \leq 2p \, E^x\{P^{X(S_n)}(M_\tau > \gamma); S_n < S, S_n < T_p\}$$
$$\leq 2p \, P^x[M_\tau \circ \theta_{S_n} > \gamma; S_n < S].$$

But almost surely $M_{\tau + S_n} = M_{S_n}(M_\tau \circ \theta_{S_n})$ and $M_{S_n} > 0$ on $\{S_n < S\}$. Therefore

$$P^x(S_n < S, S_n < T_p) \leq 2p \, P^x(M_{\tau + S_n} > 0),$$

and this last probability approaches zero as $n \to \infty$ since $S_n \uparrow S$ and $\tau > 0$. Hence we obtain Proposition 5.6.

We come now to the main result of this section.

(5.7) THEOREM. Let f be $\alpha-(X, M)$ excessive. Then

(i) $\{e^{-\alpha t} f(X_t) M_t, \mathscr{F}_t, P^\mu\}$ is a right continuous supermartingale for any μ such that $\mu(f) < \infty$.

(ii) Almost surely $t \to f(X_t)$ is right continuous on $[0, S)$ and has left-hand limits on $[0, T)$. In addition $t \to f(X_t)$ has a left-hand limit at S if $M_{S-} > 0$.

(iii) $t \to f(X_t)$ is right continuous on $[0, R)$ almost surely P^x if $x \in E_M \cup (E_M^c)^r$.

(iv) Almost surely $t \to f(X_t)$ is finite on $[u, S)$ if $f(X_u) < \infty$.

Proof. First of all in light of (5.5) and the remarks following Definition 5.1 we may assume that $\alpha = 0$. Secondly in proving (ii) and (iii) we may assume, as explained in the proof of Theorem 2.12 of Chapter II, that f is bounded. Of course, the regularity assertions in (ii) and (iii) are relative to the topology of $[0, \infty]$. Considering (iii) first of all, if $x \in (E_M^c)^r$ then $P^x(R = 0) = 1$ and so there is nothing to prove. If $x \in E_M$ and $X_0(\omega) \in E_M$, then $t \to f[X_t(\omega)]$ fails to be right continuous at a point $t_0 < R(\omega)$ only if there exist a sequence $\{t_n\}$ decreasing to t_0 and a pair of intervals I_1, I_2 of the form $[-\infty, r)$, $(s, \infty]$ (perhaps not in that order) with r and s rational and $r < s$ such that $X_{t_0}(\omega) \in f^{-1}(I_1) \cap E_M$ and $X_{t_n}(\omega) \in f^{-1}(I_2) \cap E_M$ for each $n \geq 1$. The two sets $f^{-1}(I_1) \cap E_M$ and $f^{-1}(I_2) \cap E_M$ are disjoint nearly Borel sets, and Proposition 5.2 implies that they are finely open. The proof of (iii) may now be completed by appealing to the argument which finishes the proof of 4.8 of Chapter II.

Since $P^x(S = 0) = 1$ if $x \notin E_M$, $S \leq R$, and $M_t = 0$ if $t \geq S$, it follows that $t \to f(X_t) M_t$ is right continuous on $[0, \infty)$ almost surely. Moreover

$$E^x[M_{t+s} f(X_{t+s}) \mid \mathscr{F}_t] = M_t E^{X(t)}[M_s f(X_s)]$$
$$\leq M_t f(X_t),$$

and consequently (i) is established (we are assuming $\alpha = 0$). In proving (ii) we assume that f is bounded. Then for each x, $\{M_t f(X_t); \mathscr{F}_t, P^x\}$ is a right continuous bounded supermartingale, and so a standard supermartingale theorem ((1.5) of Chapter 0) implies that $t \to f(X_t) M_t$ has left-hand limits almost surely. Therefore $t \to f(X_t)$ is right continuous and has left-hand limits almost surely on $[0, S)$. If $M_{S-} = 0$ then, by Proposition 5.6, $S = T$ and (ii) is established in this case. But if $M_{S-} > 0$ then $t \to f(X_t)$ has a left limit at $t = S$. Now define $Q_0 = 0$ and $Q_{n+1} = Q_n + S \circ \theta_{Q_n}$ for $n \geq 0$. Clearly $\{Q_n\}$ is an increasing sequence of $\{\mathscr{F}_t\}$ stopping times. Let $Q = \lim Q_n$. It is an immediate consequence of Proposition 5.3 that $Q_{n+1} \leq T$ almost surely on $\{Q_n < T\}$. We will next show that almost surely $Q \geq T$.

Now to show that $Q \geq T$ almost surely it suffices to show that $P^x(Q < T_p) = 0$ for all p and $x \in E_M$. Let p be fixed; then according to Lemma 5.4 there exists $\tau > 0$ such that $P^x(S \geq \tau) \geq 1/2p$ whenever $\varphi(x) \geq p^{-1}$. If k is a positive integer we have for $x \in E_M$ and $n \geq 1$

$$\begin{aligned} P^x\{Q_{n+1} - Q_n \geq \tau; Q_n < T_p \wedge k\} &= E^x\{P^{X(Q_n)}(S \geq \tau); Q_n < T_p \wedge k\} \\ &\geq \frac{1}{2p} P^x(Q_n < T_p \wedge k). \end{aligned}$$

Let $\Lambda_n = \{Q_{n+1} - Q_n \geq \tau; Q_n < T_p \wedge k\}$; then

$$P^x(\Lambda_n) \leq P^x[Q_{n+1} - Q_n \geq \tau; Q < \infty] + P^x[Q_n < k; Q = \infty]$$

and consequently $P^x(\Lambda_n) \to 0$ as $n \to \infty$. But

$$P^x(Q < T_p \wedge k) \leq P^x(Q_n < T_p \wedge k)$$

for each n and so letting $n \to \infty$ and then $k \to \infty$ we obtain $P^x(Q < T_p) = 0$. Therefore $Q \geq T$ almost surely.

Suppose that $t \to f(X_t)$ has left-hand limits on $[0, Q_n)$ and that either $Q_n = T$ or $t \to f(X_t)$ has a left-hand limit at $t = Q_n$. Since $Q_1 = S$ we have already seen that this is indeed the case when $n = 1$. If $Q_n < T$, then $t \to f(X_t)$ has left-hand limits on $[0, Q_n + S \circ \theta_{Q_n})$. Moreover if $M_{S-} \circ \theta_{Q_n} = 0$ then $S \circ \theta_{Q_n} = T \circ \theta_{Q_n} = T - Q_n$ and so $Q_{n+1} = T$, while if $M_{S-} \circ \theta_{Q_n} > 0$ then $t \to f(X_t)$ has a left limit at $t = Q_n + S \circ \theta_{Q_n} = Q_{n+1}$. Combining this with the results of the previous paragraph clearly yields (ii).

To prove (iv) let $A = \{f < \infty\}$ and $B = \{f = \infty\}$. If $x \in B^r \cap E_M$, then Proposition 5.2 implies that $x \in B$. Since $f(x) \geq Q_B f(x)$ one sees, just as in the proof of Theorem 2.12c of Chapter II, that $P^x(T_B < S) = 0$ for all $x \in A \cap E_M$, and hence for all $x \in A$ since $P^x(S = 0) = 1$ if $x \notin E_M$. It now follows as in Chapter II that $X_t \in A$ almost surely on (T_A, S), and this plainly yields Conclusion (iv) of (5.7). Thus the proof of Theorem 5.7 is now complete.

(5.8) REMARK. It follows from this theorem (or Proposition 5.2) that the restriction of an $\alpha - (X, M)$ excessive function to E_M is finely continuous on E_M. Another consequence is the fact that the minimum of two $\alpha - (X, M)$ excessive functions is $\alpha - (X, M)$ excessive.

We will next give a few examples to illustrate some of the various possibilities. However, inasmuch as our main results are proved under the assumption that the MF is exact, let us first give a simple criterion for exactness.

(5.9) PROPOSITION. *Let M be a right continuous MF. Then M is exact provided that for each $u > 0$ and $x \notin E_M$ one has $\lim_{t \downarrow 0} E^x\{M_{u-t} \circ \theta_t\} = 0$.*

Proof. Let $\{W^\alpha\}$ be the exact resolvent constructed from $\{V^\alpha\}$ in Theorem 4.9. According to (4.9) and (4.10) in order to show that $V^\alpha = W^\alpha$ it suffices to show that $(\beta+1)U^{\beta+1}V^1 1(x) \to 0$ as $\beta \to \infty$ for each $x \in E - E_M$. But

$$(\beta+1)U^{\beta+1}\, V^1(x) = (\beta+1)E^x \int_0^\infty e^{-(\beta+1)t} \int_0^\infty e^{-u} M_u \circ \theta_t \, du\, dt$$

$$= (\beta+1) \int_0^\infty e^{-\beta t} \int_t^\infty e^{-u}\, E^x(M_{u-t} \circ \theta_t)\, du\, dt$$

which under the hypotheses of (5.9) approaches zero as $\beta \to \infty$ if $x \notin E_M$, establishing Proposition 5.9.

We begin with an example to show that $S < T < R$ is possible. Let $E = \{0\} \cup [1, \infty)$ and let X be uniform motion to the right at speed one if $X(0) \geq 1$, while X starting from 0 remains there for an exponential holding time after which it jumps to 1 and then moves to the right at speed one. Let Q be the first hitting time of $\{1, 2\}$ and $M_t = I_{[0,Q)}(t)$. Clearly $E_M = E$ so that $R = \infty$. On the other hand $S = Q$ and it is easy to see that T is the hitting time of $\{2\}$. Thus $S < T < R$ almost surely P^0.

Next consider the following example: X is uniform motion to the right on the real line. Let $h(0) = 0$ and $h(x) = |x|^{-1}$ if $x \neq 0$ and set $M_t = \exp(-\int_0^t h(X_s)\, ds)$.* This defines a right continuous MF of X, and obviously $E - E_M = \{0\}$. It is also easy to see that M is exact (use (5.9) for example). Finally one can check without difficulty that $R = T_{\{0\}}$ (the first hitting time of $\{0\}$), while $S = T = D_{\{0\}}$ (the first entry time of $\{0\}$). Note that this shows that S need not be an exact terminal time. If $f = I_{E_M}$ then f is (X, M) excessive, but $t \to f(X_t)$ is not right continuous at $t = 0$ almost surely P^0. This shows that one may not eliminate the exceptional set of x's in (5.7iii). However in this example $t \to f(X_t)$ is right continuous on the open interval $(0, R)$ almost surely, and the reader should have no trouble in proving that this is the case in general.

Finally we give an example to show that the assertion concerning left-hand limits in Theorem 5.7 is the best that one can do in general. Let $E = \{-1, -\frac{1}{2}, -\frac{1}{3}, \ldots\} \cup [0, \infty)$. Let $\{H_n\}$ be a sequence of independent exponential holding times such that $\sum E(H_n) < \infty$. We describe the process X as follows: if X starts at $-1/n$ it remains there for a time H_n after which it jumps to $-1/(n+1)$ and remains there a time H_{n+1} and so on until it reaches 0 after which it moves to the right at unit speed. If $X(0) \geq 0$ the process moves to the right at unit speed. Note $P^x\{T_{\{0\}} < \infty\} = 1$, for each $x < 0$. Let A_t be the number of jumps of $t \to X_t$ in the interval $[0, t]$ and let $M_t = \exp(-A_t)$. It is immediate that M is a right continuous multiplicative functional and that

* Of course, we set $M_0(\omega) = 0$ if $X_0(\omega) = 0$.

$E_M = E$ which implies that M is exact and $R = \infty$. Moreover it is easy to check that $S = T = T_{\{0\}}$. Let f be defined as follows: $f(x) = 0$ if $x \geq 0$, $f(-1/n) = 1$ if n is odd, and $f(-1/n) = 2$ if n is even. The reader should have no difficulty in verifying that f is (X, M) excessive. However it is clear that $t \to f(X_t)$ does not have a left limit at T when $X(0) < 0$, although this function is everywhere right continuous.

In a certain sense the preceding examples are artificial. However it seems to be difficult to formulate simple and reasonable hypotheses that rule out such examples.

Let us return for a moment to the canonical subprocess $\hat{X}$ corresponding to M. Of course, a function f is $\alpha - (X, M)$ excessive if and only if it is α-excessive relative to $\hat{X}$. Recall that $\hat{\zeta}(\omega, \lambda) = \zeta(\omega) \wedge \lambda$ and that $\gamma(\omega, \lambda) = \lambda$. Using the notation developed above and in Section 4 we see that

$$\hat{P}^x(S < \gamma) = E^x(M_S) = 0, \tag{5.10}$$

and so it follows from Theorem 5.7 that $t \to f(\hat{X}_t)$ has left-hand limits on $[0, \hat{\zeta})$ and is right continuous on $[0, \infty)$ almost surely whenever f is α-excessive for $\hat{X}$.

In order to have a useful theory of excessive functions ($\alpha = 0$) it is necessary to impose some finiteness condition on the potential operator. See, for example, (2.19) and (4.6) of Chapter II. We will now discuss some conditions under which $V(x, B)$ has appropriate finiteness properties. These conditions are most useful when U does not have reasonable finiteness properties; for example, when X is Brownian motion in one or two dimensions.

As in the previous paragraphs X is a standard process and M is an exact MF vanishing on $[\zeta, \infty]$. We begin with the following simple result.

(5.11) Proposition. Suppose that there exists a sequence $\{g_n\}$ in $\mathscr{E}_+^*$ such that, for each n and x, $V g_n(x) < \infty$, $Vg_1 \leq Vg_2 \leq \ldots$, and $V g_n(x) \uparrow \infty$ as $n \to \infty$ for each x in E_M. Then for each $f \in \mathscr{S}(M)$ there exists a sequence $\{f_n\}$ of *bounded* functions in $\mathscr{E}_+^*$ such that Vf_n is bounded for each n and $Vf_n \uparrow f$ as $n \to \infty$.

Proof. Let $h_n = \min(Vg_n, n, f)$. Then $h_n \in \mathscr{S}(M)$ and $Q_t h_n \to 0$ as $t \to \infty$ since $h_n \leq Vg_n < \infty$. If $f_n = n[h_n - Q_{1/n} h_n]$, then $Vf_n = n \int_0^{1/n} Q_u h_n \, du$ and the reader will easily check that $\{f_n\}$ has the desired properties.

We will now formulate (following Hunt [3]) some conditions which imply the hypothesis of Proposition 5.11. As in the paragraph above (5.3) let $\varphi(x) = V^1 1(x) = E^x \int_0^\infty e^{-t} M_t \, dt$. For $0 < \beta < 1$ define $J_\beta = \{\varphi > \beta\}$. (Note that $J_{1/n}$ is what we previously called E_n.) Each J_β is finely open and nearly Borel measurable; if $0 < \beta < \gamma < 1$ then $J_\gamma \subset J_\beta \subset E_M$, and $\bigcup J_\beta = E_M$.

(5.12) DEFINITION. A set $D \in \mathscr{E}^n$ is called *special* provided (i) D is finely open, (ii) D has compact closure in E, and (iii) D is contained in some J_β, $0 < \beta < 1$.

We next introduce two conditions on M.

(D) If D is special, then $M(t + T_D \circ \theta_t) \to 0$ almost surely as $t \to \infty$.

(E) If D is special, then $V(\cdot, D)$ is bounded.

(5.13) PROPOSITION. Under (D) or (E) the hypothesis of Proposition 5.11 is satisfied.

Proof. Assume (E). Let $\{G_n\}$ be an increasing sequence of open sets with compact closures in E whose union is E. Let $D_n = G_n \cap J_{1/n}$ and $g_n = nI_{D_n}$. Since each D_n is special and $\bigcup D_n = E_M$, this sequence $\{g_n\}$ has the required properties. Assume (D). If D is special and $\varphi_D(x) = E^x\{M(T_D)\}$, then one easily checks that $\varphi_D \in \mathscr{S}(M)$. Moreover $Q_t\, \varphi_D(x) = E^x\{M(t + T_D \circ \theta_t)\} \to 0$ as $t \to \infty$, and so $V[t^{-1}(\varphi_D - Q_t\, \varphi_D)] \uparrow \varphi_D$ as $t \to 0$. But $\varphi_D = 1$ on D and E_M is a countable union of special sets, and so we can construct the desired sequence $\{g_n\}$.

Note that (D) holds whenever $M_t \to 0$ almost surely as $t \to \infty$. In particular this is the case if almost surely ζ is finite. In order to give some conditions under which (E) holds we prepare some lemmas.

(5.14) LEMMA. If $\beta < 1$, then there exist $\tau > 0$ and $\gamma < 1$ such that $E^x(M_\tau) < \gamma$ for all $x \notin J_\beta$. Moreover for all $x \in E_\Delta$ and $n \geq 1$, $E^x\{M(n\tau);\, n\tau < T_{J_\beta}\} \leq \gamma^{n-1}$.

Proof. If $x \notin J_\beta$, then for any $\tau > 0$ we have

$$\beta \geq \varphi(x) \geq E^x \int_0^\tau e^{-t}\, M_t\, dt \geq (1 - e^{-\tau})\, E^x(M_\tau).$$

Choosing τ large enough that $\beta(1 - e^{-\tau})^{-1} < 1$ proves the first assertion. For the second let $T_\beta = T_{J_\beta}$. Then for any $n \geq 1$ and x we have

$$\begin{aligned} E^x\{M[(n+1)\tau];\, (n+1)\tau < T_\beta\} &\leq E^x\{M_{n\tau}(M_\tau \circ \theta_{n\tau});\, n\tau < T_\beta\} \\ &= E^x\{E^{X(n\tau)}(M_\tau)\, M_{n\tau};\, n\tau < T_\beta\} \\ &\leq \gamma\, E^x\{M_{n\tau};\, n\tau < T_\beta\}, \end{aligned}$$

since $X(n\tau) \notin J_\beta$ if $0 < n\tau < T_\beta$. This establishes the second assertion.

(5.15) LEMMA. Let $T_\beta = T_{J_\beta}$. Then $E^x \int_0^{T_\beta} M_t\, dt$ is bounded in x. Moreover if $0 < \delta < \beta < 1$, then for all $x \notin J_\delta$ one has $E^x[M(T_\beta)] \le \delta/\beta$.

Proof. Let γ and τ be as in (5.14). Then

$$\begin{aligned} E^x \int_0^{T_\beta} M_t\,dt &\le \sum_{n=0}^{\infty} E^x\left\{\int_{n\tau}^{(n+1)\tau} M_t\,dt;\ n\tau < T_\beta\right\} \\ &\le \sum_{n=0}^{\infty} \tau\, E^x\{M(n\tau);\ n\tau < T_\beta\} \\ &\le \tau\left[1 + \sum_{n=1}^{\infty} \gamma^{n-1}\right] < \infty, \end{aligned}$$

proving the first assertion. If $x \notin J_\delta$, then

$$\begin{aligned} \delta \le \varphi(x) &= E^x \int_0^{T_\beta} e^{-t} M_t\,dt + E^x \int_{T_\beta}^{\infty} e^{-t} M_t\,dt \\ &\ge E^x\{M(T_\beta)\,[1 - e^{-T_\beta}]\} + E^x\{e^{-T_\beta} M_{T_\beta}\, \varphi(X_{T_\beta})\} \\ &\ge E^x\{M(T_\beta)\,[1 - (1-\beta)\, e^{-T_\beta}]\} \\ &\ge \beta\, E^x\{M(T_\beta)\}, \end{aligned}$$

establishing the second assertion.

(5.16) PROPOSITION. If $\delta < 1$, then $V(x, J_\delta^c)$ is a bounded function of x.

Proof. Pick β such that $\delta < \beta < 1$. Let $T = T_{J_\beta}$ and $R = T_{J_\delta^c}$. Define $T_0 = 0$ and

$$\begin{aligned} T_{2n+1} &= T_{2n} + T \circ \theta_{T_{2n}}, \\ T_{2n+2} &= T_{2n+1} + R \circ \theta_{T_{2n+1}}, \end{aligned}$$

for $n \ge 0$. It follows from (5.15) that, for $n \ge 1$,

$$E^x\{M(T_{2n+2})\} \le (\delta/\beta) E^x\{M(T_{2n})\},$$

and consequently $M(T_{2n}) \to 0$ as $n \to \infty$ almost surely. Therefore $\lim_n T_{2n} \ge S$ almost surely. Also since X_t is not in J_δ^c for $t \in (T_{2n-1}, T_{2n})$ we have, letting A be a bound for $E^x \int_0^T M_t\,dt$,

$$\begin{aligned} V(x, J_\delta^c) &\le \sum_{n=0}^{\infty} E^x \int_{T_{2n}}^{T_{2n+1}} M_t\,dt \\ &= \sum_{n=0}^{\infty} E^x\left\{M(T_{2n})\, E^{X(T_{2n})}\left[\int_0^T M_t\,dt\right]\right\} \\ &\le A\left[1 + \sum_{n=1}^{\infty} (\delta/\beta)^{n-1}\right]. \end{aligned}$$

This proves (5.16).

The next result is an immediate consequence of (5.16).

(5.17) COROLLARY. If for each special set D there exists $\beta < 1$ such that $D \cap J_\beta$ is empty, then (E) holds. In particular (E) holds if J_β is empty for some $\beta < 1$, or if each compact subset of E is disjoint from some J_β.

We close this section with a result that is applicable in many situations.

(5.18) THEOREM. Assume that (i) functions in $\mathscr{S}^\alpha$ are lower semicontinuous, (ii) the process X is such that either $U I_B(x) = 0$ for all $x \in E$ or $\Phi_B(x) > 0$ for all $x \in E$ whenever $B \in \mathscr{E}^n$, and (iii) M is such that $\varphi(x) < 1$ for some x in E. Then (E) holds.

Proof. Choose $\delta < 1$ so that $B = \{\varphi < \delta\} \cap E$ is not empty. Consequently $\Phi_B(x) = P^x(T_B < \infty) > 0$ for all $x \in E$, and so Φ_B^1 is strictly positive and lower semicontinuous. Therefore given a compact subset K of E there exists $\eta > 0$ such that $E^x(e^{-T_B}) = \Phi_B^1(x) \geq \eta$ for all x in K. Thus for such an x

$$\begin{aligned}\varphi(x) &= E^x \int_0^{T_B} e^{-t} M_t\, dt + E^x\{e^{-T_B} M(T_B)\, \varphi(X_{T_B})\} \\ &\leq E^x(1 - e^{-T_B}) + \delta E^x(e^{-T_B}) \\ &\leq 1 - \eta(1 - \delta),\end{aligned}$$

and so K is disjoint from some J_β. This establishes Theorem 5.18.

Exercises

(5.19) The following three statements are the analogs of Theorems 5.1, 5.9, and 5.11 of Chapter II. Prove each statement by adapting the proof of the corresponding result in Chapter II. In each case X is a given standard process, M an exact MF vanishing on $[\zeta, \infty]$, and f is a nonnegative function in $\mathscr{E}^n$ which vanishes off E_M.

(i) If $f \geq Q_K^\alpha f$ for all compact subsets K of E_M, then $\beta V^{\beta+\alpha} f \leq f$ for all $\beta > 0$.

(ii) Let $T_n = \inf\{t : d(X_t, X_0) > 1/n\}$ where d is a metric for E. Suppose each point of E_M is instantaneous for X and that f is finely continuous on E_M. If for each compact $K \subset E_M$ there exists an integer N such that $f(x) \geq Q_{T_n}^\alpha f(x)$ for all $x \in K$ and $n \geq N$, then $f \in \mathscr{S}^\alpha(M)$.

(iii) Let T_n be as in (ii) and assume that the restriction of f to E_M is lower semicontinuous. Suppose that for each $x \in E_M$, $f(x) \geq Q_{T_n}^\alpha f(x)$ for a sequence of values of n approaching infinity. Then $f \in \mathscr{S}^\alpha(M)$.

(5.20) Prove that if X is a standard process and M an exact MF, then for each $u > 0$ and $x \notin E_M$, $E^x\{M_{u-t} \circ \theta_t\}$ approaches 0 as $t \to 0$. Use this and Proposition 5.9 to conclude that the product of two exact MF's is exact.

(5.21) Let X be a standard process and M an exact MF. Given $f \in \mathcal{S}(M)$ with $f < \infty$ define $Y = \lim_n f(X_{T_n}) M_{T_n}$ (the limit exists by Theorem 5.7, T_n is defined just above Proposition 5.3). Define $g(x) = E^x(Y)$. Show (a) $g \in \mathcal{S}(M)$, (b) $g \le f$, (c) $g = Q_K g$ whenever K is the complement of a special set. Also show that if f is bounded, then g is the largest function satisfying (a), (b), and (c).

6. A Theorem of Hunt

In this section we are going to give a characterization of $Q_D^\alpha u$ for $u \in \mathcal{S}^\alpha(M)$ and $D \in \mathcal{E}^n$ which exhibits the relationship between the operator Q_D^α and the operation of "balayage" or "reduite" in classical potential theory (see Brelot [1]). This characterization, which is due to Hunt [2], is one of the most important results in probabilistic potential theory. Before discussing Hunt's theorem it will be necessary to establish a very useful extension of Theorem 11.2 of Chapter I.

As in previous sections $X = (\Omega, \mathcal{M}, \mathcal{M}_t, X_t, \theta_t, P^x)$ will denote a fixed standard process with state space $(E, \mathcal{E})$. For simplicity we assume that $\mathcal{M} = \mathcal{F}$ and $\mathcal{M}_t = \mathcal{F}_t$ for all $t \ge 0$. Stopping times are then $\{\mathcal{F}_t\}$ stopping times. Here is the extension of Theorem 11.2 of Chapter I mentioned above.

(6.1) THEOREM. Let $B \in \mathcal{E}^n$ and let μ be a finite measure on $\mathcal{E}^*$ such that $\mu(B - B^r) = 0$. Then there exists a decreasing sequence $\{B_n\}$ of finely open nearly Borel subsets of E, each containing B, and such that $T_{B_n} \uparrow T_B$ almost surely P^μ.

(6.2) REMARK. Note that if X is quasi-left-continuous on $[0, \infty)$, then (11.3) of Chapter I is a sharper result than (6.1). The importance of (6.1) is that it is valid for *general* standard processes.

We will break up the proof of Theorem 6.1 into several steps which we will establish as lemmas or propositions. First note that it suffices to prove (6.1) in the case $\mu(B) = 0$. To see this suppose we have established (6.1) in this case and write $\mu = \mu_1 + \mu_2$ where μ_1 is the restriction of μ to $E - B$ and μ_2 is the restriction of μ to $B \cap B^r$. Then $\mu_1(B) = 0$ and so there exists $\{B_n\}$ with the desired properties such that $T_{B_n} \uparrow T_B$ almost surely P^{μ_1}. But $T_B = T_{B_n} = 0$ almost surely P^{μ_2} and so $T_{B_n} \uparrow T_B$ almost surely P^μ. Thus we will suppose that

$\mu(B) = 0$ and, in addition, we will assume, as we may, that μ is a probability measure.

Recall that $\Phi_A^\alpha(x) = E^x(e^{-\alpha T_A}) = E^x(e^{-\alpha T_A}; T_A < \zeta)$ for any $A \in \mathscr{E}^n$. The following notation will be used in our discussion without special mention. We let $\varphi_A = \Phi_A^1$ and $\Gamma(A) = \int \varphi_A(x)\, \mu(dx) = E^\mu(e^{-T_A})$ for any $A \in \mathscr{E}^n$. We also define $\varphi(x) = E^x(e^{-\zeta})$ for $x \in E$. Both φ_A and φ are considered as functions on E and are 1-excessive. Note that $U^1 1(x) = E^x \int_0^\zeta e^{-t}\, dt = 1 - \varphi(x)$ for $x \in E$ and so φ is actually Borel measurable.

(6.3) LEMMA. Let $\{A_n\}$ and $\{B_n\}$ be two sequences of sets in $\mathscr{E}^n$ such that, for every n, $A_n \subset B_n$. Let $A = \bigcup A_n$ and $B = \bigcup B_n$. Then for each $x \in E$, $\varphi_B(x) - \varphi_A(x) \leq \sum_n [\varphi_{B_n}(x) - \varphi_{A_n}(x)]$. In particular $\Gamma(B) - \Gamma(A) \leq \sum_n [\Gamma(B_n) - \Gamma(A_n)]$.

Proof. If $T_B(\omega) < t < T_A(\omega)$ then $T_{B_n}(\omega) < t < T_{A_n}(\omega)$ for some n. Therefore

$$\varphi_B(x) - \varphi_A(x) = E^x \int_{T_B}^{T_A} e^{-t}\, dt$$

$$\leq \sum_n E^x \int_{T_{B_n}}^{T_{A_n}} e^{-t}\, dt = \sum_n [\varphi_{B_n}(x) - \varphi_{A_n}(x)],$$

and this establishes (6.3).

(6.4) LEMMA. Let $\{T_n\}$ be an increasing sequence of stopping times with $\lim_n T_n \geq \zeta$ almost surely P^μ. Let $\Lambda = \{T_n < \zeta \text{ for all } n\}$. Then

$$P^\mu\{\lim_n \varphi(X_{T_n}) < 1, \Lambda, \zeta < \infty\} = 0.$$

Proof. Since $\{e^{-T_n}\varphi(X_{T_n}), \mathscr{F}_{T_n}, P^x\}$ is a supermartingale for each $x \in E$, there exists an $L \in \mathscr{F}$ with $0 \leq L \leq 1$ and such that $e^{-T_n}\varphi(X_{T_n}) \to e^{-\zeta}L$ almost surely. Now

$$E^\mu\{e^{-\zeta}; T_n < \zeta\} = E^\mu\{e^{-T_n} e^{-(\zeta - T_n)}; T_n < \zeta\}$$
$$= E^\mu\{e^{-T_n}\varphi(X_{T_n}); T_n < \zeta\},$$

and letting $n \to \infty$ we obtain $E^\mu\{e^{-\zeta}; \Lambda\} = E^\mu\{e^{-\zeta}L; \Lambda\}$. Consequently $P^\mu(L < 1, \Lambda, \zeta < \infty) = 0$ establishing (6.4).

(6.5) PROPOSITION. Let $D \in \mathscr{E}^n$ with $\mu(D) = 0$. Suppose that for some $a < 1$, $D \subset \{\varphi < a\}$, and that $P^\mu(T_D < \zeta) = 0$. Then there exists a decreasing sequence $\{D_n\}$ of finely open Borel subsets of E, each containing D, and such that $\Gamma(D_n) \to 0$.

Proof. By Theorem 11.2 of Chapter I there exists a decreasing sequence $\{G_n\}$ of open sets containing D such that $\lim T_{G_n} \geq \zeta$ almost surely P^μ. Let $D_n = G_n \cap \{\varphi < a\}$. (Each D_n is a Borel subset of E since φ is a Borel measurable function on E.) Now $D \subset D_n \subset G_n$ for each n and so $\lim T_{D_n} \geq \zeta$ almost surely P^μ. But $\varphi[X(T_{D_n})] \leq a$ almost surely on $\{T_{D_n} < \zeta\}$ and hence, by Lemma 6.4, $P^\mu\{T_{D_n} < \zeta$ for all $n, \zeta < \infty\} = 0$. If $T_{D_n} \geq \zeta$ then $T_{D_n} = \infty$ since $D_n \subset E$, and combining this with the preceding two sentences yields $\lim T_{D_n} = \infty$ almost surely P^μ. Hence $\Gamma(D_n) \to 0$.

(6.6) COROLLARY. Let $D \in \mathscr{E}^n$ with $\mu(D) = 0$. Suppose that $P^\mu(T_D < \zeta) = 0$. Then there exists a decreasing sequence $\{D_n\}$ of finely open Borel subsets of E, each containing D, such that $\Gamma(D_n) \to 0$.

Proof. Let $A_k = D \cap \{\varphi < 1 - 1/k\}$. Let $\varepsilon > 0$. By Proposition 6.5 there exists for each k a finely open Borel set C_k with $A_k \subset C_k \subset E$ and $\Gamma(C_k) < \varepsilon/2^k$. If $C = \bigcup C_k$, then since $D = \bigcup A_k$ it follows from Lemma 6.3 that $\Gamma(C) - \Gamma(D) < \varepsilon$. By hypothesis $\Gamma(D) = 0$ and so we obtain (6.6).

(6.7) PROPOSITION. Let $A \in \mathscr{E}^n$ and assume that $P_t^1 \varphi \to \varphi$ as $t \to 0$ uniformly on A. If $\mu(A) = 0$, then there exists a decreasing sequence $\{A_n\}$ of finely open Borel subsets of E, each containing A, such that $\Gamma(A_n) \to \Gamma(A)$.

Proof. Of course, $P_t^1 \varphi \to \varphi$ pointwise on E since φ is 1-excessive. By hypothesis $\mu(A) = 0$ and so there exists a sequence $\{G_n\}$ of open sets containing A such that $(T_{G_n} \wedge \zeta) \uparrow (T_A \wedge \zeta)$ almost surely P^μ. Given $\varepsilon > 0$ choose s with $0 < s < 1$ such that $\varphi(y) - P_s^1 \varphi(y) < \varepsilon$ for all $y \in A$. If $C = \{y : \varphi(y) - P_s^1 \varphi(y) < \varepsilon\}$, then C is finely open, Borel measurable, and $A \subset C \subset E$. ($\varphi \in \mathscr{E}$ implies $P_s^1 \varphi \in \mathscr{E}$.) Define $C_n = G_n \cap C$ so that each C_n is a finely open Borel set containing A and $(T_{C_n} \wedge \zeta) \uparrow (T_A \wedge \zeta)$ almost surely P^μ. Next observe that if z is such that $\varphi(z) - P_s^1 \varphi(z) \leq \varepsilon$, in particular if $z \in C^r$, then

$$\begin{aligned}\varepsilon \geq \varphi(z) - P_s^1 \varphi(z) &= E^z(e^{-\zeta}) - E^z(e^{-\zeta}; \zeta > s)\\ &= E^z(e^{-\zeta}; \zeta \leq s) \geq e^{-s} P^z(\zeta \leq s);\end{aligned}$$

that is, $P^z(\zeta \leq s) \leq e^s \varepsilon \leq e\varepsilon$. Now

$$\begin{aligned}\lim_n [\Gamma(C_n) - \Gamma(A)] &= \lim_n E^\mu\{e^{-T_{C_n}} - e^{-T_A}\}\\ &\leq P^\mu\left\{T_{C_n} < \zeta \text{ for all } n, \lim_n T_{C_n} = \zeta < \infty\right\},\end{aligned}$$

since $(T_{C_n} \wedge \zeta) \uparrow (T_A \wedge \zeta)$ almost surely P^μ and $T_A = \infty$ if $T_A \geq \zeta$. Denote this last displayed probability by δ. If $\delta > 0$, then for any δ' with $0 < \delta' < \delta$

there exists an n such that $\gamma_n = P^\mu\{\zeta - T_{C_n} \leq s; T_{C_n} < \zeta\} > \delta'$. But

$$\gamma_n = E^\mu\{P^{X(T_{C_n})}[\zeta \leq s]; T_{C_n} < \zeta\} \leq e\varepsilon$$

since $X(T_{C_n}) \in C_n^r \subset C^r$ almost surely if $T_{C_n} < \zeta$. Hence $\delta' \leq e\varepsilon$ and so $\delta \leq e\varepsilon$. This proves that $\lim_n[\Gamma(C_n) - \Gamma(A)] \leq e\varepsilon$. Thus for each k we can find a finely open Borel set D_k with $A \subset D_k \subset E$ and such that $\Gamma(D_k) - \Gamma(A) < 1/k$. This clearly yields the conclusion of (6.7).

The following corollary is obtained from (6.7) by an appeal to Lemma 6.3 in the same manner that Corollary 6.6 was obtained from (6.5).

(6.8) COROLLARY. Let $A \in \mathscr{E}^n$ with $\mu(A) = 0$. Suppose that $A = \bigcup A_n$ where, for each n, $P_t^1 \varphi \to \varphi$ uniformly on A_n. Then given $\varepsilon > 0$ there exists a finely open Borel set D with $A \subset D \subset E$ and such that $\Gamma(D) - \Gamma(A) < \varepsilon$.

(6.9) LEMMA. Let $D, C \in \mathscr{E}^n$ with $C \subset D$. Suppose $\eta \leq 1$ and that $\varphi_D(y) \leq \eta$ for all $y \in D$. If $\int \mu(dx)\, P_D^1(x, C) = E^\mu\{e^{-T_D} I_C(X_{T_D})\} = 0$, then $\Gamma(C) \leq \eta\,\Gamma(D)$.

Proof. Obviously $T_C \geq T_D$ and $P^\mu[X_{T_D} \in C] = 0$. Therefore $T_C = T_D + T_C \circ \theta_{T_D}$ almost surely P^μ. Now the hypothesis implies that $\varphi_D(y) \leq \eta$ for all $y \in D \cup D^r$ and so

$$\begin{aligned}\Gamma(C) &= E^\mu\{e^{-T_C}\} = E^\mu\{e^{-T_D}\, \varphi_C(X_{T_D})\} \\ &\leq E^\mu\{e^{-T_D}\, \varphi_D(X_{T_D})\} \leq \eta\, \Gamma(D).\end{aligned}$$

We come now to the key step in the proof of Theorem 6.1.

(6.10) PROPOSITION. Let $B \in \mathscr{E}^n$ and suppose that, for some $\eta < 1$, $\varphi_B(y) \leq \eta$ for every $y \in B$ (and hence also for every $y \in B^r$). If $\mu(B) = 0$ then for every $\varepsilon > 0$ there exists a finely open Borel set A with $B \subset A \subset E$ and such that $\Gamma(A) - \Gamma(B) < \varepsilon$.

Proof. Let $\varepsilon > 0$ be given. Let ν be the measure on $\mathscr{E}^*$ defined by $\nu(F) = \int \mu(dx)\, P_B^1(x, F)$ and let ν_1 be the restriction of ν to B. Since $\varphi - P_t^1 \varphi \to 0$ as $t \to 0$ we may, by Egorov's theorem, find an increasing sequence $\{B_n\}$ of subsets of B such that for each n, $B_n \in \mathscr{E}$, $\varphi - P_t^1 \varphi \to 0$ as $t \to 0$ uniformly on B_n, and $\nu_1(B - \bigcup B_n) = 0$. Let $M_1 = \bigcup B_n$ and $D_1 = B - \bigcup B_n$. Since $M_1 \subset B$, $\mu(M_1) = 0$ and so by (6.8) there exists a finely open Borel set A_1 such that $M_1 \subset A_1 \subset E$ and $\Gamma(A_1) - \Gamma(M_1) < \varepsilon/4$. Furthermore $\int \mu(dx)\, P_B^1(x, D_1) = \nu(D_1) = \nu_1(D_1) = 0$ and so, by Lemma 6.9, $\Gamma(D_1) \leq \eta\, \Gamma(B) \leq \eta$. Finally $D_1 \subset B$ and hence $\varphi_{D_1}(y) \leq \varphi_B(y) \leq \eta$ for all $y \in D_1$.

Next let ν_2 be the measure $F \to \int \mu(dx)\, P^1_{D_1}(x, F)$ restricted to D_1. According to the argument just given we can find a Borel measurable subset M_2 of D_1 such that if $D_2 = D_1 - M_2$, then (i) $\int \mu(dx)\, P^1_{D_1}(x, D_2) = 0$ and (ii) there is a finely open Borel set A_2 such that $M_2 \subset A_2 \subset E$ and $\Gamma(A_2) - \Gamma(M_2) < \varepsilon/8$. By Lemma 6.9, $\Gamma(D_2) \leq \eta\, \Gamma(D_1) \leq \eta^2$. Plainly we may continue this procedure: that is, we can find a sequence $\{M_n\}$ of Borel subsets of B and a sequence $\{D_n\}$ of nearly Borel subsets of B such that (a) for each n there is a finely open Borel set A_n with $M_n \subset A_n \subset E$ and such that $\Gamma(A_n) - \Gamma(M_n) < \varepsilon/2^{n+1}$, and (b) $D_n = B - \bigcup_{k=1}^n M_k$ and $\Gamma(D_n) \leq \eta^n$. Now let $D = \bigcap_n D_n$, $M = \bigcup M_n$, and $\tilde{A} = \bigcup A_n$. Then $\tilde{A}$ is a finely open Borel subset of E containing M and, by Lemma 6.3, $\Gamma(\tilde{A}) - \Gamma(M) < \varepsilon/2$. Moreover $\Gamma(D) = 0$ since $\eta < 1$, and so $P^\mu(T_D < \zeta) = 0$. Also $\mu(D) = 0$ since $D \subset B$. Thus by Corollary 6.6 there exists a finely open Borel set C with $D \subset C \subset E$ and such that $\Gamma(C) - \Gamma(D) < \varepsilon/2$. Let $A = \tilde{A} \cup C$. Then $A \supset B$, A is a finely open Borel subset of E, and $\Gamma(A) - \Gamma(B) < \varepsilon$ by Lemma 6.3.

(6.11) PROPOSITION. Let $B \in \mathscr{E}^n$ and suppose that $\mu(B) = 0$. Then for every $\varepsilon > 0$ there exists a finely open nearly Borel set A with $B \subset A \subset E$ and such that $\Gamma(A) - \Gamma(B) < \varepsilon$.

Proof. Let $\varepsilon > 0$ be given. Choose $\eta < 1$ such that $(1/\eta - 1) < \varepsilon/2$. Let $B_1 = \{\varphi_B > \eta\}$ and $B_2 = B \cap \{\varphi_B \leq \eta\}$. The set B_2 satisfies the hypotheses of (6.10) and so there is a finely open Borel set A_2 with $B_2 \subset A_2 \subset E$ such that $\Gamma(A_2) - \Gamma(B_2) < \varepsilon/2$. Let $A = A_2 \cup B_1$. Plainly A is a finely open nearly Borel set and $B \subset A \subset E$. Now

$$\begin{aligned}\Gamma(A) = E^\mu(e^{-T_A}) &= E^\mu\{\exp[-(T_{A_2} \wedge T_{B_1})]\}\\ &= E^\mu\{e^{-T_{B_1}}; T_{B_1} < T_{A_2}\} + E^\mu\{e^{-T_{A_2}}; T_{A_2} \leq T_{B_1}\}.\end{aligned}$$

Also $T_{B_1} + T_B \circ \theta_{T_{B_1}} \geq T_B$ and so

$$\begin{aligned}E^\mu\{e^{-T_{B_1}}; T_{B_1} < T_{A_2}\} &\leq (1/\eta)\, E^\mu\{e^{-T_{B_1}}\, \varphi_B(X_{T_{B_1}}); T_{B_1} < T_{A_2}\}\\ &\leq (1/\eta)\, E^\mu\{e^{-T_B}; T_{B_1} < T_{A_2}\}.\end{aligned}$$

In addition $B_2 \subset B$ and hence

$$\begin{aligned}E^\mu\{e^{-T_{A_2}} - e^{-T_B}; T_{A_2} \leq T_{B_1}\} &\leq E^\mu\{e^{-T_{A_2}} - e^{-T_{B_2}}\}\\ &= \Gamma(A_2) - \Gamma(B_2) < \varepsilon/2.\end{aligned}$$

Therefore $E^\mu\{e^{-T_{A_2}}; T_{A_2} \leq T_{B_1}\} \leq E^\mu\{e^{-T_B}; T_{A_2} \leq T_{B_1}\} + \varepsilon/2$ and upon combining these estimates we obtain

$$\Gamma(A) = E^\mu(e^{-T_A}) \leq E^\mu(e^{-T_B}) + (1/\eta - 1) + \varepsilon/2,$$

or $\Gamma(A) \leq \Gamma(B) + \varepsilon$ by the choice of η. Thus the proof of Proposition 6.11 is complete.

Now Theorem 6.1 is an immediate consequence of (6.11). Indeed, using (6.11), we can find a decreasing sequence $\{B_n\}$ of finely open nearly Borel subsets of E, each containing B, such that $\Gamma(B_n) \to \Gamma(B)$. Let $T = \lim T_{B_n}$. Then $T \leq T_B$ and $E^\mu(e^{-T}) = \lim_n \Gamma(B_n) = E^\mu(e^{-T_B})$. Consequently $T = T_B$ almost surely P^μ, and so finally the proof of Theorem 6.1 is complete.

We are now ready to formulate and prove Hunt's fundamental theorem. We assume that M is an exact MF of X and that $M_t = 0$ whenever $t \geq \zeta$. As in Section 5, Q_t and V^α denote the semigroup and resolvent associated with (X, M).

(6.12) THEOREM. Assume that there exists a sequence $\{h_n\}$ of nonnegative functions in $b\mathscr{E}^*$ such that Vh_n is bounded for each n and $Vh_n \uparrow \infty$ on E_M. Let $f \in \mathscr{S}(M)$ and $A \in \mathscr{E}^n$. Let $\mathscr{U} = \{u \in \mathscr{S}(M) : u \geq f$ on $A\}$ and let f_A denote the lower envelope of $\mathscr{U}$; i.e., $f_A = \inf\{u : u \in \mathscr{U}\}$. Then $Q_A f \leq f_A$ and $Q_A f(x) = f_A(x)$ except possibly for those points $x \in E_M \cap (A - A^r)$.

Once again we will break up the proof into a sequence of lemmas and propositions. First note that according to Proposition 5.13 the hypothesis of (6.12) is satisfied whenever condition (D) or (E) of Section 5 holds. (Take $f = \infty$ on E_M, $f = 0$ off E_M in the statement of 5.11.) Next observe that the hypothesis of (6.12) is equivalent to the assertion that there exists an $h \in b\mathscr{E}^*_+$ such that Vh is bounded and strictly positive on E_M. Indeed if such an h exists and $h_n = nh$, then $Vh_n \uparrow \infty$ on E_M. Conversely if the hypothesis of (6.12) is satisfied and if $h = \sum_n h_n(2^n b_n)^{-1}$ where $b_n = \sup(1, \|h_n\|, \|Vh_n\|)$, then h and Vh are bounded and $Vh > 0$ on E_M. This is often a useful way to formulate the hypothesis in (6.12). Finally note that if $\alpha > 0$ then the MF, $M_t^\alpha = e^{-\alpha t} M_t$, automatically satisfies the hypothesis of (6.12) according to Proposition 5.13. As a result we have the following corollary which we state explicitly for ease of future reference.

(6.13) COROLLARY. Let $\alpha > 0$, $f \in \mathscr{S}^\alpha(M)$, and $A \in \mathscr{E}^n$. Let $f_A^\alpha = \inf\{u : u \in \mathscr{S}^\alpha(M), u \geq f$ on $A\}$. Then $Q_A^\alpha f \leq f_A^\alpha$ and $Q_A^\alpha f(x) = f_A^\alpha(x)$ except possibly for those $x \in E_M \cap (A - A^r)$.

Obviously in proving (6.12) we may assume that $A \subset E_M$ since all the functions involved vanish off E_M. Our proof of (6.12) is essentially a repetition of the original proof in Hunt [2].

(6.14) PROPOSITION. Let f, A, and f_A be as in (6.12). Then $Q_A f \le f_A$ and $Q_A f(x) = f_A(x)$ if $x \in A^r$. Here it is not necessary to assume the special hypothesis contained in the first sentence of (6.12).

Proof. If $u \in \mathscr{S}(M)$ and $u \ge f$ on A, then $u \ge f$ on $(A \cup A^r) \cap E_M$ since u and f are finely continuous on E_M. Hence $u \ge f$ on $A \cup A^r$ since both u and f vanish off E_M. Therefore $Q_A f \le Q_A u \le u$, and so $Q_A f \le f_A$. If $x \in E_M \cap A^r$, then $Q_A f(x) = f(x)$. But $f \in \mathscr{U}$ and so $f_A \le f$. Hence $Q_A f(x) = f_A(x)$ if $x \in E_M \cap A^r$, and this yields (6.14) since both f_A and $Q_A f$ vanish off E_M.

The next result is very similar to (6.4).

(6.15) PROPOSITION. Let $A \in \mathscr{E}^n$ and let $\{T_n\}$ be an increasing sequence of stopping times with limit T. Let μ be an initial measure and suppose that $P^\mu(T \ne T_A) = 0$. Then $\Phi_A^\alpha(X_{T_n}) \to 1$ almost surely P^μ on $\{T_n < T_A$ for all $n, T_A < \infty\}$. As usual $\Phi_A^\alpha(x) = E^x\{e^{-\alpha T_A}; T_A < \zeta\}$.

Proof. Since $\{e^{-\alpha T_n}\Phi_A^\alpha(X_{T_n}), \mathscr{F}_{T_n}, P^x\}$ is a nonnegative supermartingale for each x, there exists an $L \in \mathscr{F}$ with $0 \le L \le 1$ such that $e^{-\alpha T_n}\Phi_A^\alpha(X_{T_n}) \to e^{-\alpha T_A} L$ almost surely. Now

$$E^\mu\{e^{-\alpha T_n}\Phi_A^\alpha(X_{T_n}); T_n < T_A\} = E^\mu\{e^{-\alpha T_A}; T_n < T_A\}$$

and letting $n \to \infty$ we obtain

$$E^\mu\{e^{-\alpha T_A} L; \Lambda\} = E^\mu\{e^{-\alpha T_A}; \Lambda\}$$

where $\Lambda = \{T_n < T_A$ for all $n\}$. It follows from this that $L = 1$ almost surely P^μ on $\Lambda \cap \{T_A < \infty\}$, and so one obtains (6.15).

(6.16) PROPOSITION. Let $A \in \mathscr{E}^n$ and let μ be an initial measure such that $\mu(A - A^r) = 0$. Then there exists a decreasing sequence $\{A_n\}$ of sets in $\mathscr{E}^n$, each containing A, and such that the following statements hold: (i) $\lim T_{A_n} = T_A$ almost surely P^μ, (ii) $T_{A_n} = T_A$ for sufficiently large n almost surely P^μ on $\{T_A < \infty\}$, and (iii) $A_n \subset A_n^r$ for each n.

Proof. For the purposes of this proof we write $\varphi_B = \Phi_B^1$ for any $B \in \mathscr{E}^n$. Assume first of all that there is an $\eta < 1$ such that $A \subset \{\varphi_A < \eta\}$. Then A^r is empty and so $\mu(A) = 0$. By Theorem 6.1 we can find a decreasing sequence $\{B_n\}$ of finely open nearly Borel sets containing A such that $T_{B_n} \uparrow T_A$ almost surely P^μ. Let $A_n = B_n \cap \{\varphi_A < \eta\}$. Then each A_n is finely open and $\{A_n\}$ has all the properties asserted in (6.16) except possibly (ii). To see that (ii) holds let $\Lambda = \{T_{A_n} < T_A$ for all $n, T_A < \infty\}$. Then $\varphi_A(X_{T_{A_n}}) \le \eta$ almost surely on Λ, while $\varphi_A(X_{T_{A_n}}) \to 1$ almost surely P^μ on Λ according to (6.15). Consequently $P^\mu(\Lambda) = 0$, so (6.16) is proved in this case, and one may even assert that each A_n is finely open in this case.

In the general case we write $A = \bigcup_{k=0}^{\infty} B_k$ where $B_0 = \{\varphi_A = 1\} \cap A$ and $B_k = \{\varphi_A \le 1 - 1/k\} \cap A$ for $k \ge 1$. For $k \ge 1$, $B_k \subset A - A^r$ and so $\mu(B_k) = 0$. Also $\varphi_{B_k} \le \varphi_A \le 1 - 1/k$ on B_k. Thus by what was proved above we can find for each $k \ge 1$ a decreasing sequence $\{B_k^n\}$ of finely open sets in $\mathscr{E}^n$, each containing B_k, and such that almost surely P^μ, $T_k^n \uparrow T_k$ and $T_k^n = T_k$ for large enough n on $\{T_k < \infty\}$. Here we have written T_k^n for the hitting time of B_k^n and T_k for the hitting time of B_k. For each pair of positive integers k and n we can find a positive integer $m(k, n)$ such that

$$P^\mu\{T_k^{m(k,n)} < T_k;\ T_k^{m(k,n)} < n\} \le 2^{-(n+k)},$$

and we may assume that $m(k, n)$ is increasing with n. Now define

$$A_n = B_0 \cup \left(\bigcup_{k \ge 1} B_k^{m(k,n)}\right).$$

Plainly $\{A_n\}$ is a decreasing sequence of sets in $\mathscr{E}^n$, each containing A. Moreover $T_{A_n} = \min(T_{B_0}, \inf_{k\ge 1} T_k^{m(k,n)})$, and since $T_A = \inf_{k \ge 0} T_{B_k}$ we see that if $T_{A_n} < T_A$ then $T_{A_n} < T_{B_0}$ and so $T_{A_n} = \inf_{k \ge 1} T_k^{m(k,n)}$. Consequently

$$P^\mu\{T_{A_n} < T_A;\ T_{A_n} < n\} \le \sum_{k=1}^{\infty} 2^{-(n+k)} = 2^{-n}.$$

Therefore almost surely P^μ for all large n either $T_{A_n} = T_A$ or $T_{A_n} \ge n$. Hence almost surely P^μ, $\lim T_{A_n} = T_A$ and $T_{A_n} = T_A$ for sufficiently large n on $\{T_A < \infty\}$. We must still show that for each n, $A_n \subset A_n^r$. If $x \in B_0$ then x is regular for A and hence also for A_n, $n \ge 1$. Each B_k^n is finely open and so if $x \in B_k^{m(k,n)} \subset A_n$, then $x \in A_n^r$. Thus $A_n \subset A_n^r$ for each n, completing the proof of (6.16).

In light of Proposition 6.14 the proof of Theorem 6.12 will be complete as soon as the following lemma is established.

(6.17) LEMMA. The hypotheses and notation are those of (6.12). Given $\varepsilon > 0$ and $x \in E_M - A$ there exists a $u \in \mathscr{S}(M)$ such that $u \ge f$ on A and $u(x) \le Q_A f(x) + \varepsilon$.

Proof. As mentioned above we may assume that $A \subset E_M$. As a first step we suppose that there exists an $h \in \mathscr{E}_+^*$ such that Vh is bounded and $f \le Vh$ on A. Let $\{A_n\}$ be the sequence of sets constructed in (6.16) relative to the given set A and with ε_x as initial measure. Define $B_n = A_n \cap \{f < Vh + \varepsilon/2\}$. Clearly $A \subset B_n \subset E_M$ and each point of B_n is regular for B_n. Also by (6.16), $T_{B_n} \uparrow T_A$ and $T_{B_n} = T_A$ for large enough n on $\{T_A < \infty\}$. Here and in the remainder of the proof of Lemma 6.17 we omit the phrase "almost surely P^x" in those places where it is clearly appropriate. Recall that $S = \inf\{t: M_t = 0\}$. We write T for T_A and T_n for T_{B_n}. If $T < S$ then $T_n < S$ and so $f(X_{T_n}) \leq Vh(X_{T_n}) + \varepsilon/2 \leq \|Vh\| + \varepsilon/2$. Thus by the bounded convergence theorem

$$E^x\{f(X_{T_n})\, M_{T_n}; T < S\} \to E^x\{f(X_T)\, M_T; T < S\}$$
$$= Q_A f(x).$$

By Theorem (5.7i), $\lim_{t\to\infty} M_t Vh(X_t) = L$ exists almost surely, and, since Vh is bounded,

$$E^y\{L\} = \lim_{t\to\infty} E^y\{M_t Vh(X_t)\} = \lim_{t\to\infty} E^y\left\{\int_t^\infty h(X_u)\, M_u\, du\right\} = 0$$

for any y. Thus $\lim_{t\to\infty} M_t Vh(X_t) = 0$ almost surely. Next observe that

$$E^x\{f(X_{T_n})\, M_{T_n}; T = S\} = E^x\{f(X_{T_n})\, M_{T_n}; T = S < \infty, T_n < S\}$$
$$+ E^x\{f(X_{T_n})\, M_{T_n}; T = S = \infty, T_n < S\}.$$

The integral over $\{T = S < \infty, T_n < S\}$ approaches zero as $n \to \infty$ since this set decreases to the empty set and the integrand is bounded. The integral over $\{T = S = \infty, T_n < S\}$ is dominated by

$$E^x\{Vh(X_{T_n})\, M_{T_n}; T_n < \infty; T = S = \infty\} + \varepsilon/2$$

and by the above remark this integral approaches zero as $n \to \infty$. Combining these estimates we have

$$\lim_n Q_{B_n} f(x) \leq Q_A f(x) + \varepsilon/2,$$

and so for a sufficiently large value of n, $Q_{B_n} f(x) \leq Q_A f(x) + \varepsilon$. But $B_n \subset B_r^n$ and $B_n \subset E_M$. Therefore $Q_{B_n} f = f$ on $B_n \supset A$. Thus $u = Q_{B_n} f$ has the desired properties and so (6.17) is proved under the assumption that f is dominated on A by a bounded potential.

In the next step we suppose that f is finite on A. If $Q_A f(x) = \infty$, then $u = f$ has the desired properties. Therefore we suppose that $Q_A f(x) < \infty$. By hypothesis there is a sequence $\{Vh_n\}$ of bounded potentials such that $Vh_n \uparrow \infty$ on E_M. Let $A_n = A \cap \{f \leq Vh_n\}$. Then $A = \bigcup A_n$. Also $f \leq Vh_n$ on A_n. Therefore, by the proof in the previous paragraph, we can find for each n a nearly Borel set $B_n \supset A_n$ such that $B_n \subset E_M$, $B_n \subset B_n^r$, and $Q_{B_n} f(x) - Q_{A_n} f(x) \leq \varepsilon/2^n$. Let $B = \bigcup B_n \subset E_M$. Clearly each point of B is regular for B and $B \supset A$. We claim that

$$\textbf{(6.18)} \qquad Q_B f(x) - Q_A f(x) \leq \sum_{n=1}^{\infty} Q_{B_n} f(x) - Q_{A_n} f(x).$$

Assuming (6.18) for the moment, it then follows that $Q_B f(x) \leq Q_A f(x) + \varepsilon$ and that $f(x) = Q_B f(x)$ on $B \supset A$. Thus $u = Q_B f$ has the desired properties. As to (6.18) it suffices to prove that

$$(6.19) \qquad Q_B^\alpha V^\alpha g(x) - Q_A^\alpha V^\alpha g(x) \leq \sum_{n=1}^{\infty} Q_{B_n}^\alpha V^\alpha g(x) - Q_{A_n}^\alpha V^\alpha g(x)$$

for each $\alpha > 0$ and $g \in b\mathscr{E}_+^*$ with $V^\alpha g$ bounded. The left side of (6.19) is equal to

$$E^x\left\{\int_{T_B}^{T_A} e^{-\alpha T} M_t\, g(X_t)\, dt\right\} \leq E^x\left\{\sum_{n=1}^{\infty} \int_{T_{B_n}}^{T_{A_n}} e^{-\alpha t} M_t\, g(X_t)\, dt\right\}$$

since for each t, $\{T_B < t < T_A\} \subset \bigcup_{n=1}^{\infty}\{T_{B_n} < t < T_{A_n}\}$. Thus (6.19) and hence (6.18) is established. Note that (6.18) is merely an extension of Lemma 6.3. We have now established (6.17) under the assumption that f is finite on A.

We will need the following lemma when discussing the general case of (6.17).

(6.20) LEMMA. Assume the hypothesis of (6.12). Let A be a nearly Borel subset of E_M and let $\varepsilon > 0$. If $x \in E_M - A$ and $Q_A 1(x) = 0$, then there exists a $u \in \mathscr{S}(M)$ such that $u = \infty$ on A and $u(x) < \varepsilon$.

Proof. By what we have already proved applied to the function $f = I_{E_M}$, there exists for each $n \geq 1$ a $u_n \in \mathscr{S}(M)$ with $u_n \geq 1$ on A and $u_n(x) < Q_A 1(x) + \varepsilon 2^{-n} = \varepsilon 2^{-n}$. Plainly $u = \sum u_n$ has the desired properties.

We are now ready to complete the proof of (6.17) in the general case. Once again we may assume that $Q_A f(x) < \infty$. Let $B = A \cap \{f < \infty\}$ and $C = A \cap \{f = \infty\}$. Note that $f(X_{T_C}) = \infty$ almost surely on $\{T_C < S\}$. Observe that

$$\infty > Q_A f(x) \geq Q_C f(x) = E^x\{f(X_{T_C})\, M_{T_C}\},$$

and consequently $Q_C 1(x) = E^x\{M_{T_C}\} = 0$. Now by Lemma 6.20 we can find a $v \in \mathscr{S}(M)$ such that $v = \infty$ on C and $v(x) < \varepsilon/2$. On the other hand f is finite on B and so by what we have already established there exists a $g \in \mathscr{S}(M)$ with $g \geq f$ on B and $g(x) \leq Q_B f(x) + \varepsilon/2$. Let $u = g + v$. Then $u \geq f$ on A and

$$u(x) = g(x) + v(x) \leq Q_B f(x) + \varepsilon \leq Q_A f(x) + \varepsilon.$$

Thus the proof of Lemma 6.17 and hence the proof of Theorem 6.12 is finished.

In Section 1 of Chapter V we will discuss various implications and refinements of Theorem 6.12 under an additional hypothesis.

Exercises

(6.21) Let X be a standard process such that (i) there exists a nonempty polar set and (ii) the only excessive functions are constants. Show that in this situation the conclusion of (6.12) is false (take $M_t = I_{[0,\zeta)}(t)$). Use (1.5b) and (4.22) of Chapter II to show that Brownian motion in $\mathbf{R}^2$ satisfies these conditions.

(6.22) (a) Let X satisfy the hypothesis of (6.12) with $M_t = I_{[0,\zeta)}(t)$. If μ is a finite measure on $\mathscr{E}^*$ show that μU determines μ. Here μU is the measure $A \to \int \mu(dx)\, U(x, A)$ on $\mathscr{E}^*$. (Hint: let $h \in b\mathscr{E}^*_+$ be such that Uh is strictly positive and bounded. Let $g = U^1 h$. Use the resolvent equation to show that Ug is bounded. Furthermore show that g is strictly positive. Let f be a continuous function with compact support and let $f_n = f \wedge (ng)$. Show that $\beta U^\beta f_n \to f_n$ boundedly as $\beta \to \infty$ and use this to show that $\mu(f_n)$ is determined by μU.) (b) Let X and X^* be standard processes with the same state space $(E, \mathscr{E})$ and assume that both X and X^* are as in Part (a). Assume further that they have the same class of excessive functions ($\alpha = 0$). If B is a Borel set, then use (6.12) to show that $P_B f = P_B^* f$ for any excessive function f. (This makes sense since $\mathscr{S}(X) = \mathscr{S}(X^*)$.) Combine this with (a) to show that the measures $P_B(x, \cdot)$ and $P_B^*(x, \cdot)$ are identical for each $x \in E$ and $B \in \mathscr{E}^n$.

IV

ADDITIVE FUNCTIONALS AND THEIR POTENTIALS

Let $X = (\Omega, \mathcal{M}, \mathcal{M}_t, X_t, \theta_t, P^x)$ be a standard process with state space $(E, \mathscr{E})$. Let f be a nonnegative bounded Borel measurable function on E and consider

$$A_t(\omega) = \int_0^t f[X_s(\omega)]\, ds.$$

Clearly $t \to A_t(\omega)$ is continuous, nondecreasing, and $A_0 = 0$. Moreover, for each t, $A_t \in \mathscr{F}_t$. Also observe that $t \to A_t$ is constant on the interval $[\zeta, \infty]$ since $X_t = \Delta$ on this interval and $f(\Delta) = 0$. In particular $A_\infty = \int_0^\zeta f(X_t)\, dt$. Finally an elementary calculation yields the following fundamental property of A:

(*)
$$A_{t+s} = A_t + A_s \circ \theta_t.$$

The potential of f is easily expressed in terms of A,

$$Uf(x) = E^x \int_0^\infty f(X_t)\, dt = E^x(A_\infty),$$

and, more generally,

$$U^\alpha f(x) = E^x \int_0^\infty e^{-\alpha t} f(X_t)\, dt = E^x \int_0^\infty e^{-\alpha t}\, dA_t.$$

In this chapter we are going to study functionals A_t which satisfy (*) and appropriate regularity conditions. These will be called additive functionals of X. In light of the above discussion the function $u_A(x) = E^x(A_\infty)$ is called the potential of A. The main purpose of this chapter is to show that a large class of excessive functions can be represented as the potentials of additive functionals. In later chapters we will investigate further properties of additive functionals. In particular we will see that there is a strong analogy between additive functionals in the present theory and measures in classical potential theory. We will actually discuss additive functionals of (X, M) for a suitable class of multiplicative functionals M. However, we will state our definitions and theorems in terms of the process X and the multiplicative functional M rather than make any explicit use of the canonical subprocess of Section 3 of Chapter III.

1. Additive Functionals

As in the introduction, $X = (\Omega, \mathcal{M}, \mathcal{M}_t, X_t, \theta_t, P^x)$ is a fixed standard process with state space $(E, \mathcal{E})$. Throughout this chapter $M = \{M_t;\ t \geq 0\}$ will denote a fixed multiplicative functional (MF) of X. We will assume in this chapter, without special mention, *that for all ω the mapping $t \to M_t(\omega)$ is right continuous and that $M_t(\omega) = 0$ if $t \geq \zeta(\omega)$*. The reader should recall that any right continuous MF is equivalent to one having these properties. The following notation will be used throughout this chapter:

$$S = \inf\{t\colon M_t = 0\}.$$

Clearly S is a terminal time, $S \leq \zeta$, and E_M consists of those x which are irregular for S. We come now to the basic definition of this chapter.

(1.1) Definition. A family $A = \{A_t\colon t \geq 0\}$ of functions from Ω to $[0, \infty]$ is called an *additive functional* (AF) *of* (X, M) provided the following three conditions are satisfied:

(a) almost surely the mapping $t \to A_t$ is nondecreasing, right continuous, continuous at $t = S$, and satisfies $A_0 = 0$;

(b) for each t, $A_t \in \mathcal{F}_t$;

(c) for each t and s, $A_{s+t} = A_t + M_t(A_s \circ \theta_t)$ almost surely.

Of course the exceptional set in (1.1c) depends, in general, on t and s. However, in view of the right continuity of A, for each fixed $t \geq 0$, (1.1c) holds for all s almost surely. Approximating S from above by countable valued stopping times, (1.1) implies that almost surely $A_t = A_S$ for all $t \geq S$. We will sometimes write $A(t, \omega)$ for $A_t(\omega)$ and $A(t)$ for A_t.

Obviously by redefining A_t on a set $\Gamma \in \mathcal{F}$ with $P^x(\Gamma) = 0$ for all x we may assume that the regularity properties in (1.1a) hold for *all* ω. This we will do without special mention. Suppose that A satisfies all of the conditions of Definition 1.1 except that $t \to A_t$ is *not* assumed to be continuous at $t = S$. If we assume that the other conditions in (1.1a) hold for all ω, as we may, then $A_{S-} = \lim_{t \uparrow S} A_t$ exists. It is easy to check, using the fact that S is a terminal time, that if we redefine A_t so that $A_t = A_{S-}$ for $t \geq S$, then (1.1b) and (1.1c) remain valid. In particular the redefined functional satisfies all the conditions of Definition 1.1.

We define $A_\infty(\omega) = \lim_{t\to\infty} A_t(\omega)$. We may assume that this limit always exists in $[0, \infty]$. Clearly $A_\infty = A_S = A_{S-}$.

In case T is a terminal time such that $T \leq \zeta$ and $M_t = I_{[0,T)}(t)$, an AF of (X, M) will be called also an AF of (X, T). We will call an AF of (X, ζ) simply an AF of X. The reader should check that in this case (1.1c) reduces to $(*)$ of the introduction. At first reading it might be reasonable to consider only additive functionals of X. This will eliminate some technical difficulties and is easily accomplished by replacing M_t by 1 and S by ζ throughout.

(1.2) DEFINITION. Two additive functionals A and B of (X, M) are *equivalent* provided that, for each $t \geq 0$, $A_t = B_t$ almost surely on $\{S > t\}$.

In view of the right continuity of AF's, this is equivalent to the statement that almost surely $t \to A_t(\omega)$ and $t \to B_t(\omega)$ are identical functions on $[0, S(\omega))$. Clearly one may replace $[0, S(\omega))$ by $[0, \infty]$ in the preceding sentence. *Equality between AF's will always be understood to mean equivalence.*

It will be convenient for later work to introduce a stronger condition than (1.1c).

(1.3) DEFINITION. An AF, A, of (X, M) is *perfect* provided there exists a set Λ in $\mathscr{F}$ with $P^x(\Lambda) = 1$ for all x and such that $A_{t+s}(\omega) = A_t(\omega) + M_t(\omega)\,[A_s(\theta_t\omega)]$ for all t and s whenever $\omega \in \Lambda$.

Similarly we will say that an MF, M, of X is *perfect* provided there exists a set $\Lambda \in \mathscr{F}$ with $P^x(\Lambda) = 1$ for all x and such that $M_{t+s}(\omega) = M_t(\omega)\,[M_s(\theta_t\omega)]$ for all t and s whenever $\omega \in \Lambda$. We will show in Chapter V, under a slight additional hypothesis, that if M is perfect then any *continuous* AF of (X, M) is equivalent to a perfect AF of (X, M). We remark that our use of the term "perfect" is different from that of Dynkin [2].

Let M and N be two MF's of X and let B be an AF of (X, N). Let $S = \inf\{t\colon M_t = 0\}$ and $R = \inf\{t\colon N_t = 0\}$. Suppose that $f \in \mathscr{E}_+^*$ and that $\alpha \geq 0$. Frequently we will need an integral of the form

$$\textbf{(1.4)} \qquad \int_{(0,t]} f(X_u)\, e^{-\alpha u}\, M_u\, dB_u = J(f, t),$$

and so we need some discussion of its meaning and properties. In the first place if ω is such that $u \to X_u(\omega)$ is right continuous then, by an argument we have used several times already, the mapping $u \to f(X_u(\omega))$ is $\mathscr{R}^*$ measurable. ($\mathscr{R}^*$ is the σ-algebra of universally measurable subsets of $[0, \infty]$.) Consequently the integral of $f(X_u)\, e^{-\alpha u} M_u$ against the measure induced by the non-decreasing function $u \to B_u$ is defined almost surely. This is the meaning of (1.4). Observe that dB_u attributes no mass to the interval $[R, \infty)$ according to (1.1a). Next we wish to show that $J(f, t) \in \mathscr{F}_t$. This is obvious if $t = 0$ and so we assume that $t > 0$. Suppose first that f is bounded and continuous and define

$$a(n, k; \omega) = \inf_{\frac{k}{2^n} < v \le \frac{k+1}{2^n}} f(X_v(\omega))\, e^{-\alpha v} M_v(\omega),$$

$$g_n(u, \omega) = a(n, k; \omega) \quad \text{if} \quad \frac{k}{2^n} < u \le \frac{(k+1)}{2^n}$$

for $n \ge 1$, $k \ge 0$. Plainly $a(n, k) \in \mathscr{F}_{(k+1)/2^n}$. For a fixed value of n define K by $K/2^n < t \le (K+1)/2^n$. Then

$$\text{(1.5)} \qquad \int_{(0,t]} g_n(u)\, dB_u = \sum_{k<K} a(n, k)\,(B_{(k+1)/2^n} - B_{k/2^n}) + a(n, K)\,(B_t - B_{K/2^n}).$$

But the right side of (1.5) is in $\mathscr{F}_{t+2^{-n}}$, and as $n \to \infty$ the left side of (1.5) increases to $J(f, t)$. Thus the right continuity of the family $\{\mathscr{F}_t\}$ implies that $J(f, t) \in \mathscr{F}_t$. In (1.4) we may replace $(0, t]$ by $(0, \infty)$. Let us then write $J(f)$ in place of $J(f, t)$. Our discussion applies to $J(f)$ as well and shows that $J(f) \in \mathscr{F}$ if f is positive and continuous. In the extension of our results to an arbitrary $f \in \mathscr{E}_+^*$ we will simplify matters by assuming that $J(f, t)$ is bounded as a function of ω for each bounded continuous f. This assumption can be dispensed with by another passage to a limit. (Consider the integral over the interval $(0, t \wedge T_n)$ where $T_n = \inf\{u\colon B_u \ge n\}$ and then let $n \to \infty$.) It follows from this assumption in the usual way that $J(f, t) \in \mathscr{F}_t$ if $f \in \mathscr{E}_+$. Finally suppose $f \in b\mathscr{E}_+^*$. Given a finite measure μ on $\mathscr{E}$, choose $g, h \in b\mathscr{E}$ such that $g \le f \le h$ and $\nu(g) = \nu(h)$ where ν is the measure $\nu(A) = E^\mu\{J(I_A, t)\}$, $A \in \mathscr{E}$. Then $J(g, t) = J(h, t)$ almost surely P^μ and hence $J(f, t) \in \mathscr{F}_t$. The extension to arbitrary $f \in \mathscr{E}_+^*$ is by monotone convergence. All that we have said remains valid if t is replaced by an arbitrary $\{\mathscr{F}_t\}$ stopping time. We leave the details to the reader as Exercise (1.16).

(1.6) PROPOSITION. Let R be a terminal time with $S \le R \le \zeta$ almost surely, let B be an AF of (X, R), and suppose that $f \in b\mathscr{E}_+^*$. Define

$$\text{(1.7)} \qquad A_t = \int_{(0,t]} f(X_u)\, M_u\, dB_u.$$

If a.s., $\{A_t < \infty,\ t < S\} \subset \{B_t < \infty\}$ then (1.7) defines an AF of (X, M).

Proof. The supplementary hypothesis certainly holds if f is bounded away from zero. The validity of Conditions (a) and (b) of Definition 1.1 is clear. In checking (c) we must consider

$$A(t+s) - A(t) = \int_{(t,t+s]} f(X_u)\, M_u \, dB_u .$$

By a change of variable this integral becomes

$$\int_{(0,s]} f(X_{u+t})\, M_{u+t}\, dB_{u+t} .$$

This equals 0 if $t \geq S$. Using the remark following Definition 1.1 and the fact that $R \geq S$ almost surely, this last expression equals

$$M_t \int_{(0,s]} f(X_u \circ \theta_t)\, M_u \circ \theta_t \, dB_u \circ \theta_t = M_t(A_s \circ \theta_t)$$

almost surely on $\{S > t\}$, completing the proof.

The hypothesis in Proposition 1.6 that f be bounded is only for the purpose of ensuring that $t \to A_t$ be right continuous. Of course it is not needed if $t \to A_t$ is known from other considerations to be right continuous, for example if A_t is finite for all t.

The most important special case of Proposition 1.6 and the remarks following its proof is this: suppose $f \in \mathscr{E}_+^*$ and

$$A_t = \int_0^t f(X_u)\, du = \int_0^{t \wedge \zeta} f(X_u)\, du. \tag{1.8}$$

Then $\{A_t\}$ is an AF of X, at least if A_t is finite for all t or even if almost surely $t \to A_t$ has the property that $A_{t+\varepsilon} = \infty$ for all $\varepsilon > 0$ implies $A_t = \infty$.

(1.9) PROPOSITION. Let A be an AF of (X, M) and define

$$B_t = \int_{(0,t]} (M_u)^{-1}\, dA_u \tag{1.10}$$

(where we set $(M_u)^{-1} = 0$ if $M_u = 0$). Then B is an AF of (X, S).

The proof is straightforward, and so we omit it. Of course, the functionals A and B are related by (1.7) when we take f there equal to 1 on E.

(1.11) DEFINITION. An AF, A, of (X, M) is called a *strong additive functional* (or is said to have the strong Markov property) if whenever T is an $\{\mathscr{F}_t\}$ stopping time and $R \in \mathscr{M}$, $R \geq 0$, then almost surely

(1.12) $$A_{T(\omega)+R(\omega)}(\omega) = A_{T(\omega)}(\omega) + M_{T(\omega)}(\omega)\,[A_{R(\omega)}(\theta_T\,\omega)].$$

Obviously any perfect AF is a strong AF. Usually when writing expressions such as (1.12) we will omit the ω's. For example, (1.12) will be written

$$A_{T+R} = A_T + M_T[A_R(\theta_T)].$$

Observe that $A_R \circ \theta_T = A_{R\circ\theta_T}(\theta_T) \neq A_R(\theta_T)$ in general. The next result gives a simple sufficient condition for an AF to be a strong AF.

(1.13) PROPOSITION. Let A be an AF of (X, M) and suppose that M is an SMF (strong multiplicative functional). Then A is a strong additive functional.

Proof. Define an AF, B, of (X, S) by (1.10). Since B_t and A_t are related by (1.7) (with $f = 1$) and M is an SMF by hypothesis, it is clear that we need show only that for every $\{\mathscr{F}_t\}$ stopping time T and $u \geq 0$

(1.14) $$B_{T+u} = B_T + I_{[0,S)}(T)\,(B_u \circ \theta_T)$$

almost surely. One easily checks that $N_t = e^{-B_t}I_{[0,S)}(t)$ defines an MF, N, of X. Obviously $E_N = E_M$ because $B_0 = 0$ and $t \to B_t$ is right continuous. According to Proposition 4.21 of Chapter III an MF is strong if and only if it is regular. Thus if Q is any $\{\mathscr{M}_t\}$ stopping time we have for each x

$$\begin{aligned} P^x[X_Q \in E - E_N,\, N_Q > 0] &\leq P^x[X_Q \in E - E_M,\, Q < S] \\ &= P^x[X_Q \in E - E_M,\, M_Q > 0] = 0, \end{aligned}$$

and so N is regular and hence strong. From this the validity of (1.14) is immediate.

Any AF of X has the strong Markov property because the relevant MF, $M_t = I_{[0,\zeta)}(t)$, is clearly an SMF. In particular (1.8) defines a strong *AF* of X. In actual fact (1.8) defines a perfect AF of X. Moreover we can say a bit more concerning the right continuity requirement in (1.8). Let A be defined by (1.8) and let $R = \inf\{t\colon A_t = \infty\}$. Let us suppose that no point is regular for R. Then

$$P^x(A_R < \infty,\, R < \zeta) \leq E^x\{P^{X(R)}(R = 0);\, R < \zeta\} = 0.$$

From this it follows that almost surely $t \to A_t$ is continuous (in the topology of the extended reals) throughout $[0, \infty]$.

We close this section by introducing the two classes of AF's which will prove to be of fundamental importance in the remainder of this book.

(1.15) DEFINITION. An AF, A, of (X, M) is said to be *continuous* (CAF) if almost surely $t \to A_t$ is continuous. An AF, A, of (X, M) is said to be *natural* (*NAF*) if almost surely $t \to A_t$ and $t \to X_t$ have no common discontinuities.

Obviously any CAF is an NAF. Since $t \to A_t$ is constant on $[S, \infty]$ and is continuous at S, the discontinuities of $t \to A_t$ can occur only at points in $(0, S)$. The importance of the class of NAF's will emerge in the next section.

Exercises

(1.16) The notation and assumptions are those of (1.4). If T is an $\{\mathscr{F}_t\}$ stopping time and $f \in \mathscr{E}_+^*$ show that $J(f, T) = \int_{(0,T]} f(X_u)\, e^{-\alpha u} M_u\, dB_u$ is $\mathscr{F}_T$ measurable.

(1.17) An $\{\mathscr{F}_t\}$ stopping time T is called a *perfect terminal time* provided there exists a set $\Lambda \in \mathscr{F}$ with $P^x(\Lambda) = 1$ for all x and such that, for each $\omega \in \Lambda$, $T(\theta_t \omega) = T(\omega) - t$ whenever $t < T(\omega)$. Let T be a perfect terminal time such that no point in E_Δ is regular for T. Let $f \in \mathscr{E}_+^*$. Define $T_0 = 0$ and $T_{n+1} = T_n + T \circ \theta_{T_n}$ for $n \geq 0$, and define $A_t = \sum f(X_{T_n})$ where the sum is over all $n \geq 1$ such that $T_n \leq t$ (if no such n exist then $A_t = 0$). Prove that A is a perfect AF of X. [Hint: show that if $\omega \in \Lambda$ and $T_k(\omega) \leq t < T_{k+1}(\omega)$, then $t + T_n(\theta_t \omega) = T_{n+k}(\omega)$ for $k \geq 0$ and $n \geq 1$.]

(1.18) The purpose of this exercise is to illustrate the relationship between additive functionals of (X, M) as defined in this section and additive functionals of the canonical subprocess $\hat{X}$ corresponding to M. In order to avoid technical details which may obscure the basic idea we impose rather strong conditions. Assume that $\mathscr{M} = \mathscr{F}$ and $\mathscr{M}_t = \mathscr{F}_t$ for all $t \geq 0$. Let M be an MF of X and let $\hat{X}$ be the corresponding canonical subprocess. We use the notation of Section 3, Chapter III, without special mention. Let $\hat{A}$ be an AF of $\hat{X}$ such that $\hat{A}_t \in \hat{\mathscr{M}}_t$ for each t. Of course $\hat{X}$ need not be a standard process but this is of no importance in the present discussion. By (3.20) of Chapter III for each t there exists $A_t^* \in \mathscr{M}_t = \mathscr{F}_t$ such that $\hat{A}_t(\omega, \lambda) = A_t^*(\omega)$ if $t < \lambda$. Show that $\hat{A}_t(\omega, \lambda) = A_{\lambda-}^*(\omega)$ if $t \geq \lambda$. Let $B_t(\omega) = A_t^*(\omega)$ if $t < S(\omega)$ and $B_t(\omega) = A_{S-}^*(\omega)$ if $t \geq S(\omega)$. Show that $B = \{B_t\}$ is an AF of (X, S). Let $A_t = \int_{(0,t]} M_u\, dB_u$ so that $A = \{A_t\}$ is an AF of (X, M). Show that $A_t = \int_{(0,t]} M_u\, dA_u^*$. If $\mathscr{M}^*$ is the σ-algebra defined in (3.6) of Chapter III, then show that $\hat{E}^x\{\hat{A}_t \mid \mathscr{M}^*\} = A_t \circ \pi$ for each x. In other words if we "integrate out" the auxiliary variable λ in $\hat{A}_t(\omega, \lambda)$ we obtain an AF of (X, M). Conversely if A is an AF of (X, M) and B is related to A by (1.10), then putting $\hat{A}_t(\omega, \lambda) = B_t(\omega)$ if $t < \zeta(\omega, \lambda) \leq \lambda$ and $\hat{A}_t(\hat{\omega}) = \hat{A}_{\zeta-}(\hat{\omega})$ if $t \geq \zeta$ defines an AF of $\hat{X}$.

2. Potentials of Additive Functionals

As in Section 1, $X = (\Omega, \mathcal{M}, \mathcal{M}_t, X_t, \theta_t, P^x)$ will denote a fixed standard process with state space $(E, \mathcal{E})$. *From now on we will assume that M is an SMF of X*; then, according to Proposition 1.13, any AF of (X, M) is a strong AF. As in Section 1, S will always denote $\inf\{t: M_t = 0\}$, which, in the present situation, is a strong terminal time. Of course we still assume, as in Section 1, that $M_t(\omega) = 0$ if $t \geq \zeta(\omega)$ and that $t \to M_t(\omega)$ is right continuous for all ω. Consequently $S \leq \zeta$.

(2.1) Definition. Let A be an *AF* of (X, M). Suppose $\alpha \geq 0$ and $f \in \mathcal{E}_+^*$. We define the *α-potential of f relative to A, $U_A^\alpha f$,* by

$$U_A^\alpha f(x) = E^x \int_0^\infty e^{-\alpha t} f(X_t)\, dA_t. \tag{2.2}$$

It is immediate from our discussion of the integral (1.4) that $U_A^\alpha : \mathcal{E}_+^* \to \mathcal{E}_+^*$. When $f = 1$ we write u_A^α for $U_A^\alpha 1$. The function u_A^α is called the *α-potential of A.* If u_A^α is finite then the operators U_A^α are given by finite measures, which we denote by $U_A^\alpha(x, \cdot)$ as usual. We recall some notation from Chapter III: namely,

$$Q_t^\alpha f(x) = e^{-\alpha t} E^x\{f(X_t)\, M_t\},$$

$$V^\alpha f(x) = \int_0^\infty Q_t^\alpha f(x)\, dt = E^x \int_0^\infty e^{-\alpha t} f(X_t)\, M_t\, dt,$$

$$Q_T^\alpha f(x) = E^x\{e^{-\alpha T} f(X_T)\, M_T\}$$

for a stopping time T, and $\mathcal{S}^\alpha(M)$ for the set of $\alpha - (X, M)$ excessive functions. The family of operators $\{U_A^\alpha\}$ is not, in general, a resolvent. However, the following relationship is somewhat analogous to the resolvent equation. We formulate it as a proposition but leave its proof to the reader as Exercise 2.18.

(2.3) Proposition. Let $f \in \mathcal{E}_+^*$ and let $\alpha, \beta \geq 0$. If $U_A^\alpha f(x)$ and $U_A^\beta f(x)$ are finite and almost surely $A_t < \infty$ on $\{t < S\}$, then

$$U_A^\alpha f(x) - U_A^\beta f(x) = (\beta - \alpha) V^\alpha U_A^\beta f(x) = (\beta - \alpha) V^\beta U_A^\alpha f(x).$$

(2.4) Proposition. If $f \in \mathcal{E}_+^*$ and a.s., $A_t < \infty$ on $\{t < S\}$, then

(i) $U_A^\alpha f \in \mathcal{S}^\alpha(M)$.
(ii) If T is an $\{\mathcal{F}_t\}$ stopping time, then

$$Q_T^\alpha U_A^\alpha f(x) = E^x \int_{(T,\infty]} e^{-\alpha t} f(X_t)\, dA(t).$$

(iii) If $D \in \mathscr{E}^n$ and f vanishes off D, then

$$U_A^\alpha f(x) = Q_D^\alpha U_A^\alpha f(x) + E^x\{e^{-\alpha T_D} f(X_{T_D})\,(A_{T_D} - A_{T_D-})\}.$$

(iv) If A is natural, D open, and f vanishes off some compact subset of D, then $U_A^\alpha f = Q_D^\alpha U_A^\alpha f$.

Proof. The first two assertions are routine calculations with which the reader is familiar by now. Assertion (iii) is an immediate consequence of (ii). If f vanishes off a compact subset of an open set D and if $f(X_{T_D}) > 0$, then $t \to X_t$ must have a discontinuity at T_D. Consequently if A is natural, then $t \to A_t$ is continuous at T_D and so (iv) follows from (iii).

(2.5) PROPOSITION. Let A be an AF of (X, M) such that, for a fixed $\alpha \geq 0$, u_A^α is finite. Then A is natural if and only if for every $f \in \mathbf{C}_K$ and open set G containing the support of f one has

$$Q_G^\alpha U_A^\alpha f = U_A^\alpha f. \tag{2.6}$$

Proof. The necessity of this condition was established in (2.4iv). If (2.6) holds for all continuous f with compact support in G, then it also holds for $f = I_G$ by montone convergence. Since each side of (2.6) is finite it follows from (2.4iii) with $f = I_G$ that almost surely on $\{T_G < \infty\}$ either $X(T_G) \notin G$ or $A(T_G) = A(T_G-)$. Now let B be the AF of (X, S) defined in (1.10). Plainly A and B have the same points of discontinuity and so it will suffice to show that B is natural. Given $\varepsilon > 0$ define

$$T_\varepsilon = \inf\{t\colon B_t - B_{t-} > \varepsilon\}.$$

A moment's thought shows that T_ε is an $\{\mathscr{F}_t\}$ stopping time, and $\{T_\varepsilon < \infty\} = \{T_\varepsilon < S\}$ since $t \to B_t$ is constant on $[S, \infty]$ and continuous at S. It is also easy to check that T_ε is a strong terminal time. Moreover $B(T_\varepsilon) - B(T_\varepsilon-) \geq \varepsilon$ if $T_\varepsilon < \infty$. Define $T_\varepsilon^1 = T_\varepsilon$ and, for $n \geq 1$, $T_\varepsilon^{n+1} = T_\varepsilon^n + T_\varepsilon \circ \theta_{T_\varepsilon^n}$. The discontinuities of $t \to B_t$ occur only at the points T_ε^n as ε ranges over the positive rationals, and so to prove that B is natural it suffices to prove that almost surely $t \to X_t$ is continuous at T_ε when $T_\varepsilon < S$.

Suppose that, for some $\varepsilon > 0$ and $x \in E$, $P^x\{t \to X_t$ is discontinuous at $T_\varepsilon;\, T_\varepsilon < S\} > 0$. Then since E has a countable base there will exist a nonnegative $h \in \mathbf{C}_K$ and positive numbers β and r such that

$$P^x\{h(X_t) < \beta \text{ for } t \in [r, T_\varepsilon);\ h(X_{T_\varepsilon}) > \beta;\ T_\varepsilon < S\} > 0. \tag{2.7}$$

If we denote by Λ the set whose probability is being computed in (2.7) and let $G = \{h > \beta\}$, then on Λ, $T_\varepsilon = r + T_G \circ \theta_r$. Since $t \to A_t$ and $t \to B_t$ have the same discontinuities we see that $P^x(\Lambda)$ does not exceed

$$P^x\{A(r + T_G \circ \theta_r) \neq A[(r + T_G \circ \theta_r)-]; X_{T_G} \circ \theta_r \in G; r + T_G \circ \theta_r < S\}$$
$$= E^x\{P^{X(r)}[A(T_G) \neq A(T_G-); X(T_G) \in G; T_G < S]; r < S\},$$

and this last expression is zero by the first part of the proof. Thus Proposition 2.5 is established.

The main result of this section is that a natural additive functional A of (X, M) is determined by its potential u_A^α provided that the potential is finite. The next few propositions lead up to this result.

(2.8) PROPOSITION. Let A be an AF of (X, M) and let B be related to A by (1.10). Suppose for a fixed $\alpha \geq 0$, u_A^α is everywhere finite. Then for every $f \in \mathscr{E}_+^*$.

$$U_A^\alpha V^\alpha f(x) = E^x \int_0^S B_t M_t e^{-\alpha t} f(X_t)\, dt. \tag{2.9}$$

Proof. It suffices to carry out the proof for $\alpha > 0$ and nonnegative continuous f with compact support, for each side of (2.9) is a measure in f and each side is right continuous in α. Assume f and α are so. The left side of (2.9) is

$$E^x \int_0^S e^{-\alpha t} \left\{E^{X(t)} \int_0^S e^{-\alpha u} f(X_u)\, M_u\, du\right\} dA_t.$$

We may assume that the expression in braces is continuous on the right in t. Indeed, let $W^\alpha f$ be defined as in (4.9) of Chapter III. The resolvent $\{W^\alpha\}$ is exactly subordinate to $\{U^\alpha\}$ and so $W^\alpha f$ is continuous in the fine topology. But $W^\alpha f(x) = V^\alpha f(x)$ for all $x \in E_M$, while for all $t < S$, $X_t \in E_M$. Thus the expression in braces is just $W^\alpha f(X_t)$ if $t < S$. This being so we have

(2.10) $U_A^\alpha V^\alpha f(x)$

$$= \lim_{n\to\infty} E^x \sum_{k \geq 0} V^\alpha f(X_{(k+1)/2^n})\, (A_{(k+1)/2^n} - A_{k/2^n})\, e^{-\alpha(k+1)/2^n}$$

$$= \lim_{n\to\infty} E^x \sum_{k \geq 0} (A_{(k+1)/2^n} - A_{k/2^n})[M_{(k+1)/2^n}]^{-1} \int_{(k+1)/2^n}^\infty e^{-\alpha u} f(X_u)\, M_u\, du$$

where indeterminate quotients 0/0 are set equal to 0. The last limit in (2.10) is just

$$E^x \int_0^S \left\{\int_t^S e^{-\alpha u} f(X_u)\, M_u\, du\right\} dB_t.$$

The equality (2.9) now follows from an integration by parts in this last expression.

(2.11) PROPOSITION. Let B^1 and B^2 be AF's of (X, S). Suppose that, for all $t \geq 0$, $x \in E$, and $f \in b\mathscr{E}_+^*$, $E^x\{f(X_t) B_t^1 M_t\} = E^x\{f(X_t) B_t^2 M_t\} < \infty$. Then B^1 and B^2 are equivalent.

Proof. A straightforward induction argument shows that, for $0 \leq t_1 < \ldots < t_n \leq t$ and $f_1, \ldots, f_n \in b\mathscr{E}_+^*$, $E^x(\prod_{j=1}^n f_j(X_{t_j}) B_t^i M_t)$ is independent of i. Consequently for every $\Gamma \in \mathscr{F}_t$, $E^x(B_t^i M_t; \Gamma)$ is independent of i and so almost surely $B_t^1 = B_t^2$ on $\{t < S\}$.

(2.12) PROPOSITION. Let A^1 and A^2 be AF's of (X, M). Suppose that, for some fixed $\alpha \geq 0$, $u_{A^1}^\alpha < \infty$, and that, for every nonnegative continuous f with compact support in E, $U_{A^1}^\alpha f = U_{A^2}^\alpha f$. Then A^1 and A^2 are equivalent.

Proof. It follows as usual that, for every $f \in \mathscr{E}_+^*$, $U_{A^i}^\alpha f$ is independent of i. Then from Proposition 2.3 it follows that, for every $\beta \geq \alpha$, $U_{A^i}^\beta f$ is independent of i, and hence so is $U_{A^i}^\beta V^\beta f$. According to Proposition 2.8 then for all $\beta \geq \alpha$

$$E^x \int_0^\infty B_t^i M_t f(X_t) e^{-\beta t} dt = \int_0^\infty e^{-\beta t} E^x(B_t^i M_t f(X_t)) dt$$

is independent of i, where B^i is related to A^i by (1.10). Now $B_t^i M_t \leq A_t^i$ and, by hypothesis, $\infty > u_{A^i}^\alpha(x) \geq e^{-\alpha t} E^x(A_t^i)$. Consequently if f is bounded and continuous, then $E^x(B_t^i M_t f(X_t))$ is right continuous in t. From this and the uniqueness theorem for Laplace transforms it follows that $E^x(B_t^i M_t f(X_t))$ is independent of i for f bounded and continuous. The equality extends immediately to $f \in b\mathscr{E}_+^*$ because of the finiteness estimate we have just made. Proposition 2.11 now implies that B^1 and B^2 are equivalent, and hence so are A^1 and A^2.

We are now able to state the main theorem of this section, a uniqueness theorem for additive functionals.

(2.13) THEOREM. Let A and B be natural additive functionals of (X, M) and suppose that, for some fixed $\alpha \geq 0$, $u_A^\alpha < \infty$. If $u_A^\alpha = u_B^\alpha$, then A and B are equivalent.

Proof. Because of (2.12) it will suffice to show that for every nonnegative f in $\mathbf{C}_K$ we have $U_A^\alpha f = U_B^\alpha f$. Given such an f and a number $\varepsilon > 0$ let us define

$$T_\varepsilon = \inf\{t : |f(X_t) - f(X_0)| \geq \varepsilon\}$$

and $T_0 = 0$, $T_{n+1} = T_n + T_\varepsilon \circ \theta_{T_n}$ for $n \geq 0$. If $T_{n+1} < \infty$ then $|f(X_{T_{n+1}}) - f(X_{T_n})| \geq \varepsilon$ by the right continuity of the paths and the fact that f is continuous. Because the paths have left limits it follows that $\lim T_n \geq \zeta$. Now $|f(X_t) - f(X_{T_n})| \leq \varepsilon$ for all t in $[T_n, T_{n+1})$. This inequality is valid also for $t = T_{n+1}$ if the path $t \to X_t$ is continuous there. If the path is not continuous at $T_{n+1} < \infty$ or if $T_{n+1} = \infty$, the measure dA_t puts no mass at T_{n+1} almost surely by the hypothesis that A is natural. So in any event almost surely

$$\left| \int_{(T_n, T_{n+1}]} e^{-\alpha t} f(X_t)\, dA_t - f(X_{T_n}) \int_{(T_n, T_{n+1}]} e^{-\alpha t}\, dA_t \right| \leq \varepsilon \int_{(T_n, T_{n+1}]} e^{-\alpha t}\, dA_t .$$

Consequently

$$U_A^\alpha f(x) = \sum E^x \left\{ f(X_{T_n}) \int_{(T_n, T_{n+1}]} e^{-\alpha t}\, dA_t \right\} + \varepsilon L \tag{2.14}$$

where $|L| \leq u_A^\alpha(x)$. A typical summand on the right of (2.14) is

$$\begin{aligned} E^x & \left\{ f(X_{T_n}) \int_{(0, T_\varepsilon \circ \theta_{T_n}]} e^{-\alpha(t+T_n)}\, dA_{t+T_n} \right\} \\ &= E^x \left\{ e^{-\alpha T_n} f(X_{T_n})\, M_{T_n}\, E^{X(T_n)} \left[\int_{(0, T_\varepsilon]} e^{-\alpha t}\, dA_t \right] \right\}. \end{aligned}$$

Now for each $y \in E$ we have

$$E^y \int_{(0, T_\varepsilon]} e^{-\alpha t}\, dA_t = u_A^\alpha(y) - Q_{T_\varepsilon}^\alpha\, u_A^\alpha(y). \tag{2.15}$$

But all of these calculations are also valid when A is replaced by B throughout, while by hypothesis the right side of (2.15) is unchanged if one replaces A by B. Therefore the infinite sum in (2.14) is also unchanged if one replaces A by B and since ε is arbitrary it follows that $U_A^\alpha f = U_B^\alpha f$. This completes the proof of Theorem 2.13.

The example in Exercise 2.16 shows that Theorem 2.13 is *not* valid without the assumption that both A and B are natural.

Exercises

(2.16) Let X be the Poisson process on the integers with parameter 1, that is, the process constructed in Section 12 of Chapter I with $E = \{\ldots, -1, 0, 1, \ldots\}$, $\lambda(x) = 1$, and $Q(x, \cdot) = \varepsilon_{x+1}$ for all $x \in E$. Let A_t be the number of jumps of $u \to X_u$ in the interval $[0, t]$ and let $B_t = t$. Clearly A and B are additive functionals of X. Show that $u_A^\alpha = u_B^\alpha = \alpha^{-1}$ for all $\alpha > 0$.

(2.17) If A is an AF of (X, M) and $f \in \mathscr{E}_+^*$ is bounded away from zero, then $U_A^\alpha f \in \mathscr{S}^\alpha(M)$. In particular u_A^α is $\alpha - (X, M)$ excessive. Show by an example that $U_A^\alpha f$ need not be in $\mathscr{S}^\alpha(M)$ even if it is bounded ($f \in \mathscr{E}_+^*$).

(2.18) Prove Proposition 2.3.

(2.19) Let A be an AF of (X, M). The function $f_t(x) = E^x(A_t)$ is called the *characteristic* of A.

(i) Show that $f_t(x)$ has the following properties:

(a) $f_t \in \mathscr{E}_+^*$ for each $t \geq 0$,

(b) $t \to f_t(x)$ is nondecreasing, $f_0(x) = 0$, and $t \to f_t(x)$ is right continuous on $[0, u)$ provided $f_u(x) < \infty$,

(c) $f_{t+s}(x) = f_t(x) + Q_t\, f_s(x)$ for all t and s.

(ii) Show that $u_A^\alpha(x) = \alpha \int_0^\infty e^{-\alpha t} f_t(x)\, dt$ for each $\alpha \geq 0$.

(iii) If E_M is nearly Borel measurable (since we are not assuming that M is exact this is not automatic) show that, for each $t \geq 0$, f_t is nearly Borel measurable and finely lower semicontinuous on E_M and even finely continuous on E_M if f_u is finite for some $u > t$. [Hint: for a fixed t let $Y_n = A_t \wedge n$. If $g_n(x) = E^x(Y_n)$ show that $\alpha U^\alpha g_n \to g_n$ on E_M and conclude from this that f_t is nearly Borel measurable. Adapt the argument of (4.14), Chapter II, to show that g_n is finely continuous at each point of E_M and that if f_u is finite for some $u > t$, then f_t is finely upper semicontinuous at each point of E_M.]

(iv) Suppose that $f_t(x)$ is finite for all t and x and that g is a bounded continuous function on E. If A is natural show that for each t and x

$$E^x \int_{(0,t]} g(X_u)\, dA_u = \lim_n \sum_{k=0}^{n-1} Q_{kt/n}(g f_{t/n}).$$

(2.20) Let A and B be NAF's of (X, M) and assume that $E_M \in \mathscr{E}^n$. If A and B have the same finite characteristic, than A and B are equivalent. [Hint: to begin assume that A is an AF of (X, M) with $f_t(x) = E^x(A_t)$ finite, and hence $f_t \in \mathscr{E}_+^n$ and is finely continuous on E_M for each $t \geq 0$ by (2.19iii). Let $B_k = \{x: f_1(x) > k\}$ and let $T_k = T_{B_k}$. If $M^k(t) = M_t\, I_{[0, T_k)}(t)$, then show that $A^k(t) = A(t \wedge T_k)$ defines an AF of (X, M^k) which is natural if A is. In addition show that A^k has a bounded α-potential for any $\alpha > 0$. Next show that almost surely $\lim_k T_k \geq S$ and so $A^k(t) \to A(t)$ for all t as $k \to \infty$. Now let A and B be two AF's of (X, M) with the same finite characteristic $f_t(x)$. Show that if T is a bounded stopping time than $E^x(A_T) = E^x(B_T)$ for all x by approximating T from above by stopping times taking on only finitely many values. Conclude that if A^k and B^k are defined as above, then A^k and B^k have the same bounded characteristic. Finally prove the assertion of (2.20).]

(2.21) Let A be a CAF of (X, M). Let B be related to A by (1.10).

(i) Let $\varphi(x) = E^x \int_0^\infty e^{-t} M_t \exp(-B_t)\, dt$. Use (2.20i) of Chapter II to show that if $R = \inf\{t : A_t = \infty\}$, then

$$U_A^1 \varphi(x) = E^x \int_0^R e^{-t} M_t(1 - e^{-B_t})\, dt \leq 1.$$

(ii) Let h_n be the indicator function of $\{\varphi > 1/n\}$ and let $A_n(t) = \int_0^t h_n(X_u) dA_u$. Show that, for each n, A_n is a CAF of (X, M) with a bounded 1-potential and that $A_n(t) \uparrow A(t)$ as $n \to \infty$.

(iii) If, in addition, M is exact show that φ is nearly Borel measurable and finely continuous.

(2.22) Let A, B, and M be as in (2.21) with M exact. Assume that for a fixed $\alpha \geq 0$, u_A^α is finite. For $\beta \geq 0$ and $f \in \mathscr{E}_+^*$ define

$$V_A^\beta f(x) = E^x \int_0^\infty e^{-\beta B_t} e^{-\alpha t} f(X_t)\, dA_t.$$

(i) Show that, for each $\beta \geq 0$, $N_t^\beta = M_t \exp(-\beta B_t)$ is an exact MF of X and that $V_A^\beta f$ is $\alpha - (X, N^\beta)$ excessive for each $f \in \mathscr{E}_+^*$.

(ii) Show that the family $\{V_A^\beta; \beta \geq 0\}$ satisfies the resolvent equation and that $\|V_A^\beta\| \leq \beta^{-1}$ for $\beta > 0$. In particular note that $U_A^\alpha V_A^\beta = V_A^\beta U_A^\alpha = (1/\beta)[U_A^\alpha - V_A^\beta]$ for each $\beta > 0$.

(iii) Let $\{V^\gamma\}$ be the resolvent corresponding to M and let $\{W^\gamma\}$ be the resolvent corresponding to N^β. Show that $\beta U_A^\alpha W^\alpha = V^\alpha - W^\alpha = \beta V_A^\beta V^\alpha$.

3. Potentials of Continuous Additive Functionals

In this section we are going to characterize the potential of a continuous additive functional of (X, M). The following basic assumptions will be in force throughout this section. $X = (\Omega, \mathscr{M}, \mathscr{M}_t, X_t, \theta_t, P^x)$ will be a fixed standard process with state space $(E, \mathscr{E})$, and, in order to simplify the terminology, we will assume that $\mathscr{M} = \mathscr{F}$ and $\mathscr{M}_t = \mathscr{F}_t$. Therefore a stopping time will always be an $\{\mathscr{F}_t\}$ stopping time. M will be a fixed *exact* multiplicative functional of X with $M_t = 0$ if $t \geq \zeta$. As in Section 1, $S = \inf\{t: M_t = 0\}$. In the present case S is a strong terminal time and $S \leq \zeta$. We will denote the semigroup and resolvent generated by M by $\{Q_t\}$ and $\{V^\alpha\}$, respectively.

As before $u_A(x) = E^x A(\infty) = E^x A(S)$ is the potential of A. We saw in Section 2 that $A(t + T) = A(T) + M_T A_t \circ \theta_T$ almost surely whenever T is a stopping time. It follows easily from this that for $R \in \mathscr{F}_+$ and T a stopping time we have $A_R = A_T + M_T A_{R-T}(\theta_T)$ almost surely on $\{R \geq T\}$. If A is an additive functional of (X, M) with finite potential u_A and T is a stopping time, then it follows from (2.4) that

(3.1)
$$Q_T\, u_A(x) = E^x\{A(\infty) - A(T)\}$$
$$= E^x\{A(S) - A(T)\}.$$

Consequently if A is continuous and $\{T_n\}$ is an increasing sequence of stopping times with limit T then $Q_{T_n} u_A \to Q_T u_A$ as $n \to \infty$. Motivated by this fact we make the following definition.

(3.2) DEFINITION. A *finite* (X, M) excessive function f is called a *regular potential* (of (X, M)) provided that $Q_{T_n} f \to Q_T f$ whenever $\{T_n\}$ is an increasing sequence of stopping times with limit T.

We have just observed that the finite potential of a CAF of (X, M) is a regular potential of (X, M). The main purpose of this section is to prove the converse of this statement.

Let f be a fixed finite (X, M) excessive function and define for each $n \geq 1$

(3.3)
$$f_n = n \int_0^{1/n} Q_t\, f\, dt.$$

(3.4) PROPOSITION. Each f_n is (X, M) excessive and $\{f_n\}$ increases to f as $n \to \infty$.

Proof. We may write $f_n = \int_0^1 Q_{t/n} f\, dt$ and $Q_s f_n = \int_0^1 Q_{s+t/n} f\, dt$. Since f is (X, M) excessive, $Q_{s+t/n} f \leq Q_{t/n} f$ and $Q_{t/n} f$ increases to f as $n \to \infty$. This makes both assertions in the proposition obvious.

Given $\varepsilon > 0$ we define

(3.5)
$$B_n = \{x : f(x) - f_n(x) \geq \varepsilon\}.$$

Each B_n is nearly Borel measurable and, since f and f_n vanish on E_M^c, $B_n \subset E_M$. Moreover $\{B_n\}$ is a decreasing sequence of sets with empty intersection, and each B_n is finely closed in E_M. As in Section 5 of Chapter III let $S_p = \inf\{t: M_t \leq 1/p\}$; then $\{S_p\}$ is an increasing sequence of stopping times with limit S.

(3.6) PROPOSITION. Let $T_n = T_{B_n}$ and let $T = \lim T_n$. If $\lim_n Q_{T_n} f = Q_T f$, then $\lim_n P^x(T_n < S_p) = 0$ for all p and x.

Proof. Recall that if g is (X, M) excessive, then $t \to g(X_t)$ is right continuous a.s. on $[0, S)$, and so $f(X_{T_n}) - f_n(X_{T_n}) \geq \varepsilon$ almost surely on $\{T_n < S\}$. Thus for a fixed x and p we have

$$\begin{aligned}
&\frac{\varepsilon}{p} P^x[T_n < S_p \text{ for all } n] \\
&\qquad \leq \frac{1}{p} E^x\left\{\liminf_{k\to\infty}[f(X_{T_k}) - f_k(X_{T_k})];\ T_n < S_p \text{ for all } n\right\} \\
&\qquad \leq \liminf_{k\to\infty} E^x\{[f(X_{T_k}) - f_k(X_{T_k})]\, M_{T_k}\} \\
&\qquad \leq \liminf_{k\to\infty} E^x\{[f(X_{T_k}) - f_j(X_{T_k})]\, M_{T_k}\}
\end{aligned}$$

for each j. Since $Q_{T_n} f_j \geq Q_T f_j$ we obtain

$$\begin{aligned}
\frac{\varepsilon}{p} P^x[T_n < S_p \text{ for all } n] &\leq \liminf_k [Q_{T_k} f(x) - Q_T f_j(x)] \\
&= Q_T f(x) - Q_T f_j(x),
\end{aligned}$$

and letting $j \to \infty$ yields $P^x[T_n < S_p \text{ for all } n] = 0$. Since $\{T_n\}$ is an increasing sequence this evidently completes the proof of Proposition 3.6.

(3.7) REMARK. Since $f - f_n = n\int_0^{1/n}(f - Q_t f)\,dt \leq f - Q_{1/n} f$, if we let $A_n = \{x : f(x) - Q_{1/n} f(x) \geq \varepsilon\}$ then $B_n \subset A_n$. Therefore $T_{A_n} \leq T_n$ and so the conclusion of Proposition 3.6 holds whenever $\lim_n P^x[T_{A_n} < S_p] = 0$.

An obvious consequence of Proposition 3.6 is that $\lim_n T_n \geq S$ and that $\{T_n < S \text{ for all } n\}$ is contained in $\{M_{S-} = 0\}$ almost surely. It is equally clear from Definition 3.2 that $Q_t f \to 0$ as $t \to \infty$ whenever f is an (X, M) regular potential. The next result is the main step in our task of characterizing the potentials of continuous additive functionals of (X, M).

(3.8) THEOREM. *Let f be a bounded (X, M) excessive function and suppose (i) $Q_t f \to 0$ as $t \to \infty$, (ii) for each $\varepsilon > 0$, $\lim_n P^x(T_n < S_p) = 0$ for all x and p where $T_n = T_{B_n}$ and B_n is defined in (3.5). Then there exists a CAF, A, of (X, M) whose potential is f.*

REMARK. The preceding discussion shows that any bounded (X, M) regular potential satisfies the hypotheses of Theorem 3.8. Moreover A must be unique up to equivalence according to Theorem 2.13.

Proof. Let $g_n = n(f - Q_{1/n} f)$; then the fact that $Q_t f \to 0$ as $t \to \infty$ implies that $V g_n = f_n$ where f_n is defined in (3.3). We next define

(3.8 bis) $$A^n(t) = \int_0^t g_n(X_u)\, M_u\, du,$$

and clearly $A^n = \{A^n(t)\}$ is a CAF of (X, M) for each n. If u_n denotes the potential of A^n, then $u_n = Vg_n = f_n \uparrow f$ as $n \to \infty$, and so it is clear that the CAF that we are searching for is in some sense the limit of the A^n.

For each x and t

$$\begin{aligned} E^x\{A^n(\infty) \mid \mathscr{F}_t\} &= E^x\left\{\int_0^t g_n(X_u)\, M_u\, du + \int_t^\infty g_n(X_u)\, M_u\, du \mid \mathscr{F}_t\right\} \\ &= A^n(t) + M_t f_n(X_t). \end{aligned}$$

Consequently if we let $e_n(t) = A^n(t) + M_t\, f_n(X_t)$, then $\{e_n(t), \mathscr{F}_t, P^x\}$ is a nonnegative martingale for each n and x. Since $\sigma(\bigcup_t \mathscr{F}_t)^\sim = \mathscr{F}$ we have

$$e_n(\infty) = \lim_{t \to \infty} e_n(t) = E^x\{A^n(\infty) \mid \mathscr{F}\} = A^n(\infty) = A^n(S)$$

almost surely. It now follows from the extension of Kolmogorov's inequality to martingales (Doob [1, p. 315]) that for each $\delta > 0$

$$\begin{aligned} P^x\left[\sup_t |e_n(t) - e_m(t)| \geq \delta\right] &\leq \delta^{-2}\, E^x\{[e_n(\infty) - e_m(\infty)]^2\} \\ &= \delta^{-2}\, E^x\{[A^n(\infty) - A^m(\infty)]^2\}. \end{aligned}$$

We now estimate this last expectation:

$$\begin{aligned} J_{n,m} &= E^x\{[A^n(\infty) - A^m(\infty)]^2\} = E^x\left\{\left(\int_0^\infty [g_n(X_u) - g_m(X_u)] M_u\, du\right)^2\right\} \\ &= 2E^x \int_0^\infty [g_n(X_u) - g_m(X_u)] M_u \int_u^\infty [g_n(X_t) - g_m(X_t)] M_t\, dt\, du \\ &= 2E^x \int_0^\infty [g_n(X_u) - g_m(X_u)] M_u^2\, E^{X(u)} \int_0^\infty [g_n(X_t) - g_m(X_t)] M_t\, dt\, du. \end{aligned}$$

But $E^y \int_0^\infty g_n(X_t)\, M_t\, dt = f_n(y)$ and so if $n \geq m$

$$\begin{aligned} J_{n,m} &\leq 2E^x \int_0^\infty g_n(X_u)\, M_u^2 [f_n(X_u) - f_m(X_u)]\, du \\ &\leq 2E^x \int_0^\infty g_n(X_u)\, M_u^2 [f(X_u) - f_m(X_u)]\, du. \end{aligned}$$

If we let

$$H_m = \sup_{t>0} \{M_t [f(X_t) - f_m(X_t)]\}^2,$$

then from the Schwarz inequality we obtain

$$J_{n,m} \leq 2\left(E^x\{H_m\}\, E^x\left\{\left[\int_0^\infty g_n(X_u)\, M_u\, du\right]^2\right\}\right)^{1/2}.$$

However arguing as above we find

$$E^x\left\{\left[\int_0^\infty g_n(X_u)\, M_u\, du\right]^2\right\} = 2E^x \int_0^\infty g_n(X_u)\, M_u^2 E^{X(u)} \int_0^\infty g_n(X_t)\, M_t\, dt\, du$$
$$\leq 2\|f\|^2,$$

and so if $n \geq m$

$$J_{n,m} \leq 2^{3/2}\|f\|\{E^x(H_m)\}^{1/2}. \tag{3.9}$$

Of course $E^x(H_m) = 0$ if $x \notin E_M$. Recalling the definition of S_p we define

$$H_{m,p} = \sup_{0<t<S_p} \{M_t[f(X_t) - f_m(X_t)]\}^2.$$

Note that if $x \in E_M$ and $p \geq 2$, then $P^x(S_p > 0) = 1$. Also note that the suprema defining $H_{m,p}$ and H_m are almost surely unchanged if we replace the condition $t > 0$ by $t \geq 0$. Clearly $H_m \leq H_{m,p} + p^{-2}\|f\|^2$. On the other hand given $\varepsilon > 0$ and recalling the definition of T_m we may write for $x \in E_M$ and $p \geq 2$

$$E^x(H_{m,p}) = E^x(H_{m,p};\, T_m < S_p) + E^x(H_{m,p};\, T_m \geq S_p)$$
$$\leq \|f\|^2\, P^x(T_m < S_p) + \varepsilon^2.$$

It is now clear that $E^x(H_m) \to 0$ as $m \to \infty$ for all x, and hence $J_{n,m} \to 0$ as n and m approach infinity.

It is immediate from this estimate that, for all $\delta > 0$, $P^x[\sup_{t\geq 0}|e_n(t) - e_m(t)| \geq \delta] \to 0$ as $n, m \to \infty$. Also the Tchebychev inequality implies that for any $\delta > 0$ and $n > m$

$$P^x\left\{\sup_{t\geq 0}(M_t|f_n(X_t) - f_m(X_t)|) \geq \delta\right\} \leq \delta^{-2}E^x(H_m)$$

which approaches zero as $m \to \infty$. Combining these two statements we obtain

$$P^x\left[\sup_t |A^n(t) - A^m(t)| \geq \delta\right] \to 0 \quad \text{as} \quad n, m \to \infty \tag{3.10}$$

for each $\delta > 0$ and $x \in E_\Delta$. Thus $\{A_t^n\}$ converges in measure uniformly in t with respect to each P^x, and hence also with respect to P^μ whenever μ is a finite measure on $\mathscr{E}_\Delta^*$.

Let $\mathscr{P}$ be the class of all measures on $\mathscr{E}_\Delta^*$ with mass less than or equal to one. Then for each $t \in [0, \infty)$ and $\mu \in \mathscr{P}$ there exists $A_t^\mu \geq 0$ in $\mathscr{F}_t$ such that $A_t^n = A^n(t) \to A_t^\mu$ in P^μ measure as $n \to \infty$. But $E^x\{[A^n(\infty)]^2\} \leq 2\|f\|^2$ and so for each t and $\mu \in \mathscr{P}$ the family $\{A_t^n\}$ is P^μ uniformly integrable. Therefore $E^\mu\{A_t^n Y\} \to E^\mu\{A_t^\mu Y\}$ as $n \to \infty$ for all $Y \in b\mathscr{F}$. We will write A_t^x for A_t^μ when

$\mu = \varepsilon_x$. Let $0 \leq t, s < \infty$, and μ be given. Let $\mathscr{H}_s = \theta_s^{-1}\mathscr{F}$. Then by (2.7) of Chapter 0, $Y \in b\mathscr{H}_s$ if and only if there exists $Z \in b\mathscr{F}$ such that $Y = Z \circ \theta_s$. Consequently, given $Y \in b\mathscr{F}$ there is a $Z \in b\mathscr{F}$ such that $E^\mu\{Y | \theta_s^{-1}\mathscr{F}\} = Z \circ \theta_s$. Therefore if $\nu = \mu P_s$ we have

$$\begin{aligned} E^\mu\{(A_t^n \circ \theta_s)Y\} &= E^\mu\{(A_t^n \circ \theta_s)(Z \circ \theta_s)\} \\ &= E^\mu\{E^{X(s)}[A_t^n Z]\} = E^\nu\{A_t^n Z\} \\ &\to E^\nu\{A_t^\nu Z\} = E^\mu\{(A_t^\nu \circ \theta_s)(Z \circ \theta_s)\} \\ &= E^\mu\{(A_t^\nu \circ \theta_s)Y\}. \end{aligned}$$

But

$$E^\mu\{A_{t+s}^n Y\} = E^\mu\{A_s^n Y\} + E^\mu\{M_s(A_t^n \circ \theta_s)Y\},$$

and so letting $n \to \infty$ we obtain

$$A_{t+s}^\mu = A_s^\mu + M_s(A_t^\nu \circ \theta_s), \qquad \text{a.s.} \quad P^\mu, \tag{3.11}$$

where $\nu = \mu P_s$. If $Y \in b\mathscr{F}$, then $x \to E^x\{YA_t^n\}$ is $\mathscr{E}_\Delta^*$ measurable, and hence so is its limit $x \to E^x\{YA_t^x\}$. Now for a fixed t, let $Q^x = A_t^x P^x$ so that Q^x is a finite measure on $(\Omega, \mathscr{F})$ and $x \to Q^x(\Lambda)$ is $\mathscr{E}_\Delta^*$ measurable for each $\Lambda \in \mathscr{F}$. Obviously $Q^x \ll P^x$ for each x. (Also $x \to P^x(\Lambda)$ is in $\mathscr{E}_\Delta^*$ for each $\Lambda \in \mathscr{F}$}.) In view of the right continuity of the paths, it is clear that $\mathscr{F}$ is countably generated up to completion; that is, there exists a countably generated σ-algebra $\mathscr{G}$ such that $\mathscr{F} = \bigcap_\mu \mathscr{G}^{P^\mu}$ with μ ranging over $\mathscr{P}$. (If $t \to X_t(\omega)$ is right continuous for all ω, then we may take $\mathscr{G} = \mathscr{F}^\circ$.) Thus by the theorem of Doob used in the proof of (2.3), Chapter III, we can find a function $\tilde{B}_t(x, \omega)$ which is $\mathscr{E}_\Delta^* \times \mathscr{F}$ measurable and such that $\omega \to \tilde{B}_t(x, \omega)$ is a density for Q^x relative to P^x for each x. Consequently for each x

$$A_t^x(\omega) = \tilde{B}_t(x, \omega), \qquad \text{a.s.} \quad P^x.$$

Define $B_t(\omega) = \tilde{B}_t(X_0(\omega), \omega)$; then $B_t \in \mathscr{F}$. Now if $Y \in b\mathscr{F}$ then

$$E^x\{A_t^n Y\} \to E^x\{A_t^x Y\} = E^x\{B_t Y\},$$

and integrating with respect to μ we see that for each $\mu \in \mathscr{P}$

$$A_t^\mu = B_t, \qquad \text{a.s.} \quad P^\mu.$$

In particular this implies that $B_t \in \mathscr{F}_t$ for all t. Next if $\nu = \mu P_s$ then

$$P^\mu\{A_t^\nu \circ \theta_s \neq B_t \circ \theta_s\} = P^\nu\{A_t^\nu \neq B_t\} = 0,$$

and so it follows from (3.11) that for each fixed t and s

$$B_{t+s} = B_s + M_s(B_t \circ \theta_s) \tag{3.12}$$

almost surely. In addition since

$$f_n(x) - Q_t f_n(x) = E^x\{A_t^n\} \to E^x\{B_t\}$$

and $f_n \uparrow f$, we see that $f(x) - Q_t f(x) = E^x\{B_t\}$ for all x and $t < \infty$. It follows from (3.12) that $B_t \le B_s$ almost surely if $t \le s$. Now define

$$A_t = \inf_{\substack{s>t \\ s \in \mathbf{Q}}} B_s$$

where Q denotes the rationals. Then $A_t \in \mathscr{F}_{t+} = \mathscr{F}_t$ and $B_t \le A_t$ almost surely for each t. Clearly $t \to A_t$ is right continuous and nondecreasing almost surely. But for each $t < \infty$

$$\begin{aligned} E^x\{A_t\} &= \lim_{\substack{s \downarrow t \\ s \in \mathbf{Q}}} E^x\{B_s\} = \lim_{\substack{s \downarrow t \\ s \in \mathbf{Q}}} [f(x) - Q_s f(x)] \\ &= f(x) - Q_t f(x) = E^x\{B_t\}, \end{aligned}$$

and so $A_t = B_t$ almost surely for each t. Let $A_\infty = \sup_t A_t$. Then $A_\infty = \lim_{t\uparrow\infty} A_t$ almost surely and so $E^x(A_\infty) = \lim_{t\uparrow\infty} E^x\{A_t\} = \lim_{t\uparrow\infty}[f(x) - Q_t f(x)] = f(x)$. Finally

$$P^x\{A_t \circ \theta_s \ne B_t \circ \theta_s\} = E^x\{P^{X(s)}[A_t \ne B_t]\} = 0,$$

and so it follows that (3.12) holds with B replaced by A, that $f(x) = E^x\{A_\infty\}$, and that $t \to A_t(\omega)$ is almost surely right continuous. (Plainly $A_0 = 0$, a.s.; however, we do not as yet know that $t \to A_t$ is almost surely continuous at $t = S$.) We will complete the proof of Theorem 3.8 by showing that $t \to A_t$ is almost surely continuous on $[0, \infty)$.

Let $x \in E_\Delta$ be fixed; then using (3.10) one can find a sequence $\{n_k\}$, depending on x in general, such that

$$P^x\left\{\sup_t |A^{n_j}(t) - A^{n_k}(t)| \ge 2^{-k}\right\} \le 2^{-k}$$

provided $n_j \ge n_k$. It now follows by standard reasoning that $\{A^{n_k}(t, \omega)\}$ converges uniformly on $[0, \infty]$ to a finite limit almost surely P^x. Let Λ be the set of those ω's such that $\{A^{n_k}(t, \omega)\}$ converges uniformly in t on $[0, \infty]$; then $\Lambda \in \mathscr{F}$ and $P^x(\Lambda) = 1$. Let $B^x(t, \omega) = \lim_k A^{n_k}(t, \omega)$ if $\omega \in \Lambda$ and $B^x(t, \omega) = 0$ for all t if $\omega \notin \Lambda$. Thus $t \to B^x(t, \omega)$ is continuous for all ω and $A^{n_k}(t, \cdot) \to B^x(t, \cdot)$ almost surely P^x for each t. But by our previous construction if $Y \in b\mathscr{F}$ and $t < \infty$ then $E^x\{YA_t^n\} \to E^x\{YA_t\}$, and so it follows that, for each fixed $t < \infty$, $A_t = B_t^x$ almost surely P^x. Finally since $t \to A_t$ is almost surely right continuous, we see that the functions $t \to A_t(\omega)$ and $t \to B^x(t, \omega)$ agree almost surely P^x. But x was arbitrary and so $t \to A_t$ is almost surely continuous. This completes the proof of Theorem 3.8.

We now remove the restriction that f be bounded.

(3.13) Theorem. Let f be a regular (X, M) potential. Then there exists a CAF of (X, M) whose potential is f.

Proof. For each positive integer k let $R'_k = T_{\{f>k\}}$ and $R_k = \min(R'_k, S)$. Thus $\{R_k\}$ is an increasing sequence of strong terminal times. Now f is a finite (X, M) excessive function and so, for each x, $\{M_{R_k} f(X_{R_k}), \mathscr{F}_{R_k}, P^x\}$ is a nonnegative supermartingale. Hence almost surely $M_{R_k} f(X_{R_k})$ approaches a limit L as $k \to \infty$ and $E^x(L) \leq f(x)$. It is now evident that $P^x(R_k < S_p$ for all $k) = 0$ for each x and p. In particular $R_k \uparrow S$ almost surely and $R_k = S$ for large enough k on $\{M_{S-} > 0\}$ almost surely. We now define $g_k = f - Q_{R_k} f$, and since f is a regular (X, M) potential it follows that $g_k \uparrow f$ as $k \to \infty$.

Let $N^k(t) = M_t I_{[0,R_k)}(t) = M_t I_{[0,R'_k)}(t)$. In the remainder of this paragraph k will be held fixed and so we will drop it from our notation. Plainly $N = \{N_t\}$ is a right continuous MF of X. Moreover both M and $I_{[0,R')}(t)$ are exact and hence so is N according to (5.20) of Chapter III. We are now going to show that $g = f - Q_R f$ is a bounded regular (X, N) potential. Let $\{K_t\}$ denote the semigroup generated by N. First note that $E_N \subset \{x: P^x(R > 0) = 1\} \subset \{f \leq k\}$. If $R_t = \min(R, t)$, then using the fact that R is a terminal time we find

$$\begin{aligned} K_t g(x) &= E^x\{[f(X_t) - Q_R f(X_t)]M_t;\, t < R\} \\ &= E^x\{M_t f(X_t) - M_R f(X_R);\, t < R\} \\ &= Q_{R_t} f(x) - Q_R f(x) \\ &\leq f(x) - Q_R f(x) = g(x). \end{aligned}$$

If $x \in E_N$, then

$$\begin{aligned} K_t g(x) &\to E^x\{M_0 f(X_0) - M_R f(X_R);\, R > 0\} \\ &= f(x) - Q_R f(x) = g(x) \end{aligned}$$

as $t \to 0$. On the other hand if $x \notin E_N$, then it is easy to see that $g(x) = 0$. Thus g is (X, N) excessive. Moreover $g(x) \leq f(x) \leq k$ on $\{f \leq k\} \supset E_N$ and $g = 0$ off E_N. Therefore g is bounded. Finally we must show that g is a regular (X, N) potential. Let $\{T_n\}$ be an increasing sequence of stopping times with limit T. If $T'_n = T_n \wedge R$ and $T' = T \wedge R$, then $T'_n \uparrow T'$ and so

$$\begin{aligned} K_{T_n} g(x) &= E^x\{[f(X_{T_n}) - Q_R f(X_{T_n})]M_{T_n};\, T_n < R\} \\ &= E^x\{M_{T_n} f(X_{T_n}) - M_R f(X_R);\, T_n < R\} \\ &= Q_{T'_n} f(x) - Q_R f(x) \\ &\to Q_{T'} f(x) - Q_R f(x) = K_T f(x). \end{aligned}$$

Hence g is a bounded regular (X, N) potential.

We now bring the parameter k back into our notation. Since g_k is a bounded regular (X, N^k) potential there exists a CAF, A^k, of (X, N^k) whose potential is g_k. Recalling the definition of N^k we have

$$A^k(t+s) = A^k(t) + M_t\, A^k(s) \circ \theta_t$$

almost surely on $\{t < R_k\}$. If $n > k$ it is easy to check that $t \to A^n(t \wedge R_k)$ is a CAF of (X, N^k), and that

$$\begin{aligned} E^x\{A^n(R_k)\} &= E^x\{A^n(R_n) - [A^n(R_n) - A^n(R_k)]\} \\ &= g_n(x) - E^x\{A^n(R_n) - A^n(R_k);\, R_k < R_n\}. \end{aligned}$$

But $g_n = f - Q_{R_n} f$ while the term to be subtracted from it is just

$$\begin{aligned} &E^x\{M_{R_k}\, A^n(R_n) \circ \theta_{R_k};\, R_k < R_n\} \\ &\quad = E^x\{[f(X_{R_k}) - Q_{R_n} f(X_{R_k})] M_{R_k};\, R_k < R_n\} \\ &\quad = E^x\{M_{R_k} f(X_{R_k}) - M_{R_n} f(X_{R_n});\, R_k < R_n\} \\ &\quad = Q_{R_k} f(x) - Q_{R_n} f(x). \end{aligned}$$

Thus the potential relative to (X, N^k) of $A^n(t \wedge R_k)$ is g_k and so the uniqueness theorem implies that $A^n(t) = A^k(t)$ on $[0, R_k]$ almost surely. Now $R_k \uparrow S$ and so if we define $A(t) = \lim_n A^n(t)$, then this exists on $[0, S)$ almost surely and the above compatibility implies that A is continuous and finite on $[0, S)$ and that $A(t+s) = A(t) + M_t\, A(s) \circ \theta_t$ almost surely on $\{t + s < S\}$. If we now define $A(t) = \lim_{u \uparrow S} A(u)$ when $t \geq S$, then it is routine that A is a CAF of (X, M). (Of course $A(t) = 0$ for all t if $S = 0$.) Finally $A(R_n) \uparrow A(S)$ and consequently

$$\begin{aligned} E^x\{A(S)\} &= \lim_n E^x\{A(R_n)\} \\ &= \lim_n E^x\{A^n(R_n)\} \\ &= \lim_n g_n(x) = f(x). \end{aligned}$$

Thus the proof of Theorem (3.13) is complete.

Suppose that f is $\alpha - (X, M)$ excessive. Then f is (X, N) excessive where $N_t = e^{-\alpha t} M_t$. Moreover if A is a CAF of (X, N), then $B_t = \int_0^t e^{\alpha u}\, dA_u$ defines a CAF of (X, M). Consequently Theorem 3.13 implies the following result.

(3.14) COROLLARY. Let f be a finite $\alpha - (X, M)$ excessive function such that $Q_{T_n}^\alpha f \to Q_T^\alpha f$ whenever $\{T_n\}$ is an increasing sequence of stopping times with limit T. Then there exists a unique CAF, A, of (X, M) whose α-potential is f; that is,

$$f(x) = E^x \int_0^\infty e^{-\alpha t}\, dA(t).$$

Suppose now that M is a perfect MF. Then it is immediate that each of the approximating additive functionals A^n defined in (3.8bis) is perfect, and it is natural to ask if A may also be taken to be perfect. We close this section with a preliminary result in this direction. Further results on this problem will be given in Chapter V.

(3.15) DEFINITION. A function f in $\mathscr{E}_+^*$ is said to be *uniformly* (X, M) *excessive* provided that (i) f is bounded, (ii) $f \in \mathscr{S}(M)$, and (iii) $Q_t f \to f$ uniformly on E as $t \to 0$.

(3.16) THEOREM. Let M be perfect (in addition to being exact and vanishing on $[\zeta, \infty]$). Suppose that f is uniformly (X, M) excessive and that $Q_t f \to 0$ as $t \to \infty$. Then there exists a perfect CAF of (X, M) whose potential is f.

Proof. First note that f satisfies the hypotheses of Theorem 3.8 since for each $\varepsilon > 0$ the set B_n defined in (3.5) is empty if n is large enough. Thus exactly as in the proof of Theorem 3.8 one obtains the estimate (3.9). But now $H_m \le \|f - f_m\|^2 \to 0$ as $m \to \infty$ and so (3.10) may be strengthened to

$$P^x\Big[\sup_t |A^n(t) - A^m(t)| \ge \delta\Big] \to 0 \qquad \text{as} \quad n, m \to \infty$$

uniformly in x for each $\delta > 0$. Therefore one can find a sequence $\{n_k\}$ *independent* of x such that

$$P^x\Big[\sup_t |A^k(t) - A^j(t)| \ge 2^{-k}\Big] \le 2^{-k}$$

for all x provided $j \ge k$ where, for notational convenience, we have written A^k for A^{n_k}. Hence $\{A^k(t)\}$ converges uniformly for $t \in [0, \infty]$ almost surely. Let Λ be the set of those ω for which $A^k(t, \omega)$ converges for all $t \in [0, S(\omega))$. Then we define $A(t, \omega) = \lim_k A^k(t, \omega)$ if $t < S(\omega)$ and $\omega \in \Lambda$, $A(t, \omega) = \lim_{u \uparrow S(\omega)} A(u, \omega)$ for $t \ge S(\omega)$ and $\omega \in \Lambda$, and $A(t, \omega) = 0$ for all t if $\omega \notin \Lambda$. Since $\theta_t^{-1}\Lambda \cap \{S > t\} \subset \Lambda$ for each t and since $I_{[0,S)}(t)$ is also perfect, it follows that A is a perfect AF of (X, M). Clearly A is continuous and $u_A = f$. Thus Theorem 3.16 is established.

Exercises

(3.17) Let $f_t(x)$ be a nonnegative finite function defined for $t \geq 0$ and $x \in E$. Suppose that: (i) $f_t(\cdot) \in \mathscr{E}^*$ for each $t \geq 0$, (ii) $f_0(x) = 0$ and $t \to f_t(x)$ is right continuous at 0 for all x, and (iii) $f_{t+s} = f_t + Q_t f_s$ for all t and s. Show that $t \to f_t(x)$ is nondecreasing and right continuous on $[0, \infty)$. If

$$u^\alpha(x) = \alpha \int_0^\infty e^{-\alpha t} f_t(x)\, dt,$$

then show that $u^\alpha \in \mathscr{S}^\alpha(M)$. Let $\alpha \geq 0$ be fixed and suppose that whenever $\{T_n\}$ is an increasing sequence of stopping times with limit T one has $Q^\alpha_{T_n} f_t \to Q^\alpha_T f_t$ as $n \to \infty$ for almost all t (Lebesgue measure). If, in addition, $f_t(x)$ is bounded, then show that there exists a CAF, A, of (X, M) such that $f_t(x) = E^x(A_t)$ for all t and x.

(3.18) Give an example of a bounded regular potential which is not uniformly excessive. [Hint: let X be uniform motion on the line and $M_t = I_{[0,T)}(t)$ where T is the time of hitting a suitable set.]

4. Potentials of Natural Additive Functionals

In this section we are going to investigate those excessive functions which may be represented as the finite potential of a natural additive functional. The most satisfactory approach to this problem is to make use of P. A. Meyer's theory of the decomposition of supermartingales. Since Meyer [1] contains a definitive treatment of this theory there would be no purpose in our reproducing it here. Rather we will take a more constructive approach (also due to Meyer [2]) under an additional hypothesis on X (Assumption 4.1 below). However for the reader familiar with Meyer's theory we will sketch the approach based on the decomposition theorem for supermartingales in the Notes and Comments for this section. In particular *the main result of the present section, Theorem 4.22, is valid without Assumption 4.1.* The techniques of Sections 4 and 5 are of interest, but the results are not particularly essential for the remainder of the book. These two sections might be omitted at a first reading.

As in the previous section $X = (\Omega, \mathscr{M}, \mathscr{M}_t, X_t, \theta_t, P^x)$ is a fixed standard process with state space $(E, \mathscr{E})$ such that $\mathscr{M} = \mathscr{F}$ and $\mathscr{M}_t = \mathscr{F}_t$. As before, M denotes a fixed *exact* MF of X with $M_t = 0$ if $t \geq \zeta$. We will use the notation developed in the first paragraph of Section 3 without special mention.

We come now to the special assumption to be imposed on X in this section.

Suppose that $\{T_n\}$ is an increasing sequence of stopping times; we then let $\bigvee_n \mathscr{F}_{T_n}$ denote the σ-algebra $\sigma(\bigcup_n \mathscr{F}_{T_n})^{\sim}$. Our special assumption is the following:

(4.1) If $\{T_n\}$ is an increasing sequence of stopping times with limit T, then $X(T) \in (\bigvee_n \mathscr{F}_{T_n})/\mathscr{E}_\Delta$.

Note that (4.1) is certainly satisfied if X is a Hunt process, since, in that case, $X(T_n) \to X(T)$ almost surely on $\{T < \infty\}$ and $X_T = \Delta$ on $\{T = \infty\}$ while $\{T < \infty\} = \bigcup_k \bigcap_n \{T_n < k\} \in \bigvee_n \mathscr{F}_{T_n}$. Consequently $X(T) \in (\bigvee_n \mathscr{F}_{T_n})/\mathscr{E}_\Delta$. If X is only quasi-left-continuous on $[0, \zeta)$, then $X(T_n) \to X(T)$ almost surely on $\{T < \zeta\}$ while $X_T = \Delta$ on $\{T \geq \zeta\}$. But the above argument breaks down since it is not true that $\{T < \zeta\} \in \bigvee_n \mathscr{F}_{T_n}$ (see Exercise 4.34). However note that $\{T \leq \zeta\}$ is in $\bigvee_n \mathscr{F}_{T_n}$. *In the remainder of this section we will assume that (4.1) holds.* As we have seen this is no restriction if X is a Hunt process.

Before coming to the study of potentials of NAF's it will be necessary to develop certain preliminary material about stopping times. These results are useful in other situations as well.

(4.2) PROPOSITION. Let $\{T_n\}$ be an increasing sequence of stopping times with limit T. Then (under (4.1)) $\bigvee_n \mathscr{F}_{T_n} = \mathscr{F}_T$.

Proof. For the purposes of this proof let $\mathscr{G} = \bigvee_n \mathscr{F}_{T_n}$. Since $T_n \to T$, $T \in \mathscr{G}$; and $X(T) \in \mathscr{G}/\mathscr{E}_\Delta$ by assumption. Since $T_n \wedge t \uparrow T \wedge t$ and $\mathscr{F}_{T_n \wedge t} \subset \mathscr{F}_{T_n}$ for all $t \geq 0$ it also follows that $X(T \wedge t) \in \mathscr{G}/\mathscr{E}_\Delta$ for all $t \geq 0$. Let μ be a finite measure on $(E_\Delta, \mathscr{E}_\Delta)$ and let f and α with or without subscripts denote, respectively, a bounded continuous function on E_Δ and a positive constant. Suppose $H : \Omega \to \mathbf{R}$ is a finite product of the form

$$H = \prod_k \int_0^\infty e^{-\alpha_k t} f_k(X_t)\, dt. \tag{4.3}$$

If we write each integral in this product as $\int_0^T + \int_T^\infty$, then H may be written as a finite sum of products where each summand has the form

$$\prod_i \int_0^T e^{-\alpha_i t} f_i(X_t)\, dt \prod_j \int_T^\infty e^{-\alpha_j t} f_j(X_t)\, dt.$$

In view of the remarks at the beginning of this proof the product over i is $\mathscr{G}$ measurable. On the other hand the strong Markov property yields

$$E^\mu\left\{\prod_j \int_T^\infty e^{-\alpha_j t} f_j(X_t)\, dt \,\middle|\, \mathscr{F}_T\right\} = \varphi(X_T) \exp(-T \textstyle\sum \alpha_j)$$

where

$$\varphi(x) = E^x\left\{\prod_j \int_0^\infty e^{-\alpha_j t} f_j(X_t)\, dt\right\}.$$

Clearly $\varphi \in \mathscr{E}_\Delta$ and so $\varphi(X_T)\exp(-T \sum \alpha_j)$ is in $\mathscr{G}$. Of course $\mathscr{G} \subset \mathscr{F}_T$ and therefore

$$E^\mu(H \mid \mathscr{F}_T) = E^\mu(H \mid \mathscr{G}) \tag{4.4}$$

for all H of the form (4.3). In order to complete the proof of Proposition 4.2 it will certainly suffice to show that (4.4) holds whenever H is a finite product of the form $\prod_j f_j(X_{t_j})$ where $0 \le t_1 < \ldots < t_n$ and each $f_j \in \mathbf{C}(E_\Delta)$. If $f \in \mathbf{C}(E_\Delta)$ then $t \to f(X_t)$ is right continuous. Therefore given t_0 one can find a sequence $\{\gamma_n(t)\}$ of continuous functions on $[0, \infty)$ with compact support such that for each $f \in \mathbf{C}(E_\Delta)$ almost surely $\int_0^\infty \gamma_n(t) f(X_t)\, dt \to f(X_{t_0})$ boundedly as $n \to \infty$. Consequently it will suffice to show that (4.4) holds whenever H is a finite product of the form $\prod_j \int_0^\infty \gamma_j(t) f_j(X_t)\, dt$. But any such γ is a uniform limit of polynomials in e^{-t} by the Stone–Weierstrass theorem and since we know that (4.4) holds whenever H is a linear combination of products of the form (4.3) it is now evident that (4.4) also holds when $H = \prod_j \int_0^\infty \gamma_j(t) f_j(X_t)\, dt$. This completes the proof of Proposition 4.2.

Remark. Obviously (4.2) implies (4.1) and so, in fact, (4.1) and (4.2) are equivalent.

We now introduce a concept that will be very important in the remainder of this section.

(4.5) Definition. Let T be a stopping time and Λ a set in $\mathscr{F}$. Then T is said to be *accessible on* Λ if for each initial measure μ on $\mathscr{E}_\Delta$ there exists an increasing sequence $\{T_n\}$ of stopping times such that $\lim T_n = T$ on Λ and $T_n < T$ for all n on $\Lambda \cap \{T > 0\}$, both statements holding P^μ almost surely. When $\Lambda = \Omega$ we say simply that T is accessible.

If T is accessible on Λ, then the mapping $t \to X_t$ is continuous from the left (as well as from the right) at T almost surely on $\Lambda \cap \{T < \zeta\}$. This is an immediate consequence of quasi-left-continuity of the process. Meyer [5] (or [10]) has proved the remarkable converse that if X is a Hunt process, then any stopping time T is accessible on $\{t \to X_t$ is continuous at T; $T < \infty\}$. We will be concerned primarily with terminal times and will be content to prove directly that certain ones are accessible. The interested reader should consult Meyer [1] for a complete discussion of accessible stopping times.

Before coming to the main fact we shall need concerning accessibility we must introduce another concept involving terminal times.

(4.6) DEFINITION. Let T be a terminal time and define the iterates, T_n, of T by $T_0 = 0$, $T_{n+1} = T_n + T \circ \theta_{T_n}$. Then T is called *complete* if for each $k \geq 0$, $n \geq 1$, and stopping time R we have $T_{n+k} = R + T_n \circ \theta_R$ almost surely on $\{T_k \leq R < T_{k+1}\}$.

It is likely that every strong terminal time is complete. However, rather than attempt to prove a general theorem we will simply give a sufficient condition for completeness that can be checked easily in the cases of interest to us. We will give some additional discussion at the end of this section for the benefit of the interested reader (see Exercise 4.36).

(4.7) PROPOSITION. Let T be a strong terminal time and as in (4.6) let T_k, $k \geq 0$, denote the iterates of T. Suppose that for every stopping time R and every $k \geq 0$, $R + T \circ \theta_R = T_{k+1}$ almost surely on $\{T_k \leq R < T_{k+1}\}$. Then T is complete.

Proof. Suppose for the moment that T, R, and Q are any three stopping times, that $\Lambda \in \mathscr{F}$, and that $T = R + Q \circ \theta_R$ almost surely on Λ. Then (Exercise 4.35) it follows that, for every $F \in \mathscr{F}$, $F \circ \theta_T = F \circ \theta_Q \circ \theta_R$ almost surely on Λ. In particular returning to the T of the proposition we have for each $F \in \mathscr{F}$

$$F \circ \theta_T \circ \theta_{T_k} = F \circ \theta_{T_{k+1}} \tag{4.8}$$

almost surely. Now the relation $T_{n+k} = T_k + T_n \circ \theta_{T_k}$ is valid for all k if $n = 1$, by definition of the iterates. If this relation is known to hold almost surely for all k and all $n \leq m$ with a given $m \geq 1$, then using (4.8) we obtain

$$\begin{aligned} T_{m+1+k} &= T_{k+1} + T_m \circ \theta_{T_{k+1}} = T_k + T \circ \theta_{T_k} + T_m \circ \theta_T \circ \theta_{T_k} \\ &= T_k + T_{m+1} \circ \theta_{T_k} \end{aligned}$$

almost surely. So we have proved by induction that

$$T_{n+k} = T_k + T_n \circ \theta_{T_k} \tag{4.9}$$

almost surely for all n and k. Our main hypothesis is that $T_{k+1} = R + T \circ \theta_R$ almost surely on $\{T_k \leq R < T_{k+1}\}$. Using this and (4.9) twice we obtain for any n and k

$$\begin{aligned} T_{n+k} &= T_{k+1} + T_{n-1} \circ \theta_{T_{k+1}} = R + T \circ \theta_R + T_{n-1} \circ \theta_T \circ \theta_R \\ &= R + T_n \circ \theta_R \end{aligned}$$

almost surely on $\{T_k \leq R < T_{k+1}\}$ completing the proof.

We come now to the main fact that we need concerning accessibility of terminal times. Let us first establish some notation. If T is a terminal time and we define $\psi(x) = E^x(e^{-T})$ for x in E, then ψ is 1-super-mean-valued and so $\bar{\psi}(x) = \lim_{t\downarrow 0} P_t^1 \psi(x) = \lim_{t\downarrow 0} E^x\{e^{-(t+T\circ\theta_t)}\}$ is 1-excessive with $\bar{\psi} \leq \psi$. Moreover if $\psi(x) < 1$, then by the zero-one law $P^x(T > 0) = 1$ and hence $\bar{\psi}(x) = \psi(x)$. Thus $\bar{\psi} = \psi$ on $\{\psi < 1\}$. If T is a strong terminal time and Q is a stopping time, then

$$\begin{aligned} E^x\{P^{X(Q)}(T = 0);\, Q < T\} &= P^x(T - Q = 0;\, Q < T) \\ &= 0, \end{aligned}$$

and so $P^{X(Q)}(T = 0) = 0$ almost surely on $\{Q < T\}$. Therefore $\psi(X_Q) = \bar{\psi}(X_Q)$ almost surely on $\{Q < T\}$. Let $B = \{\bar{\psi} = 1\}$ and $R = T_B \wedge \zeta$. This notation will be used in the statement and proof of the next proposition.

(4.10) Proposition. Let T be a strong terminal time which is also complete. Let Q be an exact terminal time with $Q \leq R$ and suppose that, for each x,

(a) $P^x\{\psi(X_T) = 1;\, T < Q\} = 0$;

(b) $P^x\{X_{T-} \neq X_T;\, 0 < T < Q\} = 0$.

If $B_n = \{\bar{\psi} > 1 - 1/n\}$ and $R_n = T_{B_n}$, then almost surely $R_n < T$ for all n on $\{0 < T < Q\}$ and $\lim R_n = T$ on $\{R_n < T \text{ for all } n\}$. In particular T is accessible on $\{T < Q\}$.

Proof. First note that

$$\begin{aligned} E^x\{e^{-(T-R_n)};\, R_n < T\} &= E^x\{\bar{\psi}(X_{R_n});\, R_n < T\} \\ &\geq (1 - 1/n)\, P^x(R_n < T), \end{aligned}$$

and so $\lim R_n = T$ almost surely on $\{R_n < T \text{ for all } n\}$. Thus to complete the proof we must show that $R_n < T$ for all n almost surely on $\{0 < T < Q\}$. To this end suppose that, for some y and n, $P^y(0 < T < Q;\, R_n \geq T) > 0$. Since $\bar{\psi}(X_T) \leq \psi(X_T) < 1$ almost surely on $\{0 < T < Q\}$ by hypothesis, we may actually assume that for some m, $P^y(0 < T < Q;\, R_m > T) > 0$. We will derive a contradiction from this. Let $\delta = 1 - 1/m$, and $H = R_m \wedge Q$. Then $H \leq \zeta$ since $Q \leq R \leq \zeta$. Now both R_m and Q are exact and so it follows from (5.20) of Chapter III that H is also exact. As before let $T_0 = 0$ and $T_{n+1} = T_n + T\circ\theta_{T_n}$ denote the iterates of T. Next observe that (4.10a) implies that $P^x\{\psi(X_{T_n}) = 1;\, T_n < Q\} = 0$ for each x and n and hence

$$\begin{aligned} E^x\{e^{-T_{n+1}};\, T_n < H\} &= E^x\{\bar{\psi}(X_{T_n})\, e^{-T_n};\, T_n < H\} \\ &\leq \delta\, E^x\{e^{-T_n};\, T_{n-1} < H\}. \end{aligned}$$

Thus $E^x\{e^{-T_{n+1}}; T_n < H\} \le \delta^n$, and so almost surely $T_n \to \infty$ if $H = \infty$, and $T_n \ge H$ for some n if $H < \infty$. We define an AF of (X, H) by setting

$$\begin{aligned} A_t &= n, \qquad \text{if} \quad T_n \le t < T_{n+1} \text{ and } t < H, \\ &= \lim_{r \uparrow H} A_r \qquad \text{if} \quad t \ge H. \end{aligned}$$

The fact that T is complete yields readily the fact that A is indeed an AF of (X, H). The discontinuities of A occur only at the points $T_n < H$; but

$$\begin{aligned} &P^x\{X(T_n-) \ne X(T_n); T_n < H\} \\ &\quad \le E^x\{P^{X(T_{n-1})}[X(T-) \ne X(T); T < H]\} \\ &\quad = 0. \end{aligned}$$

and consequently A is natural. The 1-potential of A is

$$\begin{aligned} u_A^1(x) &= E^x \int_0^\infty e^{-t}\, dA(t) = \sum_{n=1}^\infty E^x\{e^{-T_n}; T_n < H\} \\ &\le \sum \delta^n < \infty, \end{aligned}$$

and so u_A^1 is bounded. In addition we assert that u_A^1 is a *regular* 1-potential of (X, H). Let us suppose this last statement has been established and use it to complete the proof. If u_A^1 is known to be a regular 1-potential, then according to Corollary 3.14 there is a *continuous* AF, say B, of (X, H) such that $u_A^1 = u_B^1$. Of course B is natural since it is continuous, but A is also natural and so by the uniqueness theorem, (2.13), A and B are equivalent. In particular A is continuous; but $P^y(A$ has a discontinuity$) \ge P^y(T < H) > 0$, a contradiction. Thus to complete the proof we need only show that u_A^1 is a regular 1-potential.

To this end suppose $\{J_n\}$ is an increasing sequence of stopping times with limit J. Then, with Q_t denoting the transition operators for (X, H), we have

$$Q_{J_n}^1 u_A^1(x) - Q_J^1 u_A^1(x) = E^x \int_{(J_n, J]} e^{-t}\, dA_t,$$

and to prove that this approaches 0 as $n \to \infty$ it suffices to prove that for all x and k

(4.11) $$P^x\{T_k < H, J_n < T_k \text{ for all } n, J = T_k\} = 0.$$

Since $\bar{\psi} \in \mathscr{S}^1$, $\{e^{-J_n} \bar{\psi}(X_{J_n}), \mathscr{F}_{J_n}\}$ is a positive supermartingale relative to each measure P^x and so we can define a random variable L, $0 \le L \le 1$, such that

$$e^{-J} L = \lim_n e^{-J_n} \bar{\psi}(X_{J_n})$$

almost surely P^x for each x. Because T is complete, if W is any stopping time, then $T_{n+1} = W + T \circ \theta_W$ on $\{T_n \le W < T_{n+1}\}$. Now fix x and k. Given any n, $m \ge n$, and $\Lambda \in \mathscr{F}_{J_n}$ we have

$$E^x(e^{-T_{k+1}}; \Lambda \cap \{T_k \le J_m < T_{k+1} \wedge H\})$$
$$= E^x(e^{-J_m} \bar{\psi}(X_{J_m}); \Lambda \cap \{T_k \le J_m < T_{k+1} \wedge H\}).$$

If we let $m \to \infty$ and define $\Gamma = \{T_k \le J_m < T_{k+1} \wedge H$ for all large $m\}$, then

$$E^x\{e^{-T_{k+1}}; \Lambda \cap \Gamma\} = E^x\{e^{-J}L; \Lambda \cap \Gamma\}.$$

This equality holds for all $\Lambda \in \mathscr{F}_{J_n}$ for all n and hence for all $\Lambda \in \bigvee_n \mathscr{F}_{J_n}$. But according to (4.2) this σ-algebra is simply $\mathscr{F}_J$. Thus we can let $\Lambda = \{T_{k+1} = J, J < H\}$ and obtain the conclusion that $L = 1$ almost surely P^x on

$$\Lambda \cap \Gamma = \{T_{k+1} = J,\ T_{k+1} < H,\ T_k \le J_m < T_{k+1} \text{ for all large } m\}.$$

On the other hand $\bar{\psi}(X_{J_m}) \le \delta$ if $J_m < H$ and so $L \le \delta$ on $\Lambda \cap \Gamma$. Hence $P^x(\Lambda \cap \Gamma) = 0$, which certainly implies (4.11). The proof of (4.10) is complete.

For use in Section 5 we need an additional fact concerning accessibility. As before let T be a terminal time with $T \le \zeta$ and let $\psi(x) = E^x(e^{-T})$ for x in E and $\bar{\psi} = \lim_{t \downarrow 0} P_t^1 \psi$.

(4.12) Proposition. Let T be a strong terminal time, let R_n be the hitting time of $\{\bar{\psi} > 1 - 1/n\}$, and let $\Lambda = \{R_n < T$ for all $n\}$. Then (a) $\lim R_n = T$ almost surely on Λ, (b) $\lim_{t \uparrow T} \bar{\psi}(X_t) = 1$ almost surely on $\Lambda \cap \{T < \infty\}$, and (c) if $\{Q_n\}$ is an increasing sequence of stopping times with limit Q, then almost surely $\{Q_n < T$ for all n, $Q = T < \infty\}$ is contained in Λ.

Proof. The first sentence in the proof of (4.10) establishes (a). We next prove (c). Let $L \in \mathscr{F}$ be such that $e^{-Q_n} \bar{\psi}(X_{Q_n}) \to e^{-Q}L$ almost surely as $n \to \infty$. Given positive integers k and j, a set $\Gamma \in \mathscr{F}_{Q_j}$, and $n \ge j$ we have, since $\bar{\psi}(X_{Q_n}) = \psi(X_{Q_n})$ on $\{Q_n < T\}$,

$$E^x[e^{-Q_n} \bar{\psi}(X_{Q_n}); \Gamma \cap \{Q_n < T \wedge k\}] = E^x[e^{-T}; \Gamma \cap \{Q_n < T \wedge k\}].$$

If we fix k and let $B = \{Q_n < T \wedge k$ for all $n\}$, then letting $n \to \infty$ in the above equality we obtain

$$\text{(4.13)} \qquad E^x\{e^{-Q}L; \Gamma \cap B\} = E^x\{e^{-T}; \Gamma \cap B\}.$$

This holds for all $\Gamma \in \mathscr{F}_{Q_j}$ for each j and hence by (4.2) for all $\Gamma \in \mathscr{F}_Q$. In particular $\Gamma = \{T = Q\}$ is in $\mathscr{F}_Q$. Since $L \le 1$ almost surely on B, this choice of Γ in (4.13) implies that $L = 1$ almost surely on $B \cap \Gamma$. Consequently $\bar{\psi}(X_{Q_n}) \to 1$ and so $R_n < T = Q$ for all n almost surely on $B \cap \Gamma$. If we let $k \to \infty$ we obtain (c). Finally applying the above proof to the sequence $\{R_n\}$

rather than $\{Q_n\}$ we see that $\lim_n \bar{\psi}(X_{R_n}) = 1$ on $\Lambda \cap \{T < \infty\}$ and this implies (b) since $\lim_{t \uparrow T} \bar{\psi}(X_t)$ exists almost surely on $\{T < \infty\}$.

(4.14) REMARK. If T is an exact terminal time and $\psi^\alpha(x) = E^x\{e^{-\alpha T}; T < \zeta\}$, then the above arguments show that whenever $\{Q_n\}$ is an increasing sequence of stopping times with limit Q one has $\lim_n \psi^\alpha(X_{Q_n}) = 1$ almost surely on $\{Q_n < T \text{ for all } n,\ Q = T < \infty\}$ for each $\alpha > 0$. The same conclusion then holds for $\alpha = 0$ since ψ^α increases as α decreases.

We turn next to the analysis of a particular stopping time. Let g be a finite (X, M) excessive function. Recall that almost surely the mapping $t \to g(X_t)$ is right continuous and has left-hand limits on $[0, S)$ where $S = \inf\{t: M_t = 0\}$. With this g fixed let Γ denote the set of ω such that $t \to g(X_t(\omega))$ and $t \to X_t(\omega)$ are right continuous and have left limits on $[0, S)$ and $[0, \zeta)$, respectively. Let $Y_t = g(X_t)\, M_t$; we use Y_{t-} and $g(X_t)_-$ to denote $\lim_{s \uparrow t} Y_s$ and $\lim_{s \uparrow t} g(X_s)$, respectively, whenever these limits exist. Given $\varepsilon > 0$ define

$$(4.15)\quad T(\omega) = \inf\{t < S(\omega): |Y_t - Y_{t-}| > \varepsilon M_t;\ X_t = X_{t-}\}, \qquad \omega \in \Gamma,$$

$$= 0, \qquad \omega \notin \Gamma.$$

Finally let $T(\omega) = S(\omega)$ if $\omega \in \Gamma$ and the set in braces is empty. In view of the right continuity of $t \to Y_t$ the infimum in (4.15) is actually attained. In particular $X_T = X_{T-}$ on $\{T < S\}$.

(4.16) PROPOSITION. T is a complete terminal time and T is accessible on $\{T < S\}$.

Proof. Suppose $\omega \in \Gamma$ and $a > 0$. Using the regularity properties of $Y_t(\omega)$ and $X_t(\omega)$ one checks without difficulty that $T(\omega) < a$ if and only if the following holds: either $S(\omega) < a$ or $S(\omega) \geq a$ and for some positive integer k and every positive integer n there are rational numbers r_n and s_n such that (a) $r_n < s_n < a - 1/k$, (b) $s_n - r_n < 1/n$, (c) $d[X_{r_n}(\omega), X_{s_n}(\omega)] < 1/n$, and (d) $|Y_{r_n}(\omega) - Y_{s_n}(\omega)| \geq (\varepsilon + 1/k) M_{s_n}(\omega)$, where d is a metric for E_Δ. This describes $\Gamma \cap \{T < a\}$ in terms of a countable number of sets, each of which is in $\mathscr{F}_a$. Since $P^x(\Omega - \Gamma) = 0$ for all x this proves that $\{T < a\} \in \mathscr{F}_a$. In view of the explicit description of T and the fact that M is a strong multiplicative functional one obtains immediately that T and its iterates satisfy the hypothesis of (4.7), so T is complete. Now $P^x(T = 0) = 1$ if and only if $P^x(S = 0) = 1$. Hence, in the notation of Proposition 4.10, $\{\psi < 1\} = E_M$, and so $R \geq T_{E_\Delta - E_M}$. As in Section 5 of Chapter III let $\varphi(x) = E^x \int_0^\infty e^{-t} M_t\, dt$, $E_n = \{\varphi > 1/n\}$, and $Q_n = T_{E_\Delta - E_n}$ (in Section 5 of Chapter III we used T_n in place of Q_n). According to Proposition 5.3 of Chapter III, $\lim_n Q_n \geq S$ almost surely. Now fix k and set $Q = Q_k \wedge S$. If $P^x(Q = 0) = 1$, then x is regular for Q_k or $\varphi(x) = 0$. But if $\varphi(x) = 0$ then x is regular for Q_k since φ is finely continuous. Thus $P^x(Q = 0) = 1$ implies x is regular for Q_k, and hence $t + Q \circ \theta_t \leq t + Q_k \circ \theta_t \to 0$ as $t \to 0$ almost surely P^x. Thus it follows from Proposition 5.9 of Chapter III that Q is exact. We now apply Proposition 4.10 with this $Q = Q_k \wedge S$. The hypotheses of (4.10) are obviously satisfied. Consequently with the R_n's defined there we have almost surely $R_n < T$ for all n on $\{0 < T < Q_k \wedge S\}$ and $\lim R_n = T$ on $\{R_n < T$ for all $n\}$. Finally letting $k \to \infty$ we see that T is accessible on $\{T < S\}$. We have actually proved much more: namely, that almost surely $\lim R_n \geq T$ and that $\{R_n\}$ increases to T strictly from below on $\{0 < T < S\} = \{T < S\}$.

We finally have the necessary tools at our disposal for studying the potentials of natural additive functionals. We will use the following fact repeatedly: if $f \in \mathscr{S}(M)$, Q and R stopping times, and $\Lambda \in \mathscr{F}_Q$, then for each x, $E^x\{M_Q f(X_Q); Q \leq R; \Lambda\} \geq E^x\{M_R f(X_R); Q \leq R; \Lambda\}$. (This is obvious for potentials and hence extends to elements of $\mathscr{S}(M)$.)

(4.17) DEFINITION. A finite (X, M) excessive function u is called a *natural potential* if for each x whenever $\{T_n\}$ is an increasing sequence of stopping times with limit $T \geq S$ almost surely P^x, then $\lim_n Q_{T_n} u(x) = 0$.

If A is any AF of (X, M) with a finite potential u_A, then the fact that $Q_T u_A(x) = E^x\{A(S) - A(T)\}$ implies that u_A is a natural potential. The remainder of this section is devoted to proving that any natural potential is the potential of an NAF.

(4.18) PROPOSITION. Let u be a natural potential and let μ be a measure on E_Δ such that $\int u \, d\mu < \infty$. Then the family of random variables $\{M_T u(X_T);$ T a stopping time$\}$ is P^μ uniformly integrable.

Proof. Let $R_n = \inf\{t: M_t u(X_t) > n\}$. Clearly $\{R_n\}$ is an increasing sequence of stopping times and since almost surely $M_{R_n} u(X_{R_n}) \geq n$ on $\{R_n < \infty\}$ we have

$$n P^x(R_n < S) \leq E^x\{M_{R_n} u(X_{R_n})\} \leq u(x) < \infty,$$

and so $R_n \geq S$ for large n almost surely. If T is any stopping time and $\Lambda = \{M_T u(X_T) > n\}$, then almost surely $\Lambda \subset \{R_n \leq T\}$ so

$$E^\mu\{M_T u(X_T); \Lambda\} \leq \int E^x\{M_{R_n} u(X_{R_n})\} \mu(dx). \tag{4.19}$$

But $E^x\{M_{R_n} u(X_{R_n})\} \to 0$ since u is a natural potential and the integrands are dominated by $u(x)$ with $\int u\,d\mu < \infty$, so the right side of (4.19) approaches 0. The lack of dependence on T in this estimate yields the result.

(4.20) PROPOSITION. Let u be a natural potential and $\{J_n\}$ an increasing sequence of stopping times with limit $J \leq S$ almost surely P^x. Then $E^x\{M_{J_n} u(X_{J_n}); J = S\} \to 0$ as $n \to \infty$.

Proof. Let $Y = \lim_n M_{J_n} u(X_{J_n})$. This limit exists almost surely and in fact, since $u(x) < \infty$, (4.18) implies that the convergence also takes place in $L^1(P^x)$. Of course $Y = 0$ on $\{J_n = S \text{ for some } n\}$. Let us bring in Proposition 4.12 applied to the strong terminal time S. If $\{R_n\}$ is the sequence defined there and $\Gamma = \{J_n < J \text{ for all } n\}$, then $Q_k = R_k \wedge k$ increases to S strictly from below, and so for each k

$$\begin{aligned}\lim_n E^x\{M_{J_n} u(X_{J_n});\, J = S\} &= \lim_n E^x\{M_{J_n} u(X_{J_n});\, \Gamma;\, J = S\}\\ &\leq \lim_n E^x\{M_{J_n} u(X_{J_n});\, J_n > Q_k\}\\ &\leq E^x\{M_{Q_k} u(X_{Q_k})\} \to 0\end{aligned}$$

as $k \to \infty$, since $\lim Q_k \geq S$ and u is a natural potential. The proof is complete.

(4.21) PROPOSITION. Let $\{T_n\}$ be an increasing sequence of stopping times with limit T. If $\alpha > 0$ and $f \in b\mathscr{E}_+^*$, then $\lim_n M_{T_n} V^\alpha f(X_{T_n}) = M_T V^\alpha f(X_T)$ almost surely on $\{T < \infty\}$. If g is in $\mathscr{S}^\alpha(M)$, $\alpha \geq 0$, then $\lim_n M_{T_n} g(X_{T_n}) \geq M_T\, g(X_T)$ almost surely on $\{T < \infty\}$.

Proof. If x is fixed, then $\{e^{-\alpha T_n} M_{T_n} V^\alpha f(X_{T_n}), P^x\}$ is a nonnegative supermartingale and hence $U = \lim_n e^{-\alpha T_n} M_{T_n} V^\alpha f(X_{T_n})$ exists P^x almost surely. In particular $L = \lim_n M_{T_n} V^\alpha f(X_{T_n})$ exists almost surely on $\{T < \infty\}$. If $\Lambda \in \mathscr{F}_{T_n}$ and $k \geq n$ one has

$$E^x\{e^{-\alpha T_k} M_{T_k} V^\alpha f(X_{T_k}); \Lambda\} = E^x\left\{\int_{T_k}^\infty e^{-\alpha t} M_t f(X_t)\,dt; \Lambda\right\},$$

and consequently if we subtract the corresponding expression with T_k replaced T and then let $k \to \infty$ one obtains $E^x\{U - e^{-\alpha T} M_T V^\alpha f(X_T); \Lambda\} = 0$. This then holds for all $\Lambda \in \mathscr{F}_T$ because of Proposition 4.2 and hence $U = e^{-\alpha T} M_T V^\alpha f(X_T)$ almost surely. In particular $L = M_T V^\alpha f(X_T)$ almost surely on $\{T < \infty\}$.

If g is (X, M) excessive, then g is $\alpha - (X, M)$ excessive for some positive α. Thus there exists a sequence $\{f_n\}$ of elements of $b\mathscr{E}_+^*$ such that $V^\alpha f_n \uparrow g$. Now $\lim_n M_{T_n} g(X_{T_n})$ exists almost surely on $\{T < \infty\}$, and hence almost surely on $\{T < \infty\}$

$$\begin{aligned}\lim_n M_{T_n} g(X_{T_n}) &\geq \lim_k \lim_n M_{T_n} V^\alpha f_k(X_{T_n}) \\ &= \lim_k M_T V^\alpha f_k(X_T) \\ &= M_T\, g(X_T),\end{aligned}$$

which establishes Proposition 4.21.

We are now ready to prove the main theorem on representation of natural potentials.

(4.22) THEOREM. Let u be a natural (X, M) potential. Then there exists a unique natural additive functional A of (X, M) whose potential is u.

Proof. Given $\varepsilon > 0$ let T be the stopping time defined in (4.15) with this ε and of course the (X, M) excessive function g there replaced by u. Define as usual $T_0 = 0$ and $T_{n+1} = \min(T_n + T \circ \theta_{T_n}, S)$. In the remainder of the proof we will omit the phrase "almost surely" in places where there is no doubt as to its appropriateness. Since $|[M_{T_n} u(X_{T_n})]_- - M_{T_n} u(X_{T_n})|$ exceeds εM_{T_n} it follows from Theorem 5.6 of Chapter III that $\lim T_n = S$ and in fact that $T_n = S$ for large n on $\{M_{S-} > 0\}$. We will use the following notation: $Y_n = M_{T_n} u(X_{T_n})$ and $Z_n = [M_{T_n} u(X_{T_n})]_-$. Of course $Y_n = 0$ if $T_n = S$. Furthermore, since T is a complete terminal time according to (4.16), if K is any stopping time then $Y_{n+k} = M_K(Y_k \circ \theta_K)$ and $Z_{n+k} = M_K(Z_k \circ \theta_K)$ on $\{T_n \leq K < T_{n+1}\}$. Let us define

$$A^\varepsilon(t) = \sum_{T_n \leq t} (Z_n - Y_n)$$

where of course we are setting $Z_0 = Y_0 = u(X_0)$. Also we adopt the convention $[M_S u(X_S)]_- = 0$, so that in reality we are summing only over those n for which $T_n \leq t$ and $T_n < S$. Now it follows from (4.16) and (4.21) that each of the summands $Z_n - Y_n$ is nonnegative. Suppose we are given positive numbers t and r and that $T_k \leq t < T_{k+1}$. Then $T_{k+j} \leq t + r$ if and only if $T_j \circ \theta_t \leq r$, and $T_{k+j} < S$ if and only if $T_j \circ \theta_t < S \circ \theta_t$. Since for any such j we have $Z_{k+j} - Y_{k+j} = M_t(Z_j - Y_j) \circ \theta_t$ it follows that on $\{T_k \leq t < T_{k+1}\}$, $A^\varepsilon(t + u) = A^\varepsilon(t) + M_t(A^\varepsilon(u) \circ \theta_t)$. But $\{S > t\} = \bigcup_k \{T_k \leq t < T_{k+1}\}$ and consequently $A^\varepsilon = \{A^\varepsilon(t)\}$ is an AF of (X, M). The discontinuities of A^ε occur only at the points $T_n < S$. By construction the path is continuous at all such points and so A^ε is an NAF of (X, M). Now T is accessible on $\{T < S\}$; in fact, according to the last sentence of the proof of Proposition 4.16 there is an increasing sequence $\{R_n\}$ of stopping times with limit T and such that $R_n < T$ for all n on $\{T < S\}$. Moreover $u(x) \geq E^x M_{R_n} u(X_{R_n})$ for all n and x, since u is (X, M) excessive. Also $M_{R_n} u(X_{R_n}) = 0$ if $R_n = S$ and, by (4.20), $M_{R_n} u(X_{R_n})$ approaches

0 on $\{R_n < S \text{ for all } n, R_n \to S\}$. On the remaining set, $M_{R_n} u(X_{R_n})$ approaches Z_1 and so in light of (4.18)

(4.23) $$0 \le u(x) - E^x M_{R_n} u(X_{R_n}) \to u(x) - E^x Z_1.$$

Having made these observations we continue the proof by comparing the potential $w^\varepsilon(x) = E^x A^\varepsilon(S)$ with the excessive function u. Since u is a natural potential we have $E^x M_{T_n} u(X_{T_n}) = E^x Y_n \to 0$ as $n \to \infty$ and of course $w^\varepsilon(x) = \lim_n E^x\{\sum_1^n (Z_k - Y_k)\}$. Consequently

$$\begin{aligned} u(x) - w^\varepsilon(x) &= \lim_n \left\{u(x) - E^x Y_n - \sum_1^n E^x(Z_k - Y_k)\right\} \\ &= \lim_n \left[E^x\left\{\sum_0^{n-1} (Y_k - Y_{k+1}) - \sum_1^n (Z_k - Y_k)\right\}\right] \\ &= \lim_n \sum_0^{n-1} E^x(Y_k - Z_{k+1}) \\ &= \lim_n \sum_0^{n-1} E^x\{M_{T_k}[u(X_{T_k}) - E^{X(T_k)}(Z_1)]\}. \end{aligned}$$

According to (4.23) each summand in this last expression for $u - w^\varepsilon$ is positive and so $w^\varepsilon \le u$. Let $v^\varepsilon = u - w^\varepsilon$. We are next going to show that v^ε is (X, M) excessive. Since u and w^ε are (X, M) excessive we have $Q_t v^\varepsilon \to v^\varepsilon$ as $t \to 0$, and so it is enough to show that $Q_t v^\varepsilon \le v^\varepsilon$. Let us begin by separating out the basic inequality we need. We will show that for each x, n, and t

(4.24) $$E^x\{Y_n - Z_{n+1}; T_n \le t\} \ge E^x\{[u(X_t) M_t - Z_{n+1}]; T_n \le t < T_{n+1}\}.$$

Indeed, let $u_k = Vg_k$ with u_k increasing to u (this is possible since $Q_t u \to 0$ as $t \to \infty$, u being a natural potential) and let $\{R_j\}$ be a sequence of stopping times increasing to T_{n+1} and strictly less than T_{n+1} on $\{T_{n+1} < S\}$. Moreover we may assume $R_j \ge T_n$ for all j. Then for each k and j

(4.25) $$\begin{aligned} &E^x\{u_k(X_{T_n}) M_{T_n} - u_k(X_{R_j}) M_{R_j}; T_n \le t\} \\ &= E^x\left\{\int_{T_n}^{R_j} g_k(X_s) M_s \, ds; T_n \le t\right\} \\ &\ge E^x\left\{\int_t^{R_j} g_k(X_s) M_s \, ds; T_n \le t < R_j\right\} \\ &= E^x\{u_k(X_t) M_t - u_k(X_{R_j}) M_{R_j}; T_n \le t < R_j\}. \end{aligned}$$

If in (4.25) we first let $k \to \infty$ and then $j \to \infty$ the extreme members approach the left and right sides of (4.24), and so (4.24) is established. Since $\lim T_n = S$ it is clear, writing v for v^ε, that the relationship $Q_t v \le v$ will be established if we can show that for each n

$$v(x) \geq E^x\{v(X_t)\, M_t;\ t < T_n\} + E^x\{v(X_{T_n})\, M_{T_n};\ T_n \leq t\}.$$

Obviously this is true when $n = 0$. To pass from its validity for a given value of n to the next larger one it suffices to show that

$$\text{(4.26)}\quad E^x\{v(X_{T_n})\, M_{T_n};\ T_n \leq t\} \geq E^x\{v(X_t)\, M_t;\ T_n \leq t < T_{n+1}\} + E^x\{v(X_{T_{n+1}})\, M_{T_{n+1}};\ T_{n+1} \leq t\}.$$

But we have seen that $v(x) = E^x\{\sum_0^\infty (Y_k - Z_{k+1})\}$ and so the left side of (4.26) is

$$E^x\left\{M_{T_n}\, E^{X(T_n)}\left[\sum_0^\infty (Y_k - Z_{k+1})\right];\ T_n \leq t\right\}$$

$$= E^x\left\{\sum_0^\infty [(Y_k - Z_{k+1}) \circ \theta_{T_n}]\, M_{T_n};\ T_n \leq t\right\}$$

$$= E^x\left\{\sum_n^\infty (Y_k - Z_{k+1});\ T_n \leq t\right\}.$$

In the second summand on the right of (4.26) we first express the range of integration $\{T_{n+1} \leq t\}$ as $\{T_n \leq t\} - \{T_n \leq t < T_{n+1}\}$. Bringing in the expression for v again we see that this second summand is

$$E^x\left\{\sum_{n+1}^\infty (Y_k - Z_{k+1});\ T_n \leq t\right\} - E^x\{v(X_{T_{n+1}})\, M_{T_{n+1}};\ T_n \leq t < T_{n+1}\}.$$

Consequently subtracting the right side of (4.26) from the left we obtain

$$\text{(4.27)}\quad E^x\{Y_n - Z_{n+1};\ T_n \leq t\} - E^x\{v(X_t)\, M_t - v(X_{T_{n+1}})\, M_{T_{n+1}};\ T_n \leq t < T_{n+1}\}.$$

Now $v = u - w^\varepsilon$ and $w^\varepsilon(x) = \sum_1^\infty E^x(Z_k - Y_k)$. Using this for the first occurrence of v in (4.27) and our previous expression for the second, we see that (4.27) reduces to

$$E^x\{Y_n - Z_{n+1};\ T_n \leq t\} - E^x\{u(X_t)\, M_t - Z_{n+1};\ T_n \leq t < T_{n+1}\}.$$

In view of (4.24) this is nonnegative, so (4.26) is established and we have shown that v^ε is (X, M) excessive.

Now for each $\varepsilon > 0$ we have $u = w^\varepsilon + v^\varepsilon$. It is clear from the construction that w^ε increases as ε decreases to 0. Thus $w = \lim_{\varepsilon \to 0} w^\varepsilon$ exists and is an (X, M) excessive function dominated by u. Consequently v^ε must decrease and if $v = \lim v^\varepsilon$, then it is obvious that $Q_t v \leq v$, and hence v is (X, M) excessive being the difference of two finite (X, M) excessive functions. It is also clear from the definitions and what we have already proved that $A^\varepsilon(t) - A^\eta(t)$ is an (X, M) additive functional provided $\varepsilon < \eta$. We define $A_J(t) = \lim_{\varepsilon \downarrow 0} A^\varepsilon(t)$ for $0 \leq t \leq \infty$. Now $w(x) = \lim_\varepsilon w^\varepsilon(x) = \lim_\varepsilon E^x A^\varepsilon(\infty) =$

$E^x A_J(\infty)$, and hence $A_J(\infty)$ is finite. But if $\varepsilon < \eta$ then $0 \le A^\varepsilon(t) - A^\eta(t) \le A^\varepsilon(\infty) - A^\eta(\infty)$, and therefore $A^\varepsilon(t) \to A_J(t)$ uniformly on $[0, \infty]$. This implies that A_J is right continuous on $[0, \infty]$ and continuous at S. It is now clear that A_J is an NAF of (X, M) whose potential is w.

We will complete the proof of Theorem (4.22) by showing that v is a regular (X, M) potential. Let $\{R_n\}$ be an increasing sequence of stopping times with limit R; we must then show $E^x\{M_{R_n} v(X_{R_n})\} \to E^x\{M_R v(X_R)\}$. Clearly it is no restriction to assume that $R \le S$. Now

$$0 \le E^x\{M_{R_n} v(X_{R_n}) - M_R v(X_R)\} = E^x\{M_{R_n} v(X_{R_n}); R = S\} + E^x\{M_{R_n} v(X_{R_n}) - M_R v(X_R); R < S\},$$

and Proposition 4.20 and the fact that $v \le u$ imply that the first term on the right approaches zero as $n \to \infty$. If $\Lambda = \{R_n < R \text{ for all } n\}$, then the second term approaches $E^x\{[M_R v(X_R)]_- - M_R v(X_R); \Lambda, R < S\}$. But R is accessible on $\Lambda \cap \{R < S\}$ and hence $t \to X_t$ is continuous at R on $\Lambda \cap \{R < S\}$ because of the quasi-left-continuity of X. We also claim that

$$\textbf{(4.28)} \qquad [M_R v(X_R)]_- - M_R v(X_R) = [M_R u(X_R)]_- - M_R u(X_R) - [A_J(R) - A_J(R-)]$$

on $\Lambda \cap \{R < S\}$. We will prove a statement slightly more general than this in a moment. But assuming this, the right side of (4.28) is zero since by construction the discontinuities of A_J are precisely the discontinuities of $M_t u(X_t)$ at which X_t is continuous. Consequently v is a regular (X, M) potential and so by Theorem 3.13 there exists a CAF, A_C, of (X, M) whose potential is v, and thus u is the potential of $A_J + A_C$ which is natural. Thus the proof of Theorem 4.22 will be complete as soon as (4.28) is established. We formulate this as a proposition since we will want to refer to it in the sequel.

(4.29) PROPOSITION. Let A be an (X, M) additive functional with finite potential u. Let $\{R_n\}$ be an increasing sequence of stopping times with limit R and let $\Lambda = \{R_n < R \text{ for all } n\}$. Then $\lim_n [A(R) - A(R_n)] = \lim_n M_{R_n} u(X_{R_n}) - M_R u(X_R)$ almost surely. In particular $\lim_n M_{R_n} u(X_{R_n}) = [M_R u(X_R)]_-$ and $\lim_n A(R_n) = A(R-)$ on $\Lambda \cap \{R < S\}$.

Proof. If $\Gamma \in \mathscr{F}_{R_k}$ and $n > k$, then

$$E^x\{A(R) - A(R_n); \Gamma\} = E^x\{M_{R_n} u(X_{R_n}) - M_R u(X_R); \Gamma\}.$$

Using Propositions 4.18 and 4.2 we obtain

$$\lim_n [A(R) - A(R_n)] = \lim_n M_{R_n} u(X_{R_n}) - M_R u(X_R)$$

almost surely, which establishes (4.29).

We close this section with a characterization of CAF's that is analogous to that given for NAF's in Proposition 2.5.

(4.30) THEOREM. Let A be an additive functional of (X, M) with finite potential. Then A is continuous if and only if $Q_K U_A I_K = U_A I_K$ for all compact K. Recall that $U_A f(x) = E^x \int_0^\infty f(X_t)\, dA(t)$ is the potential operator corresponding to A.

Proof. We first observe that

$$Q_K U_A I_K(x) = E^x \int_{(T_K, \infty)} I_K(X_t)\, dA(t),$$

while

$$U_A I_K(x) = E^x \int_0^\infty I_K(X_t)\, dA(t) = E^x \int_{[T_K, \infty)} I_K(X_t)\, dA(t).$$

Therefore

$$U_A I_K(x) - Q_K U_A I_K(x) = E^x\{A(T_K) - A(T_K -)\}, \tag{4.31}$$

and so if A is continuous then $U_A I_K = Q_K U_A I_K$.

We now turn to the proof of the converse. If G is open and K a compact subset of G, then $T_G \le T_K$ and so

$$Q_K U_A I_K \le Q_G U_A I_K \le U_A I_K.$$

But by hypothesis $Q_K U_A I_K = U_A I_K$ and so $Q_G U_A I_K = U_A I_K$. Hence $Q_G U_A I_G = U_A I_G$. By Proposition 2.5, or more precisely its proof, this implies that A is natural. Now given $\varepsilon > 0$ define

$$T = \inf\{t : A_t - A_{t-} > \varepsilon M_t\}$$

where as usual we set $T = S$ if the set in braces is empty. A discussion similar to that in (4.16) shows that T is a complete terminal time. Because A is natural, the path is continuous at T if $T < S$, and so, arguing as in the second part of the proof of (4.16), T is accessible on $\{T < S\}$. Somewhat more precisely, the proof of (4.12) shows that if $\psi(x) = E^x e^{-T}$, then almost surely on $\{T < S\}$, $\psi(X_t) \to 1$ as t increases to T. Of course almost surely $\psi(X_T) < 1$ on $\{T < S\}$. To complete the proof of (4.30) it will suffice to show that $P^x(T < S) = 0$ for all x, since ε in the definition of T is arbitrary. Suppose, to the contrary that, for some x, $P^x(T < S) > 0$. Because $\psi(X_T) < 1$ if $T < S$ there is an $\eta < 1$ and a compact subset K of $\{\psi < \eta\}$ such that $P^x(X_T \in K, T < S) > 0$. Let $\{T_n\}$ be a sequence of stopping times increasing to T such that almost surely $T_n < T$ for all n on $\{T < S\}$. If $\psi(X_t) \ge \eta$ for all $t \in [T_n, T)$, then $T = T_n + T_K \circ \theta_{T_n}$ on $\{X_T \in K, T < S\}$. Consequently for sufficiently large n,

$$0 < P^x\{A(T_n + T_K \circ \theta_{T_n}) - A([T_n + T_K \circ \theta_{T_n}]-) > 0\}$$
$$\leq E^x\{P^{X(T_n)}[A(T_K) - A(T_K -)] > 0\} = 0,$$

since, by (4.31), $E^y\{A(T_K) - A(T_K-)\} = 0$ for all y. This contradiction completes the proof.

REMARK. We have actually proved that if A is a NAF of (X, M), then either A is continuous or else there is a compact set K and a point $x \in E_M$ such that $t \to A_t$ is discontinuous at T_K with positive P^x probability.

Exercises

(4.32) Prove that (4.1) is satisfied if and only if $U^\alpha f(X_{T_n})$ approaches $U^\alpha f(X_T)$ almost surely on $\{T < \infty\}$ for every $\alpha > 0, f \in \mathbf{C}_K(E)$, and increasing sequence $\{T_n\}$ of stopping times (where $T = \lim T_n$).

(4.33) Prove that the following condition is sufficient for the validity of (4.1): whenever $\{T_n\}$ is an increasing sequence of stopping times with limit T, then almost surely on $\{T < \infty\}$ either $X(T_n) \to X(T)$ or $\lim X(T_n)$ does not exist in E and $T = \zeta$.

(4.34) Show that for the process considered in (9.16) of Chapter I and again in (3.18) of Chapter III condition (4.1) is not satisfied.

(4.35) Let T, R, Q be three stopping times, Λ a set in $\mathscr{F}$, and suppose that $T = R + Q \circ \theta_R$ almost surely on Λ. Prove that for every $F \in \mathscr{F}$ we have $F \circ \theta_T = F \circ \theta_Q \circ \theta_R$ almost surely on Λ. [Hint: check directly that the conclusion is valid when $F = \prod_{i=1}^n f_i(X_{t_i})$ with $f_i \in \mathscr{E}^*$. Use linearity and the MCT to conclude that it is valid for every $F \in \mathscr{F}^0$. Extend the conclusion to $F \in \mathscr{F}$ by using the definition of $\mathscr{F}$ and the strong Markov property.]

(4.36) Let T be an exact terminal time with $A = \{x : x \text{ is regular for } T\}$, let R be a stopping time and μ be a finite measure on $\mathscr{E}_\Delta$. Define $\{R_n\}$ by $R_n = (k+1)/2^n$ if $k/2^n \leq R < (k+1)/2^n$ and $R_n = \infty$ if $R = \infty$.

(a) Prove that as $n \to \infty$, $E^\mu\{f(T \circ \theta_{R_n})\, h \circ \theta_{R_n}; \Lambda\} \to E^\mu\{f(T \circ \theta_R)\, h \circ \theta_R; \Lambda\}$ for $\Lambda \in \mathscr{F}_R$, f a continuous function on $[0, \infty]$ and $h \in \mathscr{F}$ of the form $h = \prod_{i=1}^n f_i(X_{t_i})$ with each f_i bounded and continuous on E_Δ. [Hint: use the strong Markov property and the fact that (Chapter II, (4.14)) $x \to E^x\{f(T)\, h\}$ is finely continuous.]

(b) Show that $\lim_{n\to\infty} R_n + T \circ \theta_{R_n} = R + T \circ \theta_R$ in P^μ measure on $\{X_R \in A\}$.

(c) Show that if $n \geq m$, then $R_n + T \circ \theta_{R_n} = R_m + T \circ \theta_{R_m}$ almost surely on $\{R_m < R_n + T \circ \theta_{R_n}\}$, hence almost surely on $\{T \circ \theta_{R_n} \geq 2^{-m}\}$.

(d) Use (a) and the fact that $T \circ \theta_R > 0$ almost surely on $\{X_R \notin A\}$ to show that

$$\lim_{m\to\infty} \sup_{n \geq m} P^\mu\{X_R \notin A,\ T \circ \theta_{R_n} < 2^{-m}\} = 0.$$

(e) Use (d) and (c) to show that $\lim_n T \circ \theta_{R_n}$ exists in P^μ measure on $\{R < \infty,\ X_R \notin A\}$ and use (a) to identify this limit as $T \circ \theta_R$.

(*f*) Combine what we have done so far to conclude that $R_n + T \circ \theta_{R_n}$ approaches $R + T \circ \theta_R$ in P^μ measure as $n \to \infty$.

(g) Show that T is a complete terminal time. [Hint: check that the hypotheses of (4.7) are satisfied by approximating T_k and R with stopping times which take values in a fixed countable set and then using (*f*).]

(4.37) Let $\{R_n\}$ be an increasing sequence of stopping times with limit R, let T be a stopping time and let $\Lambda = \{R_n < T \text{ for all } n,\ R = T\}$. Prove that for each μ there is an increasing sequence $\{J_n\}$ of stopping times such that $J_n < T$ for all n and $\lim J_n = T$ almost surely P^μ on Λ, and such that also $J_n = \infty$ for all large n almost surely P^μ on $\Omega - \Lambda$. [Hint: use (4.2) to conclude that for each integer n there is an integer k_n and a set $\Gamma_n \in \mathscr{F}_{R_{k_n}}$ such that $P^\mu\{\Gamma_n \Delta \Lambda\} \leq 2^{-n}$. Assume (as we may) that $k_1 < k_2 < \ldots$, and define stopping times Q_n by

$$\begin{aligned} Q_n(\omega) &= R_{k_n}(\omega) \quad &&\text{if} \quad \omega \in \Gamma_n, \\ &= \infty &&\text{if} \quad \omega \notin \Gamma_n. \end{aligned}$$

Finally set $J_n = \inf_{j \geq n} Q_j$ and check that $\{J_n\}$ has the required properties.]

(4.38) Prove that if a stopping time T is accessible on each of a sequence $\{\Lambda_n\}$ of sets, then T is accessible on $\bigcup \Lambda_n$. [Hint: start by using (4.37) to conclude that if T is accessible on Λ_1 and Λ_2 then it is also accessible on $\Lambda_1 \cup \Lambda_2$.]

5. Classification of Excessive Functions

The purpose of this section is to define various special classes of excessive functions and establish some important properties of these classes. In particular we will relate them to the regular potentials and natural potentials defined in Sections 3 and 4. As usual $X = (\Omega, \mathscr{M}, \mathscr{M}_t, X_t, \theta_t, P^x)$ denotes a fixed standard process with state space $(E, \mathscr{E})$, and again in this section we assume for simplicity that $\mathscr{M} = \mathscr{F}$ and $\mathscr{M}_t = \mathscr{F}_t$ for each t. M denotes a fixed exact MF of X with $M_t = 0$ if $t \geq \zeta$, and $S = \inf\{t: M_t = 0\}$. Thus S is a strong terminal time and $S \leq \zeta$. Finally we assume *that* (4.1) *holds throughout this section*, although some of the elementary results do not depend on this assumption.

According to Proposition 4.12 we can find an increasing sequence $\{\tau_n\}$ of finite stopping times such that $\tau_n \uparrow S$ almost surely, and if $\Omega_S = \{\tau_n < S$ for all n; $S < \infty\}$ then whenever $\{R_n\}$ is an increasing sequence of stopping times with limit R we have

$$\{R_n < R \text{ for all } n;\ R = S < \infty\} \subset \Omega_S$$

almost surely. Indeed if $\bar{\psi}(x) = \lim_{t \downarrow 0} E^x\{e^{-(t + S \circ \theta_t)}\}$ for $x \in E$ and $\bar{\psi}(\Delta) = 0$ and T_n is the hitting time of $\{\bar{\psi} > 1 - 1/n\}$, then we may take τ_n to be $\min(T_n, n, S)$. We now fix such a sequence $\{\tau_n\}$ for the remainder of this section. Our definitions will be in terms of this fixed sequence. We restrict our attention to *finite* (X, M) excessive functions. This is a serious restriction since the extension to general (X, M) excessive functions can be quite delicate.

(5.1) DEFINITION. Let f be a finite (X, M) excessive function:

(i) f is *M uniformly integrable* provided that the family $\{M_T f(X_T);$ T a stopping time$\}$ is P^x uniformly integrable for all x;

(ii) f is *M-harmonic* provided $Q_{\tau_n} f = f$ for all n;

(iii) f is an *M-pseudopotential* provided that $\lim_n M_{\tau_n} f(X_{\tau_n}) = 0$ almost surely;

(iv) f is *M-regular* provided that $t \to M_t f(X_t)$ is continuous wherever $t \to X_t$ is continuous on $[0, S)$ almost surely.

Since M will be fixed throughout our discussion we will drop it from our terminology. Thus we will say that an $f \in \mathscr{S}(M)$ is harmonic or is a pseudopotential rather than M-harmonic or an M-pseudopotential. Recall that $t \to M_t f(X_t)$ is right continuous and has left-hand limits almost surely. Also $\{M_t f(X_t), \mathscr{F}_t, P^x\}$ is, for each x, a nonnegative supermartingale and so is $\{M_{R_n} f(X_{R_n}), \mathscr{F}_{R_n}, P^x\}$ whenever $\{R_n\}$ is an increasing sequence of stopping times. Thus (5.1iii) makes sense. Next observe that f is a pseudopotential if and only if

$$\lim_{t \uparrow S} M_t f(X_t) = 0 \quad \text{on} \quad \Omega_S \cup \{S = \infty\}$$

almost surely, which, using the properties of the sequence $\{\tau_n\}$, is equivalent to the statement that whenever $\{R_n\}$ is an increasing sequence of stopping times with limit R then $M_{R_n} f(X_{R_n}) \to 0$ almost surely on $\{R = S\}$. In particular this shows that the definition of a pseudopotential *does not* depend on the specific choice of the sequence $\{\tau_n\}$. By contrast the definition of a harmonic function *does* depend on the choice of the sequence $\{\tau_n\}$, as we will show by an example at the end of this section.

We now introduce some notation that will be used in the remainder of this section. Whenever f is a finite (X, M) excessive function we set $Y_t = M_t f(X_t)$ and $R_n = \inf\{t\colon Y_t > n\}$. Since $t \to Y_t$ is almost surely right continuous each R_n is a stopping time and $Y_{R_n} \geq n$ almost surely on $\{R_n < \infty\}$. More generally we write Y_T for $M_T f(X_T)$ when T is a stopping time. Finally let $\Lambda_S = \Omega_S \cup \{S = \infty\} = \{\tau_n < S \text{ for all } n\}$. Note that modulo the class of sets $\mathcal{N} = \{\Lambda \in \mathcal{F}: P^x(\Lambda) = 0 \text{ for all } x\}$ Ω_S and Λ_S do not depend on the specific choice of the sequence $\{\tau_n\}$.

(5.2) Lemma. If T is any stopping time, then $\theta_T^{-1} \Lambda_S \cap \{T < S\} = \Lambda_S \cap \{T < S\}$ almost surely.

Proof. As in the second paragraph of this section let T_n be the hitting time of $\{\bar{\psi} > 1 - 1/n\}$ and let $H_n = T_n \wedge S$ where $\bar{\psi}(x) = \lim_{t\downarrow 0} E^x\{e^{-(t+S\circ\theta_t)}\}$ for $x \in E$ and $\bar{\psi}(\Delta) = 0$. Then according to (4.12), $\Omega_S = \{H_n < S \text{ for all } n, S < \infty\}$ almost surely. Now using the fact that each H_n is a strong terminal time it is easy to see that $\theta_T^{-1}\, \Omega_S \cap \{T < S\} = \Omega_S \cap \{T < S\}$ almost surely. It is also clear that $\theta_T^{-1}\{S = \infty\} \cap \{T < S\} = \{S = \infty\} \cap \{T < S\}$ almost surely, and combining these statements yields Lemma 5.2.

(5.3) Proposition. Let f be a finite (X, M) excessive function and let $R_n = \inf\{t\colon Y_t > n\}$. Then almost surely $R_n = \infty$ for large enough n, and f is uniformly integrable if and only if $Q_{R_n} f \to 0$ as $n \to \infty$. In this case the family $\{Y_T; T \text{ a stopping time}\}$ is P^μ uniformly integrable whenever $\mu(f) < \infty$.

Proof. The only part of the conclusion not already proved in (4.18) is that if f is uniformly integrable then $Q_{R_n} f \to 0$ as $n \to \infty$. But $f(X_{R_n}) M_{R_n} = 0$ if $R_n = \infty$ and the uniform integrability allows us to interchange limits and expectations, so this part of the conclusion follows immediately from the fact that almost surely $R_n = \infty$ for large n.

(5.4) Proposition. If f is a finite (X, M) excessive function then $f = h + p$ where h is the largest uniformly integrable harmonic function dominated by f and p is a pseudopotential.

Proof. Since $\{Y_{\tau_n}, P^x\}$ is a nonnegative supermartingale for each x, $\lim_n Y_{\tau_n} = L$ exists almost surely. Recall that $\Lambda_S = \Omega_S \cup \{S = \infty\} = \{\tau_n < S \text{ for all } n\}$. Then $L = Y_S = 0$ on Λ_S^c and $L = \lim_{t\uparrow S} Y_t$ on Λ_S; that is, $L = \lim_{t\uparrow S} (I_{\Lambda_S} Y_t)$

almost surely. If T is a stopping time it is easy to see using (5.2) that $L \circ \theta_T = (M_T)^{-1}L$ almost surely on $\{T < S\}$. Define $h(x) = E^x(L) \leq f(x)$. Clearly $h \in \mathscr{E}^*$ and h vanishes off E_M. Now

$$\begin{aligned} Q_t h(x) &= E^x\{M_t L \circ \theta_t; t < S\} \\ &= E^x\{L; t < S\} \end{aligned}$$

and so $Q_t h \leq h$ and $Q_t h \to h$ as $t \to 0$. Thus h is (X, M) excessive. Moreover for each n, $\tau_n < S$ almost surely on Λ_S and hence

$$\begin{aligned} Q_{\tau_n} h(x) &= E^x\{M_{\tau_n} L \circ \theta_{\tau_n}; \tau_n < S\} \\ &= E^x\{L; \tau_n < S; \Lambda_S\} \\ &= E^x\{L; \Lambda_S\} = h(x). \end{aligned}$$

Therefore h is harmonic. Now let $R_n = \inf\{t\colon M_t\, h(X_t) > n\}$; then

$$Q_{R_n} h(x) = E^x\{L; R_n < S\}.$$

According to Proposition 5.3, $R_n = \infty$ for large enough n almost surely and hence $Q_{R_n} h(x) \to 0$ as $n \to \infty$. In view of (5.3) this implies that h is uniformly integrable. Let g be a uniformly integrable harmonic function with $g \leq f$. If $\tilde{L} = \lim_n M_{\tau_n} g(X_{\tau_n})$, then $\tilde{L} \leq L$ almost surely. Consequently, since g is uniformly integrable, $h(x) = E^x(L) \geq E^x(\tilde{L}) = \lim_n E^x\{M_{\tau_n} g(X_{\tau_n})\} = g(x)$. Thus h is the largest uniformly integrable harmonic function dominated by f.

If $L' = \lim_n M_{\tau_n} h(X_{\tau_n})$, then $L' \leq L$ almost surely since $h \leq f$. But h is a uniformly integrable harmonic function and so $E^x(L') = \lim_n Q_{\tau_n} h(x) = h(x) = E^x(L)$. Consequently $L = L'$ almost surely. Thus if we define $p = f - h$ it is obvious that $\lim_n M_{\tau_n} p(X_{\tau_n}) = 0$ almost surely. So to complete the proof of (5.4) we need show only that p is (X, M) excessive. First observe that if T is any stopping time and $T \leq \tau_n$ for some n, then $h \geq Q_T h \geq Q_{\tau_n} h = h$ and so $Q_T h = h$. Consequently $Q_T(f - h) + h = Q_T f \leq f$, and so $Q_T p \leq p$ for any such T. Given $t \geq 0$ let $T_n = t \wedge \tau_n$. We have just observed that $Q_{T_n} p \leq p$, so in particular

$$p(x) \geq E^x(p(X_t)\, M_t; t < \tau_n).$$

Since $p(X_t)\, M_t = 0$ if $t \geq \tau_n$ for all n, it follows that $p(x) \geq E^x(p(X_t)\, M_t) = Q_t\, p(x)$; that is, p is (X, M) super-mean-valued. But also p is the difference of two finite (X, M) excessive functions, so p is itself (X, M) excessive.

(5.5) PROPOSITION. (i) An (X, M) excessive function f is a uniformly integrable pseudopotential if and only if it is a natural potential in the sense of (4.17); that is, for each x whenever $\{T_n\}$ is an increasing sequence of stopping times with limit T and with $P^x(T \geq S) = 1$, then $Q_{T_n} f(x) \to 0$ as $n \to \infty$.

(ii) A uniformly integrable (X, M) excessive function f is harmonic if and only if $Q_T f = f$ for all stopping times T such that $T < S$ almost surely on Λ_S.

Proof. Let $\{T_n\}$ be as in (i). If f is a pseudopotential we have already seen in the remarks following (5.1) that $Y_{T_n} \to 0$ almost surely P^x, and so if f is uniformly integrable $E^x(Y_{T_n}) \to 0$ as $n \to \infty$. Conversely suppose f is a natural potential. Proposition 4.18 states that f is uniformly integrable, and the usual supermartingale argument implies that $L = \lim_n Y_{\tau_n}$ exists almost surely. But $\tau_n \uparrow S$ almost surely and f is uniformly integrable, hence $E^x(L) = \lim_n Q_{\tau_n} f(x) = 0$. Thus $L = 0$ almost surely and so f is a pseudopotential.

Regarding (ii) we need only show that any uniformly integrable harmonic function has the stated property. We may assume $T \leq S$ without loss of generality. If $T_n = T \wedge \tau_n$, then $f \geq Q_{T_n} f \geq Q_{\tau_n} f = f$ and so $f = Q_{T_n} f$. On the other hand since $T < S$ on Λ_S it is clear that $T_n = T$ for sufficiently large n, and using the uniform integrability of f this implies that $Q_{T_n} f \to Q_T f$ as $n \to \infty$. Hence $Q_T f = f$ and (5.5) is established.

Note that (5.5) states that within the class of uniformly integrable (X, M) excessive functions the definition of harmonicity is *independent* of the particular sequence $\{\tau_n\}$ being used.

(5.6) DEFINITION. Let f and g be (X, M) excessive functions. Then f dominates g in the *strong order*, written $f \gg g$, provided there exists an $h \in \mathscr{S}(M)$ such that $f = g + h$.

We are now going to decompose a pseudopotential into a "maximal" uniformly integrable part and a residual part which in a certain sense "contains" no uniformly integrable piece.

(5.7) PROPOSITION. Let f be a pseudopotential. Then $f = p + q$ where p is a uniformly integrable pseudopotential and q is a pseudopotential with the property that the only uniformly integrable (X, M) excessive function which q dominates in the strong order is zero. Moreover p is the largest (in the strong order) uniformly integrable (X, M) excessive function dominated in the strong order by f.

Proof. Let $R_n = \inf\{t\colon Y_t > n\}$ where $Y_t = M_t f(X_t)$ and let T_n be the hitting time of $\{f > n\}$. Clearly $\{T_n\}$ is increasing and $T_n \leq R_n$. Since $f(x) \geq E^x\{f(X_{T_n}) M_{T_n}\} \geq n\, E^x\{M_{T_n}\}$ we must have $\lim T_n \geq S$ almost surely. Now f is a pseudopotential and so the remarks following (5.1) imply that almost surely $M_{T_n} f(X_{T_n}) \to 0$ as $n \to \infty$. Finally $f \geq Q_{T_n} f \geq Q_{R_n} f$ for each n. Define

$q = \lim_n Q_{T_n} f$. Since $\{Q_{T_n} f\}$ is a decreasing sequence of (X, M) excessive functions, q is super-mean-valued relative to the semigroup $\{Q_t\}$; that is, $Q_t q \leq q$ for all $t \geq 0$. Moreover $\bar{q} = \lim_{t \to 0} Q_t q$ is (X, M) excessive and $\bar{q} \leq q$. In order to show that $q \in \mathcal{S}(M)$ we will prove that $q = \bar{q}$. Let n be fixed and let T be a stopping time such that $T \leq T_n$. Now $\{f > n\}$ is finely open and so $X(T_n)$ is regular for $\{f > n\}$ almost surely on $\{T_n < \infty\}$. Therefore $T + T_n \circ \theta_T = T_n$ almost surely on $\{T \leq T_n\} = \Omega$ and not just on $\{T < T_n\}$. Consequently

$$\begin{aligned} Q_T Q_{T_n} f(x) &= E^x\{M_{T+T_n \circ \theta_T} f(X_{T+T_n \circ \theta_T})\} \\ &= Q_{T_n} f. \end{aligned}$$

Thus if $T \leq T_k$ for some fixed k, one obtains upon letting $n \to \infty$ in the above equality $Q_T q = q$. In particular $Q_{T_n} q = q$ for each n. We are now prepared to show that $q = \bar{q}$. For a fixed $x \in E$ choose n such that $f(x) < n$. Then $P^x(T_n > 0) = 1$ and so

$$\begin{aligned} \bar{q}(x) &= \lim_{t \to 0} Q_t q(x) = \lim_{t \to 0} Q_t Q_{T_n} q(x) \\ &= \lim_{t \to 0} E^x\{M_{t+T_n \circ \theta_t} q(X_{t+T_n \circ \theta_t})\} \\ &\geq \lim_{t \to 0} E^x\{M_{T_n} q(X_{T_n}); t < T_n\} = Q_{T_n} q(x) = q(x). \end{aligned}$$

But $E = \bigcup\{f < n\}$ and hence $\bar{q} \geq q$. Therefore $\bar{q} = q$ and so $q \in \mathcal{S}(M)$. Finally q is pseudopotential since $q \leq f$.

We now define $p = f - q$. In order to show that $p \in \mathcal{S}(M)$ it suffices to show that $Q_t p \leq p$ for all $t \geq 0$, and exactly the same argument as that used in the last paragraph of the proof of Proposition 5.4 yields this fact (replace τ_n by the T_n of the present proof). Thus $p \in \mathcal{S}(M)$ and p is a pseudopotential since $p \leq f$. Now $Q_{T_n} p = Q_{T_n}(f - q) \to 0$ as $n \to \infty$. If $H_n = \inf\{t: M_t p(X_t) > n\}$, then $T_n \leq H_n$ and so $Q_{H_n} p \leq Q_{T_n} p \to 0$. Thus Proposition 5.3 implies that p is uniformly integrable. Next suppose that $q = g + h$ with $g, h \in \mathcal{S}(M)$ and g uniformly integrable. Now $Q_{T_n} g \leq g$, $Q_{T_n} h \leq h$, and $Q_{T_n} q = q$ for each n, and as a result $Q_{T_n} g = g$ and $Q_{T_n} h = h$ for each n. But $g \leq q$ and so almost surely $M_{T_n} g(X_{T_n}) \to 0$ as $n \to \infty$. Since g is uniformly integrable we obtain $g = Q_{T_n} g \to 0$ as $n \to \infty$. In other words the only uniformly integrable function in $\mathcal{S}(M)$ that q dominates in the strong order is zero. To establish the last sentence of Proposition 5.7 suppose that $f = u + v$ with $u, v \in \mathcal{S}(M)$ and u uniformly integrable. Plainly $Q_{T_n} u \to 0$ and hence $q = \lim_n Q_{T_n} f = \lim_n Q_{T_n} v \leq v$. Exactly the same argument as that used to show that $p = f - q$ is (X, M) excessive shows that $v - q$ is (X, M) excessive and consequently $f = u + v = u + (v - q) + q$. Therefore $p = u + (v - q)$ and so $p \gg u$. This completes the proof of Proposition 5.7.

REMARK. The reader should verify that q has the following property: if $u \in \mathscr{S}(M)$ and $u \geq q$ then $u - q \in \mathscr{S}(M)$; that is, $u \in \mathscr{S}(M)$ and $u \geq q$ imply $u \gg q$.

Combining Propositions 5.4 and 5.7 one readily obtains the following result.

(5.8) THEOREM. If f is a finite (X, M) excessive function, then $f = h + p + q$ where h is the largest uniformly integrable harmonic function dominated by f, p is the largest (in the strong order) uniformly integrable pseudopotential dominated in the strong order by $f - h$, and q is a pseudopotential which dominates only zero in the strong order.

We turn now to a study of regularity. We begin with the following alternate characterization of regularity.

(5.9) PROPOSITION. Let f be a finite (X, M) excessive function; then f is regular if and only if whenever $\{T_n\}$ is an increasing sequence of stopping times with limit T one has $\lim_n Y_{T_n} = Y_T$ almost surely on $\{T < S\}$.

Proof. Let us first suppose that f is regular. On the set of ω's for which $t \to X_t(\omega)$ is continuous at $T(\omega) < S(\omega)$ one has $Y_{T_n} \to Y_T$ almost surely by the definition of regularity. On the set of ω's for which $t \to X_t(\omega)$ is discontinuous at $T(\omega) < S(\omega)$ one has $T_n(\omega) = T(\omega)$ for sufficiently large n almost surely by the quasi-left-continuity of X (recall $S \leq \zeta$) and so $Y_{T_n} \to Y_T$ almost surely on this set also. Thus regularity implies the condition of Proposition 5.9. Conversely suppose f is not regular; then there exists an $\varepsilon > 0$ and a y in E such that $P^y(T^\varepsilon < S) > 0$ where T^ε is the stopping time defined by

$$T^\varepsilon = \inf\{t < S\colon |Y_t - Y_{t-}| > \varepsilon M_t \text{ and } X_t = X_{t-}\}$$

and $T^\varepsilon = S$ if the set in braces is empty. According to Proposition 4.16, T^ε is accessible on $\{T^\varepsilon < S\}$. Hence there exists an increasing sequence of stopping times $\{T_n\}$ which increases to T^ε strictly from below on $\{T^\varepsilon < S\}$ almost surely P^y, and consequently if $T = \lim_n T_n$ it is not the case that $Y_{T_n} \to Y_T$ almost surely on $\{T < S\}$. Therefore the condition in (5.9) implies regularity.

The following is just a restatement of Proposition 4.21 in the present situation.

(5.10) PROPOSITION. Let f be an (X, M) excessive function and $\{T_n\}$ an increasing sequence of stopping times with limit T. Then $\lim_n Y_{T_n} \geq Y_T$ almost surely.

(5.11) Corollary. Let f be a finite regular (X, M) excessive function. Then any (X, M) excessive function that f dominates in the strong order is itself regular.

Proof. Suppose $f = g + h$ with g and h in $\mathscr{S}(M)$ and let $\{T_n\}$ be an increasing sequence of stopping times with limit T. Now

$$\lim_n M_{T_n}[g(X_{T_n}) + h(X_{T_n})] = M_T f(X_T) = M_T[g(X_T) + h(X_T)]$$

almost surely on $\{T < S\}$ and combining this with (5.10) we can conclude that $\lim_n M_{T_n} g(X_{T_n}) = M_T\, g(X_T)$ almost surely on $\{T < S\}$ with a similar statement for h. Hence g and h are regular by Proposition 5.9.

(5.12) Corollary. Any harmonic function is regular.

Proof. Let $\{T_k\}$ be an increasing sequence of stopping times with limit T and suppose that h is harmonic. Let $Y_t = M_t\, h(X_t)$ and for fixed n let $H = T \wedge \tau_n$ and $H_k = T_k \wedge \tau_n$. Since $\tau_n \uparrow S$, in order to show that h is regular it suffices to show that $Y_{H_k} \to Y_H$ almost surely. But $\lim_k Y_{H_k} \geq Y_H$ by Proposition 5.10 and if R is any stopping time dominated by τ_n, then $Q_R\, h = h$, since h is harmonic. Therefore by Fatou's lemma

$$E^x\left\{\lim_k Y_{H_k}\right\} \leq \lim_k Q_{H_k} h(x) = h(x) = Q_H\, h(x) = E^x\{Y_H\},$$

and consequently $\lim_k Y_{H_k} = Y_H$ almost surely. Thus (5.12) is established.

Recall that in Section 3 a finite (X, M) excessive function f was called a regular potential provided $Q_{T_n} f \to Q_T f$ whenever $\{T_n\}$ is an increasing sequence of stopping times with limit T. We can now given an alternate form of this definition. Recall, however, that we are assuming (4.1) in this section.

(5.13) Proposition. A finite (X, M) excessive function is a regular potential if and only if it is uniformly integrable, regular, and a pseudopotential.

Proof. Let f be a regular potential. Then it follows from Proposition 5.5 that f is a uniformly integrable pseudopotential, and so we need only show that f is regular. Suppose $\{T_n\}$ is an increasing sequence of stopping times with limit T. Then $\lim_n Y_{T_n} \geq Y_T$ and using the uniform integrability of f

$$E^x(Y_T) = Q_T f(x) = \lim_n Q_{T_n} f(x) = E^x\left\{\lim_n Y_{T_n}\right\}.$$

Hence f is regular. Conversely if f is a uniformly integrable, regular, pseudopotential and $\{T_n\}$ and T are as above, then almost surely $Y_{T_n} \to Y_T$ on $\{T < S\}$ and $Y_{T_n} \to 0 = Y_T$ on $\{T \geq S\}$. That is, $Y_{T_n} \to Y_T$ almost surely and hence f, being uniformly integrable, is a regular potential. Thus Proposition 5.13 is established.

Let f be a finite (X, M) excessive function. By Theorem 5.8, $f = h + p + q$. Since h is harmonic it is regular, and in fact h is uniformly integrable. Now p is a natural potential and so by the proof of Theorem 4.22 there exist a unique CAF, A_C, of (X, M) and a "pure jump" NAF, A_J, of (X, M) such that $p = u_{A_C} + u_{A_J}$. Since $p_C = u_{A_C}$ is a regular potential it is regular. Clearly $p_J = u_{A_J}$ has the property that the only regular potential which p_J dominates in the strong order is zero and p_C is the largest, in the strong order, regular potential strongly dominated by p. Finally let us show that q is regular. Recall the definition of q from (5.7) and (5.8). Namely if $u = f - h$ and T_n is the hitting time of $\{u > n\}$, then $q = \lim Q_{T_n} u$, and it was shown in the course of the proof of (5.7) that $Q_{T_n} q = q$ for all n. Now using the fact that almost surely $T_n \uparrow S$ one can show that q is regular by exactly the same argument that was used in the proof of Corollary 5.12 to show that harmonic functions are regular. We summarize this discussion in the following theorem.

(5.14) THEOREM. Let f be a finite (X, M) excessive function. Then f can be written uniquely in the following manner: $f = h + p_C + p_J + q$ where p_C is the potential of a CAF of (X, M), p_J is the potential of a pure jump NAF of (X, M), h is the largest uniformly integrable harmonic function dominated by f, and q is a pseudopotential with the property that the only uniformly integrable (X, M) excessive function dominated in the strong order by q is zero. Finally h, q, and p_C are regular.

We conclude this section with an example. Let $E = \mathbf{R}^3 - \{0\}$ and let X be Brownian motion on E. Since $\{0\}$ is polar for Brownian motion in $\mathbf{R}^3$ this makes sense. Also $\zeta = \infty$ almost surely. We set $M_t = I_{[0,\zeta)}(t)$ so that $S = \zeta = \infty$ almost surely. Let $f(x) = |x|^{-1}$ where $|\cdot|$ denotes the distance from the origin. Since f is harmonic in the ordinary sense on E, it follows from Theorem 5.11 of Chapter II that f is excessive. Let τ_n be the hitting time of $\{x \in E\colon |x| > n \text{ or } |x| < n^{-1}\}$ and let $\hat{\tau}_n$ be the hitting time of $\{x \in E\colon |x| > n\}$. Both sequences are of the type under consideration; that is, they both increase to $S = \infty$ strictly from below. Clearly f is $\{\tau_n\}$-harmonic since it is harmonic in the ordinary sense, but not $\{\hat{\tau}_n\}$-harmonic. In fact $f(X_{\hat{\tau}_n}) = 1/n \to 0$. Thus f is a pseudopotential relative to $\{\hat{\tau}_n\}$ and hence also relative to $\{\tau_n\}$ since this concept does not depend on the specific sequence in question. In other words f is both harmonic and a pseudopotential relative to

$\{\tau_n\}$. In particular this implies that f can not be uniformly integrable. On the other hand it is not difficult to see that, for each x in E, $E^x\{f(X_t)^2\}$ is uniformly bounded in t. (Of course, the bound depends on x.) Consequently the family $\{f(X_t): t \geq 0\}$ is P^x uniformly integrable for each $x \in E$. Thus the P^x uniform integrability of $\{f(X_t): t \geq 0\}$ for all x does *not* imply that f is uniformly integrable. This example is due to Helms and Johnson [1].

V

FURTHER PROPERTIES OF CONTINUOUS ADDITIVE FUNCTIONALS

In this chapter we will derive and discuss a number of important properties of continuous additive functionals. We will also give some applications of these results. In many respects additive functionals are analogous to measures, and this analogy, which goes quite deep, will become clear in the light of the results of this chapter. Some of these results will be proved only under an additional regularity assumption, (1.3), on X, while others take an especially nice form under this assumption. Therefore Section 1 is devoted to a discussion of this assumption with the main line of development of the chapter beginning in Section 2. The reader whose main interest is additive functionals might proceed from Theorem 1.6 to Section 2 and refer back to the rest of Section 1 only as needed.

1. Reference Measures

Let $X = (\Omega, \mathscr{M}, \mathscr{M}_t, X_t, \theta_t, P^x)$ be a standard process with state space $(E, \mathscr{E})$ and, for ease of exposition, assume that $\mathscr{M} = \mathscr{F}$ and $\mathscr{M}_t = \mathscr{F}_t$ for all $t \geq 0$.

(1.1) DEFINITION. A measure λ on $\mathscr{E}^*$ which is a countable sum of finite measures is called a *reference measure* for X provided that a set $A \in \mathscr{E}^*$ is of potential zero if and only if $\lambda(A) = 0$.

Note that if λ is any countable sum of finite measures then there is a finite measure μ such that $\mu(A) = 0$ if and only if $\lambda(A) = 0$. Hence if there exists a reference measure, then there exists a finite reference measure.

It follows from (3.2) of Chapter II that if λ is a reference measure for X and f and g are α-excessive ($\alpha \geq 0$), then $f = g$ a.e. λ implies that f and g are identical.

(1.2) PROPOSITION. Suppose that for some fixed $\alpha > 0$ there exists a measure ξ on $(E, \mathscr{E}^*)$ which is a countable sum of finite measures and is such that given $f \in b\mathscr{E}_+^*$ with $U^\alpha f = 0$ a.e. ξ, then $U^\alpha f = 0$. Then there exists a reference measure for X.

Proof. If μ is a measure and $f \in \mathscr{E}_+^*$, then it will be convenient to write $\langle \mu, f \rangle = \mu(f)$. We may assume that ξ is finite and hence that $\langle \xi, U^\alpha 1 \rangle < \infty$; that is, the measure $\lambda = \xi U^\alpha$ is finite. Clearly $\lambda(A) = 0$ if A is of potential zero. On the other hand if $\lambda(A) = 0$ then $U^\alpha I_A = 0$ a.e. ξ and hence $U^\alpha I_A = 0$. Thus A is of potential zero. Consequently λ is a reference measure for X.

We can now state the regularity assumption that we will impose from time to time.

(1.3) There exists a reference measure for X.

When (1.3) is being assumed to hold we will explicitly say so. Note that if the functions in $\mathscr{S}^\alpha$ are lower semicontinuous for some $\alpha > 0$, then the measure $\xi = \sum_{n=1}^\infty 2^{-n} \varepsilon_{x_n}$, where $\{x_n\}$ is a countable dense subset of E, satisfies the hypotheses of (1.2), and so (1.3) is satisfied in this situation. In fact (1.3) is a very mild assumption in the sense that it is valid for practically any standard process of interest (see, however, Exercise 1.19). The following simple result begins to indicate the importance of (1.3).

(1.4) PROPOSITION. Under (1.3) any α-excessive function is Borel measurable.

Proof. It suffices to show that $U^\alpha f$ is Borel measurable for any $f \in \mathscr{E}_+^*$ and $\alpha > 0$. Let λ be a reference measure for X such that the measure $\mu = \lambda U^\alpha$ is finite. Then there exist $g, h \in \mathscr{E}_+$ such that $g \leq f \leq h$ and $\mu\{g < h\} = 0$. Therefore $U^\alpha g \leq U^\alpha f \leq U^\alpha h$ and the extreme members of this inequality agree λ a.e. and hence are identical. But $U^\alpha h$ and $U^\alpha g$ are Borel measurable and consequently so is $U^\alpha f$.

Note that if A is a nonempty set in $\mathscr{E}^*$ and A is finely open, then $\lambda(A) > 0$ for any reference measure λ. Hence if $\lambda(B) = 0$, the complement of B is finely dense in E. If M is an exact MF of X and f, g are in $\mathscr{S}^\alpha(M)$, then $\{f \neq g\}$ is finely open and nearly Borel measurable, and so if $f = g$ a.e. λ then $f = g$. Before coming to the main result of this section it is necessary to establish some notation. In the sequel M will always denote a fixed exact MF of X

which vanishes on $[\zeta, \infty]$. Recall that a function $f \in \mathscr{E}_+^*$ is said to be $\alpha - (X, M)$ super-mean-valued provided f vanishes off E_M and $Q_t^\alpha f \leq f$ for all $t \geq 0$. In this case $\bar{f} = \lim_{t \downarrow 0} Q_t^\alpha f$ exists and $\bar{f}$ is the largest function in $\mathscr{S}^\alpha(M)$ which is dominated by f. We call $\bar{f}$ the $\alpha - (X, M)$ excessive regularization of f. If $\{f_n\}$ is a decreasing sequence of $\alpha - (X, M)$ excessive functions, then it is immediate that $f = \lim_n f_n$ is $\alpha - (X, M)$ super-mean-valued, and the reader should have no difficulty in modifying the proof of Theorem 3.6 of Chapter II to show that $\{\bar{f} < f\}$ is semipolar in this case.* Since the infimum of two elements in $\mathscr{S}^\alpha(M)$ is again in $\mathscr{S}^\alpha(M)$, it follows that if $\{f_n\}$ is any countable family of functions in $\mathscr{S}^\alpha(M)$ and $f = \inf_n f_n$, then f is $\alpha - (X, M)$ super-mean-valued and $\{\bar{f} < f\}$ is semipolar.

The next two theorems are the main results of this section. Of these, Theorem 1.6 is the more important one. Both of them are due to Meyer [2].

(1.5) THEOREM. Assume (1.3) holds. Let $\mathbf{F}$ be a family of $\alpha - (X, M)$ excessive functions which is filtering upward; that is, if $f, g \in \mathbf{F}$ then there exists $h \in \mathbf{F}$ such that $f \leq h$ and $g \leq h$. Then there exists an increasing sequence $\{f_n\}$ of functions in $\mathbf{F}$ such that $\sup_n f_n = \sup\{f\colon f \in \mathbf{F}\}$. In particular $\sup\{f\colon f \in \mathbf{F}\}$ is $\alpha - (X, M)$ excessive.

Proof. First note that the last conclusion follows immediately from the first since the limit of an increasing sequence of functions in $\mathscr{S}^\alpha(M)$ is in $\mathscr{S}^\alpha(M)$. Let λ be a finite reference measure and $a = \sup\{\langle \lambda, f(1+f)^{-1}\rangle \colon f \in \mathbf{F}\}$. Since $\mathbf{F}$ is filtering upward and $t \to t(1+t)^{-1}$ is a strictly increasing continuous function from $[0, \infty]$ onto $[0, 1]$, one can find an increasing sequence $\{f_n\}$ of functions in $\mathbf{F}$ such that $\lim_n \langle \lambda, f_n(1+f_n)^{-1}\rangle = a$. Let $u = \lim f_n$. If $f \in \mathbf{F}$ then one can find an increasing sequence $\{g_n\}$ of functions in $\mathbf{F}$ such that $g_n \geq \sup(f, f_n)$ for each n. Let $g = \lim g_n$. Then $g \geq f$, $g \geq u$, and clearly $\langle \lambda, g(1+g)^{-1}\rangle = \langle \lambda, u(1+u)^{-1}\rangle$. Hence $g = u$ a.e. λ and, since both u and g are in $\mathscr{S}^\alpha(M)$, $g = u$. Consequently $f \leq u$, and since $f \in \mathbf{F}$ is arbitrary the proof is complete.

(1.6) THEOREM. Assume (1.3) holds. Let $\mathbf{F}$ be a family of $\alpha - (X, M)$ excessive functions and let $u = \inf\{f \colon f \in \mathbf{F}\}$. Then there exists a countable subset $\{f_n\}$ of $\mathbf{F}$ such that if $v = \inf_n f_n$, then $\bar{v} \leq u \leq v$. We have already remarked that $\{\bar{v} < v\}$ is semipolar.

* Observe that if B is a nearly Borel subset of E_M and no point of E_M is regular for B, then B is semipolar. To see this write $B = \bigcup_n (B \cap E_n)$ where E_n is defined above (5.3) of Chapter III and note that $B \cap E_n$ is thin for each n. Thus a subset of E_M is "semipolar relative to (X, M)" if and only if it is semipolar.

Proof. There is no loss of generality in assuming that **F** is filtering downward. Arguing as in the proof of (1.5) we can find a decreasing sequence $\{f_n\}$ of functions in **F** such that if $v = \lim_n f_n$, then for any $f \in \mathbf{F}$ we have $v \leq f$ a.e. λ, where λ is a finite reference measure. Hence $\bar{v} \leq f$, and since $f \in \mathbf{F}$ is arbitrary the proof is complete.

We may use (1.6) to give a very useful characterization of polar sets.

(1.7) PROPOSITION. Assume (1.3) and let A be a polar set. If $\alpha > 0$ then there exists a function $f \in \mathscr{S}^\alpha$ such that f is finite except on a set of potential zero and $A \subset \{f = \infty\}$. This is also true when $\alpha = 0$ provided that there is a sequence $\{h_n\}$ of functions in $\mathscr{E}_+^*$ such that Uh_n is bounded for each n and $\lim Uh_n = \infty$ on E.

Proof. We will carry out the proof under the second hypothesis. Note that it is equivalent to the statement that there is an $h \in \mathscr{E}_+^*$ such that Uh is bounded and $Uh > 0$ on E. There is no loss of generality in assuming that A is nearly Borel. Let $\mathbf{F} = \{f \in \mathscr{S} : f \geq Uh \text{ on } A\}$. The family **F** is filtering downward, and if $g = \inf\{f : f \in \mathbf{F}\}$ then by (6.12) of Chapter III, $g(x) = P_A Uh(x)$ for every $x \notin A$. But $P_A Uh = 0$ and consequently by (1.6) there is a decreasing sequence $\{f_n\}$ of functions in **F** such that $\lim f_n(x) = 0$ except for x in a Borel semi-polar set. We may assume each $f_n \leq \|Uh\|$ and so if λ is a finite reference measure then $\langle \lambda, f_n \rangle \to 0$. By passing to a subsequence we may assume that $\langle \lambda, f_n \rangle \leq 2^{-n}$ for each n. If $f = \Sigma f_n$ then $\langle \lambda, f \rangle < \infty$ and $f = \infty$ on A since $Uh > 0$. This completes the proof.

REMARK. In view of Proposition (3.5) of Chapter II this gives a complete characterization of polar sets when (1.3) holds. The name "polar" comes from this characterization in the classical case (i.e., Brownian motion), a polar set being contained in the set of "poles" of a superharmonic function.

We turn next to some applications of (1.3) (or more precisely Theorem 1.6) to the fine topology. These results are due to Doob [3] and we follow his discussion quite closely. Once again we will make it explicit when we are assuming (1.3). If A is any subset of E we say that x is *irregular* for A provided A is thin at x; that is ((3.1) of Chapter II) there exists a set $B \in \mathscr{E}^n$ such that $A \subset B$ and x is irregular for B. We say that x is *regular* for A provided that x is regular for *every* nearly Borel set containing A. Thus x is irregular for A if and only if it is not regular for A. We let A^r denote the set of all points which are regular for A. The following proposition generalizes (4.9) of Chapter II.

(1.8) PROPOSITION. Let $A \subset E$. Then (i) A is finely closed if and only if $A^r \subset A$, (ii) $(A \cup A^r)^r = A^r$, and (iii) $A \cup A^r$ is the fine closure of A.

Proof. Suppose A is finely closed. Then A is thin at each $x \in E - A$; that is, given $x \in E - A$ there exists a nearly Borel set $B \supset A$ such that x is not in B^r. Hence x is not in A^r, or $A^r \subset A$. Conversely if $A^r \subset A$ and $x \in E - A$, then x is irregular for A. Hence A is finely closed. Thus (i) is established. Let us first prove (ii) under the assumption that $A \in \mathscr{E}^n$. In this case $A^r \in \mathscr{E}^n$ and we claim that $T_A \leq T_{A^r}$ almost surely. To see this fix an x and let $\{K_n\}$ be an increasing sequence of compact subsets of A^r such that $T_{K_n} \downarrow T_{A^r}$ almost surely P^x. Then

$$\begin{aligned} P^x(T_{A^r} < T_A) &= \lim_n P^x(T_{K_n} < T_A) \\ &= \lim_n E^x\{P^{X(T_{K_n})}[T_A > 0]; T_{K_n} < T_A\} = 0, \end{aligned}$$

establishing the claim. Now $T_{A \cup A^r} = \min(T_A, T_{A^r}) = T_A$ almost surely, and so $A^r = (A \cup A^r)^r$ if A is nearly Borel measurable. In the general case $A \subset A \cup A^r$ implies that $A^r \subset (A \cup A^r)^r$. If $x \notin A^r$ then there exists a $B \in \mathscr{E}^n$ containing A such that $x \notin B^r$. But $B \cup B^r \in \mathscr{E}^n$ and contains $A \cup A^r$, and by what was proved above $x \notin (B \cup B^r)^r$. Hence x is irregular for $A \cup A^r$, or $(A \cup A^r)^r \subset A^r$. Thus (ii) holds. Next suppose that F is finely closed and $A \subset F$. Then by (i), $A \cup A^r \subset F \cup F^r = F$. On the other hand (ii) and (i) imply that $A \cup A^r$ is finely closed, and so (iii) is proved.

We now introduce a "regularization" of an arbitrary numerical function which will play an important role in our development.

(1.9) Definition. Let u be a numerical function on E. Then we define u^+ by $u^+(x) = \sup\{c : x \in \{u \geq c\}^r\}$ for each x in E.

The reader should check that if A is a subset of E then $I_A^+ = I_{A^r}$.

(1.10) Proposition. If $u \in \mathscr{E}^n$, then for each x

$$u^+(x) = \limsup_{t \downarrow 0} u(X_t) \qquad \text{a.s.} \quad P^x.$$

Proof. It is understood that $t = 0$ is excluded from the limiting procedure under consideration. Also it is not a priori clear that the limit in question is in any sense a measurable function of ω. Let $u^+(x) = a$. Then $x \in \{u \geq c\}^r$ for all $c < a$, and so P^x almost surely $u(X_t) \geq c$ for arbitrarily small strictly positive t. Hence $\limsup_{t \downarrow 0} u(X_t) \geq a$ almost surely P^x. If $c > a$, then $x \notin \{u \geq c\}^r$ and so $u(X_t) < c$ on some open interval $(0, \tau(\omega))$ almost surely P^x. Thus $\limsup_{t \downarrow 0} u(X_t) \leq a$ almost surely P^x establishing (1.10).

Note that the proof of (1.10) actually shows that the set on which the desired equality fails to hold is in $\mathscr{F}$. Also note that if u is finely continuous and nearly Borel measurable, then $u = u^+$. In particular if u is α-excessive, then $u = u^+$.

(1.11) PROPOSITION. Let $\{u_n\}$ be a decreasing sequence of α-excessive functions with limit u. Then $\bar{u} = u^+$.

Proof. Of course, $\bar{u}$ is the α-excessive regularization of u. Since $\bar{u} \leq u$ and $\bar{u} \in \mathscr{S}^\alpha$ we have $\bar{u} = \bar{u}^+ \leq u^+$. On the other hand given $\varepsilon > 0$ it was shown in the proof of (3.6) of Chapter II that $A_\varepsilon = \{u \geq \bar{u} + \varepsilon\}$ is thin. Now $u \in \mathscr{E}^n$ and so $u^+(x) = \limsup_{t \downarrow 0} u(X_t)$ almost surely P^x for each x. If x is fixed then x is not regular for A_ε, and so almost surely P^x

$$u^+(x) \leq \limsup_{t \downarrow 0} \bar{u}(X_t) + \varepsilon = \bar{u}(x) + \varepsilon,$$

which implies $u^+ \leq \bar{u}$. Thus (1.11) is proved.

The following corollary is an immediate consequence of Theorem 1.6 and Proposition 1.11.

(1.12) COROLLARY. Assume (1.3). If $\{u_i : i \in I\}$ is a family of α-excessive functions, then there exists a countable subset J of I such that if $u_I = \inf_{i \in I} u_i$ and $u_J = \inf_{i \in J} u_i$ then $u_I^+ = u_J^+ = \bar{u}_J \leq u_I \leq u_J$ and the set on which the extreme members of this inequality differ is semipolar. In particular u_I^+ is α-excessive.

In the remainder of this section we will assume that (1.3) *holds.* If $A \subset E$ and $\alpha \geq 0$ then we define

$$\mathbf{F}_A^\alpha = \{f \in \mathscr{S}^\alpha : f \geq 1 \text{ on } A\}$$

and

$$e_A^\alpha = \inf\{f : f \in \mathbf{F}_A^\alpha\}.$$

We now fix a positive value of α, say $\alpha = 1$, and write e_A and $\mathbf{F}_A$ in place of e_A^1 and $\mathbf{F}_A^1$. Note that it follows from (1.12) that $e_A^+ \in \mathscr{S}^1$ and that $\{e_A^+ < e_A\}$ is semipolar. In addition, if $B \in \mathscr{E}^n$ we write $\varphi_B(x)$ for $E^x(e^{-T_B}; T_B < \zeta)$.

(1.13) PROPOSITION. If A is a subset of E, then there is a Borel set $B \supset A$ such that $e_A^+ = \varphi_B$.

Proof. According to (1.12) we can find a decreasing sequence $\{f_n\}$ of functions in $\mathbf{F}_A$ such that if $f = \lim f_n$ then $e_A^+ = f^+ = \bar{f} \leq e_A \leq f$. Let $B_n = \{f_n \geq 1\}$ and $B = \bigcap B_n$. Clearly $f_n \geq P_{B_n}^1 f_n \geq P_{B_n}^1 1 = \varphi_{B_n} \geq \varphi_B$, and so $\varphi_B \leq \bar{f} = e_A^+$

because $\varphi_B \in \mathscr{S}^1$ and $\varphi_B \leq f$. On the other hand according to (6.13) of Chapter III if $x \notin B - B^r$ then $\varphi_B(x) = e_B(x) \geq e_A(x)$, so that $\varphi_B \geq e_A^+$ except on the semipolar set $B - B^r$. Consequently $\varphi_B \geq e_A^+$ since both functions are in $\mathscr{S}^1$, and so $\varphi_B = e_A^+$ proving (1.13).

REMARK. It is easy to see that if $D \in \mathscr{E}^n$, then $\varphi_D = e_D^+$. As a result, using the notation of (1.13), if $C \in \mathscr{E}^n$ and $C \supset A$, then $\varphi_{C \cap B} = \varphi_B$.

There are several interesting corollaries to (1.13) and the remark following it. The sets A and B are those in the statement of (1.13).

(1.14) COROLLARY. $A^r = B^r$; consequently, if A is any subset of E, then A^r is a Borel set and $A - A^r$ is semipolar.

Proof. Since $A \subset B$ it follows that $A^r \subset B^r$. If $x \notin A^r$, then there is a set $C \in \mathscr{E}^n$ such that $C \supset A$ and $\varphi_C(x) < 1$. By the remark following (1.13) we have

$$1 > \varphi_C(x) \geq \varphi_{C \cap B}(x) = \varphi_B(x)$$

and so $x \notin B^r$. Thus $A^r = B^r$. Of course $B^r = \{\varphi_B = 1\}$ and under (1.3) this is a Borel set. Finally $A - A^r = A - B^r \subset B - B^r$ which is, of course, semipolar.

REMARK. It is apparent from (1.13) and (1.14) that $A^r = \{e_A^+ = 1\}$ for any subset A of E.

(1.15) COROLLARY. If A is semipolar then A is contained in a semipolar Borel set.

Proof. It suffices to treat the case in which A is thin. If A is thin and B is a Borel set related to A as in (1.13), then by (1.14) A^r and B^r are both empty. Consequently B is thin.

REMARK. Note that the proof of (1.15) actually shows that a set which is thin at every point is itself thin. Compare this with (3.14) of Chapter II.

(1.16) COROLLARY. If u is any numerical function on E, then u^+ is Borel measurable and $\{u^+ < u\}$ is semipolar.

Proof. It is immediate from the definition of u^+ that $\{u^+ \geq c\} = \bigcap_n \{u \geq c - 1/n\}^r$ for any c. Consequently (1.14) implies that u^+ is Borel measurable. If $A = \{x\colon u(x) \geq a > b \geq u^+(x)\}$ and $x \in A^r$, then $x \in \{u \geq a\}^r$

and so $u^+(x) \geq a$; that is, $A^r \subset A^c$. Therefore (1.14) implies that $A = A - A^r$ is semipolar, and this yields the fact that $\{u^+ < u\}$ is semipolar.

We come now to the main result of this development.

(1.17) THEOREM. Let $\{u_i : i \in I\}$ be a family of finely upper semicontinuous (u.s.c.) functions on E (that is, u.s.c. in the fine topology). Then there exists a countable subset J of I such that if $u_I = \inf_{i \in I} u_i$ and $u_J = \inf_{i \in J} u_i$, then $u_J^+ \leq u_I \leq u_J$. In particular u_I and u_J differ on a semipolar set. We emphasize that (1.3) is being assumed here.

Proof. Let us suppose that we have established (1.17) in the special case in which each u_i is the indicator function of a finely closed set A_i. Since $I_A^+ = I_{A^r}$, the conclusion would read in this case that there exists a countable subset J of I such that $A_J^r \subset A_I \subset A_J$ where $A_I = \bigcap_{i \in I} A_i$ and $A_J = \bigcap_{i \in J} A_i$. We now show that this special case implies the general result. For each a, consider the family of finely closed sets $\{u_i \geq a\}$ as i ranges over I. Then there exists a countable subset J_a of I such that $\{u_{J_a} \geq a\}^r \subset \{u_I \geq a\}$ where $u_{J_a} = \inf_{i \in J_a} u_i$. Let $J = \bigcup_{a \in \mathbf{Q}} J_a$. Then $\{u_J \geq a\} \subset \{u_{J_a} \geq a\}$ and so $\{u_J \geq a\}^r \subset \{u_I \geq a\}$ for each $a \in \mathbf{Q}$. Thus if $u_J^+(x) = c$, then $x \in \{u_J \geq a\}^r \subset \{u_I \geq a\}$ for all rational $a < c$, and so $u_I(x) \geq c$. Hence $u_J^+ \leq u_I$, and consequently to complete the proof of Theorem 1.17 it will suffice to treat the above special case.

Suppose that $\{A_i; i \in I\}$ is a family of finely closed subsets of E and let $A = \bigcap_{i \in I} A_i$. Now $\{e_{A_i}^+; i \in I\}$ is a family of 1-excessive functions and so by Corollary 1.12 there exists a countable subset J of I such that if $u_J = \inf_{i \in J} e_{A_i}^+$ and $u_I = \inf_{i \in I} e_{A_i}^+$, then $u_J^+ = \bar{u}_J = u_I^+ \leq u_I$. Define $B = \bigcap_{i \in J} A_i$. Plainly $e_B^+ \leq e_{A_i}^+$ for $i \in J$ and hence $e_B^+ \leq u_J$. Consequently $e_B^+ \leq \bar{u}_J$ since $\bar{u}_J$ is the largest 1-excessive function which u_J dominates. Therefore using the remark following (1.13) we obtain

$$B^r = \{e_B^+ = 1\} \subset \{\bar{u}_J = 1\} \subset \bigcap_{i \in I} \{e_{A_i}^+ = 1\}$$

$$= \bigcap_{i \in I} A_i^r \subset A.$$

Thus the first assertion in Theorem 1.17 is established. The second follows from the first and (1.16).

REMARK. There is a result dual to (1.17) about families of finely lower semicontinuous functions. In particular if $\{B_i; i \in I\}$ is a family of finely open sets, then there exists a countable subset J of I such that $(\bigcup_{i \in I} B_i) - (\bigcup_{i \in J} B_i)$ is semipolar.

We will give a number of applications of Theorem 1.17 in the exercises. We close this section with following interesting result of Doob [3].

(1.18) PROPOSITION. Let $\mathscr{E}^f$ be the σ-algebra of fine Borel sets on E, that is, the σ-algebra generated by the finely open sets. Assume that (1.3) holds. Then $A \in \mathscr{E}^f$ if and only if there exists a Borel set B and a semipolar set N such $A = B \cup N$.

Proof. If A is finely closed, then $A = A^r \cup (A - A^r)$ where A^r is Borel and $A - A^r$ is semipolar. Let $\mathscr{D}$ be the class of all subsets of E which can be written as the union of a Borel set and a semipolar set. Clearly $\mathscr{D}$ is closed under countable unions and by the above remark contains all finely closed sets. If $A \in \mathscr{D}$ and $A = B \cup N$, then there is a Borel semipolar set $D \supset N$ according to (1.15). Consequently, letting "prime" denote complement in E, we have $A' = (B' - D) \cup [B' \cap (D - N)]$. Thus $\mathscr{D}$ is closed under complements and so $\mathscr{E}^f \subset \mathscr{D}$. Conversely any set of the form $A = B \cup N$ with $B \in \mathscr{E}$ and N semipolar is in $\mathscr{E}^f$ since a thin set is finely closed.

Exercises

(1.19) Consider the transition function $P_t(x, A) = I_A(x)$ on $(E, \mathscr{E})$. The corresponding standard process is said to be "constant in time." Prove that there is no reference measure unless E is countable.

(1.20) Assume (1.3). If A is nearly Borel show that there exists an increasing sequence $\{K_n\}$ of compact sets contained in A such that $\lim T_{K_n} = T_A$ almost surely. [Hint: use (1.5).]

(1.21) Assume (1.3) and let μ be a measure on $\mathscr{E}^*$ which vanishes on semipolar sets. Show that there exists a smallest finely closed set A such that $\mu(E - A) = 0$. The set A is called the *fine support of* μ. [Hint: use (1.17) and (1.18).]

(1.22) If A is an additive functional of (X, M), then we say that A *vanishes* on a set D provided $U_A I_D = 0$. We say that D *carries* A if A vanishes on $E - D$. Assume (1.3). Show that if A is a continuous additive functional of (X, M), then there is a smallest set F finely closed in E_M (i.e., closed in the fine topology restricted to E_M) which carries A. F is called the *fine support of* A and will be studied in some detail in Section 3.

(1.23) Suppose that there exists a finite measure λ on $\mathscr{E}^*$ such that if

$A \in \mathscr{E}^n$ and $P^x(T_A < \infty) = 0$, a.e. λ, then $P^x(T_A < \infty) = 0$ for all x. Show that X has a reference measure. [Hint: use (1.2).]

(1.24) Assume (1.3). Show that any $\alpha - (X, M)$ excessive function is Borel measurable. We are, of course, supposing that M is exact.

2. Continuous Additive Functionals

In this section we are going to establish two very important properties of continuous additive functionals under the assumption that (1.3) holds. We do not know to what extent (1.3) is necessary for the validity of these properties, but we will make crucial use of (1.3) in our discussion. As usual $X = (\Omega, \mathscr{M}, \mathscr{M}_t, X_t, \theta_t, P^x)$ denotes a fixed standard process with state space $(E, \mathscr{E})$ and M a fixed exact MF of X vanishing on $[\zeta, \infty]$. Again we will assume that $\mathscr{M} = \mathscr{F}$ and $\mathscr{M}_t = \mathscr{F}_t$ for all t. We begin with the following extension of Theorem 3.16 of Chapter IV. Recall from (1.3) of Chapter IV the definition of a perfect AF or MF.

(2.1) THEOREM. Assume (1.3) and that M is perfect. Then any CAF of (X, M) is equivalent to a perfect CAF of (X, M).

Proof. Let A be a CAF of (X, M) and let us assume to begin with that A has a bounded potential; that is, $u(x) = E^x A(\infty)$ is bounded. Let λ be a finite reference measure for X and let $U_A(x, \Gamma) = U_A I_\Gamma(x) = E^x \int_0^\infty I_\Gamma(X_t)\, dA_t$ for $\Gamma \in \mathscr{E}^*$. Plainly $U_A(x, \cdot)$ is a finite measure on $\mathscr{E}^*$ for each x. Since $\langle \lambda, u \rangle$ is finite, $\mu(\Gamma) = \int U_A(x, \Gamma)\, \lambda(dx)$ defines a finite measure on $\mathscr{E}^*$. Now $Q_t u \uparrow u$ as $t \to 0$ and so it follows from Egorov's theorem that we can find an increasing sequence $\{K_n\}$ of compact subsets of E such that (i) $Q_t u \to u$ as $t \to 0$ *uniformly* on K_n for each n and (ii) $\mu(E - K_n) \downarrow 0$ as $n \to \infty$. Clearly $U_A I_{E-K_n}$ decreases with n and $\langle \lambda, U_A I_{E-K_n} \rangle = \mu(E - K_n) \to 0$. Therefore $\{U_A I_{E-K_n}\}$ decreases to zero a.e. λ. Let $u_n = U_A I_{K_n}$. Then $u = u_n + U_A I_{E-K_n}$ and so $u_n \uparrow u$ a.e. λ. But $\lim_n u_n$ and u are both (X, M) excessive and hence $u_n \uparrow u$ everywhere. Moreover $u_{n+1} - u_n = U_A I_{(K_{n+1} - K_n)}$ is (X, M) excessive and so $u_n \ll u_{n+1}$; that is, u_{n+1} dominates u_n in the strong order (see (5.6) of Chapter IV). We next claim that each u_n is uniformly (X, M) excessive (see (3.15) of Chapter IV).

To see this first recall ((4.30) of Chapter IV) that if K is a compact subset of E, then $Q_K U_A I_K = U_A I_K$. Consequently $Q_{K_n} u_n = u_n$ for each n. Let n be fixed. Then given $\varepsilon > 0$ there exists $\delta > 0$ such that $u \le Q_t u + \varepsilon$ on K_n provided $t < \delta$. Hence on K_n we have for $t < \delta$

$$u_n + Q_t U_A I_{E-K_n} \le u_n + U_A I_{E-K_n} = u$$
$$\le Q_t u + \varepsilon = Q_t u_n + Q_t U_A I_{E-K_n} + \varepsilon,$$

and so $u_n \le Q_t u_n + \varepsilon$ on K_n if $t < \delta$. But $Q_t u_n + \varepsilon$ is (X, M) excessive and hence

$$u_n = Q_{K_n} u_n \le Q_{K_n}(Q_t u_n + \varepsilon) \le Q_t u_n + \varepsilon$$

everywhere provided $t < \delta$. Thus u_n is uniformly (X, M) excessive. If we define $u_0 = 0$, then plainly $f_n = u_n - u_{n-1}$ is (X, M) uniformly excessive for each $n \ge 1$ and $u = \sum f_n$. By Theorem 3.16 of Chapter IV for each n there is a perfect CAF, A^n, of (X, M) whose potential is f_n. If we define $B(t) = \sum A^n(t)$ for $0 \le t \le \infty$, then $E^x B(\infty) = u(x)$ and so $B(\infty) < \infty$ almost surely. Therefore $\sum A^n(t)$ converges uniformly on $[0, \infty]$ almost surely and so B is a CAF of (X, M). Clearly B is perfect. Since A and B have the same bounded potential u, they are equivalent. Thus (2.1) is proved under the assumption that A has a bounded potential.

To treat the general case we make use of (2.21) of Chapter IV which asserts that there exists a sequence $\{A^n\}$ of CAF's of (X, M) with bounded 1-potentials such that $A^n(t) \uparrow A(t)$ for all t as $n \to \infty$. Define $N_t = e^{-t} M_t$ and $B^n(t) = \int_0^t e^{-s}\, dA^n(s)$. Then N is a perfect MF of X and B^n is a CAF of (X, N) with a bounded potential. Thus by what was proved above there exist perfect CAF's, $\bar{B}^n$ of (X, N) such that B^n and $\bar{B}^n$ are equivalent for each n. If we define $\bar{A}^n(t) = \int_0^t e^s\, d\bar{B}^n(s)$, then each $\bar{A}^n$ is a perfect CAF of (X, M) equivalent to A^n. It follows easily from this that A is equivalent to a perfect CAF of (X, M), completing the proof of Theorem 2.1.

Let $\mathbf{A}(M)$ be the collection of all continuous additive functionals of (X, M). We will write simply $\mathbf{A}$ for the collection of all CAF's of X. Equality in $\mathbf{A}(M)$ is understood to be equivalence. Under the usual pointwise definitions of $A + B$ and αA for $\alpha \ge 0$ the set $\mathbf{A}(M)$ becomes a cone. This introduces an order relation "$\le$" in $\mathbf{A}(M)$ as follows: $A \le B$ provided there exists $C \in \mathbf{A}(M)$ such that $A + C = B$. The reader should observe that if $A, B \in \mathbf{A}(M)$ have *finite* α-potentials u_A^α, u_B^α, then $A \le B$ if and only if $u_A^\alpha \ll u_B^\alpha$. Given $\alpha \ge 0$ let $\mathbf{A}^\alpha(M)$ denote those elements in $\mathbf{A}(M)$ which have finite α-potentials. Then if $0 \le \alpha < \beta$, $\mathbf{A}^\alpha(M) \subset \mathbf{A}^\beta(M) \subset \mathbf{A}(M)$. We are now going to study the cone $\mathbf{A}(M)$ in some detail. However it will first be necessary to prepare some tools which will also be used in Section 3.

(2.2) LEMMA. Let $a(t)$ be a function from $[0, \infty]$ to $[0, \infty]$ which is nondecreasing and right continuous and satisfies $a(0) = 0$, $a(\infty) = \lim_{t \uparrow \infty} a(t)$. Define

$$\tau(t) = \inf\{s : a(s) > t\}$$

for $0 \leq t < \infty$ where as usual we set $\tau(t) = \infty$ if the set in braces is empty. The function τ from $[0, \infty)$ to $[0, \infty]$ is called the inverse of a. It is right continuous and nondecreasing. Define $\tau(\infty) = \lim_{t \uparrow \infty} \tau(t)$. Then

(i) $$a(s) = \inf\{t \colon \tau(t) > s\}, \qquad 0 \leq s < \infty;$$

and if f is a nonnegative Borel measurable function on $[0, \infty]$ vanishing at ∞ one has

(ii) $$\int_{(0,\infty)} f(t)\, da(t) = \int_0^\infty f[\tau(t)]\, dt.$$

Proof. It is easy to verify that $\tau(t)$ is nondecreasing and right continuous. As to (i) we first note that if $a(s) < \infty$, then $\tau[a(s)] = \inf\{u : a(u) > a(s)\} \geq s$ with equality if and only if s is a point of right increase of a; that is, $a(s + \varepsilon) > a(s)$ for all $\varepsilon > 0$. Thus $\tau[a(s + \varepsilon)] \geq s + \varepsilon > s$ if $a(s + \varepsilon) < \infty$ and so $\inf\{t \colon \tau(t) > s\} \leq a(s + \varepsilon)$ in any case. Since a is right continuous, this yields $\inf\{t \colon \tau(t) > s\} \leq a(s)$. Suppose this last inequality is strict. Then there exists an $r < a(s)$ such that $\tau(r) > s$. But the definition of τ then implies that $a(s) \leq r$. Thus (i) holds. It suffices to prove (ii) when $a(\infty) < \infty$. In this case the class of bounded f for which (ii) holds is clearly a linear space closed under increasing limits and so it suffices to prove (ii) when f is the indicator function of $[0, s]$, $s < \infty$. But then $\int_{(0,\infty)} f(t)\, da(t) = a(s) - a(0) = a(s)$. Of course, da assigns no mass to $\{0\}$ or $\{\infty\}$ and so we could just as well integrate over $[0, \infty]$. On the other hand if $t < a(s) = \inf\{u \colon \tau(u) > s\}$ then $\tau(t) \leq s$, while if $t > a(s)$ then $\tau(t) > s$. Therefore $f[\tau(t)]$ is one on $[0, a(s))$ and zero on $(a(s), \infty]$. Consequently $\int_0^\infty f[\tau(t)]\, dt = a(s)$. This completes the proof of (2.2).

Although (2.2ii) is closely related to (2.20i) of Chapter II it is not a consequence of that result.

(2.3) PROPOSITION. (i) Let M be an SMF of X vanishing on $[\zeta, \infty]$. (For this proposition only we do not assume that M is exact.) Let A be an AF of (X, M). For each t, $0 \leq t < \infty$, define

$$\tau_t(\omega) = \inf\{u \colon A_u(\omega) > t\}.$$

Then each τ_t is a stopping time. (ii) Suppose that $M_t = I_{[0,S)}(t)$ where S is a strong terminal time and that A is continuous. Then for each $u, v \geq 0$ one has

$$\tau_{u+v} = \tau_u + \tau_v \circ \theta_{\tau_u}$$

almost surely on $\{\tau_u < S\}$.

Proof. We may assume that $t \to A_t(\omega)$ is right continuous and nondecreasing for all ω. Then Lemma 2.2 implies that $\{\tau_t < u\} = \bigcup_n \{A_{u-1/n} > t\}$ and so each τ_t is a stopping time. Note that if $t \to A_t(\omega)$ is continuous, then $u \to \tau_u(\omega)$ is strictly increasing on $[0, A_\infty(\omega))$ and that $\tau_u(\omega) = \max\{t: A_t(\omega) = u\}$ provided the set in braces is not empty, while $\tau_u(\omega) = \infty$ otherwise. In particular $A[\tau_u(\omega), \omega] = u$ if $\tau_u(\omega) < \infty$. In proving (ii) we may assume that $t \to A_t(\omega)$ is continuous for all ω. Let $u, v \geq 0$ be fixed. Proposition 1.13 of Chapter IV and the right continuity of A_t imply that

$$A(\tau_u + s) = A(\tau_u) + A(s, \theta_{\tau_u})$$

for all s, almost surely on $\{\tau_u < S\}$. Since $A(\tau_u) = u$ this implies that, almost surely on $\{\tau_u < S\}$, $A(\tau_u + s) > u + v$ if and only if $A(s, \theta_{\tau_u}) > v$. Proposition 2.3 follows immediately from this and the definition of τ_t.

In general we will call τ the inverse of A. We now assume again that M is exact. We next introduce some notation. If A is a CAF of (X, M) we write $A^*(t) = \int_0^t (M_s)^{-1}\, dA_s$ so that A^* is a CAF of (X, S). For short we will write $dA_t^* = (M_t)^{-1}\, dA_t$ and $dA_t = M_t\, dA_t^*$ for this relationship. We let τ denote the inverse of A and τ^* the inverse of A^*. Since $A_t \leq A_t^*$ for all t one has $\tau_t^* \leq \tau_t$ for all t. If $f \in \mathscr{E}_+^*$ we write fA for the family of random variables $(fA)(t) = \int_0^t f(X_s)\, dA_s$. Under appropriate continuity or finiteness assumptions fA will be a CAF of (X, M). Note that $(fA)^* = fA^*$. We come now to the key step in our study of $\mathbf{A}(M)$. *We assume that* (1.3) *holds in the remainder of this section.*

(2.4) PROPOSITION. Let A be a CAF of (X, M) with finite α-potential. Then necessary and sufficient conditions that an $f \in \mathscr{S}^\alpha(M)$ have the representation $f = U_A^\alpha g$ where g is Borel measurable and $0 \leq g \leq 1$ are (i) $f \leq u_A^\alpha$ and (ii) $f(x) - Q_t^\alpha f(x) \leq E^x \int_0^t e^{-\alpha s}\, dA_s$ for all t and x.

Proof. The necessity of these conditions is immediate. By replacing M_t with $e^{-\alpha t} M_t$ and A_t with $\int_0^t e^{-\alpha s}\, dA_s$ we reduce the general case to the case $\alpha = 0$ in the usual way. Thus in proving the sufficiency it suffices to consider the case $\alpha = 0$. Now

$$Q_t(u_A - f)(x) = u_A(x) - E^x A(t) - Q_t f(x) \leq u_A(x) - f(x),$$

and so $u_A - f$ is (X, M) excessive. Since u_A is a regular (X, M) potential the relationship $f + (u_A - f) = u_A$ with both f and $u_A - f$ in $\mathscr{S}(M)$ implies that f and $u_A - f$ are regular (X, M) potentials. Hence there exist CAF's, B and C, of (X, M) such that $f = u_B$ and $u_A - f = u_C$, and the uniqueness theorem implies that $A = B + C$. Consequently if T is any stopping time

$$f(x) - Q_T f(x) = E^x B(T) \leq E^x A(T)$$

for each x. Now $A[\tau(t)] = t$ if $\tau(t) < \infty$ and $A(\infty) \leq t$ if $\tau(t) = \infty$. Since

$\tau^*(t) \le \tau(t)$ this implies that $A[\tau^*(t)] \le A[\tau(t)] \le t$ for all t. Hence $f(x) - Q_{\tau^*(t)} f(x) \le t$ for all t and x.

If $h \in \mathscr{E}_+^*$ define for $\beta \ge 0$

$$W^\beta h(x) = E^x \int_0^\infty e^{-\beta t} h(X_{\tau^*(t)}) M_{\tau^*(t)} \, dt.$$

Making use of $A^*[\tau^*(t)] = t$ if $\tau^*(t) < \infty$ and (2.2ii), this may be written

$$W^\beta h(x) = E^x \int_0^\infty e^{-\beta A^*(t)} h(X_t) \, dA_t.$$

It now follows from (2.22) of Chapter IV (or is easily verified in the present situation) that the family $\{W^\beta; \beta \ge 0\}$ is a resolvent on $b\mathscr{E}^*$ with $W^0 = U_A$. Taking Laplace transforms one obtains from $f - Q_{\tau^*(t)} f \le t$ the inequality

(2.5) $$0 \le \beta(f - \beta W^\beta f) \le 1.$$

In particular $\beta W^\beta f \to f$ as $\beta \to \infty$. Since u_A is finite we may choose a reference measure λ so that $\mu(\Gamma) = \lambda U_A(\Gamma) = \int \lambda(dx) \, U_A(x, \Gamma)$ is a finite measure on $\mathscr{E}^*$. Let $\mathbf{L}_\infty = \mathbf{L}_\infty(\mu)$ and $\mathbf{L}_1 = \mathbf{L}_1(\mu)$. Since $\mathbf{L}_\infty$ is the dual of $\mathbf{L}_1$ (as Banach spaces), a bounded set in $\mathbf{L}_\infty$ is relatively compact in the weak * topology, i.e., the topology on $\mathbf{L}_\infty$ induced by $\mathbf{L}_1$. Also $\mathbf{L}_1$ is separable since E has a countable base and so the weak * topology is metrizable on bounded subsets of $\mathbf{L}_\infty$. It now follows from (2.5) that one can find a sequence $\{\beta_n\}$ increasing to infinity such that $\beta_n(f - \beta_n W^{\beta_n} f)$ converges to an element g of $\mathbf{L}_\infty$ in the weak * topology. We may assume that g is Borel measurable and $0 \le g \le 1$. (Here we are making the usual "identification" of an element of $\mathbf{L}_\infty$ with a "function.") Let $g_n = \beta_n(f - \beta_n W^{\beta_n} f)$. Since $U_A = W^0$ on bounded functions. the resolvent equation yields

$$U_A g_n = \beta_n W^{\beta_n} f \to f$$

pointwise as $n \to \infty$. On the other hand for each x the measure $U_A(x, \cdot)$ is absolutely continuous with respect to μ, because if $\mu(\Gamma) = \int \lambda(dy) \, U_A(y, \Gamma) = 0$, then the (X, M) excessive function $U_A(\cdot, \Gamma)$ vanishes a.e. λ and hence is identically zero. For fixed x let $U_A(x, dy) = h_x(y) \, \mu(dy)$ where $h_x \in \mathbf{L}_1$. Then

$$U_A g_n(x) = \int g_n(y) \, h_x(y) \, \mu(dy)$$

$$\to \int g(y) \, h_x(y) \, \mu(dy) = U_A g(x).$$

Consequently $f = U_A g$ and the proof of Proposition 2.4 is complete.

(2.6) Corollary. If $A, B \in \mathbf{A}^\alpha(M)$, then $A \le B$ if and only if for all t and x

$$E^x \int_0^t e^{-\alpha s}\, dA_s \le E^x \int_0^t e^{-\alpha s}\, dB_s. \tag{2.7}$$

If $A, B \in \mathbf{A}(M)$, then $A \le B$ if and only if there exists a Borel measurable f with $0 \le f \le 1$ such that $A = fB$.

Proof. If $A, B \in \mathbf{A}^\alpha(M)$ and $A \le B$, then clearly (2.7) holds. If (2.7) holds then, since the left side of (2.7) is just $u_A^\alpha(x) - Q_t^\alpha u_A^\alpha(x)$, it follows from (2.4) that $u_A^\alpha = U_B^\alpha f = u_{fB}^\alpha$ with $0 \le f \le 1$. Hence $A = fB$ and so $B = A + (1 - f)B$, or $A \le B$. Note that if $f \ge 0$ is bounded and $B \in \mathbf{A}^\alpha(M)$, then $fB \in \mathbf{A}^\alpha(M)$. It remains to show that if $A, B \in \mathbf{A}(M)$ and $A \le B$, then $A = fB$ with f Borel measurable and $0 \le f \le 1$. It follows from (2.21) of Chapter IV that if $\varphi(x) = E^x \int_0^\infty e^{-t} M_t \exp(-B_t^*)\, dt$ and g_n is the indicator function of $\{1/(n+1) < \varphi \le 1/n\}$ for $n \ge 1$, then $B^n = g_n B$ has a bounded one potential for each n and $\sum g_n = I_{E_M}$. Of course each $g_n \in \mathscr{E}_+^*$. Let $A^n = g_n A$. If $A + C = B$ then $g_n A + g_n C = g_n B$ and so $A^n \le B^n$. Clearly A^n has a bounded one potential and $\sum A^n = A$. By the first part of the proof, then, there is for each n a function $h_n \in \mathscr{E}_+^*$ with $0 \le h_n \le 1$ such that $A^n = h_n g_n B$. Let λ be a finite reference measure and let μ be the measure $\mu(D) = E^\lambda \int_0^\infty e^{-t} I_D(X_t)\, dB_t$. Then $\mu(h_n g_n) = \int \lambda(dx)\, u_{A^n}^1(x) < \infty$ and so there is a function $f_n \in \mathscr{E}_+$ such that $f_n \le h_n g_n$ and $\mu(f_n) = \mu(h_n g_n)$. Now the two bounded functions $u_{f_n B}^1$ and $u_{h_n g_n B}^1$ are in $\mathscr{S}^1(M)$ and are equal almost surely λ. Hence they are identical, so by the uniqueness theorem for additive functionals $f_n B = h_n g_n B = A^n$. If we set $f = \sum f_n$ then $f \in \mathscr{E}_+$ and $A = fB$. It is obvious that $0 \le f \le 1$ and so the proof of (2.6) is complete.

(2.8) Corollary. Let A and B be in $\mathbf{A}(M)$. Then a necessary and sufficient condition that there exists a Borel measurable $f \ge 0$ such that $A = fB$ is that, for any $g \in \mathscr{E}_+$, $U_B g = 0$ implies $U_A g = 0$.

Proof. The necessity is obvious. To prove the sufficiency there is no loss of generality in assuming that A and B have bounded one potentials. If $C = A + B$, then by (2.6) there exist $f, g \in \mathscr{E}_+$, $0 \le f, g \le 1$, such that $A = fC$ and $B = gC$. Formally $A = (f/g)B$, and we will now justify this relation. Note that $U_B^1 h = 0$ if and only if $U_B h = 0$. If $N_g = \{g = 0\}$ and I_g is the indicator function of N_g, then $U_B^1 I_g = U_C^1(g I_g) = 0$ and so, using the hypothesis, $U_C^1(f I_g) = U_A^1 I_g = 0$. Hence $f = 0$ on N_g a.e. $U_C^1(x, \cdot)$ for all x. Define $h(x) = f(x)/g(x)$ if $x \notin N_g$ and $h(x) = 0$ if $x \in N_g$. Then for each x, $hg = f$ a.e. $U_C^1(x, \cdot)$ and consequently $u_A^1 = U_C^1 f = U_C^1 gh = U_B^1 h$. The uniqueness theorem now yields $A = hB$ completing the proof of (2.8).

We are now going to give an example to show that (2.4), and hence (2.6) and (2.8), are *not* valid for natural additive functionals. Let E be the following subset of the Euclidean plane where (x, y) is the generic point in $\mathbf{R}^2$:

$$E = E_1 \cup E_2 \cup E_3 \cup E_4$$

where

$$E_1 = \{(x, y): y = 0, x \leq 0\},$$
$$E_2 = \{(x, y): y = -x + 1, 0 \leq x < 1\},$$
$$E_3 = \{(x, y): y = x - 1, 0 \leq x < 1\},$$
$$E_4 = \{(x, y): y = 0; x \geq 1\}.$$

The process X is described as follows: a particle moves to the right at unit speed until reaching $(0, 0)$ which is a holding point with parameter 1 from which the particle jumps to $(0, 1)$ or $(0, -1)$ with probability $\frac{1}{2}$, respectively, and then continues to move to the right with unit speed. The reader should write down a formal definition of this process and verify that it is a Hunt process. Since excessive functions are right continuous in the obvious sense, condition (1.3) holds. If T is the hitting time of the point $(1, 0)$, then upon defining $B(t) = 0$ for $t < T$ and $B(t) = 1$ for $t \geq T$ we obtain a natural additive functional of X. Define u as follows: $u = 1$ on E_1, $u = 0$ on $E_2 \cup E_4$, and $u = 1$ on E_3. Then $A(t) = 0$ if $t < T$ and $A(T) = u(X_T)_- - u(X_T)$ if $t \geq T$ also defines a natural additive functional of X. Clearly $A \leq B$. If $A = fB$, then $A(T) - A(T-) = f(X_T)[B(T) - B(T-)] = f[(1, 0)]$ almost surely $P^{(0,0)}$. But $A(T) - A(T-) = u(X_T)_- - u(X_T)$ is *not* constant $P^{(0,0)}$ almost surely.

Exercises

(2.9) Assume (1.3.) (i) Prove that $\mathbf{A}(M)$ is a lattice under "$\leq$." (ii) Prove that $\mathbf{A}(M)$ is boundedly complete; that is, if $\{A_i\}$ is a family of elements of $\mathbf{A}(M)$ then $\inf A_i$ exists in $\mathbf{A}(M)$, and if there exists $B \in \mathbf{A}(M)$ such that $A_i \leq B$ for all i, then $\sup A_i$ exists in $\mathbf{A}(M)$. [Hint: use the fact that $\mathbf{L}_1(\mu)$ is boundedly complete under the ordering given by $f \leq g$ provided $\mu(\{f > g\}) = 0$.]

(2.10) Assume (1.3). Let M be an exact MF of X vanishing on $[\zeta, \infty]$ and such that $M_t \in \mathscr{F}^0$ for all t. Note that $M_t = I_{[0,\zeta)}(t)$ always satisfies this condition. Show that any CAF, A, of (X, M) is equivalent to a CAF, B, of (X, M) such that $B_t \in \mathscr{F}^0$ for all $t \geq 0$. [Hint: use (1.24) and imitate the proofs of (3.16) of Chapter IV and (2.1).] In particular when dealing with a CAF, A, of X we may assume without loss of generality, under (1.3), that A is perfect and that each A_t is $\mathscr{F}^0$ measurable.

(2.11) Let A be a CAF of X and let τ be the inverse of A. Assume that $t \to A_t(\omega)$ is continuous and nondecreasing for *all* ω. Then $t \to \tau_t(\omega)$ is right continuous and strictly increasing for all ω. (i) Show $\{\tau_t < a\} = \{A_a > t\}$ for all a and t, $0 \le a, t < \infty$. (ii) Show that $\hat{X} = (\Omega, \mathscr{F}, \mathscr{F}_{\tau(t)}, X_{\tau(t)}, \theta_{\tau(t)}, P^x)$ is a Markov process with right continuous paths having $(E, \mathscr{E}^*)$ as state space and whose resolvent W^α is given by $W^\alpha f(x) = E^x \int_0^\infty e^{-\alpha A(t)} f(X_t)\, dA_t$. (iii) Let $\mathscr{G}_t = \mathscr{F}_{\tau(t)}$ for $t \ge 0$ and $\mathscr{G} = \mathscr{F}$. Note that $\{\mathscr{G}_t : t \ge 0\}$ is right continuous. Show that if T is a $\{\mathscr{G}_t\}$ stopping time then τ_T ($\tau_T(\omega) = \tau_{T(\omega)}(\omega)$) is an $\{\mathscr{F}_t\}$ stopping time. Also show that $\mathscr{G}_T = \mathscr{F}_{\tau(T)}$. (iv) Show that if T is a $\{\mathscr{G}_t\}$ stopping time then $\tau_{t+T} = \tau_T + \tau_t \circ \theta_{\tau(T)}$ almost surely. Use this to show that the process $\hat{X}$ defined in (ii) is a strong Markov process. (v) If A is strictly increasing, that is, $u < v$, $u < \zeta(\omega)$, and $A_u(\omega) < \infty$ imply that $A_u(\omega) < A_v(\omega)$, show that $\hat{X}$ is normal and quasi-left-continuous and that the lifetime $\hat{\zeta}$ of $\hat{X}$ is given by $\hat{\zeta} = A_\zeta$. If, in addition, (1.3) holds use (2.10) to show that $x \to E^x f(\hat{X}_t)$ is Borel measurable for each $t \ge 0$ and $f \in b\mathscr{E}$, and hence $\hat{X}$ is a standard process. See (4.11) for further properties of $\hat{X}$. Here we have written $\hat{X}_t = X_{\tau(t)}$ for $t < \infty$, and, of course, $\hat{X}_\infty = \Delta$.

3. Fine Supports and Local Times

In this section we are going to investigate the set of time points on which a CAF, A, is increasing. Roughly speaking the situation is this: there exists a subset F of E such that $t \to A_t(\omega)$ is increasing at t_0 if and only if $X_{t_0} \in F$. (See Theorem 3.8 for the precise statement.) If (1.3) holds, then F turns out to be just the fine support of A defined in (1.22). However we will proceed somewhat more generally and will give an explicit definition of F. We will also discuss in some detail CAF's for which the corresponding F reduces to a single point x_0. Such a CAF is called a *local time* for X at x_0. We will give a simple characterization of those points x_0 which admit a local time in this sense, and will investigate the dependence of the local time on the point x_0.

As usual $X = (\Omega, \mathscr{M}, \mathscr{M}_t, X_t, \theta_t, P^x)$ denotes a fixed standard process with state space $(E, \mathscr{E})$ and with $\mathscr{M} = \mathscr{F}$, $\mathscr{M}_t = \mathscr{F}_t$ for all t. Also M denotes a fixed exact MF of X which vanishes on $[\zeta, \infty]$ and $S = \inf\{t: M_t = 0\}$. Let A be a CAF of (X, M) and let A^* be the CAF of (X, S) defined by $dA_t^* = (M_t)^{-1}\, dA_t$ where the notation is that of Section 2. In particular τ_t and τ_t^* denote the inverses of A_t and A_t^*, respectively. It will be convenient to assume that $t \to M_t(\omega)$ is right continuous and nonincreasing for all ω and that $t \to A_t(\omega)$ is continuous and nondecreasing for all ω. Define

(3.1) $$R(\omega) = \inf\{t\colon A_t(\omega) > 0\}$$

provided the set in braces is not empty and $R(\omega) = \infty$ if it is empty. Clearly R is a stopping time and $R = \sup\{t\colon A_t = 0\}$. Since $A_t(\omega) = 0$ if and only if $A_t^*(\omega) = 0$, we also have

(3.2) $$R(\omega) = \inf\{t\colon A_t^*(\omega) > 0\}$$

provided the set in braces is not empty and $R(\omega) = \infty$ if it is empty. It follows from (3.2) that if T is a stopping time, then $T + R \circ \theta_T = R$ almost surely on $\{T < R \wedge S\}$. In the notation of Section 2, $R = \tau_0 = \tau_0^*$. Finally observe that $A_R = A_R^* = 0$ since A and A^* are continuous. We next define

(3.3) $$\varphi^A(x) = E^x\{e^{-R}; R < S\}.$$

Since $R = \infty$ on $\{R \geq S\}$, it is evident that $\varphi^A(x) = E^x(e^{-R})$. It is also clear that φ^A is in $\mathscr{S}^1(M)$. We now define

(3.4) $$\text{Supp}(A) = \{x\colon \varphi^A(x) = 1\},$$

and we call Supp(A) the *support* of A. Clearly Supp(A) is nearly Borel measurable, Supp(A) $\subset E_M$, and Supp(A) is finely closed in E_M, that is, closed in the relative fine topology on E_M. Moreover Supp(A) $= \{x\colon P^x(R = 0) = 1\}$, and so intuitively Supp(A) consists of those points x such that, starting from x, $t \to A_t$ begins to increase immediately with probability one. Obviously Supp(A) = Supp(A^*).

(3.5) Proposition. Let T be the hitting time of Supp(A). Then each of the following holds almost surely: $T \leq R$, $\{R < S\} = \{T < S\}$, and $T = R$ on $\{R < S\}$. In addition each x in Supp(A) is regular for Supp(A).

Proof. First observe that $X_T \in$ Supp(A) almost surely on $\{T < S\}$. Consequently for any x

$$\begin{aligned} P^x(T < R, R < S) &= P^x(T < R, R \circ \theta_T > 0, R < S) \\ &\leq E^x\{P^{X(T)}(R > 0); T < S\} = 0. \end{aligned}$$

Similarly $P^x(T < R, T < S) = 0$. Therefore $R \leq T$ almost surely on $\{T \wedge R < S\}$. For notational convenience let $F =$ Supp(A). Now for any $t > 0$ we have

$$\begin{aligned} P^x\{R < T\} &= P^x\{A(R + t) > 0; R < T\} \\ &= E^x\{M_R\, P^{X(R)}(A_t > 0); R < T \wedge S\}. \end{aligned}$$

If $x \in F^r$ this last expression is zero since $P^x(T = 0) = 1$. If $x \notin F$ then $X_R \notin F$

almost surely P^x on $\{R < T\}$, while if $y \notin F$

$$P^y(A_t > 0) \le P^y(R < t) \to 0$$

as $t \to 0$. Thus letting $t \to 0$ we obtain $P^x(R < T) = 0$ provided $x \notin F - F^r$. Let $\psi(x) = E^x(e^{-T}; R < S)$. Then using the easily verified fact that $t + R \circ \theta_t \downarrow R$ almost surely P^x for each $x \in E_M$, one sees that ψ is in $\mathscr{S}^1(M)$. But by what was proved above, ψ and φ^A agree except possibly on the semipolar set $F - F^r$, and hence, since they are both in $\mathscr{S}^1(M)$, they are identical. It now follows that $R = T$ almost surely on $\{R < S\}$. As a result it is clear that $F - F^r$ must be empty and this yields the last assertion of (3.5). In view of what was proved above this implies that $T \le R$ almost surely. Combining this with the statement that $R \le T$ almost surely on $\{T \wedge R < S\}$ completes the proof of Proposition 3.5.

(3.6) REMARK. Note that $R = \infty$ almost surely on $\{R \ge S\}$. If $S = \zeta$ then $T = \infty$ on $\{T \ge S\}$, and so in the case $S = \zeta$ we obtain $T = R$ almost surely.

An immediate consequence of Proposition 3.5 is the fact that $A = 0$ if and only if Supp(A) is empty. We are now going to investigate the set on which $t \to A_t$ increases. To this end we define the following sets, each of which depends on ω:

(3.7)
$$\mathbf{I} = \{t\colon A(t + \varepsilon) - A(t) > 0 \text{ for all } \varepsilon > 0\}$$
$$\mathbf{J} = \{t\colon A(t + \varepsilon) - A(t - \varepsilon) > 0 \text{ for all } \varepsilon > 0\}$$
$$\mathbf{Z} = \{t < S;\ X_t \in \text{Supp}(A)\}$$
$$\mathbf{Q} = \{t < \infty;\ \tau(u) = t \text{ for some } u\}.$$

In the definition of $\mathbf{J}$ it is understood that we set $A(u) = 0$ if $u < 0$. We also define $\mathbf{I}^*$, $\mathbf{J}^*$, and $\mathbf{Q}^*$ by replacing A and τ by A^* and τ^*, respectively, in (3.7). We call $\mathbf{I}$ the set of points of *right increase* of A and $\mathbf{J}$ the set of points of *increase* of A. The reader should check that the assumed regularity of A implies that $\mathbf{J}$ is the closure of $\mathbf{I}$ and that $\mathbf{Q} = \mathbf{I}$. Of course, the same relationships hold among $\mathbf{I}^*$, $\mathbf{J}^*$, and $\mathbf{Q}^*$. Finally the assumed regularity of M implies that $\mathbf{I} = \mathbf{I}^*$ and $\mathbf{J} = \mathbf{J}^*$; hence, $\mathbf{Q} = \mathbf{Q}^*$. We come now to our main result.

(3.8) THEOREM. Almost surely $\mathbf{I} \subset \mathbf{Z} \subset \mathbf{J}$.

Proof. If T is the hitting time of Supp(A), then almost surely we have

$$\{\omega\colon \mathbf{Z} \not\subset \mathbf{J}\} \subset \bigcup_{r<t} \{A(t) - A(r) = 0,\ r + T \circ \theta_r < t < S\}$$

where the union is over all pairs (r, t) of rationals with $0 \le r < t$. But for each x

$$P^x\{A(t) - A(r) = 0, r + T \circ \theta_r < t < S\} \le E^x\{P^{X(r)}[A(t-r) = 0, T < t - r < S]; r < S\}$$

and Proposition 3.5 implies that this last expression is zero. To show that $\mathbf{I} \subset \mathbf{Z}$ we first observe that if $t \in \mathbf{I}$ and $t \notin \mathbf{Z}$, then $t < S$ and $\varphi^A(X_t) < 1$. Using the right continuity of $u \to \varphi^A(X_u)$ on $[0, S)$ we then have

$$\{\mathbf{I} \not\subset \mathbf{Z}\} \subset \bigcup_{r<q} \{A_q - A_r > 0, r + T \circ \theta_r \ge q, r < S\}$$

almost surely where the union is over all pairs (r, q) of rationals with $0 \le r < q$. As before, Proposition 3.5 implies that the above union has P^x measure zero for all x. Hence Theorem 3.8 is established.

(3.9) Corollary. Let g^A be the indicator function of Supp(A). Then $A = g^A A$.

Proof. For any ω, A induces a measure $dA_t(\omega)$ on the Borel sets of $[0, \infty)$ whose support is $\mathbf{J}(\omega)$. Consider an ω such that $\mathbf{I}(\omega) \subset \mathbf{Z}(\omega) \subset \mathbf{J}(\omega)$. Now $\mathbf{J}(\omega) - \mathbf{I}(\omega)$ is countable and so, $t \to A_t(\omega)$ being continuous, the set $\mathbf{I}(\omega)$ carries all of the mass of $dA_t(\omega)$. Therefore

$$\begin{aligned} A_t(\omega) &= \int_{[0,t]\cap \mathbf{I}(\omega)} dA_s(\omega) \\ &= \int_{[0,t]\cap \mathbf{I}(\omega)} g^A[X_s(\omega)]\, dA_s(\omega) \\ &= \int_0^t g^A[X_s(\omega)]\, dA_s(\omega). \end{aligned}$$

In light of (3.8) this proves the equality (i.e., equivalence) of A and $g^A A$.

(3.10) Corollary. Supp(A) is the smallest nearly Borel subset of E_M which is finely closed in E_M and on whose complement A vanishes.

Proof. Recall that A vanishes on a set D provided $U_A I_D = 0$ or, equivalently, $I_D A = 0$. A certainly vanishes on the complement of Supp(A) in view of (3.9). To complete the proof of (3.10) it suffices to show that if D is nearly Borel and finely open and A vanishes on D, then $D \cap$ Supp(A) is empty. Suppose $x \in D \cap$ Supp(A). Since A vanishes on D, $A = I_{D^c} A$, and this and the continuity of A imply $A(T_{D^c}) = \int_0^{T_{D^c}} I_{D^c}(X_t)\, dA_t = 0$. Consequently $T_{D^c} \le R$.

But $x \in D \cap \text{Supp}(A)$ implies $P^x(R = 0) = 1$ and $P^x(T_{D^c} > 0) = 1$. This contradiction establishes (3.10).

We leave it to the reader as Exercise 3.33 to show that if (1.3) holds, then $\text{Supp}(A)$ is the fine support of A as defined in (1.22). We are now going to study a very special situation, but one which is of importance in the applications. We begin with the following special case of Corollary 2.8 which is easily verified directly even when (1.3) is not assumed to hold. We leave the details to the reader as Exercise 3.32.

(3.11) PROPOSITION. Let A and B be CAF's of (X, M) and suppose that $\text{Supp}(B)$ is countable. Then a necessary and sufficient condition that $A = fB$ for some $f \in \mathscr{E}_+$ is that if $g \in \mathscr{E}_+$ and $U_B g = 0$ then $U_A g = 0$. One may choose f so that it vanishes off $\text{Supp}(B)$.

In the remainder of this section we will deal with CAF's A of X. In this case $\text{Supp}(A)$ is finely closed and, according to (3.6), $T = R$ almost surely, where the notation is that of (3.5).

(3.12) DEFINITION. Let x_0 be a fixed point of E. Then a CAF, A, of X is called a *local time for X at x_0* provided that $\text{Supp}(A) = \{x_0\}$.

(3.13) THEOREM. A necessary and sufficient condition that there exists a local time for X at x_0 is that x_0 be regular for $\{x_0\}$. Let T be the hitting time of $\{x_0\}$. Then if A is any local time for X at x_0, there exists a positive constant k such that $u_A^1(x) = k\, E^x(e^{-T})$ for all x. Moreover A is equivalent to a perfect CAF of X.

Proof. If there exists a CAF, A with $\text{Supp}(A) = \{x_0\}$, then x_0 is regular for $\{x_0\}$ according to Proposition 3.5. Conversely suppose x_0 is regular for $\{x_0\}$ and define $\psi(x) = E^x(e^{-T})$ where $T = T_{\{x_0\}}$. Then $\psi \in \mathscr{S}^1$ and $\psi(x_0) = 1$. Now $X_T = x_0$ almost surely on $\{T < \infty\}$ and so

$$\begin{aligned} P_T^1\, \psi(x) &= E^x\{e^{-T} E^{X(T)}(e^{-T})\} \\ &= \psi(x_0)\, \psi(x) = \psi(x). \end{aligned}$$

Given $\varepsilon > 0$ there exists a $\delta > 0$ such that $P_t^1\, \psi(x_0) \le \psi(x_0) \le P_t^1\, \psi(x_0) + \varepsilon$ whenever $t \le \delta$. But $P_T^1(x, dy)$ is concentrated on $\{x_0\}$, and consequently if $t \le \delta$ we have

$$\psi = P_T^1 \psi \le P_T^1(P_t^1 \psi + \varepsilon) \le P_t^1 \psi + \varepsilon.$$

Therefore ψ is uniformly 1-excessive. Also $P_t^1 \psi \to 0$ as $t \to \infty$ since ψ is bounded, and so it follows from Theorem 3.16 of Chapter IV that there exists a *perfect* CAF, B, of X such that $u_B^1(x) = \psi(x)$. We will assume, as we may, that $t \to B_t(\omega)$ is continuous and nondecreasing for all ω.

We will next show that $\text{Supp}(B) = \{x_0\}$; that is, B is a local time for X at x_0. Let $R = \inf\{t : B_t > 0\}$. Then

$$\psi(x) = P_T^1 \psi(x) = E^x \int_T^\infty e^{-t} \, dB_t,$$

and so $E^x \int_0^T e^{-t} \, dB_t = 0$ for all x. Consequently $T \leq R$ almost surely. Now if $x \neq x_0$ then $P^x(T > 0) = 1$ and so x is not in $\text{Supp}(B)$ since $R \geq T$ almost surely. But $\text{Supp}(B)$ is not empty because B is not zero ($u_B^1(x_0) = \psi(x_0) = 1$). Hence $\text{Supp}(B) = \{x_0\}$. In particular $R = T$ almost surely.

To complete the proof of Theorem 3.13 it will suffice to show that if A is any local time for X then there exists a constant $k > 0$ such that $A = kB$. If $g \in \mathscr{E}_+$ and $U_B g = 0$, then $g(x_0)$ must be zero. Consequently $U_A g = 0$ since $\text{Supp}(A) = \{x_0\}$. Therefore according to (3.11) there exists an $f \in \mathscr{E}_+$ such that $A = fB$. Let $k = f(x_0)$. Then obviously $A = kB$ completing the proof of Theorem 3.13.

Theorem 3.13 states that when a local time at x_0 exists it is unique up to a multiplicative constant. Therefore whenever x_0 is regular for $\{x_0\}$ we define *the local time of X at x_0* to be the unique CAF, L, of X satisfying $u_L^1(x) = E^x(e^{-T})$ for every x. Here $T = T_{\{x_0\}}$. We assume, as we may, that L is perfect and that $t \to L_t(\omega)$ is continuous and nondecreasing for all ω. We let $\tau_t(\omega) = \tau(t, \omega)$ be the functional inverse to L. It follows from Proposition 2.3 that for each $u, v \geq 0$ one has

$$\tau_{u+v} = \tau_u + \tau_v \circ \theta_{\tau_u} \tag{3.14}$$

almost surely on $\{\tau_u < \zeta\} = \{\tau_u < \infty\}$. But (3.14) is obviously valid if $\tau_u = \infty$. Thus (3.14) holds almost surely and moreover the exceptional set may be chosen independent of u and v since L is perfect. Therefore (3.14) remains valid if u and v are replaced by nonnegative random variables U and V. In addition $u \to \tau_u(\omega)$ is strictly increasing for all ω on $[0, L_\infty)$ and Theorem 3.8 implies that $X(\tau_u) = x_0$ almost surely on $\{\tau_u < \infty\} = \{u < L_\zeta = L_\infty\}$. Finally note that $\tau_u < \infty$ for all $u < \infty$ if and only if $L_\infty = \infty$.

We are now going to investigate the stochastic properties of τ_t considered as a function of t. We will use the notation developed above without special mention. For notational simplicity we write for $\alpha > 0$

$$u^\alpha(x) = u_L^\alpha(x) = E^x \int_0^\infty e^{-\alpha t} \, dL_t, \tag{3.15}$$

so that in our previous notation ($T = T_{\{x_0\}}$)

$$u^1(x) = \psi(x) = E^x(e^{-T}).$$

Observe that

$$(3.16) \qquad u^\alpha(x) = E^x \int_T^\infty e^{-\alpha t}\, dL_t = E^x\{e^{-\alpha T}\, u^\alpha(X_T)\}$$
$$= u^\alpha(x_0)\, E^x(e^{-\alpha T}).$$

We are now in a position to compute the Laplace transform of the distribution of τ_t.

(3.17) THEOREM. For each x, t, and $\alpha > 0$

$$E^x\{e^{-\alpha\tau(t)}\} = E^x(e^{-\alpha T}) \exp[-t/u^\alpha(x_0)].$$

Proof. Fix $\alpha > 0$. Then

$$(3.18) \qquad u^\alpha(x) = E^x \int_0^\infty e^{-\alpha t}\, dL_t = \int_0^\infty E^x\{e^{-\alpha\tau(t)}\}\, dt.$$

Define $f(t) = E^{x_0}\{e^{-\alpha\tau(t)}\}$. But $X_{\tau(t)} = x_0$ if $\tau(t) < \infty$ and so

$$f(t+s) = E^{x_0}\{e^{-\alpha\tau(t+s)}\}$$
$$= E^{x_0}\{e^{-\alpha\tau(t)}\, E^{X[\tau(t)]}(e^{-\alpha\tau(s)})\}$$
$$= f(t)\, f(s).$$

Since $t \to f(t)$ is bounded and right continuous, $f(t) = e^{-kt}$ for some $k \geq 0$. By (3.18), $\int_0^\infty f(t)\, dt = u^\alpha(x_0)$ and so $k = [u^\alpha(x_0)]^{-1}$. Now for a general x we have

$$E^x\{e^{-\alpha\tau(t)}\} = E^x\{\exp[-\alpha(\tau_0 + \tau_t \circ \theta_{\tau_0})]\}$$
$$= E^x\{e^{-\alpha\tau(0)}\}\, E^{x_0}\{e^{-\alpha\tau(t)}\}$$
$$= E^x\{e^{-\alpha\tau(0)}\} \exp[-t/u^\alpha(x_0)],$$

and since $\tau_0 = T$ this completes the proof of (3.17).

The fact that $X[\tau(t)] = x_0$ almost surely on $\{\tau(t) < \infty\}$ has as a consequence the fact that, roughly speaking, the process $\{\tau(t)\colon t \geq 0\}$ has stationary independent increments. The following result is the precise statement.

(3.19) PROPOSITION. If $0 = t_0 < t_1 < \ldots < t_n$ and if $B_1, \ldots, B_n$ are Borel subsets of $[0, \infty)$, then

$$P^{x_0}\{\tau(t_j) - \tau(t_{j-1}) \in B_j;\, j = 1, \ldots, n;\, \tau(t_n) < \infty\} = \prod_{j=1}^n P^{x_0}\{\tau(t_j - t_{j-1}) \in B_j\}.$$

Proof. Recall that $\tau_0 = \inf\{t: L(t) > 0\}$ and so $P^{x_0}[\tau(0) = 0] = 1$. Thus the above assertion is obvious when $n = 1$ since B_1 is a subset of $[0, \infty)$. Let $\Lambda_n = \bigcap_{j=1}^{n} \{\tau(t_j) - \tau(t_{j-1}) \in B_j\}$. Then using (3.14) and the fact that $X[\tau(t)] = x_0$ on $\{\tau(t) < \infty\}$ we find

$$\begin{aligned} P^{x_0}[\Lambda_n; \tau(t_n) < \infty] &= P^{x_0}[\Lambda_{n-1}; \tau(t_n - t_{n-1}) \circ \theta_{\tau(t_{n-1})} \in B_n; \tau(t_{n-1}) < \infty] \\ &= P^{x_0}[\tau(t_n - t_{n-1}) \in B_n]\, P^{x_0}[\Lambda_{n-1}; \tau(t_{n-1}) < \infty], \end{aligned}$$

and so Proposition 3.19 follows by induction.

Note that Theorem 3.17 implies that $P^{x_0}[\tau(t) < \infty] = 1$ if and only if $u_L(x_0) = \lim_{\alpha \to 0} u^\alpha(x_0) = \infty$. In this case the process $\{\tau(t); P^{x_0}\}$ has stationary independent increments in the ordinary sense.

Let $Y = \{Y(t); t \geq 0\}$ be a real-valued stochastic process defined over some probability space $(\Omega^*, \mathscr{G}, P)$. Then Y is called a *subordinator* provided Y has right continuous paths, $Y(0) = 0$, and Y has *stationary independent nonnegative* increments. It is known that for such a process $E\{e^{-\alpha Y(t)}\} = e^{-tg(\alpha)}$ for all $t \geq 0$ and $\alpha \geq 0$ where

$$\textbf{(3.20)} \qquad g(\alpha) = b\alpha + \int_0^\infty (1 - e^{-\alpha u})\, \nu(du).$$

In (3.20), b is a nonnegative constant and ν is a Borel measure on $(0, \infty)$ satisfying $\int_0^\infty u(1 + u)^{-1}\, \nu(du) < \infty$. It is easy to see that $b = \lim_{\alpha \to \infty} \alpha^{-1} g(\alpha)$. The function g is called the *exponent* of Y and ν is called the *Lévy measure* of Y. Conversely given such a b and ν there exists a subordinator whose exponent is given by (3.20). In this connection, see (2.18) and (2.19) of Chapter I. For a general discussion of subordinators, see Blumenthal and Getoor [1], Bochner [1], or Ito and McKean [1, p. 31]. (What we call a subordinator Ito and McKean call a *homogeneous differential process with increasing* (and right continuous) *paths*.)

According to (3.15), $\alpha \to u^\alpha(x_0)$ is a decreasing function and so $\gamma = \lim_{\alpha \to 0} [u^\alpha(x_0)]^{-1}$ exists with $0 \leq \gamma < \infty$. This notation is used in the statement of the next theorem, which gives the stochastic structure of $\tau(t)$ under P^{x_0}.

(3.21) THEOREM. There exists a subordinator Y and a nonnegative random variable Z defined on some probability space $(\Omega^*, \mathscr{G}, P)$ such that (i) $\{Y_t; t \geq 0\}$ and Z are independent, (ii) $P(Z \geq t) = e^{-\gamma t}$ for all $t \geq 0$, and (iii) if one defines $\tau^*(t) = Y(t)$ for $t < Z$ and $\tau^*(t) = \infty$ for $t \geq Z$, then $\{\tau^*(t), P\}$ and $\{\tau(t), P^{x_0}\}$ are stochastically equivalent, i.e., they have the same finite-dimensional distributions. Moreover the exponent g of Y is given by $g(\alpha) = [u^\alpha(x_0)]^{-1} - \gamma$.

REMARKS. In fact $\{\tau^*(t), P\}$ and $\{\tau(t), P^{x_0}\}$ are temporally homogeneous Markov processes (in the sense of Section 2 of Chapter I) with values in $(\mathbf{T}, \mathscr{B}(\mathbf{T}))$, where $\mathbf{T} = [0, \infty]$, having the same transition function. If $\gamma = 0$ then $\{\tau(t), P^{x_0}\}$ is itself a subordinator with exponent $g(\alpha) = [u^\alpha(x_0)]^{-1}$.

Proof of (3.21). Let $\mathbf{R}^+ = [0, \infty)$. For each $t > 0$ define a measure μ_t on $\mathscr{B}(\mathbf{R}^+)$ by $\mu_t(B) = P^{x_0}[\tau(t) \in B]$. A straightforward computation making use of (3.19) yields $\mu_{t+s} = \mu_t * \mu_s$ for all $t, s > 0$ where "$*$" denotes the ordinary convolution of measures on $\mathbf{R}$. Since $\tau(t) \to 0$ as $t \downarrow 0$ almost surely P^{x_0}, it is clear that $\mu_t \to \varepsilon_0$ weakly as $t \to 0$. Finally using (3.17) we see that

$$\mu_t(\mathbf{R}^+) = \lim_{\alpha \to 0} \exp[-t/u^\alpha(x_0)] = e^{-\gamma t}.$$

Therefore if we define $\nu_t = e^{\gamma t} \mu_t$, the family $\{\nu_t; t > 0\}$ is a semigroup (under convolution) of probability measures on $\mathbf{R}^+$ such that $\nu_t \to \varepsilon_0$ weakly as $t \to 0$. Under these circumstances it follows from (9.14) of Chapter I that there exists a subordinator Y on a probability space $(\Omega^*, \mathscr{G}, P)$ such that $P[Y(t) \in B] = \nu_t(B)$ for all $t > 0$ and $B \in \mathscr{B}(\mathbf{R}^+)$. Next observe that

$$\begin{aligned} E(e^{-\alpha Y(t)}) &= \int e^{-\alpha u}\, \nu_t(du) = e^{\gamma t} \int e^{-\alpha u}\, \mu_t(du) \\ &= e^{\gamma t} E^{x_0}(e^{-\alpha \tau(t)}) = \exp\{-t([u^\alpha(x_0)]^{-1} - \gamma)\}, \end{aligned}$$

and so the exponent of Y is $[u^\alpha(x_0)]^{-1} - \gamma$. We may assume without loss of generality that there exists a nonnegative random variable Z on $(\Omega^*, \mathscr{G}, P)$ that is independent of Y and such that $P(Z \geq t) = e^{-\gamma t}$ for all $t \geq 0$. Now define τ^* as in the statement of (3.21). Then for any Borel subset B of $\mathbf{R}^+$ and $t > 0$ we have

$$\begin{aligned} P[\tau^*(t) \in B] &= P[Y(t) \in B, Z > t] \\ &= e^{-\gamma t} P[Y(t) \in B] \\ &= e^{-\gamma t} \nu_t(B) \\ &= P^{x_0}[\tau(t) \in B]. \end{aligned}$$

Combining this with (3.19) and the fact that Y has stationary independent increments one easily obtains the stochastic equivalence of $\{\tau(t), P^{x_0}\}$ and $\{\tau^*(t), P\}$ since $P[\tau^*(0) = 0] = P^{x_0}[\tau(0) = 0] = 1$. This completes the proof of Theorem 3.21.

The local time can be a powerful tool in studying the "local" properties of X. In many situations it reduces the study of $\mathbf{Z} = \{t \colon X_t = x_0\}$ to a study of $\mathbf{Q} = \{t < \infty \colon \tau(u) = t \text{ for some } u\}$, and, in view of Theorem 3.21, the set $\mathbf{Q}$

is much more tractable than **Z**. However we will not go into such applications in this book. The interested reader should consult Blumenthal and Getoor [3]. Also Ito and McKean [1, Ch. 6] contains a very deep study of the local times for one-dimensional diffusion processes. We will return to the subject of local times in Chapter VI. There we will discuss a number of examples.

We next turn to a discussion of how the local time, L^x, of X at x varies with x. These results, in the present context, are due to Meyer [8], and our discussion will follow his closely. Many deep results are known about $(x, t) \to L^x(t)$ in various special situations. See, for example, Trotter [1], McKean [1], Ray [3], and Boylan [1].

Let a and b be fixed points in E each satisfying the condition of Theorem 3.13. Let T_a and T_b denote the hitting times of $\{a\}$ and $\{b\}$ and let L^a and L^b denote the local times at a and b (for X). Let $B_t = L_t^a - L_t^b$. Then $B = \{B_t\}$ has all the properties of a CAF of X except that $t \to B_t$ is of bounded variation on each finite interval rather than nondecreasing and, of course, B_t may have arbitrary sign. Let us define a family of operators on the space of bounded complex-valued $\mathscr{E}^*$ measurable functions as follows:

(3.22)
$$W^\lambda f(x) = E^x \int_0^\infty e^{i\lambda B_t} e^{-t} f(X_t)\, dB_t$$

where λ is a real parameter. Since $|dB_t| \leq dL_t^a + dL_t^b$ each W^λ is a bounded linear operator. The next lemma is essentially (2.22ii) of Chapter IV.

(3.23) LEMMA. For each real λ, $i\lambda W^0 W^\lambda = W^\lambda - W^0 = i\lambda W^\lambda W^0$.

Proof. This is obvious if $\lambda = 0$, and so we consider only the case $\lambda \neq 0$. If $f \in b\mathscr{E}^*$ then (4.14) of Chapter II implies that $W^\lambda f$ is nearly Borel measurable and finely continuous. Let f be a bounded continuous function on E. Then making use of the right continuity of $t \to W^\lambda f(X_t)$ the following formal calculations are easily justified.

$$\begin{aligned} W^0 W^\lambda f(x) &= E^x \int_0^\infty e^{-t}\, W^\lambda f(X_t)\, dB_t \\ &= E^x \int_0^\infty e^{-t} \int_0^\infty e^{i\lambda[B_{t+u} - B_t]} f(X_{t+u})\, e^{-u}\, dB_{t+u}\, dB_t \\ &= E^x \int_0^\infty e^{-i\lambda B_t} \int_t^\infty e^{i\lambda B_u} f(X_u)\, e^{-u}\, dB_u\, dB_t \\ &= E^x \int_0^\infty e^{i\lambda B_u} f(X_u)\, e^{-u} \int_0^u e^{-i\lambda B_t}\, dB_t\, dB_u . \end{aligned}$$

But B_t is continuous and so the integration on t yields $(1/i\lambda)[1 - e^{-i\lambda B_u}]$.

(See Exercise 3.42.) Thus the first identity in (3.23) is proved. The second is proved by a similar calculation.

If g^{ab} is the indicator function of the two-point set $\{a, b\}$, then clearly $W^\lambda g^{ab} f = W^\lambda f$ for each real λ; that is, the (complex) measures $W^\lambda(x, \cdot)$ are supported by $\{a, b\}$. Let $\psi_a(x) = E^x(e^{-T_a})$ and $\psi_b(x) = E^x(e^{-T_b})$. If f is a function on E we write $\langle f \rangle$ for the column vector $\begin{pmatrix} f(a) \\ f(b) \end{pmatrix}$. Using this notation and the properties of L^a and L^b we may write $\langle W^0 f \rangle = \mathbf{W}^0 \langle f \rangle$ where $\mathbf{W}^0$ is the matrix

$$\mathbf{W}^0 = \begin{pmatrix} 1 & -\psi_b(a) \\ \psi_a(b) & -1 \end{pmatrix}. \tag{3.24}$$

(Recall that $\psi_a(a) = \psi_b(b) = 1$.) Also, for general λ, $\langle W^\lambda f \rangle = \mathbf{W}^\lambda \langle f \rangle$ where $\mathbf{W}^\lambda$ is a square matrix of order two. In this notation (3.23) becomes $i\lambda \mathbf{W}^0 \mathbf{W}^\lambda = \mathbf{W}^\lambda - \mathbf{W}^0$ and so $(I - i\lambda \mathbf{W}^0)\mathbf{W}^\lambda = \mathbf{W}^0$. But the determinant of $(I - i\lambda \mathbf{W}^0)$ is given by $1 + \lambda^2[1 - \psi_a(b)\,\psi_b(a)] \geq 1$, and so $I - i\lambda \mathbf{W}^0$ is invertible. Consequently

$$I + i\lambda \mathbf{W}^\lambda = (I - i\lambda \mathbf{W}^0)^{-1} = (1 + \lambda^2 \gamma^2)^{-1} \begin{pmatrix} 1 + i\lambda & -i\lambda\,\psi_b(a) \\ i\lambda\,\psi_a(b) & 1 - i\lambda \end{pmatrix} \tag{3.25}$$

$$\gamma^2 = 1 - \psi_a(b)\,\psi_b(a).$$

We will use these calculations to make the following basic estimate.

(3.26) LEMMA. For each $x \in E$ let θ^x be the measure defined on the Borel sets of the real line by $\theta^x(\Gamma) = \int_0^\infty P^x(B_t \in \Gamma)\, e^{-t}\, dt$. Then for each $\delta > 0$, $\theta^x(\{u\colon |u| \geq \delta\}) \leq e^{-\delta/\gamma}$.

Proof. First observe that integrating by parts we have

$$\begin{aligned} i\lambda W^\lambda 1(x) &= i\lambda E^x \int_0^\infty e^{i\lambda B_t}\, e^{-t}\, dB_t \\ &= E^x \int_0^\infty e^{i\lambda B_t}\, e^{-t}\, dt - 1 \\ &= \int_{-\infty}^\infty e^{i\lambda u}\, \theta^x(du) - 1, \end{aligned}$$

and so

$$\hat{\theta}^x(\lambda) = \int_{-\infty}^\infty e^{i\lambda u}\, \theta^x(du) = 1 + i\lambda W^\lambda 1(x).$$

Therefore (3.25) yields

$$\hat{\theta}^a(\lambda) = \frac{1 + i\lambda[1 - \psi_b(a)]}{1 + \lambda^2\gamma^2}, \qquad \hat{\theta}^b(\lambda) = \frac{1 - i\lambda[1 - \psi_a(b)]}{1 + \lambda^2\gamma^2}.$$

Inverting these Fourier transforms one sees that $\theta^a(du)$ and $\theta^b(du)$ have densities with respect to Lebesgue measure given, respectively, by

$$\frac{d\theta^a}{du} = \frac{1}{2\gamma}\left[1 - \frac{\eta'}{\gamma}\right] Y(-u)\, e^{u/\gamma} + \frac{1}{2\gamma}\left[1 + \frac{\eta'}{\gamma}\right] Y(u)\, e^{-u/\gamma}$$

$$\frac{d\theta^b}{du} = \frac{1}{2\gamma}\left[1 + \frac{\eta}{\gamma}\right] Y(-u)\, e^{u/\gamma} + \frac{1}{2\gamma}\left[1 - \frac{\eta}{\gamma}\right] Y(u)\, e^{-u/\gamma}$$

where $\eta' = 1 - \psi_b(a)$, $\eta = 1 - \psi_a(b)$, and $Y(u) = 1$ for $u \geq 0$, $Y(u) = 0$ for $u < 0$. One now obtains the conclusion of (3.26) when $x = a$ or b by a simple computation. (There actually is equality in this case.)

Let $T = T_a \wedge T_b$ be the hitting time of the two-point set $\{a, b\}$. Then for any x, since $dB_t = 0$ on $[0, T]$ and $B_T = 0$, we have

$$\begin{aligned}
\hat{\theta}^x(\lambda) - 1 &= i\lambda W^\lambda 1(x) = i\lambda E^x \int_T^\infty e^{-i\lambda B_t}\, e^{-t}\, dB_t \\
&= i\lambda\, E^x\left\{e^{-T}\, E^{X(T)} \int_0^\infty e^{-i\lambda B_t}\, e^{-t}\, dB_t\right\} \\
&= i\lambda\, p_a(x)\, W^\lambda 1(a) + i\lambda\, p_b(x)\, W^\lambda 1(b),
\end{aligned}$$

where we have written $p_a(x) = E^x(e^{-T}; T = T_a)$ and $p_b(x) = E^x(e^{-T}; T = T_b)$. Thus

$$\hat{\theta}^x(\lambda) = 1 - p_a(x) - p_b(x) + p_a(x)\, \hat{\theta}^a(\lambda) + p_b(x)\, \hat{\theta}^b(\lambda),$$

and consequently

(3.27) $$\theta^x = [1 - p_a(x) - p_b(x)]\varepsilon_0 + p_a(x)\, \theta^a + p_b(x)\, \theta^b,$$

where ε_0 is unit mass at the origin. The conclusion of (3.26) for general x now follows from the results in the cases $x = a$ and $x = b$. Of course (3.27) may also be obtained by a simple direct computation beginning with the definition of the measure θ^x.

We now apply Lemma 3.26 to obtain the following result.

(3.28) PROPOSITION. Let N and δ be positive numbers. Then for any x,

$$P^x\left\{\sup_{0 \leq t \leq N} |L_t^a - L_t^b| > 2\delta\right\} \leq 2\, e^N\, e^{-\delta/\gamma}$$

where the notation is that used above.

Proof. Let $T = \inf\{t: |B_t| > 2\delta\}$. Obviously T is a stopping time. Let $f_\delta(u) = 1$ if $|u| < \delta$ and $f_\delta(u) = 0$ if $|u| \geq \delta$, and let $g_\delta = 1 - f_\delta$. Then for a fixed x we have

$$E^x(e^{-T}) = E^x \int_T^\infty e^{-t}\, dt = E^x \int_T^\infty e^{-t}[f_\delta(B_t) + g_\delta(B_t)]\, dt$$

$$\leq E^x \int_0^\infty e^{-t}\, g_\delta(B_t)\, dt + E^x \int_T^\infty e^{-t} f_\delta(B_t)\, dt.$$

By (3.26) the integral involving g_δ is equal to $\theta^x(\{u: |u| \geq \delta\}) \leq e^{-\delta/\gamma}$. If $t > T$ and $|B_t| < \delta$, then, since $|B_T| \geq 2\delta$, one must have $\delta < |B_t - B_T| = |B_{t-T}(\theta_T)|$. Therefore

$$E^x \int_T^\infty e^{-t} f_\delta(B_t)\, dt \leq E^x \int_T^\infty e^{-t}\, g_\delta[B_{t-T}(\theta_T)]\, dt$$

$$= E^x\left\{e^{-T} \int_0^\infty e^{-t}\, g_\delta[B_t \circ \theta_T]\, dt\right\}$$

$$\leq E^x\left\{e^{-T} \int_{|u| \geq \delta} \theta^{X(T)}(du)\right\} \leq e^{-\delta/\gamma},$$

and so $E^x(e^{-T}) \leq 2e^{-\delta/\gamma}$. Now for any N

$$e^{-N}\, P^x(T \leq N) \leq E^x(e^{-T};\, T \leq N) \leq 2e^{-\delta/\gamma}.$$

Finally

$$P^x\left[\sup_{0 \leq t \leq N} |B_t| > 2\delta\right] \leq P^x[T \leq N]$$

$$\leq 2e^N\, e^{-\delta/\gamma},$$

proving (3.28), since $B_t = L_t^a - L_t^b$.

(3.29) COROLLARY. Let a be a fixed point in E and suppose that the local time for X exists at all points b in some neighborhood of a. If $\lim_{b \to a} \psi_b(a) = 1$, then for any N, $\sup_{0 \leq t \leq N} |L_t^a - L_t^b|$ approaches zero in P^x probability as $b \to a$ for all x.

Proof. One first observes that $\lim_{b \to a} \psi_a(b) = 1$ also. (See Exercise 3.36.) Hence (3.29) is an immediate consequence of (3.28) since $\gamma = [1 - \psi_a(b)\psi_b(a)]^{1/2} \to 0$ as $b \to a$.

We will close this section by applying Proposition 3.28 to obtain a result which is essentially due to Boylan [1]. Again we follow Meyer [8]. We will assume that E is an interval of the real line $\mathbf{R}$ since this seems to be the only case of interest, and for simplicity we will actually assume that $E = \mathbf{R}$–the extension to the more general situation being obvious. Thus we assume that X is a standard process with state space $(\mathbf{R}, \mathscr{B}(\mathbf{R}))$ and we further assume that, for each $x \in \mathbf{R}$, x is regular for $\{x\}$ so that the local time L^x exists for all x. As above T_x is the hitting time of $\{x\}$ and $\psi_x(y) = E^y(e^{-T_x})$. Finally we assume that there exists a monotone increasing function h on $[0, \infty]$ with $h \le 1$, $\lim_{u \downarrow 0} h(u) = 0$, and with $\sum_n n[h(2^{-n})]^{1/2} < \infty$ and such that for each integer $M > 0$ there exists a K for which $1 - \psi_x(y) \le K\, h(|x - y|)$ whenever x and y are in $[-M, M]$.

(3.30) THEOREM. Under the above assumptions it is possible to choose the local time L^x in such a manner that $(t, x) \to L_t^x(\omega)$ is continuous for all ω and L^x is perfect for all x.

Proof. To begin we may assume that for each x, L^x is perfect and that $t \to L_t^x(\omega)$ is continuous for all ω. Let $\mathbf{D}$ be the set of all dyadic rationals. For each pair of positive integers (N, M) consider the map $\Phi_\omega^{N,M}$ from $\mathbf{D} \cap [-M, M]$ to $\mathbf{C}[0, N]$ which assigns to each $a \in \mathbf{D} \cap [-M, M]$ the continuous function $t \to L_t^a(\omega)$ on $[0, N]$. Theorem 3.30 will be established once the following statement is proved.

(3.31) For almost all ω, $\Phi_\omega^{N,M}$ is uniformly continuous on $\mathbf{D} \cap [-M, M]$.

Indeed let Ω_0 be in $\mathscr{F}$ with $P^x(\Omega_0) = 1$ for all x and such that, for each $\omega \in \Omega_0$, $\Phi_\omega^{N,M}$ is uniformly continuous for all N and M. Then for each $\omega \in \Omega_0$, $\Phi_\omega^{N,M}$ can be extended to a continuous map from $[-M, M]$ to $\mathbf{C}[0, N]$, and this clearly defines a continuous function $\bar{L}_t^x(\omega)$ of the pair (t, x), $t \ge 0$, $x \in \mathbf{R}$. If we set $\bar{L}_t^x(\omega) = 0$ for all (t, x) if $\omega \in \Omega - \Omega_0$, then $(t, x) \to \bar{L}_t^x(\omega)$ is continuous for all ω. We next claim that, for each x, $\bar{L}^x$ is a perfect CAF of X. To see this let Ω_1 be in $\mathscr{F}$ with $P^x(\Omega_1) = 1$ for all x and such that $L_{t+s}^a(\omega) = L_t^a(\omega) + L_s^a(\theta_t \omega)$ for all $\omega \in \Omega_1$, $t, s \ge 0$, and $a \in \mathbf{D}$. If $\omega \in \Omega_0 \cap \Omega_1$ then for a fixed t, $L_s^a(\theta_t \omega) = L_{t+s}^a(\omega) - L_t^a(\omega)$ for all $a \in \mathbf{D}$ and $s \ge 0$, and so $\theta_t \omega \in \Omega_0$. If $\omega \in \Omega_0 \cap \Omega_1$, $t, s \ge 0$, and x are fixed, then we can find a sequence $\{a_n\}$ of elements in $\mathbf{D}$ such that $L^{a_n}(\omega) \to \bar{L}^x(\omega)$ and $L^{a_n}(\theta_t \omega) \to \bar{L}^x(\theta_t \omega)$ in $\mathbf{C}[0, t + s + 1]$. This clearly implies that $\bar{L}_{t+s}^x(\omega) = \bar{L}_t^x(\omega) + \bar{L}_s^x(\theta_t \omega)$ and so $\bar{L}^x$ is a perfect CAF. Finally (3.29) implies that, for each x, L^x and $\bar{L}^x$ are equivalent, which establishes Theorem 3.30.

Thus it remains to prove (3.31). Suppose $\lambda > (2K)^{1/2} \log 2$, and let $\delta_n = \lambda n[h(2^{-n})]^{1/2}$. If $a = i2^{-n}$ and $b = (i + 1)2^{-n}$ are in $\mathbf{D} \cap [-M, M]$, then $\gamma = [1 - \psi_a(b)\, \psi_b(a)]^{1/2} \le [2K\, h(2^{-n})]^{1/2}$. Now using (3.28) we obtain

$$P^x\left\{\sup_{0\le t\le N} |L_t^{i2^{-n}} - L_t^{(i+1)2^{-n}}| > 2\delta_n\right\} \le 2\, e^N\, e^{-\delta_n \gamma^{-1}}$$
$$\le 2\, e^N \exp[-\lambda n/(2K)^{1/2}].$$

We next define

$$\rho_n = P^x\left\{\sup_{-M2^n\le i\le M2^n}\ \sup_{0\le t\le N} |L_t^{i2^{-n}} - L_t^{(i+1)2^{-n}}| > 2\delta_n\right\}.$$

Then it follows from the estimate we have just made that

$$\rho_n \le 2M2^n \cdot 2\, e^N \exp[-\lambda n/(2K)^{1/2}].$$

Consequently $\sum \rho_n < \infty$ because of our choice of λ. It now follows from the Borel–Cantelli lemma that for almost all ω there exists an integer ν such that

$$\sup_{-M2^n\le i\le M2^n}\ \sup_{0\le t\le N} |L_t^{i2^{-n}} - L_t^{(i+1)2^{-n}}| \le 2\delta_n$$

for all $n \ge \nu$. Finally this implies (3.31). Indeed if $a, b \in \mathbf{D} \cap [-M, M]$ and $|a - b| < 2^{-m}$, then we may write

$$a = a_0 + p_1 2^{-m-1} + p_2 2^{-m-2} + \cdots$$
$$b = b_0 + q_1 2^{-m-1} + q_2 2^{-m-2} + \cdots$$

where each p_i and q_j has the value zero or one and only finitely many are different from zero, and where a_0 and b_0 are of the form $a_0 = i2^{-m}$, $b_0 = j2^{-m}$ with $j = i - 1$, i, or $i + 1$. Consequently

$$\sup_{0\le t\le N} |L_t^a - L_t^b| \le \sup_{0\le t\le N} \left(|L_t^a - L_t^{a_0}| + |L_t^{a_0} - L_t^{b_0}| + |L_t^{b_0} - L_t^b|\right)$$
$$\le \sum_{n=m}^{\infty} 2\delta_n + 2\delta_m + \sum_{n=m}^{\infty} 2\delta_n$$

provided $m \ge \nu$, and since $\sum \delta_n < \infty$ this establishes (3.31). Thus the proof of (3.30) is complete.

Exercises

(3.32) Give a proof of Proposition 3.11. [Hint: use (3.9) to show that it suffices to consider the case in which Supp(B) consists of a single point x_0. If u_B^1 is finite, then $f(x_0) = u_A^1(x_0)/u_B^1(x_0)$, $f(x) = 0$ if $x \ne x_0$ has the desired properties.]

(3.33) Assume (1.3). If A is a CAF of (X, M) show that Supp(A) is the fine support of A as defined in (1.22).

(3.34) Let L be the local time for X at x_0 and ν be the Lévy measure of the corresponding subordinator Y constructed in Theorem 3.21. Show that $\nu(\mathbf{R}^+) < \infty$ if and only if x_0 is a holding point. [Hint: use the fact that $\nu[(a, \infty)]$ is the expected number of jumps of Y in unit time which exceed a in magnitude. See Blumenthal and Getoor [3, p. 60].]

(3.35) Let x_0 be a holding point for X. Show that the local time for X at x_0 exists and that $\mathbf{Z} = \mathbf{I}$ in this case. See (3.7) for the definitions of $\mathbf{I}$, $\mathbf{Z}$, and $\mathbf{J}$. If X has continuous paths and the local time for X exists at x_0, then show that $\mathbf{Z} = \mathbf{J}$. Finally observe that $\mathbf{I} = \mathbf{J}$ if and only if x_0 is a trap. Thus Theorem 3.8 cannot be improved in general. Lamperti [1] has shown that almost surely $\mathbf{Z} = \mathbf{J}$ for a wide class of processes with independent increments.

(3.36) Let $\psi_a(b) = E^b(e^{-T_a})$. Suppose $\psi_a(a) = 1$. Show that if $\{b_n\}$ is a sequence of points and $\psi_{b_n}(a) \to 1$, then $T_{b_n} + T_a \circ \theta_{T_{b_n}} \to 0$ in P^a probability. Conclude that in this situation $\psi_a(b_n) \to 1$. Here T_x is the hitting time of $\{x\}$.

(3.37) Let X be Brownian motion in $\mathbf{R}$. According to (3.16) of Chapter II the local time at x_0 exists for all x_0. Let $x_0 = 0$ and $T = T_{\{0\}}$. By (3.18) of Chapter II, $E^x(e^{-T}) = e^{-|x|}$. Use (2.3) of Chapter IV and the explicit form of the Gauss kernel to show that $u^\alpha(0) = (\alpha)^{-1/2}$. Conclude that $\{\tau(t), P^{x_0}\}$ is a stable subordinator of index $\frac{1}{2}$ in this case. See (2.19) of Chapter I. Show that the hypothesis of Theorem 3.30 is satisfied in this case.

(3.38) Let A be a CAF of (X, M) with a finite α-potential for some fixed $\alpha \geq 0$. Let D be a nearly Borel subset of E_M which is finely closed in E_M. Show that $Q_D^\alpha u_A^\alpha = u_A^\alpha$ if and only if $D \supset \text{Supp}(A)$. This gives another characterization of $\text{Supp}(A)$ when $A \in \mathbf{A}^\alpha$.

(3.39) Let x_0 be a holding point for X and assume that x_0 is not a trap. Let Q and T be the hitting times of $E_\Delta - \{x_0\}$ and $\{x_0\}$, respectively. Then there exists a constant λ_0, $0 < \lambda_0 < \infty$, such that $P^{x_0}(Q > t) = e^{-\lambda_0 t}$ for all $t \geq 0$. See (8.18) of Chapter I. Let f be the indicator function of $\{x_0\}$ and define $B_t = \int_0^t f(X_u)\, du$. Let $k = (1 + \lambda_0)(1 - \gamma_0)$ where $\gamma_0 = E^{x_0}\{e^{-Q} E^{X(Q)}(e^{-T})\}$. Show that $L_t = kB_t$ is the local time for X at x_0. Thus in this case the local time at x_0 is just a constant multiple of the Lebesgue measure of $\{u \leq t: X_u = x_0\}$—the actual time X spends at x_0.

(3.40) Let x_0 be a fixed point in E and assume that $\{x_0\}$ is not polar and that x_0 is not regular for $\{x_0\}$, i.e., $\{x_0\}$ is thin but not polar. Let $T = T_{\{x_0\}}$ and $T_0 = 0$, $T_{n+1} = T_n + T \circ \theta_{T_n}$ for $n \geq 0$. Prove that $T_n \to \infty$ almost surely. Define A_t to be the number of visits to $\{x_0\}$ by X in the interval $[0, t]$; that is,

$A_t = n$ if $T_n \leq t < T_{n+1}$. Show that if $\Phi^\alpha(x) = E^x(e^{-\alpha T})$, then $U_A^\alpha f(x) = f(x_0)\,[1 - \Phi^\alpha(x_0)]^{-1}\,\Phi^\alpha(x)$ for $\alpha > 0$. Thus we *might* call an appropriate multiple of A the *local time at* x_0. Show by an example that A need not be natural.

(3.41) Let X satisfy the assumptions of Theorem 3.30 and also condition (1.3). If ξ is a reference measure for X, then show that there exists a non-negative Borel function g on $\mathbf{R}$ such that for each $D \in \mathscr{B}(\mathbf{R})$ one has almost surely

$$\int_0^t I_D(X_u)\,du = \int_D L_t^x\, g(x)\, \xi(dx)$$

for all t. The left side is the "amount of time" the process spends in D up to time t, and so, for all t, $x \to L_t^x$ is a continuous density for this "occupation time" with respect to the measure $g\xi$ for almost all ω. [Hint: we may assume without loss of generality that ξ is finite. Show that $L_t = \int L_t^x\, \xi(dx)$ is a CAF of X with a finite one potential. If $A_t = t \wedge \zeta$, then by computing one potentials and using (2.8) show that $A = gL$ for some $g \in \mathscr{B}(\mathbf{R})_+$. Show that this g has the desired property.] Compute g explicitly if X is Brownian motion in $\mathbf{R}$ and ξ is Lebesgue measure.

(3.42) Let b be a continuous function of bounded variation on $[0, a]$ with $b(0) = 0$ and let f be a continuous function on $\mathbf{R}$. Show that $\int_0^a f[b(t)]\, db(t) = \int_0^{b(a)} f(t)\, dt$. [Hint: first consider the case in which b is absolutely continuous.]

(3.43) Let A be a CAF of X and suppose that $t \to A_t(\omega)$ is continuous for all ω. Let τ be the inverse of A. Then if T is a stopping time, $A[\tau(T)] = T$ if $\tau(T) < \infty$. Show that $\tau[A(T)] = T$ almost surely on $\{X_T \in \operatorname{Supp}(A)\}$.

4. Balayage of Additive Functionals

Let X be a fixed standard process with state space $(E, \mathscr{E})$ and let A be a CAF of X. It is an immediate consequence of (3.8) and the remarks preceding it that the "time changed" process $\hat{X}$ defined in (2.11) "lives" on $D = \operatorname{Supp}(A)$ in the sense that almost surely $X_{\tau(t)} \in D$ for all $t < \hat{\zeta} = A_\zeta$. In some sense $\hat{X}$ is the process X "sampled" when it is in D. Therefore it is often useful to know whether or not a particular set D is the support of a CAF. If A is in $\mathbf{A}^\alpha$ and $D \in \mathscr{E}^n$ and if $P_D^\alpha u_A^\alpha = u_B^\alpha$ for some $B \in A^\alpha$, then it is not unreasonable to hope that $\operatorname{Supp}(B) = D$ at least if $\operatorname{Supp}(A) \supset D$. (See (4.8) for the precise statement.) Therefore we will begin by studying the sets D for which such a B exists and the relationship between A and B.

Suppose for the moment that A is an NAF of X with a finite potential u_A. If $D \in \mathscr{E}^n$ then $v = P_D u_A$ is obviously a natural potential and so by Theorem 4.22 of Chapter IV there exists a unique NAF, B, of X such that $v = u_B$. We will call B the *projection* (or *balayage*) of A on D and write A_D for B. (We will always write $A_D(t)$ or $A_D(t, \omega)$ and never use t as a subscript on A_D.) As explained above, our main interest is in the projection of CAF's and so we will discuss CAF's directly rather than as a specialization of NAF's. Most of the results of this section are due to M. Motoo [2].

Let X be as above with $\mathscr{M} = \mathscr{F}$ and $\mathscr{M}_t = \mathscr{F}_t$ for all t. Let M be a fixed exact MF of X with $M_t = 0$ for $t \geq \zeta$. As in Section 2, $\mathbf{A}(M)$ denotes the set of all CAF's of (X, M) and $\mathbf{A}^\alpha(M)$ those elements in $\mathbf{A}(M)$ having finite α-potentials.

(4.1) Definition. Let $A \in \mathbf{A}(M)$ and let $D \in \mathscr{E}^n$. If there exists a $B \in \mathbf{A}^\alpha(M)$ such that $u_B^\alpha = Q_D^\alpha u_A^\alpha$, then B is called the *α-projection* (or *balayage*) of A on D.

Note that if such a B exists it is necessarily unique. We will denote it by A_D^α; in particular, we will write A_D for A_D^0. Also by Corollary 3.14 of Chapter IV, A_D^α exists if and only if $Q_D^\alpha u_A^\alpha$ is a regular $\alpha - (X, M)$ potential; that is, $Q_D^\alpha u_A^\alpha$ is finite and, whenever $\{T_n\}$ is an increasing sequence of stopping times with limit T, $Q_{T_n}^\alpha Q_D^\alpha u_A^\alpha \to Q_T^\alpha Q_D^\alpha u_A^\alpha$.

(4.2) Definition. A set $D \in \mathscr{E}^n$ is called *M-projective* provided that, for each $\alpha \geq 0$ and $A \in \mathbf{A}^\alpha(M)$, A_D^α exists. We say that D is *projective* if it is M-projective for $M_t = I_{[0,\zeta)}(t)$.

(4.3) Proposition. A set $D \in \mathscr{E}^n$ is M-projective if and only if whenever $\{T_n\}$ is an increasing sequence of stopping times with limit T we have $\lim_n(T_n + T_D \circ \theta_{T_n}) = T + T_D \circ \theta_T$ almost surely on $\{\lim_n(T_n + T_D \circ \theta_{T_n}) < S\}$.

Proof. Let $A \in \mathbf{A}^\alpha(M)$ and $v = Q_D^\alpha u_A^\alpha$. Clearly v is a finite $\alpha - (X, M)$ excessive function, and, as we noted earlier, $v = u_B^\alpha$ with $B \in \mathbf{A}^\alpha(M)$ if and only if $Q_{T_n}^\alpha v \to Q_T^\alpha v$ whenever $\{T_n\}$ is an increasing sequence of stopping times with limit T. Now $T_n + T_D \circ \theta_{T_n}$ increases to a limit R, and in any event $R \leq T + T_D \circ \theta_T$. If, for the moment, T is any stopping time, then

$$Q_T^\alpha v(x) = Q_T^\alpha Q_D^\alpha u_A^\alpha(x) = E^x \int_{T+T_D \circ \theta_T}^{S} e^{-\alpha t} \, dA_t.$$

Thus in the present case we have

$$\lim_n Q_{T_n}^\alpha v(x) - Q_T^\alpha v(x) = E^x \int_{R \wedge S}^{(T+T_D \circ \theta_T) \wedge S} e^{-\alpha t} \, dA_t. \tag{4.4}$$

Consequently if $R = T + T_D \circ \theta_T$ almost surely on $\{R < S\}$, then the right side of (4.4) equals 0. Suppose, on the contrary, that for some $\{T_n\}$ we have $P^x(R < T + T_D \circ \theta_T, R < S) > 0$ for some x. If we let $A_t = \int_0^t M_s\, ds$ then $A \in \mathbf{A}^\alpha(M)$ provided $\alpha > 0$, and $t \to A_t$ is strictly increasing on $[0, S)$. Consequently the right side of (4.4) is strictly positive. This completes the proof of (4.3).

Since $S \leq \zeta$ it follows from (4.3) that if D is projective then D is M-projective for all M. Also note that if we let $\Lambda_t = (t \wedge \zeta)$, then the proof of (4.3) actually shows that D is projective if and only if Λ_D^α exists for some $\alpha > 0$.

(4.5) Proposition. Let $D \in \mathscr{E}^n$ and assume that $(\bar{D} - D^r) \cap E_M$ is polar where $\bar{D}$ is the closure of D in E (or E_Δ). Then D is M-projective.

Proof. We will verify that the condition in (4.3) holds. Let T_n, T, and R be as in the proof of (4.3) and let $R_n = T_n + T_D \circ \theta_{T_n}$. One easily sees that $\{R < T + T_D \circ \theta_T, R < S\}$ is contained in

$$\Lambda = \{T_n < T \text{ for all } n,\ T_D \circ \theta_{T_n} \to 0,\ T_D \circ \theta_T > 0,\ T < S\}.$$

But on Λ, $R_n \uparrow T$ and so $X(R_n) \to X(T)$ almost surely on Λ. Since $X(R_n) \in \bar{D}$ this implies that $X(T) \in \bar{D}$ almost surely on Λ. Thus if $x \in E_M$ we have

$$\begin{aligned} P^x(\Lambda) &\leq P^x\{T_D \circ \theta_T > 0,\ X(T) \in \bar{D},\ 0 < T < S\} \\ &= E^x\{P^{X(T)}(T_D > 0);\ X(T) \in \bar{D},\ 0 < T < S\} = 0, \end{aligned}$$

since $(\bar{D} - D^r) \cap E_M$ is polar. Consequently $R = T + T_D \circ \theta_T$ almost surely on $\{R < S\}$ and so Proposition 4.5 is established.

Let $A \in \mathbf{A}(M)$ and $D \in \mathscr{E}^n$ and asssume that, for some fixed $\alpha \geq 0$, A_D^α exists. We are going to study the relationships among D, Supp(A), and Supp(A_D^α). It is possible for D and Supp(A_D^α) to be disjoint even if D is closed and M-projective and $A \in \mathbf{A}^\alpha(M)$. See Exercise 4.14. However we have the following simple result.

(4.6) Proposition. Let A and D be as above and assume that $(D - D^r) \cap E_M$ is polar. Then Supp$(A_D^\alpha) \subset D^r$.

Proof. If $(D - D^r) \cap E_M$ is polar then obviously $T_{D^r} = T_D$ almost surely on $\{T_D < S\}$ and so $Q_{D^r}^\alpha f = Q_D^\alpha f$ for all $f \in \mathscr{E}_+^*$. Let $B = A_D^\alpha$ and let $v = u_B^\alpha = Q_D^\alpha u_A^\alpha$. Suppose that there is a point $x \in$ Supp(B) such that $x \notin D^r$. Then $P^x(T_{D^r} > 0) = 1$ so that $P^x(B(T_{D^r}) > 0) = 1$ and hence $v(x) - Q_{D^r}^\alpha v(x) = E^x \int_0^{T_{D^r}} e^{-\alpha t}\, dB(t) > 0$. Now $T_{D^r} + T_D \circ \theta_{T_{D^r}} = T_{D^r}$ almost surely and so

$Q^{\alpha}_{D^r} Q^{\alpha}_D f = Q^{\alpha}_{D^r} f$ for all $f \in \mathscr{E}^*_+$. From this and the definition of v it follows that $v = Q^{\alpha}_{D^r} v$. This contradiction completes the proof.

(4.7) PROPOSITION. Let A and D satisfy the conditions above (4.6). Then $D^r \cap \text{Supp}(A) \subset \text{Supp}(A^{\alpha}_D)$.

Proof. If $B = A^{\alpha}_D$ and $R^* = \inf\{t: B_t > 0\}$, then $u^{\alpha}_B = Q^{\alpha}_{R^*} u^{\alpha}_B = Q^{\alpha}_{R^*} Q^{\alpha}_D u^{\alpha}_A$. Now if $x \in D^r$ then $u^{\alpha}_A(x) = Q^{\alpha}_D u^{\alpha}_A(x) = u^{\alpha}_B(x)$, and so $u^{\alpha}_A(x) = Q^{\alpha}_{R^*} Q^{\alpha}_D u^{\alpha}_A(x)$. It follows from this that $A[R^* + T_D \circ \theta_{R^*}] = 0$ and hence $R^* + T_D \circ \theta_{R^*} \leq R$ almost surely P^x where, as before, $R = \inf\{t: A_t > 0\}$. If, in addition, $x \in \text{Supp}(A)$, then $P^x(R = 0) = 1$, and consequently $P^x(R^* = 0) = 1$. Therefore $x \in \text{Supp}(B)$ and (4.7) is established.

The following corollary is an immediate consequence of (4.6) and (4.7).

(4.8) COROLLARY. Let A and D be as in (4.6) and assume, in addition, that that $E_M \cap D^r \subset \text{Supp}(A)$. Then $\text{Supp}(A^{\alpha}_D) = D^r \cap E_M$.

We next study the manner in which A^{α}_D varies with α. Let $\Lambda_t = \int_0^t M_s \, ds$. Then Λ is a CAF of (X, M) and $\Lambda \in \mathbf{A}^{\alpha}(M)$ for all $\alpha > 0$. If $M_t = I_{[0,\zeta)}(t)$, then $\Lambda_t = (t \wedge \zeta)$.

(4.9) PROPOSITION. Let $A \in \mathbf{A}^{\alpha}(M)$, $D \in \mathscr{E}^n$, and suppose that A^{α}_D exists. Then A^{β}_D exists for any $\beta > \alpha$. Let $f_{\alpha}(x) = E^x \int_0^{T_D} e^{-\alpha t} \, dA_t$. Then $f_{\alpha}\Lambda \in \mathbf{A}^{\beta}(M)$, $(f_{\alpha}\Lambda)^{\beta}_D$ exists, and $A^{\alpha}_D = A^{\beta}_D + (\beta - \alpha)(f_{\alpha}\Lambda)^{\beta}_D$.

Proof. Since D is fixed in this proposition we will write A^{α} for A^{α}_D. Fix $\beta > \alpha$. Then $u^{\beta}_A \leq u^{\alpha}_A < \infty$ and $u^{\alpha}_{A^{\alpha}} \leq u^{\alpha}_{A^{\alpha}} < \infty$. Now $f_{\alpha} = u^{\alpha}_A - Q^{\alpha}_D u^{\alpha}_A \leq u^{\alpha}_A$ and so $f_{\alpha} < \infty$. Also by (2.3) of Chapter IV we have $u^{\alpha}_A = u^{\beta}_A + (\beta - \alpha) V^{\beta} u^{\alpha}_A$ and hence $V^{\beta} f_{\alpha} < \infty$. Therefore $(f_{\alpha}\Lambda) \in \mathbf{A}^{\beta}(M)$. We next compute the β-potential of A^{α}. Using (2.3) of Chapter IV we have

$$\begin{aligned} u^{\beta}_{A^{\alpha}} &= u^{\alpha}_{A^{\alpha}} + (\alpha - \beta) V^{\beta} u^{\alpha}_{A^{\alpha}} \\ &= Q^{\alpha}_D u^{\alpha}_A + (\alpha - \beta) V^{\beta} Q^{\alpha}_D u^{\alpha}_A . \end{aligned}$$

Also a straightforward computation yields

$$Q^{\alpha}_D - Q^{\beta}_D = (\beta - \alpha)[V^{\beta} - Q^{\beta}_D V^{\beta}] Q^{\alpha}_D ,$$

and combining this with the above equality we obtain

$$u^{\beta}_{A^{\alpha}} = Q^{\beta}_D u^{\alpha}_A - (\beta - \alpha) Q^{\beta}_D V^{\beta} Q^{\alpha}_D u^{\alpha}_A .$$

Using (2.3) of Chapter IV once again the above becomes

$$u^{\beta}_{A^{\alpha}} = Q^{\beta}_{D}u^{\beta}_{A} + (\beta - \alpha)Q^{\beta}_{D}V^{\beta}(u^{\alpha}_{A} - Q^{\alpha}_{D}u^{\alpha}_{A}) \tag{4.10}$$
$$= Q^{\beta}_{D}u^{\beta}_{A} + (\beta - \alpha)Q^{\beta}_{D}V^{\beta}f_{\alpha}.$$

It follows from this that $Q^{\beta}_{D}u^{\beta}_{A}$ and $Q^{\beta}_{D}V^{\beta}f_{\alpha}$ are regular $\beta - (X, M)$ potentials since $u^{\beta}_{A^{\alpha}}$ is. Thus A^{β}_{D} and $(f_{\alpha}\Lambda)^{\beta}_{D}$ exist. Finally (4.10) and the uniqueness theorem yield the conclusion of Proposition 4.9.

In the remainder of this section we will consider only CAF's of X; that is, we assume that $M_t = I_{[0,\zeta)}(t)$. If $D = \text{Supp}(A)$ for some $A \in \mathbf{A}$, then $D \in \mathscr{E}^n$ and $D = D^r$. It is natural to ask if any such set D is the support of a CAF. We are unable to answer this question at present. However if, in addition, $\bar{D} - D^r$ is polar, then the answer is in the affirmative. For example if $\Lambda_t = t \wedge \zeta$ then $\text{Supp}(\Lambda) = E$, and so it follows from (4.5) and (4.8) that Λ^{α}_{D} exists for each $\alpha > 0$ and that $\text{Supp}(\Lambda^{\alpha}_{D}) = D$. Also $E^x \int_0^{T_D} e^{-\alpha t}\, d\Lambda_t \le \alpha^{-1}$. Consequently (4.9) implies that $\Lambda^{\beta}_{D} \le \Lambda^{\alpha}_{D} \le (\beta/\alpha)\Lambda^{\beta}_{D}$ whenever $0 < \alpha < \beta$. In particular if (1.3) holds then $\Lambda^{\beta}_{D} = g\Lambda^{\alpha}_{D}$ and $\Lambda^{\alpha}_{D} = g^{-1}\Lambda^{\beta}_{D}$ where $g \in \mathscr{E}$ and $\alpha/\beta \le g \le 1$. If D consists of just one point x_0, then Λ^{α}_{D} is a multiple of the local time at x_0. For general sets D subject to the above conditions one might call any of the functionals Λ^{α}_{D} a *local time* on D for X. However, at present there seems to be no uniqueness theorem which would give an intrinsic definition of local time on D.

We close this section with the following result which is often useful in discussing the time changed process $\hat{X}$.

(4.11) PROPOSITION. Let A be a CAF of X and let $\hat{X}$ be the time changed process defined in (2.11). Let $D = \text{Supp}(A)$ and suppose that D is projective. Then $\hat{X}$ is quasi-left-continuous. Note that by (4.5) the hypothesis on D is certainly satisfied if $\bar{D} - D$ is polar, where $\bar{D}$ is the closure of D in E.

Proof. Let $\{\hat{T}_n\}$ be an increasing sequence of stopping times for $\hat{X}$ with limit $\hat{T}$. By (2.11), $T_n = \tau(\hat{T}_n)$ and $T = \tau(\hat{T})$ are stopping times for X. Moreover $T_n \uparrow R \le T$. If $\hat{T} < \hat{\zeta} = A_{\zeta}$ then $T < \zeta$, and so $X_{\tau(\hat{T}_n)} = X_{T_n} \to X_R$ almost surely on $\{\hat{T} < \hat{\zeta}\}$. The proof of (4.11) will be complete provided we show that $R = T$ almost surely on $\{\hat{T} < \hat{\zeta}\}$. In the remainder of the proof we omit the phrase "almost surely" in those places where it is clearly appropriate. Now $X(T_n) = X[\tau(\hat{T}_n)] \in D$ on $\{T_n < \zeta\}$ and so $T_n + T_D \circ \theta_{T_n} = T_n$ on $\{\hat{T} < \hat{\zeta}\}$ since $D = D^r$. Letting $n \to \infty$ and using the fact that D is projective we see that $R = R + T_D \circ \theta_R$ and hence $X_R \in D$ on $\{\hat{T} < \hat{\zeta}\}$. Therefore by (3.43), $\tau(A_R) = R$ on $\{\hat{T} < \hat{\zeta}\}$. On the other hand $A_R = \lim_n A(T_n) = \lim_n \hat{T}_n = \hat{T}$, and so $R = \tau(\hat{T}) = T$ on $\{\hat{T} < \hat{\zeta}\}$ completing the proof of (4.11).

REMARK. If (1.3) holds and A is a CAF of X and if $D = \text{Supp}(A)$ is closed in E, then according to (2.10), (2.11), (3.8), and (4.11) we may regard $\hat{X}$ as a standard process with state space $(D, \mathscr{B}(D))$.

Exercises

(4.12) The assumptions are those of Proposition 4.9. Show that $\text{Supp}(A_D^\beta) = \text{Supp}(A_D^\alpha)$.

(4.13) Let A be a CAF of X and let $\hat{X}$ be the corresponding time changed process. (i) If X satisfies (1.3), then there exists a finite measure μ such that a set Γ has potential zero relative to $\hat{X}$ if and only if $\mu(\Gamma) = 0$. [Hint: use (2.21) of Chapter IV to show that λU_A^1 is a σ-finite measure carried by $D = \text{Supp}(A)$ whenever λ is a finite reference measure for X. Take μ to be a finite measure equivalent to λU_A^1.] (ii) If Γ is a nearly Borel (relative to X) subset of D show that $P^x[\hat{T}_\Gamma < \infty] = P^x[T_\Gamma < \infty]$ for all x. (iii) Use (ii) to show that a subset Γ of D is nearly Borel relative to X if and only if it is nearly Borel relative to $\hat{X}$.

(4.14) Consider the following process: $E = (-\infty, 0] \cup [1, \infty)$ and starting from any point $x \geq 1$ one translates to the right at unit speed, 0 is a holding point with parameter 1 from which one jumps to 1, and starting from $x < 0$ one translates to the right at unit speed until reaching zero. Let $D = \{1\}$. Show D is projective even though $D - D^r = \{1\}$ is not polar. [Hint: use (4.3).] Show that $\text{Supp}(\Lambda_D^\alpha) = \{0\}$ for any $\alpha > 0$ where $\Lambda_t = t$.

(4.15) Let X be uniform motion to the right on $\mathbf{R}$. Let $D = [0, 1)$. Show that D is *not* projective although $D = D^r$. Show that there is an $A \in \mathbf{A}^0$ such that $\text{Supp}(A) = D$.

(4.16) Let X be as in (4.15). Let $D = [0, 1) \cup [2, 3)$ and let $A_t = \int_0^t I_D(X_s)\, ds$. Show that the corresponding time changed process $\hat{X}$ is *not* quasi-left-continuous.

5. Processes with Identical Hitting Distributions

Let $X = (\Omega, \mathscr{M}, \mathscr{M}_t, X_t, \theta_t, P^x)$ be a standard process with state space $(E, \mathscr{E})$. Let A be a CAF of X which is strictly increasing and finite on $[0, \zeta)$ and let τ be the inverse of A. If $\hat{X} = (\Omega, \mathscr{F}, \mathscr{F}_{\tau(t)}, X_{\tau(t)}, \theta_{\tau(t)}, P^x)$ then, according to (2.11), $\hat{X}$ is a standard process with state space $(E, \mathscr{E})$, at least if

(1.3) holds. Moreover it is easy to see that X and $\hat{X}$ have the same hitting distributions; that is, for all $D \in \mathscr{E}_\Delta$ and $x \in E_\Delta$, $P_D(x, \cdot) = \hat{P}_D(x, \cdot)$. The purpose of this section is to prove the converse of this statement. We now formulate this converse precisely.

Let X be as above with $\mathscr{M} = \mathscr{F}$ and $\mathscr{M}_t = \mathscr{F}_t$. Let $\tilde{X} = (\tilde{\Omega}, \tilde{\mathscr{M}}, \tilde{\mathscr{M}}_t, \tilde{X}_t, \tilde{\theta}_t, \tilde{P}^x)$ be another standard process with the same state space $(E, \mathscr{E})$ and with $\tilde{\mathscr{M}} = \tilde{\mathscr{F}}$, $\tilde{\mathscr{M}}_t = \tilde{\mathscr{F}}_t$. For typographical convenience we will omit the tilde "$\sim$" in those places where it is clearly applicable. For example in place of $\tilde{E}^x\{f(\tilde{X}_{\tilde{T}_B}); \tilde{T}_B < \tilde{\zeta}\}$ we will write simply $\tilde{E}^x\{f(X_{T_B}); T_B < \zeta\}$. We can now state the main result of this section.

(5.1) THEOREM. Let X and $\tilde{X}$ be as above, and suppose that, for each compact subset K of E_Δ, $P_K(x, \cdot) = \tilde{P}_K(x, \cdot)$ for all x. Then there exists a CAF, A, of X which is strictly increasing and finite on $[0, \zeta)$ such that if τ is the inverse of A and $\hat{X}$ is defined as in the preceding paragraph, then $\hat{X}$ and $\tilde{X}$ are equivalent.

We will break up the proof of Theorem 5.1 into a number of steps. Roughly speaking, the content of Theorem 5.1 is that if X and $\tilde{X}$ have the same hitting distributions (as in the hypothesis of (5.1)), then they have the same "set-theoretic" paths but they move along these paths with different "speeds."

First note that X and $\tilde{X}$ must have the same traps; for simplicity we assume that there are no traps except Δ. Second, it is clear that the hypothesis implies that $P_B(x, \cdot) = \tilde{P}_B(x, \cdot)$ for every Borel set B in E_Δ since the hitting time of B can be approximated by the hitting times of compact subsets of B. Finally if $B \in \mathscr{E}_\Delta$ and $x \notin B$, then x is regular for B relative to $X(\tilde{X})$ if and only if $P_B(x, \cdot) = \varepsilon_x$ $(\tilde{P}_B(x, \cdot) = \varepsilon_x)$. It follows from this and (4.3) of Chapter II that X and $\tilde{X}$ induce the *same* fine topology on E, and so we may speak of *the* fine topology without confusion.

Let G be a Borel subset of E and suppose that $G^c = E_\Delta - G$ is finely open. Let $T(\tilde{T})$ be the hitting time of G^c for $X(\tilde{X})$. Then $M_t = I_{[0,T)}(t)$ $(\tilde{M}_t = I_{[0,\tilde{T})}(t))$ defines an exact perfect MF of $X(\tilde{X})$ vanishing on $[\zeta, \infty]$ $([\tilde{\zeta}, \infty])$. If E_G denotes the set of points not regular for G^c, then $E_M = E_{\tilde{M}} = E_G$ and E_G is a finely open Borel set contained in G. The fact that E_G is a Borel set is not completely obvious but follows from the following assertion. See Exercise 5.30.

(5.1 bis) If B is a finely open Borel set, then $x \to P^x[T_B < t]$ is Borel measurable for each t.

Indeed (5.1 bis) implies that $x \to E^x(e^{-T})$ is Borel measurable and so $E_G =$

$\{x : E^x(e^{-T}) < 1\}$ is Borel measurable. This notation will be used without special mention in the next few paragraphs.

(5.2) LEMMA. If D is a Borel subset of E_G, then for each $f \in b\mathscr{E}$ and $x \in E$

$$E^x\{f(X_{T_D}); T_D < T\} = \tilde{E}^x\{f(X_{T_D}); T_D < T\}.$$

Proof. It suffices to prove this when D is compact and x is in E_G. Since D is a compact subset of E_G and $(G^c)^r = (E_G)^c$, $P^x[T = T_D < \infty] = 0$. Therefore if $x \in E_G$ and $F = D \cup G^c$ we have

$$P_D f(x) = E^x\{f(X_{T_D}); T < T_D\} + E^x\{f(X_{T_D}); T_D < T\}$$

$$= \int_{E_G^c} P_F(x, dy)\, P_D f(y) + E^x\{f(X_{T_D}); T_D < T\}.$$

Since X and $\tilde{X}$ have the same hitting distributions this yields Lemma 5.2.

We will write (X, T) and $(\tilde{X}, \tilde{T})$ in place of (X, M) and $(\tilde{X}, \tilde{M})$, respectively. It follows from (5.2) and Dynkin's theorem (5.19i) of Chapter III that (X, T) and $(\tilde{X}, \tilde{T})$ have the same excessive functions; that is, $\mathscr{S}(T) = \tilde{\mathscr{S}}(\tilde{T})$. Indeed if $f \in \tilde{\mathscr{S}}(\tilde{T})$ then (5.2) and Dynkin's theorem imply that $\alpha V^\alpha f \leq f$ for all $\alpha > 0$ where V^α is the resolvent corresponding to (X, T). In addition f is finely continuous on E_G and so $f \in \mathscr{S}(T)$.

A set G as above ($G \in \mathscr{E}$ and G^c finely open) is called an *exit set* provided $\tilde{P}^x(T < \infty) = 1$ for all x. Since $\tilde{P}^x(T < \infty) = \tilde{P}_{G^c}(x, E_\Delta)$ it follows that whenever G is an exit set $P^x(T < \infty) = 1$ for all x also. We now assume that G is an exit set. In what follows, $Q_t(\tilde{Q}_t)$ and $V^\alpha(\tilde{V}^\alpha)$ will denote the semigroup and resolvent associated with (X, T) $((\tilde{X}, \tilde{T}))$, respectively. We next define

$$\text{(5.3)} \qquad u(x) = \tilde{E}^x(1 - e^{-T}), \qquad g(x) = 1 - u(x) = \tilde{E}^x(e^{-T}).$$

It is immediate that $0 < u < 1$ on E_G, and $u \in \tilde{\mathscr{S}}(\tilde{T})$. Consequently u is (X, T) excessive. We come now to the key step in our construction.

(5.4) PROPOSITION. There exists a CAF, A, of (X, T) such that $u = u_A$.

Proof. According to Theorem 3.13 of Chapter IV it suffices to show that u is a regular (X, T) potential. It is clear from the definition that u is a regular $(\tilde{X}, \tilde{T})$ potential. In fact a simple computation shows that $u = \tilde{V}g$. Suppose for the moment that the following statement has been established.

(5.5) For each fixed $x \in E_G$ there exists an increasing sequence $\{u_n\}$ of regular (X, T) potentials such that $\lim u_n(x) = u(x)$ and, for each n, $u - u_n \in \mathscr{S}(T)$.

Then u must be a regular (X, T) potential. Indeed let $\{R_n\}$ be an increasing sequence of stopping times with limit R and let x be fixed. Let $\{u_n\}$ be the sequence whose existence is assumed in (5.5) for this fixed x. Then $Q_{R_n}u_k \to Q_R u_k$ as $n \to \infty$ and $Q_{R_n} u(x) - Q_{R_n} u_k(x) \leq u(x) - u_k(x) \to 0$ as $k \to \infty$, uniformly in n. Therefore

$$\lim_n Q_{R_n} u(x) = \lim_n \lim_k Q_{R_n} u_k(x)$$

$$= \lim_k Q_R u_k(x) \leq Q_R u(x).$$

But $\lim_n Q_{R_n} u(x) \geq Q_R u(x)$, and so to prove (5.4) it suffices to establish (5.5).

Let x be fixed and let μ be the measure

$$\mu(\Gamma) = \tilde{V}I_\Gamma g(x) = \tilde{E}^x \int_0^T I_\Gamma(X_t)\, g(X_t)\, dt.$$

Then $\mu(E) = \mu(E_G) = u(x) \leq 1$. Let $\varphi_t(x) = P^x(T < t)$ and $\tilde{\varphi}_t(x) = \tilde{P}^x(T < t)$. Then φ_t and $\tilde{\varphi}_t$ decrease to zero pointwise on E_G as $t \downarrow 0$, and so by Egorov's theorem we can find an increasing sequence $\{D_n\}$ of compact subsets of E_G such that, on each D_n, φ_t and $\tilde{\varphi}_t$ approach zero uniformly as $t \downarrow 0$ and $\mu(E_G - D_n) \downarrow 0$. Define $u_n = \tilde{V}I_{D_n}g$. Then $u_n \leq u$ and $u - u_n \in \tilde{\mathscr{S}}(\tilde{T}) = \mathscr{S}(T)$. If $F = \bigcup D_n$, then $\mu(E_G - F) = 0$ and so $u_n(x) \uparrow \tilde{V}I_F\, g(x) = \tilde{V}g(x) = u(x)$ by the definition of μ. Thus (5.5) will be established if we can show that each u_n is a regular (X, T) potential. Fix n and let $D = D_n$, $h = u_n$. Thus $h = \tilde{V}I_D g \leq 1$ where D is compact, and φ_t and $\tilde{\varphi}_t$ approach zero uniformly on D. We must show that h is a regular (X, T) potential. It is, of course, clear that h is a regular $(\tilde{X}, \tilde{T})$ potential.

First observe that $h = \tilde{Q}_D h = Q_D h$. Let $h_n = n \int_0^{1/n} Q_t h\, dt$. Then according to (3.4) of Chapter IV each $h_n \in \mathscr{S}(T)$ and $h_n \uparrow h$. Next fix $\varepsilon > 0$ and let $R_n(\tilde{R}_n)$ be the hitting time by $X(\tilde{X})$ of $B_n = \{x\colon h(x) - h_n(x) \geq \varepsilon\}$, and let $R = \lim R_n$, $\tilde{R} = \lim \tilde{R}_n$. Finally let $S_n = R_n + T_D \circ \theta_{R_n}$, $\tilde{S}_n = \tilde{R}_n + \tilde{T}_D \circ \tilde{\theta}_{\tilde{R}_n}$ and $S = \lim S_n$, $\tilde{S} = \lim \tilde{S}_n$. Now given $\eta > 0$ there exists $t > 0$ such that $P^y(T < t) = \varphi_t(y) < \eta$ for all $y \in D$, and so for any $z \in E_G$

$$P^z[S_n < T \text{ for all } n,\ S = T] \leq \lim_n P^z[T - S_n < t,\ S_n < T]$$

$$= \lim_n E^z\{P^{X(S_n)}(T < t);\ S_n < T\} < \eta$$

since $X(S_n) \in D$ almost surely on $\{S_n < T\}$. Therefore $P^z(S_n < T \text{ for all } n) = P^z(S < T)$ for all $z \in E_G$ (and hence in E). This argument of course yields the same statement with P^z replaced by $\tilde{P}^z$. It now follows from the quasi-left-continuity of X and $\tilde{X}$ that, for each z, $Q_{S_n}(z, \cdot) \to Q_S(z, \cdot)$, $\tilde{Q}_{\tilde{S}_n}(z, \cdot) \to \tilde{Q}_{\tilde{S}}(z, \cdot)$ weakly as $n \to \infty$. But $Q_{S_n} = Q_{R_n} Q_D = \tilde{Q}_{\tilde{R}_n} \tilde{Q}_D = \tilde{Q}_{\tilde{S}_n}$ and so $Q_S =$

$\tilde{Q}_{\tilde{S}}$. Consequently $Q_{R_n}h = Q_{S_n}h = \tilde{Q}_{\tilde{S}_n}h \to \tilde{Q}_{\tilde{S}}h = Q_S h$ and since $S \geq R$ we obtain $\lim_n Q_{R_n}h \leq Q_R h$. On the other hand $\lim_n Q_{R_n}h \geq Q_R h$ and so we have $Q_{R_n}h \to Q_R h$. Also for any z, $Q_t\, h(z) \leq P^z(t < T) \to 0$ as $t \to \infty$ since $P^z(T < \infty) = 1$. Therefore (3.6) and (3.8) of Chapter IV imply that h is a regular (X, T) potential, and so (5.5) and hence (5.4) are proved.

(5.6) LEMMA. Let A be the CAF from (5.4) and let D be a Borel subset of E_G. If $R = T_D \wedge T$, $\tilde{R} = \tilde{T}_D \wedge \tilde{T}$, then $E^x(A_R) = \tilde{E}^x\{e^{-T}(e^R - 1)\}$.

Proof. Since $E_G^c = (G^c)^r$, $T - R = T \circ \theta_R$ almost surely, and so, by (1.13) of Chapter IV, $A_T = A_R + A_T \circ \theta_R$ almost surely on $\{R < T\}$. In addition $A_T \circ \theta_R = 0$ almost surely on $\{R = T\}$. Therefore

$$\begin{aligned} E^x(A_R) &= u(x) - E^x\{u(X_R)\} = u(x) - \tilde{E}^x\{u(X_R)\} \\ &= \tilde{E}^x\{1 - e^{-T} - 1 + e^{-(T-R)}\} \\ &= \tilde{E}^x\{e^{-T}(e^R - 1)\}. \end{aligned}$$

(5.7) LEMMA. Let A be as in (5.4). Then almost surely $t \to A_t$ is strictly increasing on $[0, T)$.

Proof. Obviously it suffices to show that $\mathrm{Supp}(A) = E_G$. Let $F = \mathrm{Supp}(A)$ and suppose that $x \in E_G - F$. If $R = T \wedge T_F$, $\tilde{R} = \tilde{T} \wedge \tilde{T}_F$, then since $E_G - F$ is finely open $\tilde{P}^x(R > 0) = 1$. But $A_R = 0$ and so we obtain from (5.6)

$$0 = E^x(A_R) = \tilde{E}^x\{e^{-T}(e^R - 1)\} > 0,$$

and this contradiction establishes (5.7).

The next result is the fundamental property of A for our construction.

(5.8) PROPOSITION. Let A be as in (5.4) and let $f \in b\mathscr{E}^*$. Then $U_A f = \tilde{V}(fg)$ where g is defined in (5.3).

Proof. If $f = 1$ we have already seen that $u_A = u = \tilde{V}g \leq 1$. Consequently it suffices to establish (5.8) when f is continuous and has compact support. Given such an f and $\varepsilon > 0$ define $T_\varepsilon = \inf\{t\colon |f(X_t) - f(X_0)| \geq \varepsilon\}$ and then let $T_0 = 0$, $T_{n+1} = T_n + T_\varepsilon \circ \theta_{T_n}$ for $n \geq 0$. Let $\tilde{T}_\varepsilon$ and $\tilde{T}_n$ be defined similarly relative to $\tilde{X}$. Then arguing as in the proof of Theorem 2.13 of Chapter IV we find that

$$U_A f(x) = \sum E^x\{f(X_{T_n})\,[u(X_{T_n}) - Q_{T_\varepsilon}u(X_{T_n})]\} + \varepsilon\, L(x)$$

where $|L(x)| \leq \|f\|$, and similarly

$$\tilde{V}f\, g(x) = \sum \tilde{E}^x\{f(X_{T_n})\,[u(X_{T_n}) - \tilde{Q}_{T_\varepsilon}\, u(X_{T_n})]\} + \varepsilon\, \tilde{L}(x)$$

where $|\tilde{L}(x)| \le \|f\|$. If $B_x = \{y\colon |f(y) - f(x)| \ge \varepsilon\}$, then $Q_{T_\varepsilon}(x, \cdot) = Q_{B_x}(x, \cdot) = \tilde{Q}_{B_x}(x, \cdot) = \tilde{Q}_{T_\varepsilon}(x, \cdot)$, and so, since ε is arbitrary, the proof will be complete provided we show that $\mu_n = \tilde{\mu}_n$ where $\mu_n(\tilde{\mu}_n)$ is the distribution of $X(T_n)$ $(\tilde{X}(\tilde{T}_n))$ under $P^x(\tilde{P}^x)$. We have just observed that this is true when $n = 1$, while, for any $\Gamma \in \mathscr{E}$, $\mu_{n+1}(\Gamma) = \int \mu_n(dz)\, Q_{T_\varepsilon}(z, \Gamma)$ with a similar statement for $\tilde{\mu}_{n+1}$. As a result $\mu_n = \tilde{\mu}_n$ for all n and so (5.8) is established.

(5.9) LEMMA. Let A be as in (5.4) and define $B_t = \int_0^t [g(X_u)]^{-1}\, dA_u$. Then B_T is almost surely finite and B is a CAF of (X, T) which is strictly increasing on $[0, T)$.

REMARK. By (5.8), $U_B f = \tilde{V}f$ for all $f \in \mathscr{E}_+^*$.

Proof. The last two assertions follow immediately from the first, (5.7), and the fact that $g < 1$ on E_G. In order to show that B_T is almost surely finite it suffices to show that almost surely $t \to g(X_t)$ is bounded away from zero on $[0, T)$. To this end let T_n be the hitting time of $\{x\colon g(x) < 1/n\}$, $R = \lim T_n$, and let $\delta = P^x[T_n < T$ for all $n]$ where x is a fixed point in E_G. Since u is a regular (X, T) potential, $Q_{T_n}u \to Q_R u$. But $u = 1 - g$ and so $Q_{T_n}\, u(x) \ge (1 - 1/n)\, P^x(T_n < T)$. Letting $n \to \infty$ this yields

$$\delta \le E^x\{u(X_R);\, R < T\} \le P^x(R < T) \leqslant \delta,$$

and consequently $\delta = P^x(R < T)$. Therefore $E^x\{u(X_R);\, R < T\} = P^x(R < T)$ and since u is strictly less than one on E_G this implies that $\delta = P^x(R < T) = 0$. Thus (5.9) is established.

We now interrupt our main development in order to prove the following uniqueness result which we need.

(5.10) PROPOSITION. Let $\{R^\alpha;\, \alpha \ge 0\}$ and $\{W^\alpha;\, \alpha \ge 0\}$ be two families of nonnegative bounded linear operators on $b\mathscr{E}^*$, each of which satisfies the resolvent equation. If $W^0 = R^0$ then $W^\alpha = R^\alpha$ for all $\alpha \ge 0$.

Proof. One sees easily that $\|W^\alpha\| \le \|W^0\|$ and $\|R^\alpha\| \le \|R^0\|$ for all $\alpha \ge 0$. Let $M = \|W^0\| = \|R^0\|$. For any α and β, $R^\alpha[I - (\beta - \alpha)R^\beta] = R^\beta$ and so

$$R^\alpha = \sum_{n=0}^{\infty} (\beta - \alpha)^n (R^\beta)^{n+1}$$

provided $|\beta - \alpha| < M^{-1}$. A similar statement holds for W^α and so letting $\beta = 0, \frac{1}{2}M^{-1}, M^{-1}, \ldots$ we obtain Proposition 5.10.

It is worthwhile to point out what we have achieved so far. Suppose that not only is $\tilde{P}^x(T < \infty) = 1$ for all x, but that actually $\tilde{E}^x(T)$ is a bounded function of x. Then it is clear that $E^x \int_0^T f(X_t)\, dB_t = \tilde{V}f(x)$ for all x and positive f, where B is the CAF of (X, T) from Lemma 5.9. Let τ be the inverse of B. Since B_t is strictly increasing and finite on $[0, T)$, τ_t is *continuous* and strictly increasing on $[0, B_T)$, $\tau_t = \infty$ if $t \geq B_T$, and τ_t increases to T strictly from below as t increases to B_T. Of course if these statements are to hold for all ω we must assume that $t \to B_t(\omega)$ is continuous and strictly increasing on $[0, T(\omega))$ for all ω, and this we may do. Now it follows from (2.3) that $\hat{X}_G = (\Omega, \mathscr{F}, \mathscr{F}_{\tau(t)}, X_{\tau(t)}, \theta_{\tau(t)}, P^x)$ may be regarded as a Markov process with right continuous paths and state space $(E_G, \mathscr{E}_G^*)$, where $\mathscr{E}_G^*$ is the trace of $\mathscr{E}^*$ on E_G. We claim that this process and $(\tilde{X}, \tilde{T})$ are equivalent; that is, for all $x \in E_G$, $t \geq 0$, and $\Gamma \in \mathscr{E}^*$

$$P^x\{X[\tau(t)] \in \Gamma\} = \tilde{P}^x[X_t \in \Gamma;\, t < T].$$

Indeed, if we define $W^\alpha f(x) = E^x \int_0^\infty e^{-\alpha t} f(X_{\tau(t)})\, dt$ then by the right continuity of the processes it suffices to show that $W^\alpha f = \tilde{V}^\alpha f$ for all α and all continuous f with compact support. Now it follows from (2.22) of Chapter IV or is easily checked in the present case that $\{W^\alpha;\, \alpha \geq 0\}$ satisfies the resolvent equation. But $W^0 f(x) = E^x \int_0^T f(X_t)\, dB_t$ and this is equal to $\tilde{V}f(x)$. Consequently by (5.10) $W^\alpha f = \tilde{V}^\alpha f$ for all α.

Note that this establishes Theorem 5.1 in the special case in which $\tilde{E}^x(\zeta)$ is bounded in x, for in this case we may take $G = E$ so that $T = \zeta$ and $\tilde{T} = \tilde{\zeta}$. In the general case according to the preceding paragraph we know that "locally" (that is, on any exit set for which $\tilde{E}^x(T)$ is bounded) $\tilde{X}$ and X run with the appropriate "clock" are equivalent. It remains to "piece together" these local results in order to obtain Theorem 5.1 in the general case. The procedure for this "piecing together" is obvious enough in outline, but the details are rather involved. We will carry out the complete argument because some of the techniques are of interest and may be useful in other contexts.

To start with we need a compatability result. Let G_1 and G_2 be exit sets, T_1 and T_2 the hitting times of G_1^c and G_2^c, and B^1 and B^2 the CAF's of (X, T_1) and (X, T_2), respectively, defined in (5.9). The resolvent corresponding to $(\tilde{X}, \tilde{T}_j)$ will be denoted by $\tilde{V}_j^\alpha$, $j = 1, 2$.

(5.11) PROPOSITION. Almost surely $B_t^1 = B_t^2$ on the interval $[0, T_1 \wedge T_2]$.

Proof. Since $G_1 \cap G_2$ is an exit set and $T_1 \wedge T_2$ is the hitting time of $(G_1 \cap G_2)^c$ it will suffice to treat the case in which $G_1 \subset G_2$, so that $T_1 \leq T_2$. Define $B_t = B_t^2$ if $t \leq T_1$ and $B_t = B_{T_1}^2$ if $t > T_1$. Then B is a CAF of (X, T_1), and we wish to show that B and B^1 are equivalent. Since $g = g_1 g_2$ is strictly positive on E_{G_1} and $U_{B^1}g \le \tilde{V}_1 g_1 < \infty$ it will suffice to show that $U_{B^1}g = U_B g$, because of the uniqueness theorem for CAF's. We compute

$$U_{B^2}g(x) = U_B g(x) + E^x \int_{T_1}^{T_2} g(X_t)\, dB_t^2$$
$$= U_B g(x) + P_{T_1} U_{B^2} g(x)$$

so that $U_B g = \tilde{V}_2 g - \tilde{P}_{T_1} \tilde{V}_2 g = \tilde{V}_1 g = U_{B^1} g$, establishing the proposition.

We need some discussion of hitting times before proceeding. The arguments used here have appeared many times already, so we will not separate off these facts as propositions. We will call a set $G \subset E$ a *strong exit set* (for $\tilde{X}$) if G is a Borel set, G^c is finely open, and $\tilde{E}^x(T_{G^c})$ is bounded in x. Let x be any point of E. By assumption $\tilde{X}$ has no traps and so there is an open set N with compact closure such that $x \in N$ and $\tilde{P}^x(T_{\bar{N}^c} < \infty) > 0$. Since $\bar{N}^c$ is open it follows that $v(x) = \int_0^\infty e^{-t}\, \tilde{P}_t(x, \bar{N}^c)\, dt > 0$. The function v is Borel measurable. Given $\varepsilon > 0$ let $G = \bar{N} \cap \{v \ge \varepsilon\}$. Then $G^c = \bar{N}^c \cup \{v < \varepsilon\}$ which is Borel and finely open. In addition G is a strong exit set. Indeed if $v(z) \ge \varepsilon$ and b is such that $e^{-b} < \varepsilon/2$, then

$$\varepsilon \le v(z) \le \tilde{P}^z(T_{\bar{N}_i^c} \le b) + e^{-b}$$

and so $\tilde{P}^z(T_{G^c} \le b) \ge \tilde{P}^z(T_{\bar{N}^c} \le b) \ge \varepsilon/2$. Then for any z

$$\tilde{P}^z\{T_{G^c} > (n+1)b\} = \tilde{E}^z\{\tilde{P}^{X(nb)}(T_{G^c} > b);\ T_{G^c} > nb\}$$
$$\le \tilde{P}^z(T_{G^c} > nb)\{1 - \varepsilon/2\}$$

because $v(\tilde{X}_{nb}) \ge \varepsilon$ if $\tilde{T}_{G^c} > nb$ and $nb < \tilde{\zeta}$. So for every z, $\tilde{E}^z(T_{G^c}) \le b \sum_n (1 - \varepsilon/2)^n + b$, proving the assertion. Compare this with (10.25) of Chapter I. Now let $\{N_i\}$ be a countable base for the topology of E consisting of open sets with compact closures. If $v_i(x) = \tilde{U}^1 I_{\bar{N}_i^c}(x) = \int_0^\infty e^{-t}\, \tilde{P}_t(x, \bar{N}_i^c)\, dt$ and $W_{ij} = \bar{N}_i \cap \{v_i \ge 1/j\}$, then what we have just said implies that each W_{ij} is a strong exit set and furthermore $\bigcup_{i,j} W_{ij} = E$.

Next let G be any exit set and let $\varphi(x) = E^x(e^{-T_{G^c}})$. If $\eta < 1$ define $K_\eta = \{x\colon \varphi(x) < \eta\}$. Then each K_η is a finely open Borel set and $E_G = \bigcup_n K_{1-1/n}$. Fix $\eta < 1$ and define

$$T_0 = 0$$
$$T_{2n+1} = T_{2n} + T_{G^c} \circ \theta_{T_{2n}}$$
$$T_{2n+2} = T_{2n+1} + T_{K_\eta} \circ \theta_{T_{2n+1}}.$$

The sequence $\{T_n\}$ of stopping times is increasing, and for any x and $n \ge 1$

$$E^x(e^{-T_{2n+1}}; T_{2n} < \zeta) \le E^x\{e^{-T_{2n}}\varphi(X_{T_{2n}}); T_{2n} < \zeta\}$$
$$\le \eta\, E^x\{e^{-T_{2n}}; T_{2n} < \zeta\}$$

because $\varphi(X_{T_{2n}}) \le \eta$ if $T_{2n} < \zeta$ and $n \ge 1$. It follows that $\lim T_n = \infty$ almost surely (relative to X). We will use these notions in the next proposition.

(5.12) PROPOSITION. Let G be an exit set, let $T = T_{G^c}$, and let A be a CAF of (X, T) such that $E^x A(T)$ is bounded in x. Let $K = \{x: E^x(e^{-T}) < \eta\}$ for a fixed $\eta < 1$. Then there is a CAF $\tilde{A}$ of X such that

$$E^x \int_0^T f(X_t)\, dA_t = E^x \int_0^T f(X_t)\, d\tilde{A}_t$$

for every x and every $f \in \mathscr{E}_+^*$ which vanishes off K.

Proof. Let

$$B_t = \int_0^{t \wedge T} I_K(X_t)\, dA_t$$

so that B is a CAF of (X, T). Define

$$\psi(x) = E^x \int_0^T e^{-t}\, dB_t$$

$$z(x) = E^x\left\{\sum_{n=0}^{\infty} e^{-T_{2n}}\, \psi(X_{T_{2n}})\right\}$$

where the sequence $\{T_n\}$ is the one constructed in the paragraph before (5.12). Note that ψ is bounded and that $z(x) \le \|\psi\| \sum_{n=0}^{\infty} E^x(e^{-T_{2n}}; T_{2n} < \zeta) \le \|\psi\|(1 + \sum_0^{\infty} \eta^n)$. Thus z is also bounded. The main part of the proof is in showing that z is a regular 1-potential of X. We will break up the argument into steps. Also we will omit the phrase "almost surely" where it is clearly applicable.

(a) $z = P_K^1 z$.

Proof. We will write R in place of T_K. Now

$$P_K^1 z(x) = E^x\{e^{-R} z(X_R)\}$$
$$= \sum_{n=0}^{\infty} E^x\{e^{-(T_{2n}\circ\theta_R + R)}\psi(X(T_{2n}\circ\theta_R + R))\}.$$

Let us break up each summand into an integral over $\{R < T_1\}$ and one over $\{R \ge T_1\}$. Now observe that for $n \ge 1$, $T_{2n}\circ\theta_R + R = T_{2n}$ on $\{R < T_1\}$ while for $n \ge 0$, $T_{2n}\circ\theta_R + R = T_{2(n+1)}$ on $\{R \ge T_1\}$. If we use these facts we find that

$$P_K^1 z(x) = E^x(e^{-R}\,\psi(X_R);\, R < T_1) + \sum_{n=1}^{\infty} E^x(e^{-T_{2n}}\,\psi(X_{T_{2n}})),$$

and so

$$z(x) - P_K^1 z(x) = \psi(x) - E^x(e^{-R}\,\psi(X_R);\, R < T_1).$$

This last difference is 0 because $R = T_K$, $T_1 = T$, and B vanishes off K.

(b) If J is any Borel set, then $P_{K \cup J}^1 z = z$.

Proof. Since K is finely open we have $R = R \circ \theta_{T_{K \cup J}} + T_{K \cup J}$, where as above $R = T_K$. Consequently using (a)

$$P_{K \cup J}^1 z = P_{K \cup J}^1 P_K^1 z = P_K^1 z = z.$$

(c) If J is any compact set, then $P_J^1 z \le z$.

Proof. Let $S = T_J + R \circ \theta_{T_J}$ where $R = T_K$. Then $X(S) \in K \cup K^r$ on $\{S < \infty\}$, and so if $S < \infty$ then $T_{2k} \le S < T_{2k+1}$ for some k. Now from the definition of z it is immediate that

$$z(x) = \psi(x) + E^x\{e^{-T_2}\, z(X_{T_2})\}. \tag{5.13}$$

But $S + T_{2n} \circ \theta_S = T_{2n}$ on $\{S < T_1\}$ for $n \ge 1$ and $\{S < T_1\} \in \mathscr{F}_{T_2}$. Therefore

$$\begin{aligned} &E^x\{e^{-S}\, z(X_S);\, S < T_1\} \\ &\quad = E^x\{e^{-S}\,\psi(X_S);\, S < T_1\} + E^x\left\{\sum_1^{\infty} e^{-T_{2n}}\,\psi(X_{T_{2n}});\, S < T_1\right\} \\ &\quad = E^x\{e^{-S}\,\psi(X_S);\, S < T_1\} + E^x\{e^{-T_2}\, z(X_{T_2});\, S < T_1\}, \end{aligned}$$

and using (5.13) and the fact that $\psi \in \mathscr{S}^1(T)$ we obtain

$$\begin{aligned} &E^x\{e^{-S}\, z(X_S);\, S < T_1\} + E^x\{e^{-T_2}\, z(X_{T_2});\, S \ge T_1\} \\ &\qquad \le \psi(x) + E^x\{e^{-T_2}\, z(X_{T_2})\} = z(x). \end{aligned} \tag{5.14}$$

We make the induction hypothesis

$$z(x) \ge E^x\{e^{-S}\, z(X_S);\, S < T_{2n}\} + E^x\{e^{-T_{2n}}\, z(X_{T_{2n}});\, S \ge T_{2n}\} \tag{5.15}$$

which, by (5.14), is valid when $n = 1$ since S lies in some interval $[T_{2k}, T_{2k+1})$. The second summand on the right in (5.15) may be written

$$E^x\{e^{-S}\, z(X_S);\, S = T_{2n}\} + E^x\{e^{-T_{2n}}\, z(X_{T_{2n}});\, S > T_{2n}\}.$$

Now using (5.14) and the fact that $T_{2n} + S \circ \theta_{T_{2n}} = S$ on $\{T_{2n} < S\}$ we obtain

$$\begin{aligned}
E^x\{e^{-T_{2n}} z(X_{T_{2n}}); S > T_{2n}\} \\
&\geq E^x\{e^{-T_{2n}} E^{X(T_{2n})}[e^{-S} z(X_S); S < T_1]; S > T_{2n}\} \\
&\quad + E^x\{e^{-T_{2n}} E^{X(T_{2n})}[e^{-T_2} z(X_{T_2}); S \geq T_1]; S > T_{2n}\} \\
&= E^x\{e^{-S} z(X_S); T_{2n} < S < T_{2n+1}\} \\
&\quad + E^x\{e^{-T_{2n+2}} z(X_{T_{2n+2}}); S \geq T_{2n+1}\}.
\end{aligned}$$

But $\{T_{2n} < S < T_{2n+1}\} = \{T_{2n} < S < T_{2n+2}\}$ and $\{S \geq T_{2n+1}\} = \{S \geq T_{2n+2}\}$, and combining these facts we obtain (5.15) with n replaced by $n + 1$. Thus (5.15) holds for all n, and since $\lim T_n = \infty$ this yields $z \geq P_S^1 z$. But $P_S^1 z = P_J^1 P_K^1 z = P_J^1 z$ by (a) and so (c) is proved.

(d) $z \in \mathscr{S}^1$.

Proof. In view of (c) and Dynkin's theorem ((5.3) of Chapter II) it will suffice to show that $\liminf_{t \downarrow 0} P_t^1 z(x) \geq z(x)$ for all x. Suppose x is not regular for K. Then almost surely P^x, $t + R \circ \theta_t = R$ for t sufficiently small, and since $z = P_K^1 z$ this yields

$$\begin{aligned}
\lim_{t \to 0} P_t^1 z(x) &= \lim_{t \to 0} E^x\{e^{-(t + R \circ \theta_t)} z(X_{t + R \circ \theta_t})\} \\
&= E^x\{e^{-R} z(X_R)\} = P_K^1 z(x) = z(x).
\end{aligned}$$

Suppose on the other hand that x is regular for K. Then in particular $P^x(t < T) \to 1$ as $t \to 0$ and so

$$\begin{aligned}
\textbf{(5.16)} \quad P_t^1 z(x) &\geq E^x\{e^{-t} z(X_t); t < T_1\} \\
&= E^x\{e^{-t} \psi(X_t); t < T_1\} + E^x\{e^{-T_2} z(X_{T_2}); t < T_1\}.
\end{aligned}$$

Since ψ is $1 - (X, T)$ excessive this approaches $\psi(x) + E^x\{e^{-T_2} z(X_{T_2})\} = z(x)$ as $t \to 0$. The proof of (d) is complete.

(e) z is a regular 1-potential.

Proof. We must show that if $\{S_n\}$ is any increasing sequence of stopping times with limit S, then $P_{S_n}^1 z \to P_S^1 z$. A glance at the development in Section 3 of Chapter IV shows that we need consider only the case in which $S_n = T_{B_n}$ where $\{B_n\}$ is a decreasing sequence of sets in $\mathscr{E}^n$. In particular each S_n will then be a strong terminal time and hence so will S. Also in checking $P_{S_n}^1 z(x) \to P_S^1 z(x)$ we may assume that $P^x(S_n > 0) = 1$, for if $S_n = 0$ almost surely P^x for all n there is nothing to prove. Now fix x and let

$$\begin{aligned}
a_{nk} &= E^x\{e^{-S_n} z(X_{S_n}); T_k < S_n \leq T_{k+1}\} \\
a_k &= E^x\{e^{-S} z(X_S); T_k < S \leq T_{k+1}\}
\end{aligned}$$

so that $P^1_{S_n} z(x) = \sum_k a_{nk}$ and $P^1_S z(x) = \sum_k a_k$. It will suffice to show that, for each k, $a_{nk} \to a_k$ as $n \to \infty$ because $\sum_{k \geq N} a_{nk} \leq \|z\| \, E^x(e^{-T_N}) \to 0$ as $N \to \infty$. Suppose first that k is even, say $k = 2j$. Now if Q is any strong terminal time, then on $\{T_{2j} < Q \leq T_{2j+1}\}$ we have $Q = T_{2j} + Q \circ \theta_{T_{2j}}$ and $Q + T_2 \circ \theta_Q = T_{2j+2}$, and so using $z(x) = \psi(x) + E^x(e^{-T_2} z(X_{T_2}))$ we obtain for any such Q

$$
\begin{aligned}
(5.17) \qquad & E^x\{e^{-Q} z(X_Q);\, T_{2j} < Q \leq T_{2j+1}\} \\
&= E^x\{e^{-Q} \psi(X_Q);\, T_{2j} < Q,\, Q \circ \theta_{T_{2j}} \leq T \circ \theta_{T_{2j}}\} \\
&\quad + E^x\{e^{-Q} E^{X(Q)}(e^{-T_2} z(X_{T_2}));\, T_{2j} < Q \leq T_{2j+1}\} \\
&= E^x\{e^{-T_{2j}} E^{X(T_{2j})}(e^{-Q} \psi(X_Q);\, Q \leq T);\, T_{2j} < Q\} \\
&\quad + E^x\{e^{-T_{2j+2}} z(X_{T_{2j+2}});\, T_{2j} < Q \leq T_{2j+1}\}.
\end{aligned}
$$

Now in (5.17) we may replace Q by either S_n or S. Observe that the set $\{T_{2j} < S_n\}$ approaches $\{T_{2j} < S\}$ and that $\{T_{2j} < S_n \leq T_{2j+1}\}$ approaches $\{T_{2j} < S \leq T_{2j+1}\}$ as $n \to \infty$. Furthermore for any y we have

$$
\begin{aligned}
E^y(e^{-S_n} \psi(X_{S_n});\, S_n \leq T) &= E^y(e^{-S_n} \psi(X_{S_n});\, S_n < T) \\
&\to E^y(e^{-S} \psi(X_S);\, S < T) \\
&= E^y(e^{-S} \psi(X_S);\, S \leq T),
\end{aligned}
$$

because $S_n \to S$ and ψ, being the 1-potential of a CAF of (X, T), is a regular (X, T) potential. It follows that $a_{n,2j} \to a_{2j}$ as $n \to \infty$. Next consider the case in which $k = 2j + 1$. Here we use the fact that $z = P^1_K z$ to obtain

$$
a_{n,2j+1} = E^x\{e^{-(S_n + T_K \circ \theta_{S_n})} z(X(S_n + T_K \circ \theta_{S_n}));\, T_{2j+1} < S_n \leq T_{2j+2}\}
$$

with the same expression for a_{2j+1} upon replacing S_n with S. Now on $\{T_{2j+1} < S_n \leq T_{2j+2}\} \cap \{T_{2j+1} < S \leq T_{2j+2}\}$ we have $S_n + T_K \circ \theta_{S_n} = S + T_K \circ \theta_S$. From this and the fact that $S_n \uparrow S$ it is immediate that $a_{n,2j+1} \to a_{2j+1}$ as $n \to \infty$. This establishes (e).

Finally we can complete the proof of (5.12). Since z is a regular 1-potential there is, according to (3.14) of Chapter IV a CAF $\tilde{A}$ of X such that

$$
z(x) = E^x \int_0^\infty e^{-t} \, d\tilde{A}_t.
$$

Now $D_t = \int_0^{t \wedge T} d\tilde{A}_s$ defines a CAF of (X, T) and $E^x \int_0^T e^{-t} \, dD_t = z(x) - E^x\{e^{-T_1} z(X_{T_1})\}$. But ψ vanishes off E_G, and using (a) this last expression is $z(x) - E^x\{e^{-T_2} z(T_2)\} = \psi(x)$. Hence D and B are equivalent, and so $E^x \int_0^T f(X_t) \, dB_t = E^x \int_0^T f(X_t) \, d\tilde{A}_t$ for all x and $f \in \mathscr{E}^*_+$. This completes the proof of (5.12) because $\int_0^T f(X_t) \, dB_t = \int_0^T f(X_t) \, dA_t$ if f vanishes off K.

We will now specialize to the case in which G is a strong exit set (for $\tilde{X}$)

and A is the finite, continuous, strictly increasing on $[0, T_{G^c}]$, CAF of (X, T_{G^c}) satisfying $E^x \int_0^{T_{G^c}} f(X_t)\, dA_t = \tilde{V}f(x)$ for all x and $f \in \mathscr{E}_+^*$. With this A, pick a K as in (5.12) and let $\tilde{A}$ be the CAF of (5.12) going with this A and K. Now let H be a strong exit set (for $\tilde{X}$) and let C_t be the CAF of (X, T_{H^c}) which satisfies

$$E^x \int_0^{T_{H^c}} f(X_t)\, dC_t = \tilde{E}^x \int_0^{T_{H^c}} f(X_t)\, dt.$$

(5.18) PROPOSITION. $E^x \int_0^{T_{H^c}} f(X_t)\, dC_t = E^x \int_0^{T_{H^c}} f(X_t)\, d\tilde{A}_t$ for all x and all positive f vanishing off K.

Proof. There is no loss of generality in assuming that f vanishes off H as well. According to (5.11) we have $\int_0^t f(X_s)\, dC_s = \int_0^t f(X_s)\, dA_s$ on $[0, T]$ where T is the hitting time of $H^c \cup G^c$. Let R be the hitting time of $(K \cap H) \cup H^c$ and let

$$\begin{aligned} T_0 &= 0 \\ T_{2n+1} &= T_{2n} + T \circ \theta_{T_{2n}} \\ T_{2n+2} &= T_{2n+1} + R \circ \theta_{T_{2n+1}}. \end{aligned}$$

Clearly $T_0 \le T_1 \le \dots$, $\int_{T_{2n+1}}^{T_{2n+2}} f(X_t)\, dC_t = 0$ for all n, and furthermore $\lim T_n = T_{H^c}$ because for $n \ge 1$, $T_{2n+1} = T_{2n} + T_{G^c} \circ \theta_{T_{2n}}$ on $\{T_{2n+1} < T_{H^c}\}$ and so

$$E^x(e^{-T_{2n+1}}; T_{2n+1} < T_{H^c}) \le \eta\, E^x(e^{-T_{2n}}; T_{2n} < T_{H^c})$$

where $\eta < 1$ is the constant appearing in the definition of K; that is, $E^x\{e^{-T_{G^c}}\} < \eta$ for all $x \in K$. As a result

$$\begin{aligned} E^x \int_0^{T_{H^c}} f(X_t)\, dC_t &= \sum_{n=0}^{\infty} E^x \int_{T_{2n}}^{T_{2n+1}} f(X_t)\, dC_t \\ &= \sum_{n=0}^{\infty} E^x \left\{ E^{X(T_{2n})} \int_0^T f(X_t)\, dC_t \right\} \\ &= \sum_{n=0}^{\infty} E^x \left\{ E^{X(T_{2n})} \int_0^T f(X_t)\, dA_t \right\} \\ &= \sum_{n=0}^{\infty} E^x \left\{ E^{X(T_{2n})} \int_0^T f(X_t)\, d\tilde{A}_t \right\} \\ &= \sum_{n=0}^{\infty} E^x \int_{T_{2n}}^{T_{2n+1}} f(X_t)\, d\tilde{A}_t \\ &= E^x \int_0^{T_{H^c}} f(X_t)\, d\tilde{A}_t. \end{aligned}$$

This completes the proof of (5.18).

Given a strong exit set G we will write A^G for the CAF of (X, T_{G^c}) satisfying

(5.19)
$$E^x \int_0^{T_{G^c}} f(X_t)\, dA_t^G = \tilde{E}^x \int_0^{T_{G^c}} f(X_t)\, dt$$

for all x and $f \geq 0$. The next proposition gives the desired synthesis of the various A^G.

(5.20) Proposition. There is a CAF, A, of X such that for every strong exit set G, every x, and every $f \in \mathscr{E}_+^*$

$$E^x \int_0^{T_{G^c}} f(X_t)\, dA_t^G = E^x \int_0^{T_{G^c}} f(X_t)\, dA_t.$$

Furthermore A is strictly increasing and finite almost surely on $[0, \zeta)$.

Proof. Let $\{(K_i, G_i); i \geq 1\}$ be a family of pairs of sets, each of which satisfies these conditions: (1) G_i is a strong exit set; (2) K_i is a finely open Borel set, $K_i \subset G_i$, and for some $\eta_i < 1$, $E^x(e^{-T_{G_i^c}}) < \eta_i$ for every $x \in K_i$. Also assume that $\bigcup K_i = E$. We have seen that such a family always exists. Let $\tilde{A}^i$ be a CAF of X such that $E^x \int_0^{T_{G_i^c}} f(X_t)\, dA_t^{G_i} = E^x \int_0^{T_{G_i^c}} f(X_t)\, d\tilde{A}_t^i$ for all x and all positive f vanishing off K_i. The existence of such $\tilde{A}^i$ is the content of (5.12). Now disjoint the K_i: $J_1 = K_1$, $J_i = K_i - \bigcup_{j=1}^{i-1} K_j$, and define the CAF's, A^i, by

$$A_t^i = \int_0^t I_{J_i}(X_s)\, d\tilde{A}_s^i.$$

Finally, set

$$A_t = \sum_i A_t^i$$

so that A satisfies the measurability and additivity conditions of an AF of X. It is also clear $t \to A_t$ is everywhere left continuous and is right continuous at t_0 if $A_{t_0+\varepsilon} < \infty$ for some $\varepsilon > 0$. Moreover A is constant on $[\zeta, \infty]$. Let f be a function in $\mathscr{E}_+^*$ and let G be any strong exit set. Then $f = \sum f I_{J_i}$, $f I_{J_i}$ vanishes off K_i, and so by (5.18)

(5.21)
$$\begin{aligned} E^x \int_0^{T_{G^c}} f(X_t)\, dA_t^G &= \sum_i E^x \int_0^{T_{G^c}} f I_{J_i}(X_t)\, dA_t^G \\ &= \sum_i E^x \int_0^{T_{G^c}} f I_{J_i}(X_t)\, d\tilde{A}_t^i \\ &= \sum_i E^x \int_0^{T_{G^c}} f(X_t)\, dA_t^i \\ &= E^x \int_0^{T_{G^c}} f(X_t)\, dA_t. \end{aligned}$$

It follows from (5.21) that $A(T_{G^c}) = A^G(T_{G^c}) < \infty$ whenever G is a strong exit set and so $\lim_{t \downarrow 0} A_t = 0$. If $S = \inf\{t\colon A_{t+} > A_t\}$, then by the above discussion $A_S < \infty$ and $A_{S+\varepsilon} = \infty$ for all $\varepsilon > 0$ when $S < \infty$. But $A_{S+\varepsilon} = A_S + A_\varepsilon \circ \theta_S$ and this leads to a contradiction unless $S = \infty$. Thus $t \to A_t$ is everywhere continuous. It is now clear that $A_t = A_t^G$ on $[0, T_{G^c}]$ for any strong exit set G, and this certainly implies that $t \to A_t$ is strictly increasing on $[0, R)$ where $R = \inf\{t\colon A_t = \infty\}$. Of course in this discussion we have been omitting the qualifying phrase "almost surely."

In order to complete the proof of Proposition 5.20 it will suffice to show that $R \geq \zeta$. This requires an argument that will also be needed later on and so we formulate it as a lemma. Coming to this, let G be any *open* subset of E, $L = \bar{G}^c$, and $G_\varepsilon = \bar{G} \cap \{\tilde{\varphi}_L \geq \varepsilon\}$ where $\tilde{\varphi}_L(x) = \tilde{E}^x(e^{-T_L})$. Then G_ε^c is a finely open Borel set by (5.1 bis). If b is so large that $e^{-b} < \varepsilon/2$, then $\tilde{P}^z(T_L \leq b) \geq \varepsilon - e^{-b} > \varepsilon/2$ whenever $z \in G_\varepsilon$, and so it follows from the familiar iteration estimate that G_ε is a strong exit set for each $\varepsilon, 0 < \varepsilon < 1$. Now let $T_n(\tilde{T}_n)$ be the hitting time of $G_{1/n}^c = L \cup \{\tilde{\varphi}_L < 1/n\}$ by $X(\tilde{X})$. Then $T_n(\tilde{T}_n)$ increases to a limit $T(\tilde{T})$. If $Q(\tilde{Q})$ denotes the hitting time of $L \cup \{\tilde{\varphi}_L = 0\}$ by $X(\tilde{X})$, then obviously $T \leq Q$ and $\tilde{T} \leq \tilde{Q}$. We can now state the fact which we require. We continue to omit the phrase "almost surely."

(5.22) LEMMA. $T = Q$ and $\tilde{T} = \tilde{Q}$.

Before coming to the proof of (5.22) let us use it to complete the proof of (5.20). First observe that if $\varphi_L(x) = E^x(e^{-T_L})$, then because of the identity of the hitting distributions $\{\varphi_L = 0\} = \{\tilde{\varphi}_L = 0\}$. Now for each n, $G_{1/n}$ is a strong exit set and hence by what we have already established $A(T_n)$ is finite. Consequently $t \to A_t$ is finite on $[0, T)$. Let $R_n = \inf\{t\colon A_t \geq n\}$. Since $t \to A_t$ is continuous, $A(R_n) = n$ and $\{R_n\}$ increases to R strictly from below. Suppose that for some x, $P^x(R < \zeta) > 0$. Then by an argument similar to that which follows the proof of (5.11) we can find an open set $G \subset E$ such that $P^x[X_R \in G \cap \{\varphi_L > 0\}] > 0$ where as above $L = \bar{G}^c$. Lemma 5.22 and the ensuing discussion applies then to this open set G. Since $X_{R_n} \to X_R$ on $\{R < \zeta\}$ we must have $X_t \in G$ for all t in $[R_n, R]$ provided n is sufficiently large. If $\varphi_L(y) = 0$, then $P^y(T_L = \infty) = 1$ and so $P^y[\varphi_L(X_t) > 0] = P^y[T_L \circ \theta_t < \infty] = 0$ for all t. Hence if S is the hitting time of $\{\varphi_L = 0\}$, then $\varphi_L(X_{S+t}) = 0$ on $\{S < \infty\}$. Consequently $R < S$ if $\varphi_L(X_R) > 0$ and $R < \zeta$ and so $X_t \in G \cap \{\varphi_L > 0\}$ for all t in $[R_n, R]$ for all sufficiently large n provided $X_R \in G \cap \{\varphi_L > 0\}$. Therefore there exists an n such that $P^x[R < R_n + Q \circ \theta_{R_n}] > 0$ where Q is the hitting time of $L \cup \{\varphi_L = 0\}$. Hence $E^x\{P^{X(R_n)}(R < Q)\} > 0$, but this is a contradiction since $t \to A_t$ is finite on $[0, T)$ which, by (5.22), is the same as $[0, Q)$. Thus the proof of (5.20) will be complete as soon as we establish Lemma 5.22.

Proof of (5.22). We will first show that $\tilde{T} = \tilde{Q}$. Since L is open, Q and $\tilde{Q}$ are also the hitting times of $L^r \cup \{\tilde{\varphi}_L = 0\} = L^r \cup \{\varphi_L = 0\} = \{\varphi_L = 0 \text{ or } 1\} = \{\tilde{\varphi}_L = 0 \text{ or } 1\}$ by X and $\tilde{X}$, respectively. If $\tilde{X}(\tilde{T}_n) \in L^r \cup \{\tilde{\varphi}_L = 0\}$ for some n, then $\tilde{T}_m = \tilde{T} = \tilde{T}_n = \tilde{Q}$ for all $m \geq n$. On the other hand if $\tilde{X}(\tilde{T}_n) \notin L^r$ then $\tilde{\varphi}_L(\tilde{X}_{\tilde{T}_n}) \leq 1/n$. Since $\tilde{T} + \tilde{T}_L \circ \tilde{\theta}_{\tilde{T}} \geq \tilde{T}_n + \tilde{T}_L \circ \tilde{\theta}_{\tilde{T}n}$ we have for any x

$$\tilde{E}^x\{e^{-T}\, \tilde{\varphi}_L(X_T);\, X_{T_n} \notin L^r\} = \tilde{E}^x\{e^{-(T+T_L \circ \theta_T)};\, X_{T_n} \notin L^r\}$$

$$\leq \tilde{E}^x\{e^{-T_n}\, \tilde{\varphi}_L(X_{T_n});\, X_{T_n} \notin L^r\} \leq \frac{1}{n}.$$

Now letting $n \to \infty$ we see that $\tilde{\varphi}_L(\tilde{X}_{\tilde{T}}) = 0$ on $\{\tilde{T} < \infty, \tilde{X}_{\tilde{T}_n} \notin L^r \text{ for all } n\}$. Thus $\tilde{T} = \tilde{Q}$ almost surely $\tilde{P}^x$ for any x.

We come now to the proof that $Q = T$. Recall that $L^r = \{\varphi_L = 1\} = \{\tilde{\varphi}_L = 1\}$. Fix δ and η such that $0 < \delta < \eta < 1$ and let $D = \{x\colon \delta \leq \tilde{\varphi}_L(x) \leq \eta\}$. Let $S(\tilde{S})$ be the hitting time of $D \cup L^r \cup \{\tilde{\varphi}_L = 0\}$ by $X(\tilde{X})$. We first claim that for all x

$$\textbf{(5.23)} \qquad \tilde{E}^x\{\tilde{P}^{X(T_n)}[X_S \in D;\, Q < \infty]\} \to 0$$

as $n \to \infty$. As above if $\tilde{\varphi}_L(y) = 0$ then $\tilde{\varphi}_L(\tilde{X}_t) = 0$ for all t almost surely $\tilde{P}^y$. As a result $\tilde{X}_{\tilde{Q}}$ is regular for $L^r \cup \{\tilde{\varphi}_L = 0\}$. In particular $\tilde{Q} + \tilde{S} \circ \tilde{\theta}_{\tilde{Q}} = \tilde{Q}$. Therefore we can write

$$\begin{aligned} &\tilde{E}^x\{\tilde{P}^{X(T_n)}[X_S \in D;\, Q < \infty]\} \\ &\quad = \tilde{E}^x\{\tilde{P}^{X(T_n)}[X_S \in D;\, Q < \infty];\, T_n < Q\} \\ &\quad \leq \tilde{P}^x\{X(T_n + S \circ \theta_{T_n}) \in D;\, Q < \infty\}. \end{aligned}$$

But $t \to e^{-t}\, \tilde{\varphi}_L(\tilde{X}_t)$ is a bounded right continuous supermartingale with respect to $\tilde{P}^x$ for each x and so $t \to \tilde{\varphi}_L(\tilde{X}_t)$ has left-hand limits on $[0, \infty)$. Now $\tilde{T}_n + \tilde{S} \circ \tilde{\theta}_{\tilde{T}_n} < \tilde{Q}$ if $\tilde{X}(\tilde{T}_n + \tilde{S} \circ \tilde{\theta}_{\tilde{T}_n}) \in D$ and so the last displayed probability does not exceed

$$\tilde{P}^x\{\delta \leq \varphi_L(X_t) \leq \eta \text{ for some } t \in [T_n, Q),\, T_n < Q < \infty\},$$

and this must approach zero as $n \to \infty$ because $\tilde{T}_n \to \tilde{Q}$, $\tilde{\varphi}_L(\tilde{X}_{\tilde{T}_n}) \leq 1/n$, and $t \to \tilde{\varphi}_L(\tilde{X}_t)$ has a left-hand limit at $\tilde{Q}$ when $\tilde{Q} < \infty$. Thus (5.23) is established.

Next observe that since $\tilde{S} + \tilde{Q} \circ \tilde{\theta}_{\tilde{S}} = \tilde{Q}$ $(S + Q \circ \theta_S = Q)$ one has

$$\tilde{P}^y[X_S \in D;\, Q < \infty] = \int_D \tilde{P}_{\tilde{S}}(x, dz)\, \tilde{P}_{\tilde{Q}}(z, E_\Delta)$$

for each y. Now using the fact that X and $\tilde{X}$ have the same hitting distributions it follows from (5.23) (since Q, S, T_n are all hitting times) that

(5.24) $$\lim_n E^x\{P^{X(T_n)}[X_S \in D, Q < \infty]\} = 0$$

for all x. Now suppose that for some $x, P^x\{0 < \varphi_L(X_T) < 1\} > 0$. Since $0 < \varphi_L < 1$ if and only if $0 < \tilde{\varphi}_L < 1$ there must then exist δ and η, $0 < \delta < \eta < 1$, such that if $D = \{\delta \le \tilde{\varphi}_L \le \eta\}$, then $P^x\{X_T \in D\} > 0$. Consequently

$$\begin{aligned} E^x\{e^{-Q}; X_T \in D\} &\ge E^x\{e^{-(T + T_L \circ \theta_T)}; X_T \in D\} \\ &= E^x\{e^{-T}\, \varphi_L(X_T); X_T \in D\} > 0, \end{aligned}$$

since $T < \infty$ if $X_T \in D$; and so $P^x[Q < \infty; X_T \in D] > 0$. Finally observe that if $1/n < \delta$ then $T_n < T$ when $X_T \in D$ and hence for such n we have

$$\begin{aligned} P^x[Q < \infty; X_T \in D] &= P^x\{Q < \infty, X_T \in D, T_n < T < Q\} \\ &\le P^x\{Q \circ \theta_{T_n} < \infty, S \circ \theta_{T_n} < Q \circ \theta_{T_n}; T_n < T\} \\ &\le E^x\{P^{X(T_n)}[S < Q; Q < \infty]\} \\ &= E^x\{P^{X(T_n)}[X_S \in D; Q < \infty]\} \to 0 \end{aligned}$$

as $n \to \infty$ by (5.24). This is a contradiction and so $\varphi_L(X_T) = 0$ or 1 almost surely and this implies that $Q \le T$. Thus Lemma 5.22 and consequently Proposition 5.20 are established.

We are finally ready to complete the proof of Theorem 5.1. The CAF, A, is the one we have just constructed in Proposition 5.20 and τ is the inverse of A. The process $\hat{X}$ is of course the "time changed" process $\hat{X}_t = X_{\tau(t)}$.

(5.25) PROPOSITION. Let G be a strong exit set and $T = T_{G^c}$. Then for $f \in b\mathscr{E}^*$, $\alpha > 0$, and $x \in E$

$$E^x\{e^{-\alpha A(T)} f(X_T)\} = \bar{E}^x\{e^{-\alpha T} f(X_T)\}.$$

Proof. Let x be fixed and then denote the left and right sides of the desired equality by $\mathbf{L}(f)$ and $\mathbf{R}(f)$, respectively. Clearly it suffices to show that $\mathbf{L}(f) = \mathbf{R}(f)$ for $f \in \mathbf{C}_K$. For such an f, $t \to P_T f(X_t)$ is right continuous by (4.14) of Chapter II and so the following manipulations are easily justified:

$$\begin{aligned} P_T f(x) - \mathbf{L}(f) &= \alpha E^x \int_0^T e^{-\alpha A(t)} f(X_T)\, dA_t \\ &= \alpha E^x \int_0^T e^{-\alpha A(t)}\, P_T f(X_t)\, dA_t \\ &= \alpha W^\alpha P_T f(x) \end{aligned}$$

where, as in the discussion preceding (5.11), $\{W^\alpha; \alpha \ge 0\}$ is the resolvent of the process $X_{\tau(t)}$ terminated when it first leaves G; that is, $W^\alpha f(x) = E^x \int_0^T e^{-\alpha A(t)} f(X_t)\, dA_t$. Similarly $\tilde{P}_T f(x) - \mathbf{R}(f) = \alpha \tilde{V}^\alpha \tilde{P}_T f(x)$ where $\{\tilde{V}^\alpha\}$ is the resolvent of $(\tilde{X}, \tilde{T})$. Now $P_T = \tilde{P}_T$ by the identity of the hitting distributions, while the identity of W^α and $\tilde{V}^\alpha$ has been established in the discussion preceding (5.11). Consequently $\mathbf{L}(f) = \mathbf{R}(f)$, and the proof of (5.25) is complete.

REMARK. An important consequence of (5.25) is that the joint distribution of (A_T, X_T) under P^x is the same as the joint distribution of $(\tilde{T}, \tilde{X}_{\tilde{T}})$ under $\tilde{P}^x$. Also the equality

$$E^x \int_0^T e^{-\alpha A_t} f(X_t)\, dA_t = E^x \int_0^T e^{-\alpha t} f(X_t)\, dt \tag{5.26}$$

is simply the equality $W^\alpha = \tilde{V}^\alpha$ which we established prior to (5.11).

Proof of Theorem 5.1. We must show that for all positive $f \in \mathscr{E}^*$, $\alpha > 0$, and $x \in E$

$$E^x \int_0^\infty e^{-\alpha t} f(X_{\tau(t)})\, dt = \tilde{E}^x \int_0^\infty e^{-\alpha t} f(X_t)\, dt. \tag{5.27}$$

Of course the left side of (5.27) is simply

$$E^x \int_0^\infty e^{-\alpha A_t} f(X_t)\, dA_t,$$

and we will use the latter expression instead.

Let G be a strong exit set and K a finely open Borel subset of G such that, for some $\eta < 1$, $E^x(e^{-T_{G^c}}) < \eta$ and $\tilde{E}^x(e^{-T_{G^c}}) < \eta$ for all $x \in K$. Let $\tilde{\varphi}_K(x) = \tilde{E}^x(e^{-T_K})$ and $\varphi_K(x) = E^x(e^{-T_K})$. Pick a number $\delta > 0$ and let $J = \{\tilde{\varphi}_K \ge \delta\}$. From past discussions it is clear that $J \cap K^c$ is a strong exit set. With this δ fixed, define $R = T_{G^c \cup J^c}$, $S = T_{K \cup J^c}$, and

$$T_0 = 0$$

$$T_{2n+1} = T_{2n} + R \circ \theta_{T_{2n}}$$

$$T_{2n+2} = T_{2n+1} + S \circ \theta_{T_{2n+1}}.$$

The identity of the hitting distributions readily yields the fact that $\tilde{P}^x(T_n = T_{J^c}) = P^x(T_n = T_{J^c})$ for all x and n. On the other hand the condition on K implies that the event $\{T_n < T_{J^c}$ for all n, $\lim T_n < \infty\}$ has probability 0 under each P^x and a similar statement holds for the process $\tilde{X}$. Consequently $\lim T_n = T_{J^c}$ almost surely for X and $\tilde{X}$. We next observe that, for each $g \in b\mathscr{E}^*$, $\alpha > 0$, $x \in E$, and $n \ge 0$,

(5.28) $$E^x\{e^{-\alpha A(T_n)}g(X_{T_n})\} = \tilde{E}^x\{e^{-\alpha T_n}g(X_{T_n})\}.$$

Indeed by (5.25) this is valid when T_n is replaced by R or S, and so (5.28) follows by an easy induction, whose details we omit. Now suppose $f \in b\mathscr{E}^*$ and f vanishes off K. We will prove (5.27) for such an f. Since $f(X_t)$ vanishes for $T_{2n+1} \le t < T_{2n+2}$ we have

$$\begin{aligned} E^x \int_0^{T_{J^c}} e^{-\alpha A(t)} f(X_t)\, dA_t &= \sum_{n=0}^{\infty} E^x \int_{T_{2n}}^{T_{2n+1}} e^{-\alpha A(t)} f(X_t)\, dA_t \\ &= \sum_{n=0}^{\infty} E^x\left\{e^{-\alpha A(T_{2n})}\, E^{X(T_{2n})} \int_0^R e^{-\alpha A(t)} f(X_t)\, dA_t\right\} \\ &= \sum_{n=0}^{\infty} \tilde{E}^x\left\{e^{-\alpha T_{2n}}\, \tilde{E}^{X(T_{2n})} \int_0^R e^{-\alpha t} f(X_t)\, dt\right\} \\ &= \tilde{E}^x \int_0^{T_{J^c}} e^{-\alpha t} f(X_t)\, dt. \end{aligned}$$

Now recall that J depends on a positive constant δ. Write R_δ for the hitting time of $\{\tilde{\varphi}_K < \delta\}$ so that we already have shown

(5.29) $$E^x \int_0^{R_\delta} e^{-\alpha A(t)} f(X_t)\, dA_t = \tilde{E}^x \int_0^{R_\delta} e^{-\alpha t} f(X_t)\, dt.$$

As δ decreases to 0, R_δ increases to a stopping time R, and, arguing as in the proof of Lemma 5.22 one sees that R is the hitting time of $\{\tilde{\varphi}_K = 0\} = \{\varphi_K = 0\}$. By monotone convergence and the fact that A is continuous we may replace R_δ by R in (5.29). But then we may replace R by ∞ in (5.29) because f vanishes off K and neither process hits K at a time exceeding R. In other words, we have proved (5.27) for any f vanishing off such a K. But this establishes (5.27), for E can be covered with a countable family of sets such as K.

The proof of (5.1) is complete.

Exercises

(5.30) Prove (5.1 bis). [Hint: if $\Lambda = \bigcup_{r<t, r\in\mathbf{Q}} \{X_r \in B\}$ show that $P^x[T_B < t] = P^x(\Lambda)$ for all x.]

VI

DUAL PROCESSES AND POTENTIAL THEORY

1. Dual Processes

In this chapter we will study Markov processes for which an "adjoint" or "dual" Markov process exists. This additional structure will allow us to introduce and study some notions from classical potential theory. In particular we will define potentials of measures and study their relationship with potentials of additive functionals. As before $X = (\Omega, \mathcal{M}, \mathcal{M}_t, X_t, \theta_t, P^x)$ will be a fixed standard process with state space $(E, \mathscr{E})$. In the first part of this section only the potential operators U^α associated with X will play a role.

(1.1) DEFINITION. A family $\{V^\alpha; \alpha > 0\}$ of positive linear operators from $b\mathscr{E}^*$ to $b\mathscr{E}^*$ is called a *resolvent* if it satisfies:

(i) $\|V^\alpha\| \leq 1/\alpha$, $\alpha > 0$;

(ii) $V^\alpha f(x)$ is a measure in f for each $x \in E$ and $\alpha > 0$;

(iii) $V^\alpha - V^\beta = (\beta - \alpha) V^\alpha V^\beta$, for all $\alpha, \beta > 0$.

Of course Conditions (i) and (iii) are simply Conditions (4.2i) and (4.2iii) from Chapter III. Condition (ii) is satisfied by any resolvent subordinate to the resolvent $\{U^\alpha; \alpha > 0\}$ of X.

Given any resolvent $\{V^\alpha; \alpha > 0\}$, not necessarily subordinate to U^α, we define $\alpha - V$ *supermedian* and $\alpha - V$ *excessive* functions as in (4.5) of Chapter III. The reader will have no difficulty in checking that the propositions of Section 4, Chapter III, most notably Proposition 4.6, are valid in the present more general situation. We will use them without further elaboration.

One fact not explicitly mentioned in Chapter III is that f is $\alpha - V$ excessive if and only if it is $\beta - V$ excessive for all $\beta > \alpha$.

(1.2) Definition. Let $\{U^\alpha; \alpha > 0\}$ and $\{\hat{U}^\alpha; \alpha > 0\}$ be any two resolvents on $b\mathscr{E}^*$ and let ξ be a σ-finite measure on $\mathscr{E}^*$. Then we say that $\{U^\alpha\}$ and $\{\hat{U}^\alpha\}$ are *in duality relative to* ξ provided that for each $\alpha > 0$ there is a nonnegative function $u^\alpha \in \mathscr{E}^* \times \mathscr{E}^*$ with the following properties:

(i) the function $x \to u^\alpha(x, y)$ is α-excessive relative to the resolvent $\{U^\alpha\}$ for each $y \in E$ and $\alpha > 0$;

(ii) the function $y \to u^\alpha(x, y)$ is α-excessive relative to the resolvent $\{\hat{U}^\alpha\}$ for each $x \in E$ and $\alpha > 0$;

(iii) $U^\alpha f(x) = \int u^\alpha(x, y) f(y)\, \xi(dy)$ and $\hat{U}^\alpha f(y) = \int u^\alpha(x, y) f(x)\, \xi(dx)$ for all $\alpha > 0$, $f \in b\mathscr{E}^*$ and x and y in E.

Remark. As the notation suggests, $\{U^\alpha; \alpha > 0\}$ usually will be the resolvent of a standard process X. However we will not introduce this assumption for the moment.

Suppose that $\{U^\alpha\}$ and $\{\hat{U}^\alpha\}$ are in duality relative to ξ. In most discussions the measure ξ will be fixed and expressions like "almost everywhere" are to be regarded as defined relative to ξ. Frequently in writing integrals against ξ we will write dx in place of $\xi(dx)$. We will write $\langle f, g\rangle$ for $\int f(x)\, g(x)\, dx$. The functions u^α are by definition nonnegative but they may perhaps take on the value ∞. We will write $f\hat{U}^\alpha$ rather than $\hat{U}^\alpha f$ for the action of the operator $\hat{U}^\alpha$ on the function f, and we will write $\hat{U}^\alpha(A, x)$ for the measures associated with these operators. Thus we have the expressions

$$f\,\hat{U}^\alpha(y) = \int f(x)\, \hat{U}^\alpha(dx, y) = \int f(x)\, u^\alpha(x, y)\, dx.$$

If $G(x, y)$ is a positive or bounded function in $\mathscr{E}^* \times \mathscr{E}^*$ we write $G\,\hat{U}^\alpha(x, y)$ for $\int G(x, z)\, \hat{U}^\alpha(dz, y)$ and $U^\alpha G(x, y)$ for $\int U^\alpha(x, dz)\, G(z, y)$. With this guide the reader should have no trouble unscrambling similar notation not specifically mentioned here.

(1.3) Proposition. Let $\{V^\alpha\}$ be a resolvent and θ a measure such that $V^\alpha(x, \cdot)$ is absolutely continuous with respect to θ for each $\alpha > 0$ and $x \in E$. If f and g are $\alpha - V$ excessive and $f = g$ a.e. θ, then f and g are identical.

Proof. The hypotheses imply that $\beta V^{\alpha+\beta} f = \beta V^{\alpha+\beta} g$ for each $\beta > 0$. The conclusion then follows from Definition 4.5 of Chapter III.

An obvious consequence of (1.3) is that the functions u^α appearing in Definition 1.2 are uniquely determined.

Let $\{U^\alpha\}$ and $\{\hat{U}^\alpha\}$ be in duality relative to ξ. We will write $\mathscr{S}^\beta$ ($\hat{\mathscr{S}}^\beta$) for the class of all functions which are β-excessive relative to $\{U^\alpha\}$ ($\{\hat{U}^\alpha\}$). By (4.6) of Chapter III, $U^\alpha f \in \mathscr{S}^\alpha$ and $f\hat{U}^\alpha \in \hat{\mathscr{S}}^\alpha$ whenever $f \in \mathscr{E}^*_+$. If μ is a (nonnegative) measure on $\mathscr{E}^*$, then we define $U^\alpha\mu(x) = \int u^\alpha(x, y)\, \mu(dy)$ and $\mu\hat{U}^\alpha(y) = \int \mu(dx)\, u^\alpha(x, y)$. Note that $\beta U^{\alpha+\beta} U^\alpha \mu(x) = \int \beta U^{\alpha+\beta} u^\alpha(x, y)\, \mu(dy)$ and since $x \to u^\alpha(x, y)$ is in $\mathscr{S}^\alpha$ for each y it is clear that $U^\alpha\mu \in \mathscr{S}^\alpha$. Similarly $\mu\hat{U}^\alpha \in \hat{\mathscr{S}}^\alpha$.

We will now give a condition which guarantees that a pair of resolvents $\{U^\alpha\}$ and $\{\hat{U}^\alpha\}$ are in duality relative to a given measure ξ.

(1.4) THEOREM. Let $\{U^\alpha\}$ and $\{\hat{U}^\alpha\}$ be resolvents on $b\mathscr{E}^*$ and let ξ be a σ-finite measure on $\mathscr{E}^*$. Then $\{U^\alpha\}$ and $\{\hat{U}^\alpha\}$ are in duality relative to ξ if and only if the following two conditions are satisfied: (a) The measures $U^\alpha(x, \cdot)$ and $\hat{U}^\alpha(\cdot, x)$ are absolutely continuous with respect to ξ for each $\alpha > 0$ and $x \in E$, (b) $\langle f, U^\alpha g\rangle = \langle f\hat{U}^\alpha, g\rangle$ for each $\alpha > 0$ and all $f, g \in \mathscr{E}^*_+$. Moreover if (a) and (b) hold then (in the notation of Definition 1.2) for $0 < \alpha \le \beta$ we have

(1.5)
$$u^\alpha(x, y) = u^\beta(x, y) + (\beta - \alpha)U^\alpha u^\beta(x, y)$$
$$= u^\beta(x, y) + (\beta - \alpha)u^\beta \hat{U}^\alpha(x, y).$$

Proof. If $\{U^\alpha\}$ and $\{\hat{U}^\alpha\}$ are in duality relative to ξ, then the validity of (a) and (b) are immediate. Furthermore, in this event the resolvent equation $U^\alpha = U^\beta + (\beta - \alpha)U^\alpha U^\beta = U^\beta + (\beta - \alpha)U^\beta U^\alpha$ implies that for each x the first equation in (1.5) is valid for almost all y (relative to ξ, of course). But each term in this equation is, as a function of y, in $\hat{\mathscr{S}}^\beta$. Consequently, by Proposition 1.3, the first equality in (1.5) is valid for all y. The second equality follows in a similar manner.

So now we turn our attention to proving that conditions (a) and (b) imply duality. As before we write $\hat{U}^\alpha(A, x)$ for the measures associated with $\hat{U}^\alpha$ and use the notation described above. By hypothesis, $U^\alpha(x, \cdot)$ is absolutely continuous with respect to ξ and, for each $A \in \mathscr{E}$, $U^\alpha(x, A)$ is in $\mathscr{E}^*_+$ as a function of x. Since $\mathscr{E}$ is countably generated we can find a nonnegative function $w^\alpha \in \mathscr{E}^* \times \mathscr{E}^*$ such that

(1.6)
$$U^\alpha f(x) = \int w^\alpha(x, y) f(y)\, dy$$

for all $f \in \mathscr{E}^*_+$ (see the proof of (2.3) of Chapter III). For the moment let us fix α and x and denote by w the function $y \to w^\alpha(x, y)$. Using Condition (b) and (1.6) we obtain for any $f \in \mathscr{E}^*_+$

$$\langle w\hat{U}^{\alpha+\beta}, f\rangle = \langle w, U^{\alpha+\beta} f\rangle = U^\alpha U^{\alpha+\beta} f(x).$$

By the resolvent equation $U^\alpha U^{\alpha+\beta} f \le \beta^{-1} U^\alpha f$. Thus $\beta\langle w\hat{U}^{\alpha+\beta}, f\rangle \le U^\alpha f(x)$ for all $f \in \mathscr{E}_+^*$, and this implies that

$$\beta w^\alpha\, \hat{U}^{\alpha+\beta}(x, z) \le w^\alpha(x, z) \tag{1.7}$$

for almost all z. The measure $\hat{U}^{\alpha+\gamma}(\cdot, y)$ is absolutely continuous with respect to ξ and so it follows from (1.7) that

$$\beta w^\alpha \hat{U}^{\alpha+\gamma}\, \hat{U}^{\alpha+\beta}(x, y) = \beta w^\alpha \hat{U}^{\alpha+\beta}\, \hat{U}^{\alpha+\gamma}(x, y) \le w^\alpha\, \hat{U}^{\alpha+\gamma}(x, y) \tag{1.8}$$

for all x and y. Thus the function $y \to w^\alpha \hat{U}^{\alpha+\gamma}(x, y)$ is α supermedian relative to the resolvent $\{\hat{U}^\alpha\}$, and so, according to (4.6) of Chapter III, as $\beta \to \infty$ the left side of (1.8) increases to a limit $L_\gamma(x, y)$ which is in $\hat{\mathscr{S}}^\alpha$ as a function of y and is $\mathscr{E}^* \times \mathscr{E}^*$ measurable as a function of (x, y). Let $f \in \mathscr{E}_+^*$ and suppose f is bounded. Then applying hypothesis (b) and the resolvent equation we obtain

$$\begin{aligned}\langle \beta w \hat{U}^{\alpha+\gamma} \hat{U}^{\alpha+\beta}, f\rangle &= \langle w, \beta U^{\alpha+\gamma} U^{\alpha+\beta} f\rangle \\ &= \left\langle w, \frac{\beta}{\beta-\gamma}\,(U^{\alpha+\gamma} f - U^{\alpha+\beta} f)\right\rangle,\end{aligned}$$

and so letting $\beta \to \infty$

$$\langle L_\gamma, f\rangle = \langle w, U^{\alpha+\gamma} f\rangle$$

where L_γ is the function $y \to L_\gamma(x, y)$ with x fixed. Using the definition of w we see that

$$\langle \gamma L_\gamma, f\rangle = \gamma U^\alpha U^{\alpha+\gamma} f(x) = U^\alpha f(x) - U^{\alpha+\gamma} f(x), \tag{1.9}$$

and so $\gamma_1 L_{\gamma_1}(x, y) \le \gamma_2 L_{\gamma_2}(x, y)$ for almost all y if $\gamma_1 \le \gamma_2$. But then this inequality is valid for all y by the fact that $y \to L_\gamma(x, y)$ is in $\hat{\mathscr{S}}^\alpha$. Define the function u^α by

$$u^\alpha(x, y) = \lim_{\gamma\to\infty} \gamma\, L_\gamma(x, y).$$

As a function of y, $u^\alpha(x, y)$ is in $\hat{\mathscr{S}}^\alpha$ since it is an increasing limit of functions in $\hat{\mathscr{S}}^\alpha$. Also it follows from (1.9) that

$$\int u^\alpha(x, y) f(y)\, dy = U^\alpha f(x)$$

for all $f \in \mathscr{E}_+^*$.

We have now defined the functions u^α on $E \times E$ for all $\alpha > 0$. All that remains is to check that $x \to u^\alpha(x, y)$ is in $\mathscr{S}^\alpha$ and that this function is a density relative to ξ for the measure $\hat{U}^\alpha(\cdot, y)$. Let f and g be in $\mathscr{E}_+^*$. By what we have already proved, and Fubini's theorem,

$$\langle f\hat{U}^\alpha, g\rangle = \langle f, U^\alpha g\rangle = \int\left\{\int dx\, f(x)\, u^\alpha(x, y)\right\} g(y)\, dy,$$

and so for each f

$$f\hat{U}^\alpha(y) = \int dx\, f(x)\, u^\alpha(x, y)$$

for almost all y. Operating on each side of this relationship with $\hat{U}^{\alpha+\beta}$ we obtain

$$\beta f\hat{U}^\alpha\, \hat{U}^{\alpha+\beta}(y) = \beta \int dx\, f(x)\, u^\alpha\, \hat{U}^{\alpha+\beta}(x, y)$$

for all y. Now $f\hat{U}^\alpha \in \mathscr{S}^\alpha$ and $y \to u^\alpha(x, y) \in \mathscr{S}^\alpha$ for all x and so letting $\beta \to \infty$ we have by (4.6) of Chapter III and the monotone convergence theorem

$$f\hat{U}^\alpha(y) = \int dx\, f(x)\, u^\alpha(x, y)$$

for all y, and of course also for all α. Thus for each y and α, $x \to u^\alpha(x, y)$ is the desired density function. Finally, for each y we have that for all x

$$\beta u^\alpha\, \hat{U}^{\alpha+\beta}(x, y) = \beta \int u^\alpha(x, z)\, u^{\alpha+\beta}(z, y)\, dz;$$

that is, $x \to \beta u^\alpha\, \hat{U}^{\alpha+\beta}(x, y)$ is $U^\alpha f(x)$ where $f(z) = \beta\, u^{\alpha+\beta}(z, y)$. Hence it is in $\mathscr{S}^\alpha$. But as $\beta \to \infty$ this increases to $u^\alpha(x, y)$ and so $x \to u^\alpha(x, y)$ is, for each y, in $\mathscr{S}^\alpha$. This completes the proof of Theorem 1.4.

We now assume that $\{U^\alpha\}$ is the resolvent of our fixed standard process X and that $\{U^\alpha\}$ and a resolvent $\{\hat{U}^\alpha\}$ are in duality relative to a measure ξ. The functions u^α are those of Definition 1.2.

Before coming to the main point of this section we will make a few observations based on what we have already derived. These will be used later on without special reference. First of all, by (1.3) of this chapter, and (1.2) of Chapter V, X has a reference measure. Hence by (1.4) of Chapter V for each α and y the function $x \to u^\alpha(x, y)$ is Borel measurable. If $\alpha \leq \beta$ then by (1.5) we have $u^\alpha(x, y) \geq u^\beta(x, y)$. As β decreases to α, $u^\beta(x, y)$ increases to a function $\varphi(x, y)$ and for each x we have $u^\alpha(x, y) = \varphi(x, y)$ for almost all y because by monotone convergence $\int \varphi(x, y) f(y)\, dy$ is equal to $U^\alpha f(x)$. But this implies that u^α and φ are identical because each is in $\mathscr{S}^\alpha$ as a function of y. Another application of (1.5) shows that if β increases to α then $u^\beta(x, y)$ decreases to $u^\alpha(x, y)$ for all (x, y) such that $u^\beta(x, y) < \infty$ for some $\beta < \alpha$. As α decreases to 0, $u^\alpha(x, y)$ increases to a function $u(x, y)$ which is excessive in x for each y and 0-excessive, relative to $\{\hat{U}^\alpha\}$, in y for each x. In addition we have by monotone convergence

$$Uf(x) = E^x \int_0^\infty f(X_t)\, dt = \int u(x, y) f(y)\, dy \quad \text{for } f \in \mathscr{E}_+^*.$$

Of course it is entirely possible that u is identically infinite, but if Uh is finite for some strictly positive h then for each x, $u(x, y) < \infty$ for almost all y. The transformation $\hat{U} = \hat{U}^0$ can always be defined by monotoneity also—$f\hat{U} = \lim_{\alpha \to 0} f\hat{U}^\alpha$ for $f \in \mathscr{E}_+^*$; the relationship $\langle f\hat{U}, h\rangle = \langle f, Uh\rangle$ shows that if Uh is finite for some strictly positive h, then $f\hat{U}$ is almost everywhere finite for some strictly positive f. For each y and $\alpha > 0$, $u^\alpha(x, y) < \infty$ for almost all x; but by (3.5) of Chapter II this implies that $\{x: u^\alpha(x, y) = \infty\}$ is polar. This observation will be used many times. Finally, note that (1.5) was stated for $\alpha \leq \beta$. This was done merely to ensure that the sum is defined. Of course it is valid for any α and β provided the sums on the right side are defined. In particular for any α and β we have

$$U^\alpha u^\beta(x, y) = u^\beta \hat{U}^\alpha(x, y)$$

for any (x, y) such that $u^\gamma(x, y) < \infty$ where $\gamma = \alpha \vee \beta$.

Given a measure μ on $\mathscr{E}$ we write μP_t for the measure

$$\mu P_t(A) = \int \mu(dx) P_t(x, A)$$

and μU^α for the measure

$$\mu U^\alpha(A) = \int \mu(dx)\, U^\alpha(x, A) = \int \mu(dx) \int_A u^\alpha(x, y)\, dy.$$

Warning. Distinguish carefully between the *measure* μU^α and the *function* $U^\alpha \mu(x) = \int u^\alpha(x, y)\, \mu(dy)$.

(1.10) DEFINITION. A measure μ on $\mathscr{E}$ is α-excessive if μ is σ-finite and $\mu P_t^\alpha \leq \mu$ for all $t \geq 0$.

Note that this definition makes sense for any standard process and in no way depends upon the existence of the dual resolvent $\{\hat{U}^\alpha\}$.

If μ is α-excessive then obviously μP_t^α increases setwise as $t \downarrow 0$ to a measure ν with $\nu \leq \mu$. We will show that $\nu = \mu$. First note that, for $\beta > 0$, $\mu U^{\alpha+\beta}(A) \leq \int_0^\infty e^{-\beta t}\, \mu(A)\, dt$, and so if $\mu(A) < \infty$ then $\mu U^{\alpha+\beta}(A) < \infty$. Fix such an A. Then for any $\beta > 0$ and $\varepsilon > 0$ we have

$$\mu\{x: U^{\alpha+\beta}(x, A) > \varepsilon\} \leq (\beta\varepsilon)^{-1}\, \mu(A) < \infty.$$

Let $\{A_n\}$ be an increasing sequence of sets whose union is E and such that $\mu(A_n) < \infty$ for all n, and let $B_n = \{x: U^{\alpha+\beta}(x, A_n) > 1/n\}$. Then each B_n is a finely open Borel set, $\bigcup B_n = E$ and $\mu(B_n) < \infty$. If f is any positive continuous function then $\liminf_{t\downarrow 0} P_t^\alpha(f I_{B_n}) \geq f I_{B_n}$ and so

$$\begin{aligned} \nu(f I_{B_n}) &= \lim_{t\downarrow 0} \mu P_t^\alpha(f I_{B_n}) \\ &= \lim_{t\downarrow 0} \mu(P_t^\alpha(f I_{B_n})) \geq \mu(f I_{B_n}). \end{aligned}$$

Consequently $\mu = \nu$ on B_n and so $\mu = \nu$ on E. The reader will verify easily that, given a measure η, the measure ηU^β is β-excessive if it is σ-finite. It follows then from arguments like those in Section 2 of Chapter II that a σ-finite measure μ is α-excessive if and only if $\beta\mu U^{\beta+\alpha} \leq \mu$ for all $\beta > 0$; in particular, if μ is α-excessive then $\beta\mu U^{\beta+\alpha}$ increases setwise to μ as $\beta \to \infty$.

The next proposition establishes an important relationship between excessive measures and excessive functions.

(1.11) PROPOSITION. A measure μ is α-excessive if and only if $\mu(dy) = f(y)\,dy$ for some $f \in \mathscr{S}^\alpha$ which is almost everywhere finite.

Proof. Given such an f, the measure μ is certainly σ-finite. Moreover the relationship

$$\beta\mu U^{\beta+\alpha}(A) = \int_A \left\{\int \beta f(x)\, u^{\beta+\alpha}(x, y)\, dx\right\} dy$$

shows that $\beta\mu U^{\beta+\alpha}$ is absolutely continuous, with derivative $\beta f \hat{U}^{\beta+\alpha}$. Since $f \in \mathscr{S}^\alpha$ this implies that $\beta\mu U^{\beta+\alpha} \leq \mu$; that is, μ is α-excessive. Conversely suppose we are given an α-excessive μ. Given $\beta > 0$ define the function f_β by

$$f_\beta(y) = \beta \int \mu(dx)\, u^{\alpha+\beta}(x, y).$$

Clearly f_β is a density with respect to ξ for the measure $\beta\mu U^{\alpha+\beta}$ and so f_β is finite almost everywhere. It is also evident that $f_\beta \in \mathscr{S}^{\alpha+\beta}$. We next assert that, in fact, $f_\beta \in \mathscr{S}^\alpha$ for each $\beta > 0$. Indeed, given $\lambda > 0$ and any y such that $f_\beta(y) < \infty$ it is the case that $u^{\alpha+\beta}(x, y) < \infty$ for almost all x relative to μ and so, in particular, $u^{\alpha+\beta}\, \hat{U}^{\alpha+\lambda}(x, y) = U^{\alpha+\lambda}\, u^{\alpha+\beta}(x, y)$ for this y and almost all x (μ). Consequently, since $\lambda\mu U^{\alpha+\lambda}$ increases to μ as $\lambda \uparrow \infty$,

$$\begin{aligned}
\lambda f_\beta\, \hat{U}^{\alpha+\lambda}(y) &= \lambda\beta \int \mu(dx)\, u^{\alpha+\beta}\, \hat{U}^{\alpha+\lambda}(x, y) \\
&= \lambda\beta \int \mu(dx)\, U^{\alpha+\lambda}\, u^{\alpha+\beta}(x, y) \\
&\uparrow \beta \int \mu(dx)\, u^{\alpha+\beta}(x, y) = f_\beta(y) \quad \text{as} \quad \lambda \uparrow \infty.
\end{aligned}$$

In particular, for each $\lambda > 0$, $\lambda f_\beta \hat{U}^{\alpha+\lambda} \leq f_\beta$ almost everywhere, and hence everywhere since both of these functions are in $\mathscr{S}^\gamma$ where $\gamma = \max(\alpha + \lambda, \alpha + \beta)$. Consequently f_β is $\alpha - \hat{U}$ supermedian and so $g_\beta = \lim_{\lambda\to\infty} \lambda f_\beta \hat{U}^{\alpha+\lambda}$ is in $\mathscr{S}^\alpha$. But by the above computation $g_\beta = f_\beta$ almost everywhere, and hence everywhere. Thus $f_\beta \in \mathscr{S}^\alpha$. Since $\beta\mu U^{\alpha+\beta}$ increases with β it follows that $f_{\beta_1} \leq f_{\beta_2}$ almost everywhere if $\beta_1 \leq \beta_2$. But the inequality is then valid everywhere because $f_\beta \in \mathscr{S}^\alpha$. Finally as $\beta \to \infty$, f_β increases to a function $f \in \mathscr{S}^\alpha$. Since $\beta\mu U^{\beta+\alpha}$ increases to μ it is obvious that $\mu(dy) = f(y)\,dy$. This completes the proof.

(1.12) Corollary. The measure ξ is excessive.

Proof. $\xi(dy) = 1\,dy$ where 1 denotes the function on E identically equal to 1. Of course $\beta 1\hat{U}^\beta \le 1$ and increases as $\beta \to \infty$ to a function h which is excessive relative to $\{\hat{U}^\alpha\}$. Now if f is a positive bounded continuous function on E then $\lim_{\beta\to\infty} \beta U^\beta f = f$. Consequently using Fatou's lemma we obtain

$$\langle h, f\rangle = \lim_{\beta\to\infty} \langle \beta 1 \hat{U}^\beta, f\rangle = \lim_{\beta\to\infty} \int \beta U^\beta f(x)\,dx$$

$$\ge \int f(x)\,dx,$$

and so $h = 1$ almost everywhere. Thus $\xi(dy) = h(y)\,dy$ and so by (1.11), ξ is excessive.

(1.13) Remark. It is now evident that ξ itself is a reference measure for X. Indeed, on the one hand if $\xi(A) = 0$ then $U^\alpha(x, A) = \int_A u^\alpha(x, y)\,\xi(dy) = 0$ for all x, while on the other hand if A is of potential zero then $0 = \beta\xi U^\beta(A) \uparrow \xi(A)$.

Before proceeding let us repeat that our basic data is a standard process X with resolvent $\{U^\alpha\}$ and a resolvent $\{\hat{U}^\alpha\}$ which is in duality with $\{U^\alpha\}$ relative to a measure ξ. We are now going to assume that $\{\hat{U}^\alpha\}$ *is also the resolvent of a standard process* $\hat{X} = (\hat{\Omega}, \hat{\mathscr{M}}, \hat{\mathscr{M}}_t, \hat{X}_t, \hat{\theta}_t, \hat{P}^x)$ *with the same state space* $(E, \mathscr{E})$. This assumption will be in force throughout the rest of the chapter. The processes X and $\hat{X}$ are said to be in duality relative to ξ. Potential-theoretic objects defined in terms of $\hat{X}$ will be designated by the prefix co-. For example a function $f \in \mathscr{S}^\alpha$ will now also be called α-coexcessive and the function $g\hat{U}^\alpha$ will be called the α-copotential of g. We will denote by $\{\hat{P}_t\}$ the semigroup of transition operators for $\hat{X}$, and, in keeping with our previous notation, will write $f\hat{P}_t$ for the action of $\hat{P}_t$ on f, and, $\hat{P}_t(A, x)$ for the associated measures. If A is a set, we will write $\hat{T}_A$ for the time that $\hat{X}$ hits A and (when A is nearly Borel relative to $\hat{X}$) $\hat{P}_A(\Gamma, x) = \hat{P}^x[\hat{X}(\hat{T}_A) \in \Gamma,\ \hat{T}_A < \infty]$ for the associated hitting measure. For example, to say that a Borel set A is copolar is the same as saying $\hat{P}^x(\hat{T}_A < \infty) = 0$ for all $x \in E$. With this much introduction the reader should have no difficulty interpreting notation not specifically mentioned here. Finally, in keeping with our previous convention, for each α we extend $u^\alpha(x, y)$ to $E_\Delta \times E_\Delta$ by setting it equal to zero when either argument equals Δ.

(1.14) REMARK. It is of course possible to impose restrictions on the resolvent $\{\hat{U}^\alpha\}$ which will guarantee the existence of a corresponding standard process $\hat{X}$. For example if $f\hat{U}^\alpha \in \mathbf{C}_0(E)$ for every $\alpha > 0$ and $f \in \mathbf{C}_0(E)$ and if also $\alpha f\hat{U}^\alpha \to f$ pointwise as $\alpha \to \infty$ for every $f \in \mathbf{C}_0(E)$, then the Hille–Yosida theorem yields a strongly continuous semigroup $\{\hat{P}_t\}$ of operators on $\mathbf{C}_0(E)$ such that $f\hat{U}^\alpha = \int_0^\infty f\hat{P}_t\, e^{-\alpha t}\, dt$. The discussion in Section 9 of Chapter I then shows that $\{\hat{P}_t\}$ is the semigroup of transition operators of a standard process $\hat{X}$. The process is in fact quasi-left-continuous on $[0, \infty)$. However we will postpone imposing further restrictions on the resolvents until the need arises.

Clearly, either process, X or $\hat{X}$, may be regarded as basic. Consequently some of the statements we have already made have a valid dual version. For example, (1.11) may be extended to read "a measure μ is α-coexcessive if and only if $\mu(dy) = f(y)\, dy$ for some $f \in \mathscr{S}^\alpha$ where f is almost everywhere finite." In particular the measure ξ is coexcessive. Also for each x and α the function $u^\alpha(x, y)$ is Borel measurable in y because ξ is a reference measure for $\hat{X}$. Finally let us mention that, for each x and $\alpha > 0$, $\{y: u^\alpha(x, y) = \infty\}$ is copolar.

Recall that if μ is a measure we have defined the *function* $U^\alpha\mu(x) = \int u^\alpha(x, y)\, \mu(dy)$, and that $U^\alpha\mu \in \mathscr{S}^\alpha$. Plainly $U^\alpha\mu$ is just the density of the *measure* $\hat{U}^\alpha\mu(\Gamma) = \int \hat{U}^\alpha(\Gamma, x)\, \mu(dx)$ relative to ξ. In particular if $U^\alpha\mu$ is almost everywhere finite then $\hat{U}^\alpha\mu$ is α-coexcessive. Similarly the *function* $\mu\hat{U}^\alpha$ is the density of the *measure* μU^α relative to ξ. Because of this duality between excessive measures for one of our processes and excessive functions for the other we will in the future discuss only excessive functions. The following proposition is essentially (6.22a) of Chapter III. However we will give a proof since it is of basic importance in this chapter.

(1.15) PROPOSITION. Let μ be a measure. If $U^\alpha\mu$ is almost everywhere finite, then $U^\alpha\mu$ determines μ.

Proof. Suppose μ and ν are measures such that, for some α, $U^\alpha\mu = U^\alpha\nu < \infty$ a.e. The second equality in (1.5) then shows that $U^\beta\mu = U^\beta\nu$, a.e., and hence everywhere, for all $\beta \geq \alpha$. Let $\beta > 0$ be such that $U^\beta\mu = U^\beta\nu < \infty$ a.e., and let h be a strictly positive bounded function such that $\langle h, U^\beta\mu\rangle < \infty$. Note that if $g = h\hat{U}^\beta$ then g is strictly positive on E, and $\mu(g) = \langle h, U^\beta\mu\rangle < \infty$. Now suppose f is a continuous function on E with $0 \leq f \leq 1$. Then for every $\gamma > 0$

$$\begin{aligned}\mu[\gamma(fg)\, \hat{U}^{\beta+\gamma}] &= \langle \gamma fg, U^{\beta+\gamma}\mu\rangle = \langle \gamma fg, U^{\beta+\gamma}\nu\rangle \\ &= \nu[\gamma(fg)\, \hat{U}^{\beta+\gamma}].\end{aligned}$$

Since $g \in \mathscr{S}^\beta$ it follows that fg is cofinely continuous and so as $\gamma \to \infty$, $\gamma(fg)\,\hat{U}^{\beta+\gamma} \to fg$. Furthermore $\gamma(fg)\,\hat{U}^{\beta+\gamma} \le \gamma g \hat{U}^{\beta+\gamma} \le g$, so letting $\gamma \to \infty$ and applying the dominated convergence theorem we obtain $\mu(fg) = \nu(fg)$ for all bounded continuous f. This implies that the finite measures $g\mu$ and $g\nu$ are the same, and so $\mu = \nu$ since $g > 0$.

We come now to the key technical fact of this chapter.

(1.16) THEOREM. Let A be a Borel subset of E_Δ. Then for each $\alpha \ge 0$, $P_A^\alpha u^\alpha(x, y) = u^\alpha \hat{P}_A^\alpha(x, y)$ for all $(x, y) \in E_\Delta \times E_\Delta$.

Proof. By monotoneity it suffices to consider the case $\alpha > 0$. Let $\alpha > 0$ be fixed. Then we must show that

$$\textbf{(1.17)} \qquad \int P_A^\alpha(x, dz)\, u^\alpha(z, y) = \int u^\alpha(x, z)\, \hat{P}_A^\alpha(dz, y)$$

for all (x, y) and all Borel sets A. First note that both sides of (1.17) reduce to zero if either x or y equals Δ. Next let G be an open subset of E_Δ and let f be a bounded nonnegative Borel function which vanishes off G. Then $P_G^\alpha U^\alpha f = U^\alpha f$. If $g \in b\mathscr{E}$ then

$$\langle g, P_G^\alpha U^\alpha f\rangle = \langle g, U^\alpha f\rangle = \langle g\hat{U}^\alpha, f\rangle.$$

But $\hat{P}_G^\alpha(\cdot, x) = \varepsilon_x$ if x is in G and so $g\hat{U}^\alpha \hat{P}_G^\alpha = g\hat{U}^\alpha$ on G. Consequently since f vanishes off G, $\langle g\hat{U}^\alpha, f\rangle = \langle g\hat{U}^\alpha \hat{P}_G^\alpha, f\rangle$. As a result

$$\int \left\{\int P_G^\alpha(x, dz)\, u^\alpha(z, y)\right\} f(y)\, dy = \int \left\{\int u^\alpha(x, z)\, \hat{P}_G^\alpha(dz, y)\right\} f(y)\, dy$$

for almost all x, and hence for all x since both sides are α-excessive functions of x. Consequently for each fixed x, (1.17) (with $A = G$) holds for almost all y in $G \cap E$, and hence for all y in G since both sides are α-coexcessive functions of y and G is open. That is $P_G^\alpha u^\alpha(x, y) = u^\alpha \hat{P}_G^\alpha(x, y)$ on $E_\Delta \times G$, and by duality this also holds on $G \times E_\Delta$. For a fixed y let $v(x) = P_G^\alpha u^\alpha(x, y)$. Then $v(x) = u^\alpha \hat{P}_G^\alpha(x, y)$ on $G \cup G^r$. But $P_G^\alpha P_G^\alpha = P_G^\alpha$ and $\hat{P}_G^\alpha \hat{P}_G^\alpha = \hat{P}_G^\alpha$ since G is open, and so for each fixed x

$$\begin{aligned} v(x) = P_G^\alpha\, v(x) &= \int P_G^\alpha(x, dz)\, u^\alpha \hat{P}_G^\alpha(z, y) \\ &= \int P_G^\alpha\, u^\alpha(x, z)\, \hat{P}_G^\alpha(dz, y). \end{aligned}$$

However for fixed x, $P_G^\alpha u^\alpha(x, z) = u^\alpha \hat{P}_G^\alpha(x, z)$ for z in G and hence also if z is coregular for G since both expressions are α-coexcessive in z. Now the measure $\hat{P}_G^\alpha(\cdot, y)$ is carried by G and the points coregular for G, and combining these observations this last integral reduces to $u^\alpha \hat{P}_G^\alpha(x, y)$. Thus we have proved (1.17) for open sets A.

In the general case it clearly suffices to show that

$$\iint g(x)\, P_A^\alpha\, u^\alpha(x, y) f(y)\, dx\, dy = \iint g(x)\, u^\alpha \hat{P}_A^\alpha(x, y) f(y)\, dx\, dy$$

whenever f and g are nonnegative bounded functions in $\mathscr{E}^*$ with $\xi(f)$ and $\xi(g)$ finite. Let $\nu(dx) = g(x)\, dx$ and $\mu(dx) = f(x)\, dx$. Then the desired equality takes the form

$$\textbf{(1.18)} \qquad E^\nu\{e^{-\alpha T_A}\, U^\alpha f(X_{T_A})\} = \hat{E}^\mu\{e^{-\alpha \hat{T}_A}\, g\hat{U}^\alpha(\hat{X}_{\hat{T}_A})\}$$

and by the first part of the argument we know that this holds whenever A is open. Let ${}^rA$ denote the points which are coregular for A. Then $A - A^r$ and $A - {}^rA$ are semipolar and cosemipolar, respectively, and hence both have ξ measure zero. Therefore we can find a decreasing sequence of open sets $\{G_n\}$ containing A such that $T_{G_n} \uparrow (T_A \wedge \zeta)$ almost surely P^ν and $\hat{T}_{G_n} \uparrow (\hat{T}_A \wedge \hat{\zeta})$ almost surely $\hat{P}^\mu$. But $U^\alpha f$ and $g\hat{U}^\alpha$ are regular α-potentials for X and $\hat{X}$, respectively and so the validity of (1.18) when A is replaced by G_n implies the validity of (1.18) for general A. Thus we have established Theorem 1.16.

REMARK. For typographical convenience we will omit the hat "^" in those places where it is clearly appropriate. For example we will write $\hat{E}^\mu\{e^{-\alpha T_A} g\, \hat{U}^\alpha(X_{T_A})\}$ for the expression appearing on the right side of (1.18).

We will give several simple consequences of (1.16). Deeper consequences will appear in later sections.

(1.19) PROPOSITION. (i) A set is polar if and only if it is copolar. (ii) A set is semipolar if and only if it is cosemipolar.

Proof. If A is polar then it follows from (1.7) of Chapter V that A is contained in polar Borel set B. Let $\alpha > 0$. Then using (1.16), $\int u^\alpha(x, z)\, \hat{P}_B^\alpha(dz, y) = P_B^\alpha\, u^\alpha(x, y) = 0$. If, for a fixed y, ν is the measure $\hat{P}_B^\alpha(\cdot, y)$, then this says that $U^\alpha \nu = 0$ and so, by (1.15), $\nu = 0$. As a result B, and hence A, is copolar. Dually a copolar set is polar.

In proving that a semipolar set A is cosemipolar we may assume that A is a thin Borel set (see the proof of (1.15) of Chapter V) and that $A \subset E$. Let $\eta < 1$ and let $B = A \cap \{\Phi_A^1 \le \eta\}$. (As usual $\Phi_A^1(x) = E^x\{e^{-T_A}; T_A < \zeta\}$.) Then $(P_B^1)^n \Phi_B^1 \le \eta^{n-1}$ and $\Phi_B^1 \ge P_B^1 U^1 1$. Consequently as $n \to \infty$, $(P_B^1)^n U^1 1 \downarrow 0$. Let $g \in b\mathscr{E}$ be strictly positive and satisfy $\xi(g) < \infty$. Let $\hat{f}_n = g\hat{U}^1(\hat{P}_B^1)^n$; then $\{\hat{f}_n\}$ is a decreasing sequence of 1-coexcessive functions. If $\hat{f} = \lim \hat{f}_n$ then using (1.16) one has

$$\xi(\hat{f}_n) = \langle g, (P_B^1)^n U^1 1\rangle \downarrow 0$$

as $n \to \infty$, and so $\hat{f} = 0$ almost everywhere. Therefore the regularization of $\hat{f}$ is the zero function, and hence, according to (3.6) of Chapter II $\{\hat{f} > 0\}$ is cosemipolar.* If x is coregular for B, then $\hat{P}_B^1(\cdot, x) = \varepsilon_x$ and so, for each n, $\hat{f}_n(x) = g\,\hat{U}^1(x) > 0$. Thus if ${}^rB$ denotes the set of points coregular for B, then ${}^rB \subset \{\hat{f} > 0\}$ and hence ${}^rB$ is cosemipolar. But $B = (B - {}^rB) \cup (B \cap {}^rB)$ and since $B - {}^rB$ is cosemipolar ((3.3) of Chapter II), it follows that B, and hence A, is cosemipolar. By duality this yields (1.19).

REMARK. It is *not* true that a set is thin if and only if it is cothin (see Exercise 1.21.)

From now on we will speak just of polar sets or semipolar sets. For example if $\alpha > 0$ both $\{x: u^\alpha(x, y) = \infty\}$ and $\{y: u^\alpha(x, y) = \infty\}$ are polar (or copolar). If A is any set, then ${}^rA$ denotes the set of points which are coregular for A. It follows from (1.14) of Chapter V and (1.19) that ${}^rA$ is Borel and that $A - {}^rA$ is semipolar. Also (1.18) of Chapter V and (1.19) imply that the σ-algebra $\mathscr{E}^f$ of fine Borel sets and the σ-algebra $\hat{\mathscr{E}}^f$ of cofine Borel sets coincide. The following is a closely related result. However it is not particularly important since in the present case excessive functions are Borel measurable.

(1.20) PROPOSITION. A set is nearly Borel relative to X if and only if it is nearly Borel relative to $\hat{X}$; that is, $\mathscr{E}^n = \hat{\mathscr{E}}^n$.

Proof. Let $B \in \mathscr{E}^n$. Then there exist Borel sets B_1 and B_2 with $B_1 \subset B \subset B_2$ and such that $P^\xi(T_{B_2 - B_1} < \infty) = 0$. Consequently $P^x(T_{B_2 - B_1} < \infty) = 0$ for all x and so $B_2 - B_1$ is polar. Since B is universally measurable, given an initial measure μ one can find Borel sets A_1 and A_2 such that $A_1 \subset B \subset A_2$ and $\mu(A_2 - A_1) = 0$. Let $C_1 = A_1 \cup B_1$ and $C_2 = A_2 \cap B_2$. Then $C_1 \subset B \subset C_2$ and $C_2 - C_1$, being contained in $B_2 - B_1$, is polar. Since $\mu(C_2 - C_1) = 0$, $\hat{P}^\mu\{\hat{X}_t \in C_2 - C_1 \text{ for some } t \geq 0\} = 0$, and hence B is nearly Borel relative to $\hat{X}$. Thus (1.20) is established.

Of course it is now obvious that the conclusion of Theorem 1.16 is valid for nearly Borel sets as well as for Borel sets.

Exercises

(1.21) Let X be translation to the *right* at unit speed in **R** and let $\hat{X}$ be translation to the *left* at unit speed in **R**. Show that X and $\hat{X}$ are in duality

* In fact (3.2) (a much more elementary result than (3.6) of Chapter II) implies that $\{\hat{f} > 0\}$ is copolar.

relative to Lebesgue measure. Compute $u^\alpha(x, y)$ explicitly. Describe the fine and cofine topologies. Exhibit a thin set which is not cothin. Note that the transition functions $P_t(x, A)$ and $\hat{P}_t(A, x)$ are *not* absolutely continuous with respect to Lebesgue measure.

(1.22) Let X be a Hunt process with state space $(\mathbf{R}^N, \mathscr{B}(\mathbf{R}^N))$ whose transition function $P_t(x, A)$ has the form (2.12) of Chapter I. Let $\hat{\mu}_t(A) = \mu_t(-A)$ where $\{\mu_t\}$ is the semigroup of measures appearing in (2.12) of Chapter I. Show that $f\hat{P}_t(x) = \int f(x + y)\, \hat{\mu}_t(dy)$ defines a transition function on $\mathbf{R}^N$ and that there exists a Hunt process $\hat{X}$ on $\mathbf{R}^N$ with this transition function. If $\mu(A) = \int_0^\infty e^{-t}\mu_t(A)\, dt$ is absolutely continuous with respect to (N-dimensional) Lebesgue measure ξ show that X and $\hat{X}$ are in duality relative to ξ. If ψ_R denotes the real part of the function ψ appearing in (2.13) of Chapter I, then show that a sufficient condition that μ be absolutely continuous with respect to ξ is that, for some $\alpha > 0$, $\int [\alpha + \psi_R(x)]^{-1}\, dx < \infty$. Whenever X is a process in $\mathbf{R}^N$ with transition function of the form (2.12) of Chapter I, $\hat{X}$ will always denote the process decribed above.

(1.23) (Special cases of (1.22).) If X is the symmetric stable process in $\mathbf{R}^N$ of index α with $0 < \alpha < N$, then show that

$$u(x, y) = \Gamma\left(\frac{N - \alpha}{2}\right)\left[2^\alpha \pi^{N/2}\Gamma\left(\frac{\alpha}{2}\right)\right]^{-1} |y - x|^{\alpha - N}.$$

See (1.7) of Chapter II. If X is the one-sided stable process in $\mathbf{R}$ of index β with $0 < \beta < 1$, then show that

$$\begin{aligned} u(x, y) &= [\Gamma(\beta)]^{-1}\, (y - x)^{\beta - 1} \qquad &\text{if} \quad x < y, \\ &= 0 &\text{if} \quad x \geq y. \end{aligned}$$

See (1.6) of Chapter II.

(1.24) Let $E = \mathbf{Z}$, the integers, and let $\mathscr{E}$ be all subsets of E. Let q be a probability measure on E. Define $Q(x, A) = \sum_{y \in A} q(y - x)$ and $\hat{Q}(A, x) = \sum_{y \in A} q(x - y)$. Let $\lambda(x) = \lambda$ where $0 < \lambda < \infty$. Using the notation of Section 12, Chapter I, let X and $\hat{X}$ be the regular step processes constructed from (Q, λ) and $(\hat{Q}, \lambda)$, respectively. Show that X and $\hat{X}$ are in duality relative to the counting measure ξ on E; that is, $\xi(A)$ is the cardinality of A. If $q(1) = 1$ and $q(x) = 0$ for $x \neq 1$, describe the processes X and $\hat{X}$.

(1.25) Let X and $\hat{X}$ be in duality relative to ξ. (i) Let A be any subset of E. Show that $A^r - {}^rA$ and ${}^rA - A^r$ are semipolar. [Hint: let $F = {}^rA - A^r$. Then F is a Borel set and $F \cup {}^rF \subset {}^rA$. Let B be a Borel set containing A such that

$B^r = A^r$. See (1.13) and (1.14) of Chapter V. Let $\alpha > 0$ and let $\Phi_F^\alpha(x) = E^x(e^{-\alpha T_F}; T_F < \zeta) = P_F^\alpha 1(x)$. Let $\{f_n\}$ be a sequence of bounded functions such that $U^\alpha f_n \uparrow 1$. Now use (1.16) to show that $\Phi_F^\alpha = P_B^\alpha \Phi_F^\alpha$ and conclude from this that $\Phi_F^\alpha < 1$ on F. Use this to show that F is semipolar.] (ii) Use (i) to show that if A is any subset of E then the fine closure and the cofine closure of A differ by a semipolar set. Also show that the fine interior and the cofine interior of A differ by a semipolar set.

(1.26) Let X and $\hat{X}$ be in duality relative to ξ. (i) Let μ be a measure and B be a cofinely open Borel set carrying μ. If $M = \sup_{x \in B} U^\alpha \mu(x)$, then show that $U^\alpha \mu \leq M$ everywhere. (Here $\alpha \geq 0$.) If μ does not charge semipolar sets and K is the support of μ, then show that $\|U^\alpha \mu\| = \sup_{x \in K} U^\alpha \mu(x)$. (ii) If the fine and cofine topologies coincide and μ is any measure with support K, then show that $\|U^\alpha \mu\| = \sup_{x \in K} U^\alpha \mu(x)$. Note that this last result is applicable whenever X and $\hat{X}$ are equivalent—for example, Brownian motion or the *symmetric* stable processes. (iii) Let X be the one-sided stable process in $\mathbf{R}$ of index β, $0 < \beta < 1$. Exhibit a finite measure μ such that $U\mu = 0$ on the support of μ and such that $U\mu$ is unbounded.

(1.27) Suppose X and $\hat{X}$ are in duality relative to a measure ξ. Let A be an open subset of E and suppose $A^c = (A^c)^r = {}^r(A^c)$. (a) Show that (X, T_{A^c}) and $(\hat{X}, \hat{T}_{A^c})$ are in duality relative to the restriction of ξ to A. (b) Show that X and $\hat{X}$ are equivalent if and only if $u^\alpha(x, y) = u^\alpha(y, x)$ for all x and y. (c) Show that if X and $\hat{X}$ are equivalent, then (X, T_{A^c}) and $(\hat{X}, \hat{T}_{A^c})$ are equivalent also.

2. Potentials of Measures

In this section we are going to establish the analog of the classical Riesz decomposition theorem; that is, under suitable assumptions an excessive function f will be written in the form $f = U\mu + h$ where h is in some sense "harmonic." We assume that X and $\hat{X}$ are standard processes with the same state space $(E, \mathscr{E})$ and that they are in duality relative to a fixed σ-finite measure ξ. The notation and terminology of Section 1 will be used without special mention. In addition the following smoothness and boundedness conditions will be imposed throughout this section except in certain propositions where explicitly stated otherwise.

(2.1) If $\alpha > 0$ and $f \in b\mathscr{E}^*$ vanishes outside a compact subset of E, then $f\hat{U}^\alpha(y) = \int u^\alpha(x, y) f(x)\, dx$ is a continuous function of y on E. Of course $f\hat{U}^\alpha$ is bounded.

(2.2) If $f \in b\mathscr{E}^*$ vanishes outside a compact set, then

$$f\hat{U}(y) = \int u(x, y) f(x)\, dx, \qquad Uf(x) = \int u(x, y) f(y)\, dy$$

are bounded functions of y and x, respectively, and in addition $f\hat{U}$ is continuous on E.

We will assume both (2.1) and (2.2) and state and prove theorems about excessive functions ($\alpha = 0$). However, if only (2.1) holds, then obviously for any $\alpha > 0$ the α-subprocess ((3.17) of Chapter III) satisfies both (2.1) and (2.2). Consequently all of our results are valid for α-excessive functions when $\alpha > 0$ just under condition (2.1). These results often will be used in this form in later sections.

Before proceeding we will draw some elementary consequences of (2.1) which do *not* depend on (2.2). An immediate consequence of (2.1) is that α-coexcessive functions are lower semicontinuous for any $\alpha \geq 0$. See (2.6) and (2.16) of Chapter II. Also (2.1) implies that ξ is a Radon measure; that is, ξ is finite on compact subsets of E. To see this let $\{A_n\}$ be an increasing sequence of Borel sets with $E = \bigcup A_n$ and $\xi(A_n) < \infty$. We may also assume that each A_n has compact closure in E. Then $y \to \hat{U}^1(A_n, y)$ is continuous for each n. Let y_0 be fixed and choose n so that $\hat{U}^1(A_n, y_0) > 0$. Consequently there exist a neighborhood G of y_0 and an $\eta > 0$ such that $\hat{U}^1(A_n, y) \geq \eta$ on G. Now ξ is coexcessive and so

$$\eta\, \xi(G) \leq \int_G \hat{U}^1(A_n, y)\, \xi(dy) \leq \xi(A_n) < \infty.$$

Thus every point has a neighborhood on which ξ is finite and so ξ is a Radon measure. Here is another useful consequence of (2.1). Let K be a compact subset of E and G a neighborhood of K. Then for any $\alpha \geq 0$, $y \to \hat{U}^\alpha(G, y)$ is lower semicontinuous and strictly positive on K, and so $\inf_{y \in K} \hat{U}^\alpha(G, y) > 0$.

In the present situation the various finiteness assumptions on excessive functions and potentials of measures take a particularly nice form. Recall that $f \in \mathscr{E}^*$ is said to be *locally integrable* provided $\int_K |f|\, d\xi < \infty$ for all compact $K \subset E$.

(2.3) **Proposition.** Let f be an excessive function. Then the following conditions are equivalent under (2.1): (Condition 2.2 is not assumed for this proposition.)

(i) f is locally integrable;
(ii) f is finite a.e.;
(iii) f is finite except on a polar set.

Proof. The equivalence of (ii) and (iii) follows from Proposition 3.5 of Chapter II and the fact that ξ is a reference measure for X. Plainly (i) implies (ii) and so we need only show that (ii) implies (i). Assume (ii) and let y_0 be a fixed point in E. Then there exists an $x_0 \in E$ such that $u^1(x_0, y_0) > 0$ and $f(x_0) < \infty$. Since $y \to u^1(x_0, y)$ is lower semicontinuous one can find an $\eta > 0$ and a neighborhood G of y_0 such that $u^1(x_0, y) \geq \eta$ for all y in G. Consequently

$$\infty > f(x_0) \geq \int u^1(x_0, y) f(y)\, dy \geq \eta \int_G f(y)\, dy.$$

Thus f is integrable over some neighborhood of each fixed point y_0 and hence f is locally integrable.

(2.4) PROPOSITION. Let μ be a measure on $\mathscr{E}^*$. Then $U\mu$ is locally integrable if and only if $\hat{P}_K \mu$ is a finite measure for each compact subset K of E. In this case μ is finite on compact subsets of E.

Proof. Suppose $\hat{P}_K \mu$ is a finite measure for compact K. Given a compact K, $\int_K U\mu\, d\xi = \int (I_K \hat{U})\, d\mu$. Now $I_K \hat{U} = I_K \hat{U} \hat{P}_K$ and $\int (I_K \hat{U} \hat{P}_K)\, d\mu = \int (I_K \hat{U})\, d\hat{P}_K \mu$ which is finite because $\hat{P}_K \mu$ is a finite measure and $I_K \hat{U}$ is bounded. Next suppose that $U\mu$ is locally integrable. Let K be compact, G a neighborhood of K with compact closure, and $\eta = \inf_{y \in K} \hat{U}(G, y) > 0$. Then

$$\infty > \int_G U\mu(x)\, dx = \int \hat{U}(G, y)\, \mu(dy) \geq \eta\, \mu(K),$$

and so μ is finite on compact subsets of E. In particular $U\hat{P}_K \mu = P_K U\mu \leq U\mu$, and so if $U\mu$ is locally integrable then $\hat{P}_K \mu$ is finite on compacts and hence is finite since it is supported by K. Consequently the proof of Proposition 2.4 is complete.

Let $\mathbf{M}$ denote the set of all measures μ on $\mathscr{E}^*$ such that $U\mu$ is locally integrable. This will turn out to be the appropriate class of measures for our discussion. In particular, according to (1.15) and (2.3), $U\mu$ determines μ when μ is in $\mathbf{M}$. Finally (2.4) implies that any finite measure is in $\mathbf{M}$ and that any measure in $\mathbf{M}$ with compact support in E is finite.

We will introduce some terminology before stating the next proposition. Recall from Section 4 of Chapter III that an $f \in \mathscr{E}_+^*$ is called supermedian if $\alpha U^\alpha f \leq f$ for all $\alpha > 0$ and that in this situation $\alpha U^\alpha f$ increases as $\alpha \to \infty$ to an excessive function $\bar{f}$, the regularization of f. Moreover $\bar{f}$ is the largest excessive function dominated by f and $\xi(\{\bar{f} < f\}) = 0$. A function $g \in \mathscr{E}_+^*$ will be called *nearly supermedian* if, for each $\alpha > 0$, $\alpha U^\alpha g \leq g$, a.e. Let $D_\alpha = \{\alpha U^\alpha g > g\}$ and $D = \bigcup D_\alpha$, the union being over all rational α. Define

$f = g$ on $E - D$ and $f = \infty$ on D. Then $g \le f$, $g = f$ a.e., so that $U^\alpha f = U^\alpha g$ for all α, and $\alpha U^\alpha f \le f$ if α is rational. It follows immediately that $\alpha U^\alpha f \le f$ for all α, so f is supermedian. Consequently $\alpha U^\alpha g$ increases to an excessive function $\bar{g}$ which we again call the regularization of g, and $\bar{g} = g$ a.e. Moreover $\bar{g}$ is the largest excessive function dominated almost everywhere by g.

(2.5) THEOREM. Let $\{\mu_n\}$ be a sequence of measures in $\mathbf{M}$ and suppose that there exists a locally integrable function f such that $U\mu_n \le f$ for all n. Then the following conclusions hold.

(i) $\{\mu_n\}$ contains at least one weakly convergent subsequence $\{\nu_n\}$. If ν is the weak limit of $\{\nu_n\}$, then $\nu \in \mathbf{M}$.

(ii) Let $\{\nu_n\}$ and ν be as in (i) and assume that $U\nu_n$ converges to a function g a.e.; then g is nearly supermedian and $\bar{g} \ge U\nu$. If, in addition for each compact subset K of E and each $\varepsilon > 0$ there exists a compact subset J of E such that $\int_{J^c} \hat{U}(K, x)\, \nu_n(dx) < \varepsilon$ for all n, then $\bar{g} = U\nu$.

Proof. In order to show that $\{\mu_n\}$ contains a weakly convergent subsequence it suffices to show that $\{\mu_n(K)\}$ is bounded in n for each compact subset K of E. For such a K let G be a compact neighborhood of K and let $\eta = \inf_{x \in K} \hat{U}(G, x)$. Then

$$\infty > \int_G f\, d\xi \ge \int_G U\mu_n\, d\xi = \int \hat{U}(G, x)\, \mu_n(dx) \ge \eta\, \mu_n(K).$$

Now let $\{\nu_n\}$ be a weakly convergent subsequence with limit ν. If h is a bounded nonnegative function vanishing off a compact subset of E, then $h\hat{U}$ is in $\mathbf{C}_+$ and so, for any continuous function k with compact support and $0 \le k \le 1$,

$$\int k(h\hat{U})\, d\nu = \lim_n \int k(h\hat{U})\, d\nu_n \le \langle h, f \rangle < \infty.$$

Consequently

$$\langle h, U\nu \rangle = \int h\hat{U}\, d\nu \le \langle h, f \rangle$$

and hence (i) is established. As for (ii), using the same notation as above we obtain with the aid of the bounded convergence theorem

$$\begin{aligned} \langle h, g \rangle = \lim_n \langle h, U\nu_n \rangle &\ge \lim_n \int k(h\hat{U})\, d\nu_n \\ &= \int k(h\hat{U})\, d\nu. \end{aligned}$$

Letting k run through a sequence in $(\mathbf{C}_K)_+$ increasing to 1 we obtain $\langle h, g \rangle \ge \int h\hat{U}\, d\nu = \langle h, U\nu \rangle$. Hence $g \ge U\nu$ a.e. Finally using Fatou's lemma we have almost everywhere

$$g = \lim U\nu_n \geq \liminf \alpha U^\alpha U\nu_n$$
$$\geq \alpha\, U^\alpha(\lim U\nu_n) = \alpha U^\alpha g,$$

and so g is nearly supermedian. Clearly $\bar{g} \geq U\nu$ since $g \geq U\nu$ almost everywhere.

We have now established all of (2.5) except for the last assertion. Let K be a compact subset of E and given $\varepsilon > 0$ let J be a compact set in E such that $\int_{J^c} \hat{U}(K, x)\, \nu_n(dx) < \varepsilon$ for all n. Let k be a continuous function with compact support such that $0 \leq k \leq 1$, and $k = 1$ on J. Then

$$\int_K g\, d\xi = \lim \int_K U\nu_n\, d\xi = \lim \int \hat{U}(K, y)\, \nu_n(dy)$$
$$\leq \int k(y)\, \hat{U}(K, y)\, \nu(dy) + \varepsilon \leq \int_K U\nu\, d\xi + \varepsilon.$$

Since ε is arbitrary this implies that $\langle I_K, g\rangle \leq \langle I_K, U\nu\rangle$. On the other hand $U\nu \leq g$ a.e. and so $U\nu = g$ a.e. on K. Since K is arbitrary, $U\nu = g$ a.e., and so $U\nu = \bar{g}$, completing the proof of (2.5).

REMARK. The assumption in the last sentence of (2.5) certainly holds whenever there is a fixed compact subset J of E containing the supports of all the μ_n.

We now introduce some notation that will be used in the rest of this section. Let $\{K_n\}$ be an increasing sequence of compact subsets of E such that $K_n \subset K_{n+1}^0$ for each $n \geq 1$ and $E = \bigcup K_n$. Here K_n^0 denotes the interior of K_n. If $T_n = T_{K_n^c}$ then clearly $T_n \uparrow \zeta$ almost surely, and if $\hat{T}_n = \hat{T}_{K_n^c}$ then $\hat{T}_n \uparrow \hat{\zeta}$ almost surely also. If f is an excessive function, then $\{P_{T_n} f\}$ decreases to a limit g as $n \to \infty$ and this limit is obviously independent of the particular sequence $\{K_n\}$ used—subject to the above conditions of course. It will be convenient to write $\lim_{K \uparrow E} P_{K^c} f$ for this limit g in order to emphasize its independence of the sequence $\{K_n\}$. Obviously g is super-mean-valued and it differs from its regularization $\bar{g}$ on a semipolar set, by Theorem 3.6 of Chapter II. Of course $g = \inf_K P_{K^c} f$ where the infimum is over all compact subsets K of E.

REMARK. In fact, by (3.20) of Chapter II, $\bar{g} = g$ except possibly on the set $\{g = \infty\}$, and when f is locally integrable this set is polar since it is contained in $\{f = \infty\}$.

Suppose that $f = U\mu$ with $\mu \in \mathbf{M}$ and let h be the indicator function of a compact subset of E. Then

$$\infty > \langle h, P_{T_n} U\mu \rangle = \langle h, U\hat{P}_{T_n}\mu \rangle$$
$$= \hat{E}^{\mu} \int_{T_n}^{\zeta} h(X_t)\, dt,$$

and this last expression approaches zero as $n \to \infty$ since $\hat{T}_n \uparrow \hat{\zeta}$ almost surely. Consequently in this case $g = 0$ almost everywhere and so $\bar{g} = 0$. It now follows from the above remark that $g = 0$ on $\{U\mu < \infty\}$. The exceptional set cannot be eliminated in general. See Exercise 2.17.

The following corollaries of Theorem 2.5 are perhaps the most useful forms of this theorem for our later applications.

(2.6) Corollary. Let $\{\mu_n\}$ be a sequence of measures in $\mathbf{M}$ such that the sequence $\{U\mu_n\}$ is increasing and suppose that $f = \lim U\mu_n$ is locally integrable. Suppose further that either the supports of all the μ_n are contained in some fixed compact subset of E or that $\lim_{K \uparrow E} P_{K^c} f = 0$, a.e. Then $\{\mu_n\}$ converges weakly to a measure μ in $\mathbf{M}$ and $f = U\mu$.

Proof. Let us show first that under these hypotheses the condition in the last sentence of Theorem 2.5 is satisfied. It is certainly satisfied if the supports of all the μ_n are contained in some fixed compact subset of K. So suppose J is a compact subset of E and $y \in J^c$. Then for any x

$$P_{J^c}\, u(x, y) = \int u(x, z)\, \hat{P}_{J^c}(dz, y) = u(x, y).$$

Consequently if K and J are compact subsets of E

$$\int_{J^c} \hat{U}(K, y)\, \mu_n(dy) = \int_K dx \int_{J^c} u(x, y)\, \mu_n(dy)$$
$$= \int_K dx \int_{J^c} P_{J^c}\, u(x, y)\, \mu_n(dy)$$
$$\leq \int_K dx\, P_{J^c}\, U\mu_n(x) \leq \int_K P_{J^c} f(x)\, dx,$$

and under the second assumption in the second sentence of (2.6) this last integral approaches zero as J runs through the sequence $\{K_n\}$. Thus in either case the condition in the last sentence of Theorem 2.5 is satisfied.

Since $\{U\mu_n\}$ is increasing, f is excessive. If $\{\nu_n\}$ is a weakly convergent subsequence of $\{\mu_n\}$ with limit μ, then $f = \lim U\nu_n$. Since f is excessive, Theorem 2.5 and the preceding paragraph imply that $f = \bar{f} = U\mu$. If ν is the limit of any other weakly convergent subsequence of $\{\mu_n\}$, then by the above argument $U\nu = f = U\mu$. Consequently $\mu = \nu$ and so the entire sequence $\{\mu_n\}$ must converge weakly to μ.

(2.7) Corollary. Let $\{\mu_n\}$ be a sequence of measures in $\mathbf{M}$ such that

$\{U\mu_n\}$ decreases. Then $\{\mu_n\}$ converges weakly to a measure μ in **M** and $\{U\mu_n\}$ decreases to $U\mu$ except on a semipolar set.

Proof. Since $U\mu_1$ is locally integrable and $\lim_{K\uparrow E} P_{K^c}U\mu_1 = 0$ a.e., the argument in the first paragraph of the proof of (2.6) shows that Theorem 2.5 is applicable. Let $f = \lim U\mu_n$. Then under the present assumptions f is super-mean-valued and $f = \bar{f}$ except on a semipolar set by Theorem 3.6 of Chapter II. If $\{\nu_n\}$ is a weakly convergent subsequence of $\{\mu_n\}$ with limit μ, then $f = \lim U\nu_n$ and, by Theorem 2.5, $\bar{f} = U\mu$. It now follows as in the proof of (2.6) that the entire sequence $\{\mu_n\}$ converges weakly to μ.

REMARK. If in (2.6) or (2.7) all of the measures μ_n have their supports contained in some fixed compact subset of E, then the limit measure μ is finite since it is in **M** and has compact support.

We come now to one of the important results of this section.

(2.8) THEOREM. Let B be a Borel set with compact closure in E and let f be a locally integrable excessive function. Then $P_B f = U\mu$ where μ is a finite measure concentrated on the union of B and the set of points coregular for B.

Proof. Under the present assumptions one can find an increasing sequence $\{Ug_n\}$ of bounded potentials with limit f. Let $\mu_n = g_n\xi$. Then each μ_n is in **M**, and $P_B U\mu_n = U\hat{P}_B\mu_n$. Let $\nu_n = \hat{P}_B\mu_n$. Then $\{U\nu_n\}$ increases to $P_B f$ and each ν_n is a measure in **M** which is concentrated on $\bar{B}$. Since $U\nu_n \le f$, Corollary 2.6 is applicable, and hence there exists a finite measure μ concentrated on $\bar{B}$ such that $P_B f = U\mu$. It remains to show that μ is concentrated on $B \cup {}^rB$. To this end let G be a compact neighborhood of B. By what we have proved so far, $P_G f = U\nu$ where ν is a finite measure concentrated on G. But $P_B P_G = P_B$ and so

$$P_B f = P_B P_G f = P_B U\nu = U\hat{P}_B\nu.$$

Hence $\mu = \hat{P}_B\nu$ and this representation of μ clearly shows that μ is concentrated on $B \cup {}^rB$. Thus Theorem 2.8 is established.

We are going to derive a very useful consequence of Theorem 2.8 before proceeding to the main result of this section.

(2.9) PROPOSITION. For this proposition we assume only (2.1) and not (2.2). Let A and B be Borel sets with $B \subset A$ and assume that $B \subset {}^rA$. (This is certainly satisfied if A is a neighborhood of B.) Then $T_B \circ \theta_{T_A} = 0$ almost surely on $\{T_A = T_B < \infty\}$.

Proof. Let $\alpha > 0$. Since $T_A \le T_B$ the conclusion of (2.9) is equivalent to the statement that $T_A + T_B \circ \theta_{T_A} = T_B$ almost surely, which in turn is equivalent to $P_A^\alpha P_B^\alpha 1 = P_B^\alpha 1$. In light of (1.20) of Chapter V (or (10.19) of Chapter I applied to the measures ε_x and $P_A^\alpha(x, \cdot)$) it suffices to prove this when B is compact. In this case $P_B^\alpha 1 = U^\alpha \pi_B^\alpha$ where π_B^α is a finite measure concentrated on B, according to Theorem 2.8. (Note that as explained above we are applying (2.8) when $\alpha > 0$ assuming only condition (2.1).) Since $B \subset {}^r A$, $\hat{P}_A^\alpha \pi_B^\alpha = \pi_B^\alpha$ and so

$$P_A^\alpha P_B^\alpha 1 = P_A^\alpha U^\alpha \pi_B^\alpha = U^\alpha \hat{P}_A^\alpha \pi_B^\alpha = U^\alpha \pi_B^\alpha = P_B^\alpha 1,$$

completing the proof of (2.9).

REMARK. In view of the remark following the proof of (1.20), the sets A and B appearing in (2.8) and (2.9) need only be nearly Borel measurable. As mentioned before, this is not of much interest in the present situation. Note also that (2.9) can be stated in the form $P_A^\alpha P_B^\alpha = P_B^\alpha$ for all $\alpha \ge 0$. As a result if G is a neighborhood of B, then $P_G^\alpha P_B^\alpha = P_B^\alpha = P_B^\alpha P_G^\alpha$ for all $\alpha \ge 0$. Yet another way of formulating the conclusion of (2.9) is that $X(T_A) \in B^r$ almost surely on $\{T_A = T_B < \infty\}$.

(2.10) PROPOSITION. A locally integrable excessive function f is the potential, $U\mu$, of a measure μ in $\mathbf{M}$ if and only if $\lim_{K \uparrow E} P_{K^c} f = 0$ a.e.

Proof. We have already seen that if $\mu \in \mathbf{M}$ then $U\mu$ satisfies this condition. Conversely suppose f satisfies the hypotheses of (2.10). By (2.2), $U(x, K)$ is a bounded function of x if K is compact, and so according to (2.19) of Chapter II there is a sequence $\{g_n\}$ of bounded functions such that Ug_n is bounded for each n and Ug_n increases to f. Let μ_n be the measure $g_n \xi$. It is then an immediate consequence of (2.6) that μ_n converges weakly to a measure μ in $\mathbf{M}$ and that $U\mu = f$.

We are now ready to state and prove the analog of the Riesz decomposition theorem.

(2.11) THEOREM. Let f be a locally integrable excessive function. Then f has a unique representation of the form $f = U\mu + h$ where $\mu \in \mathbf{M}$ and h is an excessive function with the property that $P_D h = h$ whenever D is the complement of a compact subset of E.

Proof. Suppose f has two such representations—$f = U\mu_1 + h_1 = U\mu_2 + h_2$. Operating on this last equality by P_{K^c} and letting $K \uparrow E$ we find, since $P_{K^c} U\mu \to 0$ a.e. if $\mu \in \mathbf{M}$, that $h_1 = h_2$, a.e. and hence everywhere. Therefore

$U\mu_1 = U\mu_2$ and this implies that $\mu_1 = \mu_2$. Thus we have established the uniqueness of such a representation. We will next show that $f = g + h$ where h is as in (2.11) and g is an excessive function with the property that $\lim_{K \uparrow E} P_{K^c}\, g = 0$ a.e. By Proposition 2.10 this will complete the proof of Theorem 2.11.

Let $\{K_n\}$ be our fixed increasing sequence of compacts with $K_n \subset K_{n+1}^0$ and $\bigcup K_n = E$. Define $h_1 = \lim_n P_{K_n^c} f$. Then h_1 is super-mean-valued and $h_1 \leq f$. If J is any Borel set with compact closure in E, then $\bar{J} \subset K_n^0$ for all large n and $P_{J^c} P_{K_n^c} = P_{K_n^c}$ for such n. Therefore $P_{J^c}\, h_1(x) = h_1(x)$ at each x satisfying $f(x) < \infty$. Next define $g_1(x) = f(x) - h_1(x)$ if $f(x) < \infty$ and $g_1(x) = \infty$ if $f(x) = \infty$. Clearly $f = g_1 + h_1$, and $\lim_{K \uparrow E} P_{K^c} g_1 = 0$ on $\{f < \infty\}$. Moreover if J is any Borel set with compact closure in E, then $P_{J^c} g_1 \leq g_1$ everywhere. Now let K be a fixed compact subset of E and let $J = K^c \cap K_n$. Then $\bar{J}$ is compact and $J^c = K \cup K_n^c$ so that $P_{K \cup K_n^c}\, g_1 \leq g_1$ for each n. Let T and T_n be the hitting times of K and K_n^c, respectively. Then

$$E^x\{g_1(X_T);\ T \leq T_n\} \leq P_{K \cup K_n^c}\, g_1(x) \leq g_1(x),$$

and letting $n \to \infty$ we obtain $P_K g_1 \leq g_1$ since $T_n \uparrow \zeta$ almost surely. Thus, by Theorem 5.1 of Chapter II, g_1 is supermedian. Let $g = \lim_{\alpha \to \infty} \alpha U^\alpha g_1$ and $h = \lim_{\alpha \to \infty} \alpha U^\alpha h_1$. Then g and h are excessive and $f = g + h$. Since $g \leq g_1$, $\lim_{K \uparrow E} P_{K^c} g = 0$ on $\{f < \infty\}$; in particular, this limit is zero almost everywhere. Finally we have already observed in the remark preceding (2.6) that $h = h_1$ except on a polar set, and hence $P_{K^c} h = P_{K^c} h_1$ except on a polar set, whenever K is a Borel set. In particular if K is compact then we already know that $P_{K^c} h_1 = h_1$ a.e., so that $P_{K^c} h = h$ a.e. and hence everywhere because h and $P_{K^c} h$ are excessive. The proof of (2.11) is complete.

We will next formulate and prove another very important theorem due to Hunt, which complements the result in Section 6 of Chapter III. If f is excessive and $B \in \mathscr{E}$ we define $\mathscr{U}(f, B)$ to be the family of all excessive functions g which dominate f on a *neighborhood* (depending on g) of B and f^B to be the infimum of the functions in $\mathscr{U}(f, B)$.

(2.12) THEOREM. Let f be a locally integrable excessive function and $B \in \mathscr{E}$. Then f^B coincides, except possibly at the points outside B where f is infinite and at the points of $B - B^r$, with the supremum of the potentials $U\mu$ where μ is a finite measure carried by a compact subset of B and $U\mu \leq f$ everywhere. There is a sequence of such measures μ_n such that $\{U\mu_n\}$ increases to the supremum in question. This supremum is itself the potential, $U\nu_B$, of a measure ν_B in **M** which is concentrated on $B \cup {}^rB$, whenever either f is the potential of a measure in **M**, or B has compact closure in E. Finally if B is open f^B and the supremum in question coincide everywhere.

Proof. The proof of this theorem is rather long and so we will proceed in steps. First of all if $g \geq f$ on a neighborhood of B, then $g \geq P_B g \geq P_B f$ everywhere and so $f^B \geq P_B f$. Of course $f^B \leq f$. Now if G is a neighborhood of B and $g \geq f$ on G, then $g \geq P_G g \geq P_G f$ everywhere and $P_G f = f$ on G. Consequently for any $B \in \mathscr{E}$

$$f^B = \inf\{P_G f\colon G \text{ open}, G \supset B\}. \tag{2.13}$$

It is, of course, no restriction in (2.13) to assume that $G \subset E$. In the remainder of this proof all sets are understood to be subsets of E unless explicitly mentioned otherwise. An immediate consequence of (2.13) is that $f^G = P_G f$ whenever G is open. If $x \in B^r$ then $f(x) \geq f^B(x) \geq P_B f(x) = f(x)$, and so $f = f^B = P_B f$ on B^r.

If μ is a finite measure with compact support contained in B such that $U\mu \leq f$ and if $g \in \mathscr{U}(f, B)$ dominates f on a neighborhood G of B, then $g \geq P_G g \geq P_G U\mu = U\hat{P}_G \mu = U\mu$ since G is also a neighborhood of the support of μ. Thus the supremum in question nowhere exceeds f^B. Let $\{K_n\}$ be an increasing sequence of compact sets contained in B such that, for all x, $T_{K_n} \downarrow T_B$ almost surely P^x. (See (1.20) of Chapter V for the existence of such a sequence.) Certainly $P_{K_n} f \leq P_B f$ while by the right continuity of $t \to f(X_t)$ and Fatou's lemma we have $\liminf_{n \to \infty} P_{K_n} f \geq P_B f$. Therefore $\lim_n P_{K_n} f = P_B f$ and so $\lim_n P_{K_n} f(x) = f^B(x)$ at any point x such that $P_B f(x) = f^B(x)$, in particular at any point $x \in B^r$. Finally, according to Theorem 2.8, $P_{K_n} f = U\mu_n$ where μ_n is a finite measure carried by K_n. Thus we have proved the assertion in the last sentence of Theorem 2.12 and that part of the assertion in the second sentence dealing with points inside B. Note also for future reference that if G is open and K a compact subset of G, then $P_K f \leq f^K \leq P_G f = f^G$, and so it follows from what we have proved that $f^G = \sup\{f^K\colon K \text{ a compact subset of } G\}$.

Next suppose that B is compact and let $\{G_n\}$ be a decreasing sequence of open sets with compact closures $\bar{G}_n$ in E and such that $\bar{G}_{n+1} \subset G_n$, $B = \bigcap G_n$. Then $\{P_{G_n} f\}$ decreases to a super-mean-valued function $f_0 \geq P_B f$. But if G is any neighborhood of B then $G \supset G_n$ for all large n and so, by (2.13), $f_0 = f^B$. Now by Theorem 2.8 $P_{G_n} f = U\mu_n$ where μ_n is a finite measure with support contained in $\bar{G}_n \subset \bar{G}_1$, and so, by Corollary 2.7, $\{\mu_n\}$ converges weakly to a measure μ and $U\mu = f_0 = f^B$ a.e. In particular $U\mu$ is the regularization of f^B, and, according to (3.20) of Chapter II, $f^B(x) = U\mu(x)$ at each x outside B at which f^B is finite. Moreover the support of μ is contained in $\bar{G}_n$ for each n, and hence μ is carried by B. Obviously $U\mu \leq f$ and so the second sentence of Theorem 2.12 is proved whenever B is a compact subset of E. Note that we have actually proved more in this case: namely, there exists a measure μ with support in B such that $U\mu \leq f$ and $U\mu(x) = f^B(x)$ at any $x \in B^c$ at which $f^B(x) < \infty$.

We will complete the proof by using Choquet's extension theorem for capacities (Theorem 10.6 of Chapter I). Let x be a fixed point at which $f(x) < \infty$ and define $\varphi(K) = f^K(x)$ for all compact subsets K of E. Then $\varphi(K)$ is finite, $\varphi(K) \leq \varphi(L)$ if $K \subset L$, and it follows from (2.13) that φ is right continuous, that is, satisfies condition (10.5ii) of Chapter I. Let K and L be compact and $\varepsilon > 0$. Using (2.13) we can choose open sets G and H containing K and L, respectively, such that $P_G f(x) + P_H f(x) \leq \varphi(K) + \varphi(L) + \varepsilon$. Since $G \cup H$ and $G \cap H$ are open sets containing $K \cup L$ and $K \cap L$, respectively, in order to show that $\varphi(K \cap L) + \varphi(K \cap L) \leq \varphi(K) + \varphi(L)$ it will suffice to show that $P_{G \cup H} f + P_{G \cap H} f \leq P_G f + P_H f$. For this it is enough to consider the case in which f is the bounded potential of a bounded nonnegative function g. But then

$$P_{G \cup H} f(x) - P_H f(x) = E^x \int_{T_G \wedge T_H}^{T_H} g(X_t)\, dt$$

$$\leq E^x \int_{T_G}^{T_{G \cap H}} g(X_t)\, dt = P_G f(x) - P_{G \cap H} f(x).$$

Thus we have checked that φ is a Choquet capacity on the class of compact subsets of E. Also it follows from (2.13) and the remark at the end of the second paragraph of the proof that $\varphi(B) = f^B(x)$ for all Borel sets where now φ has been extended to $\mathscr{E}$ by Theorem 10.6 of Chapter I. Finally suppose in addition that $x \notin B$. Given $\varepsilon > 0$ we can find, using Choquet's theorem, a compact subset K of B such that $f^B(x) - f^K(x) < \varepsilon$. But by what was proved in the preceding paragraph there exists a measure $\mu \in \mathbf{M}$ with support in K such that $U\mu(x) = f^K(x)$ and $U\mu \leq f$. Thus the proof of the second sentence of Theorem 2.12 is complete.

In completing the proof of Theorem 2.12 we will need the following fact which is of interest in itself.

(2.14) PROPOSITION. Let μ_1 and μ_2 be measures in $\mathbf{M}$. Then the smallest excessive function dominating both $U\mu_1$ and $U\mu_2$ is the potential $U\nu$ of a measure $\nu \in \mathbf{M}$ whose support is contained in the union of the supports of μ_1 and μ_2.

Proof. Let $\mu = \mu_1 + \mu_2$. Then $\mu \in \mathbf{M}$ and $U\mu$ dominates both $U\mu_1$ and $U\mu_2$. Consider the collection $\mathscr{V}$ of all excessive functions f such that f dominates $U\mu_1$ and $U\mu_2$ and f is dominated by $U\mu$, and let u denote the infimum of this collection. Then according to Theorem 1.6 of Chapter V we can find a decreasing sequence $\{f_n\}$ of elements of $\mathscr{V}$ such that if $v = \inf_n f_n$ then $\bar{v} \leq u \leq v$. Moreover, since any excessive function dominated by the potential of a measure in $\mathbf{M}$ is itself the potential of measure in $\mathbf{M}$, $\bar{v} = U\nu$ with $\nu \in \mathbf{M}$.

Clearly $\bar{v}$ dominates both $U\mu_1$ and $U\mu_2$ almost everywhere and hence everywhere, and so $\bar{v}$ is the smallest excessive function dominating $U\mu_1$ and $U\mu_2$. Let G be any open set containing the supports of μ_1 and μ_2. Then $P_G\bar{v} \geq P_G U\mu_j = U\hat{P}_G \mu_j = U\mu_j$ for $j = 1, 2$, so $P_G\bar{v}$ dominates $U\mu_1$ and $U\mu_2$. Consequently $\bar{v} = P_G\bar{v}$, and so $U\nu = P_G U\nu = U\hat{P}_G\nu$. Therefore $\nu = \hat{P}_G\nu$ and hence the support of ν must be contained in the union of the supports of μ_1 and μ_2. This establishes Proposition 2.14.

We return now to the proof of Theorem 2.12. Let $\mathscr{W}$ be the collection of all $U\mu$ where μ is a finite measure with compact support in B and $U\mu \leq f$. Proposition 2.14 implies that $\mathscr{W}$ is filtering upward, and so by Theorem 1.5 of Chapter V we can find an increasing sequence $\{U\mu_n\}$ of potentials in $\mathscr{W}$ such that $\sup_n U\mu_n = \sup\{U\mu\colon U\mu \in \mathscr{W}\}$. This establishes the third sentence of Theorem 2.12.

It remains only to check the fourth sentence of Theorem 2.12. Let u be the supremum in question, which, in view of the preceding paragraph, is excessive. We also know that $u = f^B$ except possibly on a semipolar set. Suppose that $f = U\nu$ with $\nu \in \mathbf{M}$. Let ν_1 be the restriction of ν to B and $\nu_2 = \nu - \nu_1$. By (2.13) and Theorem 1.6 of Chapter V we can find a decreasing sequence $\{G_n\}$ of open sets containing B such that if $v = \lim_n P_{G_n} f$ then $\bar{v} \leq f^B \leq v$, and, since $\nu_2(B) = 0$ we may also assume that $(\hat{T}_{G_n} \wedge \hat{\zeta}) \uparrow (\hat{T}_B \wedge \hat{\zeta})$ almost surely $\hat{P}^{\nu_2}$. Now $P_{G_n} U\nu_2 = U\hat{P}_{G_n}\nu_2$ decreases to a super-mean-valued function g and if $h \geq 0$ is a bounded function vanishing outside a compact set, then

$$\langle h, U\hat{P}_{G_n}\nu_2\rangle = \hat{E}^{\nu_2} \int_{T_{G_n}}^{\zeta} h(X_t)\, dt$$

$$\to \hat{E}^{\nu_2} \int_{T_B}^{\zeta} h(X_t)\, dt = \langle h, U\hat{P}_B \nu_2\rangle.$$

Therefore $g = U\hat{P}_B\nu_2$ a.e. But $B \subset G_n$ and so

$$\begin{aligned} P_{G_n} f &= P_{G_n} U\nu_1 + P_{G_n} U\nu_2 \\ &= U\nu_1 + U\hat{P}_{G_n}\nu_2, \end{aligned}$$

and letting $n \to \infty$ we obtain $v = U\nu_1 + g$. But $v = f^B = u$ a.e., and $g = U\hat{P}_B\nu_2$ a.e. Therefore $u = U\nu_1 + U\hat{P}_B\nu_2$ a.e., and hence everywhere. Thus $u = U\nu_B$ where $\nu_B = \nu_1 + \hat{P}_B\nu_2$, and clearly ν_B is carried $B \cup {}^rB$. Finally if B has compact closure in E and G is a compact neighborhood of B, then $P_G f = U\nu$ for some $\nu \in \mathbf{M}$ by Theorem 2.8, and the same argument as above shows that $u = U\nu_B$ for $\nu_B = \nu_1 + \hat{P}_B\nu_2$ where again ν_1 is the restriction of ν to B and $\nu_2 = \nu - \nu_1$. Thus, at last, the proof of Theorem 2.12 is complete.

We close this section by investigating the relationship between $P_B f$ and f^B when $B \in \mathscr{E}$. The notation is that of (2.12). Since condition (2.2) certainly implies the hypothesis of Theorem 6.12 of Chapter III (with $M_t = I_{[0,\zeta)}(t)$) we know that $P_B f$ agrees with the infimum of the family of all excessive functions dominating f on B except possibly at the points of $B - B^r$, and so there must be a close relationship between $P_B f$ and f^B. In the course of the proof of (2.12) we showed that $P_B f = f^B = f$ on B^r and $P_G f = f^G$ if G is open. Since f^B is only super-mean-valued and $P_B f$ is excessive we cannot hope to show that they are equal in general. The most one could hope for is that $P_B f$ is the regularization of f^B and this is certainly the case if $f^B = P_B f$, a.e. The next result characterizes those excessive functions f for which this is true. The following concept is used in its statement. A locally integrable excessive function f is said to be *admissible* if whenever K is a compact subset of E there exists a decreasing sequence $\{G_n\}$ of open sets containing K and such that $P_{G_n} f \downarrow P_K f$ a.e.

(2.15) PROPOSITION: (i) Let f be a locally integrable excessive function. Then for each $B \in \mathscr{E}$, $f^B = P_B f$, a.e. if and only if f is admissible. In this case $f^B = P_B f$ except possibly on $(B - B^r) \cup (B^c \cap \{f = \infty\})$. (ii) If $f = U\mu$ with $\mu \in \mathbf{M}$, then f is admissible if and only if $\mu(K - {}^rK) = 0$ for all compact subsets K of E. In this case, for all $B \in \mathscr{E}$, $P_B f = U\nu_B$ where ν_B is the measure defined in Theorem 2.12.

Proof. Suppose f is admissible, K is compact, and $\{G_n\}$ is a decreasing sequence of open sets containing K such that $P_{G_n} f \downarrow P_K f$ a.e. Clearly we may assume that no point of G_n is distant from K by more than $1/n$ and so if G is any open neighborhood of K, then $G_n \subset G$ for large n. Since by (2.13), $f^K = \inf P_G f$ as G ranges over such neighborhoods we have $f^K = \lim P_{G_n} f$. Consequently $P_K f = f^K$ a.e. and $P_K f$ is the regularization of f^K. Thus by (3.20) of Chapter II, $f^K = P_K f$ on $K^c \cap \{f^K < \infty\} \supset K^c \cap \{f < \infty\}$. We now assert that, for any $B \in \mathscr{E}$, $P_B f = f^B$ except perhaps on $(B - B^r) \cup (B^c \cap \{f = \infty\})$. Indeed we have already observed that $P_B f(x) = f^B(x)$ if $x \in B^r$. Now suppose $x \in B^c \cap \{f < \infty\}$. Then by the proof of (2.12) there is a sequence $\{K_n\}$ of compact subsets of B such that $f^{K_n}(x) \to f^B(x)$. We may also suppose the K_n are such that $T_{K_n} \downarrow T_B$ a.s. P^x, so that $P_{K_n} f(x) \to P_B f(x)$. By what was proved above $P_{K_n} f(x) = f^{K_n}(x)$ for all n, and so $P_B f(x) = f^B(x)$.

Conversely if K is compact we can find a decreasing sequence $\{G_n\}$ of open sets containing K such that $P_{G_n} f \downarrow f^K$. Indeed $f^K = \inf P_G f$ as G ranges over the open neighborhoods of K and so we have merely to choose a sequence $\{G_n\}$ which is ultimately contained in any such G. Now if $P_K f = f^K$ a.e. and $P_{G_n} f \downarrow f^K$, then $P_{G_n} f \downarrow P_K f$ a.e. Consequently if, for every compact K, $P_K f = f^K$ a.e. then f is admissible. This proves assertion (i) in (2.15).

Coming to assertion (ii), let K be a compact set and let $G_n = \{x: \rho(x, K) < 1/n\}$, where $\rho(x, K)$ is the distance from x to K. It is obvious that $P_{H_n} f \downarrow P_K f$ a.e. for some sequence $\{H_n\}$ of open sets containing K if and only if $P_{G_n} f \downarrow P_K f$ a.e. Since $\{P_{G_n} f\}$ decreases, this will occur if and only if

$$\textbf{(2.16)} \qquad \langle h, P_{G_n} U\mu \rangle \downarrow \langle h, P_K U\mu \rangle$$

whenever $h \in b\mathscr{E}^*$ is positive with compact support. Now for any set J

$$\langle h, P_J U\mu \rangle = \hat{E}^{\mu} \int_{T_J}^{\zeta} h(X_t)\, dt$$

and this makes it obvious that (2.16) holds for all relevant h if and only if $\hat{T}_{G_n} \wedge \hat{\zeta} \uparrow \hat{T}_K \wedge \hat{\zeta}$ almost surely $\hat{P}^{\mu}$. Now this will happen if and only if $\mu(K - {}^rK) = 0$, and so $U\mu$ is admissible if and only if $\mu(K - {}^rK) = 0$ whenever K is compact. Finally when $f = U\mu$ with $\mu \in \mathbf{M}$ then, by (2.12), $f^B = U\nu_B$ a.e. and if f is also admissible then $U\nu_B = P_B f$ a.e., and hence everywhere. This completes the proof.

REMARK. It is easy to see that $\mu(K - {}^rK) = 0$ for all compact subsets K of E if and only if $\mu(B) = 0$ for every semipolar set B.

Exercises

(2.17) Let $A \subset \mathbf{R}^4$ be the set of points, all of whose coordinates are rational, and let ν be a probability measure carried by A and such that $\nu(\{q\}) = \nu(\{-q\}) > 0$ for all $q \in A$. Let $g_t(x, y)$ be the transition density for Brownian motion in $\mathbf{R}^4$ (see (2.17) of Chapter I). Define ν_0 to be unit mass at $0 \in \mathbf{R}^4$ and $\nu_{k+1} = \nu * \nu_k$ for $k \geq 0$. Let $\mu_t = e^{-t} \sum_{k=0}^{\infty} (t^k/k!)\nu_k$.

(a) Show that $\{\mu_t; t \geq 0\}$ is a convolution semigroup of probability measures on $\mathbf{R}^4$.

(b) Show that $p_t(x, y) = \int g_t(x - q, y)\, \mu_t(dq)$ defines a transition density with respect to Lebesgue measure for a Hunt process X in $\mathbf{R}^4$.

(c) Show that $p_t(x, y) = p_t(y, x)$ so that this process is in duality with itself relative to Lebesgue measure. Show that conditions (2.1) and (2.2) are satisfied. [Hint: show that $p_t(x, y) \leq t^{-2}$.]

(d) Show that if $x - y \in A$ then $p_t(x, y) \geq kt^{-1}$ $(k > 0)$ for small t and conclude that $u(x, y) = \infty$ whenever $x - y \in A$.

(e) Observe that $\nu \in \mathbf{M}$ and use (d) to show that $P_G U\nu(x) = \infty$ for all $x \in A$ whenever G is open and nonempty. Conclude that the exceptional set in the discussion preceding (2.6) cannot in general be eliminated.

(2.18) Let B be a thin Borel set. Show that T_B is accessible on $\{T_B < \zeta\}$.

[Hint: given an initial measure μ with $\mu(B) = 0$ let $\{G_n\}$ be a decreasing sequence of neighborhoods of B such that $T_{G_n} \wedge \zeta \uparrow T_B \wedge \zeta$ a.e. P^μ. Use (2.9) to conclude that $P^\mu(T_{G_n} = T_B < \infty) = 0$ for all n. For the general case use the first part to conclude that if K is a compact subset of B^c then T_B is accessible on $\{T_K < T_B < \zeta\}$. Finally use the fact that there is an increasing sequence $\{K_n\}$ of compact subsets of B^c such that $P^x(T_{K_n} \downarrow 0) = 1$ for all x.]

(2.19) Let X be the process (Z, T_{A^c}) where Z is Brownian motion in $\mathbf{R}^n$ and $A = \{x: |x| < 1\}$ (see (1.27)). Show that in this case the decomposition (2.11) is the classical Riesz decomposition of a positive superharmonic function into the sum of a harmonic function and the potential of a measure.

3. Potentials of Additive Functionals

In this section we are going to study the relationship between the potentials of measures in $\mathbf{M}$ and the potentials of additive functionals. In particular we will characterize those measures $\mu \in \mathbf{M}$ having the property that $U\mu$ is either a natural potential or a regular potential as defined in Chapter IV. The assumptions in this section are the same as those in Section 2. Thus X and $\hat{X}$ are standard processes in duality relative to ξ, and (2.1) and (2.2) are assumed to hold. As in Section 2 we will state and prove our results in the case $\alpha = 0$. However these results are valid also for positive α under (2.1) alone.

(3.1) THEOREM. Let μ be a measure in $\mathbf{M}$ and let A be an NAF of X such that $U\mu = u_A$. Then for any $f \in \mathscr{E}_+^*$ we have $U_A f = U(f\mu)$.

Proof. Recall that $u_A(x) = U_A 1(x) = E^x A(\infty)$. Also

$$U(f\mu)(x) = \int u(x, y) f(y)\, \mu(dy).$$

Let x be a fixed point in E at which $U\mu(x) < \infty$. Since both sides of the desired equality are excessive functions it will suffice to establish the desired equality for such x. On the other hand, with such an x fixed, $U_A f(x)$ and $U(f\mu)(x)$ are finite measures in f and so it will be enough to establish the equality when $f = I_G$ where G is an open set with compact closure in E whose boundary has μ measure zero. Let $g = I_{E-G}$ and let $\varphi = U_A f$ and $\psi = U_A g$. Since φ and ψ are dominated by $u_A = U\mu$ it follows from (2.10) that $\varphi = U\nu$ and $\psi = U\lambda$ where ν and λ are measures in $\mathbf{M}$. Now

$$P_G\, \varphi(x) = E^x \int_{(T_G, \infty)} f(X_t)\, dA_t,$$

and, since $f(X_t)$ vanishes on $[0, T_G)$,

$$\varphi(x) - P_G\,\varphi(x) = E^x\{f(X_{T_G})\,[A_{T_G}) - A(T_G-)]\}.$$

If A has a jump at $T_G < \infty$, then since A is natural the path X_t must be continuous at T_G, and consequently $X(T_G)$ is in G^c. Therefore $\varphi = P_G\varphi$ and so $Uv = P_G Uv = U\hat{P}_G v$. But this implies that $v = \hat{P}_G v$, and hence v is carried by $\bar{G}$. Coming to ψ, let H be an open neighborhood of the closed set $E - G$. Then

$$P_H\,\psi(x) = E^x \int_{(T_H,\infty)} g(X_t)\,dA_t,$$

and since $g(X_t)$ vanishes on $[0, T_{E-G}) \supset [0, T_H)$ one obtains, as above, $\psi = P_H\psi$. As a result $\lambda = \hat{P}_H\lambda$ and so λ is carried by $\bar{H}$. But H was an arbitrary neighborhood of $E - G$ and hence $E - G$ must carry λ.

Next observe that

$$U\mu = u_A = U_A(f + g) = \varphi + \psi = U(v + \lambda),$$

and so $\mu = v + \lambda$. But v is carried by $\bar{G}$, λ is carried by $E - G$, and the boundary of G has μ measure zero. Plainly this yields $v = f\mu$ and $\lambda = g\mu$. Therefore $U_A f = \varphi = Uv = U(f\mu)$, and so Theorem 3.1 is established.

We are next going to characterize those measures $\mu \in \mathbf{M}$ such that $U\mu$ is a natural potential. In the course of the discussion we will need the following fact which is of considerable interest in itself.

(3.2) Proposition. Let X be a standard process—no assumptions about the existence of a dual process or even the existence of a reference measure are made. Let $\{f_n\}$ be a decreasing sequence in $\mathcal{S}^\alpha$ $(\alpha \geq 0)$ with limit f and suppose that $f = 0$ except on a set of potential zero. Then $f = 0$ except on a polar set.

Proof. The argument is the same for all α and so for convenience we assume that $\alpha = 0$. If T is any $\{\mathcal{F}_t\}$ stopping time, then $P_T f_n \leq f_n$, and so by the bounded convergence theorem $P_T f(x) \leq f(x)$ at any x for which $f(x) < \infty$. In particular $P_T f = f$ on $\{f = 0\}$. Let $\varepsilon > 0$ and let K be a compact subset of $\{f \geq \varepsilon\}$. Then if $f(x) = 0$ we have

$$0 = f(x) \geq P_K f(x) \geq \varepsilon P^x(T_K < \infty),$$

and so $P^x(T_K < \infty) = 0$. But $\{f > 0\}$ is of potential zero and since the excessive function $x \to P^x(T_K < \infty)$ vanishes except on this set it follows that, for all x, $P^x(T_K < \infty) = 0$. Therefore K is polar and this implies that $\{f > 0\}$ is polar proving (3.2).

(3.3) Remark. If X has a reference measure ξ, then Proposition 3.2 may be restated as follows: a decreasing sequence $\{f_n\}$ in $\mathscr{S}^\alpha$ which approaches zero a.e. must, in fact, approach zero except on a polar set.

(3.4) Theorem. Let μ be a measure in $\mathbf{M}$. Then necessary and sufficient conditions that $U\mu$ be a natural potential are (i) $U\mu$ is everywhere finite and (ii) μ charges no polar set.

Proof. If $U\mu$ is a natural potential, then $U\mu$ is finite by Definition 4.17 of Chapter IV. On the other hand if μ charges some polar set then μ must charge a compact polar set K. Let ν be the restriction of μ to K. Then $U\nu \leq U\mu$ and $U\nu$ is a natural potential. Let $\{G_n\}$ be a decreasing sequence of open sets containing K such that $\bigcap \bar{G}_n = \bigcap G_n = K$. Since ν is supported by K, $P_{G_n} U\nu = U\hat{P}_{G_n}\nu = U\nu$. But if $x \notin K$ then $\lim T_{G_n} \geq \zeta$ almost surely P^x, and so by the definition of natural potential $P_{G_n} U\nu(x) \to 0$. Consequently $U\nu = 0$ except possibly on the polar set K and so $U\nu$ vanishes. Therefore $\nu = 0$ and so μ charges no polar set. This argument used only the definition of natural potential If one makes use of the theory developed in Chapter IV, then we know that there exists an NAF, A, of X such that $U\mu = u_A$. If D is a polar Borel set, then by (3.1)

$$U(I_D \mu)(x) = U_A I_D(x) = E^x \int_0^\infty I_D(X_t)\, dA_t,$$

and this last expression is clearly zero. This gives an alternate proof of the fact that μ does not charge polar sets.

Next suppose that μ does not charge polar sets and that $U\mu$ is finite. Given x let $\{T_n\}$ be an increasing sequence of stopping times such that $\lim T_n \geq \zeta$ almost surely P^x. Since $z \to u(z, y)$ is excessive, it follows that for each fixed y and z, $P_{T_n} u(z, y)$ decreases with n. If $f \in b\mathscr{E}_+^*$ and vanishes off a compact subset of E, then

$$P_{T_n} U f(x) = E^x \int_{T_n}^{\zeta} f(X_t)\, dt \to 0$$

as $n \to \infty$. Therefore $P_{T_n} u(x, \cdot)$ decreases to zero a.e. But for each n, $y \to P_{T_n} u(x, y)$ is coexcessive and so, by (3.3), $P_{T_n} u(x, \cdot)$ decreases to zero except on a polar set. Since μ doesn't charge polar sets and $U\mu$ is finite, we see that $P_{T_n} U\mu(x) = \int P_{T_n} u(x, y)\, \mu(dy)$ decreases to zero. Consequently $U\mu$ is a natural potential and Theorem 3.4 is proved.

(3.5) Theorem. Let μ be a measure in $\mathbf{M}$. Then necessary and sufficient conditions that $U\mu$ be a regular potential are that (i) $U\mu$ is everywhere finite and (ii) μ charges no semipolar set.

Proof. If $U\mu$ is a regular potential, then $U\mu$ is finite and there exists, by Theorem 3.13 of Chapter IV, a CAF, A of X such that $U\mu = u_A$. Any semipolar set is contained in a Borel semipolar set D, and by (3.1) we have

$$U(I_D \mu)(x) = U_A I_D(x) = E^x \int_0^\infty I_D(X_t)\, dA_t .$$

Since D is semipolar X_t is in D for at most countably many values of t almost surely, and since A_t is continuous the measure dA_t induced by A_t on $[0, \infty]$ vanishes on countable sets. Therefore the last displayed expression is zero, and consequently μ charges no semipolar set.

Conversely suppose $U\mu$ is finite and μ charges no semipolar set. Let $u = U\mu$. If K is a compact subset of E and μ_K is the restriction of μ to K, then since $K - {}^rK$ is semipolar it follows that $P_K U\mu_K = U\hat{P}_K \mu_K = U\mu_K$. Now let λ be a measure equivalent to ξ such that $\int u\, d\lambda < \infty$, and let η be the measure given by $\eta(\Gamma) = \int_\Gamma \lambda\hat{U}\, d\mu$ for $\Gamma \in \mathscr{E}^*$. Then $\eta(E) = \int u\, d\lambda < \infty$. Since $P_t u \uparrow u$ as $t \to 0$ we can find, using Egorov's theorem, an increasing sequence $\{K_n\}$ of compact subsets of E such that (i) u is bounded on each K_n, (ii) $P_t u \to u$ uniformly on each K_n as $t \to 0$, and (iii) $\eta(E - K_n) \downarrow 0$ as $n \to \infty$. For each n let μ_n be the restriction of μ to K_n and ν_n be the restriction of μ to $E - K_n$. Then $U\nu_n$ decreases with n and since $\int U\nu_n\, d\lambda = \eta(E - K_n)$ it follows that $U\nu_n$ decreases to zero a.e. (λ). But $u = U\mu_n + U\nu_n$ and hence $U\mu_n$ increases to u a.e. (λ), which implies that $\lim U\mu_n = u$ everywhere since both u and $\lim_n U\mu_n$ are excessive. Also $U\mu_{n+1} - U\mu_n = U\mu_n'$ where μ_n' is the restriction of μ to $K_{n+1} - K_n$. Therefore the sequence $\{U\mu_n\}$ is increasing in the strong order; that is, $U\mu_{n+1} - U\mu_n$ is excessive for each n.

We next claim that, for each n, $u_n = U\mu_n$ is uniformly excessive. We already have observed that $P_{K_n} u_n = u_n$ for each n. Since $u_n \le u$ and u is bounded on K_n this implies that each u_n is bounded. Now fix n. Then given $\varepsilon > 0$ there exists a $\delta > 0$ such that $u \le P_t u + \varepsilon$ on K_n provided $t < \delta$. Hence on K_n we have for $t < \delta$

$$u_n + P_t U\nu_n \le u_n + U\nu_n = u \le P_t u + \varepsilon = P_t u_n + P_t U\nu_n + \varepsilon.$$

and so $u_n \le P_t u_n + \varepsilon$ on K_n. But $P_t u_n + \varepsilon$ is excessive and so

$$u_n = P_{K_n} u_n \le P_{K_n}(P_t u_n + \varepsilon) \le P_t u_n + \varepsilon$$

everywhere provided $t < \delta$. Consequently u_n is uniformly excessive. If we define $u_0 = 0$, then plainly $f_n = u_n - u_{n-1}$ is uniformly excessive for each n and $u = \sum f_n$. By (3.4), u is a natural potential and so, for each n, $P_t f_n \le P_t u \to 0$ as $t \to \infty$. Therefore according to Theorem 3.16 of Chapter IV there exists for each n a CAF, A^n, of X such that $f_n = u_{A^n}$. If we define $A = \sum A^n$, then $E^x(A_\infty) = u(x) < \infty$ and so $\sum A_t^n$ converges uniformly on $[0, \infty]$

almost surely. As a result A is a CAF of X whose potential is $u = U\mu$, and consequently $U\mu$ is a regular potential. This establishes Theorem 3.5.

REMARK. The reader should note that the proof of the second half of Theorem 3.5 is essentially the same as the proof of Theorem 2.1. of Chapter V.

The following decomposition complements Theorems 3.4 and 3.5.

(3.6) PROPOSITION. Let μ be a measure in **M**. Then μ can be written uniquely in the form $\mu = \mu_1 + \mu_2 + \mu_3$ where μ_1 is carried by a polar (Borel) set, μ_2 is carried by a semipolar (Borel) set but charges no polar set, and μ_3 charges no semipolar set. In particular μ_1, μ_2, and μ_3 are carried by disjoint Borel sets.

Proof. Recall that any polar set is contained in a polar Borel set. Let μ_B denote the restriction of μ to B whenever B is a Borel set. Define

$$\mathscr{P} = \{U\mu_B\colon B \text{ a polar Borel set}\}.$$

Clearly $\mathscr{P}$ is filtering upward and so by Theorem 1.5 of Chapter V we can find an increasing sequence $\{U\mu_{B_n}\}$ in $\mathscr{P}$ such that $\sup U\mu_{B_n} = \sup \mathscr{P}$. Let $B = \bigcup B_n$ and $\mu_1 = \mu_B$. Plainly B is a polar Borel set and $\mu - \mu_1$ charges no polar set. One next constructs μ_2 in a similar manner starting from the measure $\mu - \mu_1$, and then $\mu_3 = \mu - \mu_1 - \mu_2$. Moreover it is evident from this construction that this decomposition is unique. (The uniqueness may also be established very easily by a direct argument.) Thus Proposition 3.6 is proved.

An immediate consequence of (3.6) and (2.11) is the following decomposition which should be compared with Theorem 5.14 of Chapter IV. A locally integrable excessive function f can be written uniquely in the form $f = U\mu_1 + U\mu_2 + U\mu_3 + h$ where μ_1, μ_2, and μ_3 are as in Proposition 3.6 and h has the property that $P_D h = h$ whenever D is the complement of a compact subset of E.

4. Capacity and Related Topics

In this section we will develop a theory of capacity which in the classical case reduces to the ordinary Newtonian capacity. We will also explore the situation in which every semipolar set is polar—this additional hypothesis being necessary if one wishes to obtain some of the strongest results of classical potential theory. We will assume throughout this section that X

and $\hat{X}$ are standard processes in duality relative to ξ, that (2.1) and (2.2) hold, and, in addition, that X satisfies the same smoothness hypotheses as $\hat{X}$. To be more explicit we assume:

(4.1) If $\alpha > 0$ and $f \in b\mathscr{E}^*$ vanishes off a compact subset of E, then $U^\alpha f(x) = \int u^\alpha(x, y) f(y)\, dy$ is continuous on E.

(4.2) With f as in (4.1), $Uf(x) = \int u(x, y) f(y)\, dy$ is continuous on E.

We separate (4.1) and (4.2) just as we did (2.1) and (2.2) because the results which we will establish hold for strictly positive α just under (2.1) and (4.1). Again, however, we will formulate and prove these results mainly in the case $\alpha = 0$. Under these assumptions there is a complete duality between X and $\hat{X}$ and hence the results of Sections 2 and 3 apply to $\hat{X}$ as well as X. In particular the elements of $\mathscr{S}^\alpha$ $(\alpha \geq 0)$ are lower semicontinuous (only (4.1) is needed for this fact). Also (4.1) implies that condition (4.1) of Chapter IV holds. To see this suppose that $\{T_n\}$ is an increasing sequence of stopping times with limit T. Let $\alpha > 0$ and $f \in \mathbf{C}_K$, $f \geq 0$ be fixed. Then there exists a nonnegative random variable $L \in \mathscr{F}$ such that $e^{-\alpha T_n} U^\alpha f(X_{T_n}) \to e^{-\alpha T} L$ almost surely. Clearly $L \geq U^\alpha f(X_T)$ almost surely. On the other hand

$$
\begin{aligned}
E^x\{e^{-\alpha T_n} U^\alpha f(X_{T_n})\} &= E^x \int_{T_n}^{\infty} e^{-\alpha t} f(X_t)\, dt \\
&\to E^x \int_{T}^{\infty} e^{-\alpha t} f(X_t)\, dt \\
&= E^x\{e^{-\alpha T} U^\alpha f(X_T)\},
\end{aligned}
$$

and therefore $\lim_n U^\alpha f(X_{T_n}) = U^\alpha f(X_T)$ almost surely on $\{T < \infty\}$. By (4.32) of Chapter IV this implies that (4.1) of Chapter IV holds. Consequently all of the results of Sections 4 and 5 of Chapter IV are applicable to X and $\hat{X}$.

As in Section 1 of Chapter V we let e_B denote the lower envelope of the family of all excessive functions which dominate 1 on B, where B is an arbitrary subset of E. In the present situation e_B is also the lower envelope of the family of all excessive functions which dominate 1 on some (variable) neighborhood of B. To see this let e_B^* denote this second infimum for the moment. Clearly $e_B \leq e_B^*$. On the other hand if $f \in \mathscr{S}$ and $f \geq 1$ on B, then $\{f > 1 - \varepsilon\}$ is an open set containing B since f is lower semicontinuous. Thus for any ε with $0 < \varepsilon < 1$ the excessive function $(1 - \varepsilon)^{-1} f$ dominates 1 on a neighborhood of B and so $(1 - \varepsilon) e_B^* \leq f$. Letting $\varepsilon \downarrow 0$ we obtain $e_B^* \leq f$, and hence $e_B^* \leq e_B$. Thus $e_B = e_B^*$ in the present situation. As a result we may apply either Theorem 6.12 of Chapter III or Theorem 2.12 to e_B when B is a Borel set to obtain the following statement.

(4.3) PROPOSITION. Let B be a Borel set. Then $e_B = P_B 1$ except perhaps on $B - B^r$, and e_B agrees except perhaps on $B - B^r$ with the supremum of all potentials $U\mu$ where $U\mu \le 1$ and μ has compact support contained in B. Since this supremum is excessive it agrees everywhere with $P_B 1$. Finally if B has compact closure $P_B 1 = U\pi_B$ where π_B is a finite measure carried by $B \cup {}^rB$.

An immediate, but important, consequence of (4.3) is the fact that if $U\mu$ is bounded then μ charges no polar set. Indeed if ν is the restriction of μ to a compact polar set K and $\nu \ne 0$, then $\eta = \nu/\|U\nu\|$ is a measure on K with $U\eta \le 1$ and $U\eta \ne 0$. Consequently $P_K 1 \ge U\eta$ and this is a contradiction since K is polar. This also shows that a set B is polar if and only if $U\mu$ is unbounded whenever μ is a finite nonzero measure with compact support contained in B.

Let B be a Borel set with compact closure in E. Then by (4.3) or (2.8) we have, for each $\alpha \ge 0$, $\Phi_B^\alpha = P_B^\alpha 1 = U^\alpha \pi_B^\alpha$ and $\hat{\Phi}_B^\alpha = 1\hat{P}_B^\alpha = \hat{\pi}_B^\alpha \hat{U}^\alpha$ where π_B^α and $\hat{\pi}_B^\alpha$ are finite measures carried by $B \cup {}^rB$ and $B \cup B^r$, respectively. The measure π_B^α is called the *α-capacitary measure* of B, and $\hat{\pi}_B^\alpha$ the *α-cocapacitary measure of* B. As usual when $\alpha = 0$ we will drop it entirely from our notation and terminology. If A is a Borel set with compact closure in E and if $B \subset A^r$ then $U\pi_A = 1$ on $B \cup B^r$ and so $P_B 1 = P_B U\pi_A$. Hence $\pi_B = \hat{P}_B \pi_A$, and similarly $\hat{\pi}_B = \hat{\pi}_A P_B$ if $B \subset {}^rA$. On the other hand if $B \subset {}^rA$ and $B \subset A$ then $P_A P_B = P_B$ by (2.9) and so $\pi_B = \hat{P}_A \pi_B$. Similarly $\hat{\pi}_B = \hat{\pi}_B P_A$ if $B \subset A$ and $B \subset A^r$. All of these relationships are obviously valid for positive α also.

(4.4) PROPOSITION. If B is a Borel set with compact closure in E, then the measures π_B^α and $\hat{\pi}_B^\alpha$ have the same mass.

Proof. Let G be a neighborhood of $\bar{B}$ having compact closure in E. Then using the above remarks we have

$$\hat{\pi}_B(E) = \hat{\pi}_B(\bar{B}) = \int \hat{\pi}_B(dx)\, U\pi_G(x)$$

$$= \int \hat{\pi}_G\, P_B(dx)\, U\pi_G(x) = \iint \hat{\pi}_G(dx)\, u\hat{P}_B(x, y)\, \pi_G(dy)$$

$$= \int \hat{\pi}_G\, \hat{U}(y)\, \pi_B(dy) = \pi_B(E),$$

which proves (4.4) since the same computation is valid for $\alpha > 0$.

(4.5) DEFINITION. The common value $\pi_B^\alpha(E) = \hat{\pi}_B^\alpha(E)$ is called the (*natural*) *α-capacity of* B. We denote it by $C^\alpha(B)$.

In the following discussion we assume that $\alpha = 0$. However the results obviously are valid for arbitrary $\alpha \geq 0$. Let B denote a Borel set with compact closure in E. If K is a compact subset of B and μ is carried by K, then $\mu(K) \leq C(B)$ provided $U\mu \leq 1$. Indeed by (4.3), $U\mu \leq U\pi_B$ and so if $\{f_n\}$ is a sequence of bounded nonnegative functions such that $\{f_n \hat{U}\}$ increases to 1 we have

$$\mu(K) = \lim_n \int (f_n \hat{U})\, d\mu = \lim_n \langle f_n, U\mu \rangle$$

$$\leq \lim_n \langle f_n, U\pi_B \rangle = \lim_n \int (f_n \hat{U})\, d\pi_B = C(B).$$

On the other hand by (1.20) of Chapter V we can find an increasing sequence $\{K_n\}$ of compact subsets of B such that $U\pi_{K_n} = P_{K_n} 1 \uparrow P_B 1 = U\pi_B$. It then follows from (2.6) that $\{\pi_{K_n}\}$ converges weakly to π_B and consequently $C(K_n) \uparrow C(B)$. Thus we have shown that $C(B) = \sup \mu(E)$ where the supremum is over all measures μ with support in B such that $U\mu \leq 1$ (or dually $\mu\hat{U} \leq 1$). In particular there is an increasing sequence $\{K_n\}$ of compact subsets of B such that $C(K_n) \uparrow C(B)$. It is also clear that $C(B) = 0$ if and only if B is polar. So far we have defined the capacity, $C(B)$, of a set B only when B has compact closure in E. If B is any Borel subset of E we define the capacity, $C(B)$, of B to be the supremum of $C(K)$ as K varies over all compact subsets of B. This definition agrees with the previous one if B has compact closure. We leave it to the reader as Exercise 4.14 to check that $K \to C(K)$ defines a Choquet capacity on the compact subsets of E. Obviously its extension to $\mathscr{E}$ is just C. Moreover C is countably subadditive. See Exercise 4.14. Since a set is polar if and only if every compact subset of it is polar, it remains true that a Borel set B is polar if and only if $C(B) = 0$.

Suppose now that $B \in \mathscr{E}$ and that there exists a measure π_B such that $\Phi_B = U\pi_B$—here B need not have compact closure in E. Let us show that it is still the case that $C(B)$ is the total mass of π_B and that π_B is carried by $B \cup {}^rB$. As in the preceding paragraph we can find an increasing sequence $\{K_n\}$ of compact subsets of B such that $U\pi_{K_n} \uparrow U\pi_B$, and, in addition, we may suppose that $C(K_n) \uparrow C(B)$. If $\{f_n\}$ is a sequence of bounded nonnegative functions such that $\{f_n \hat{U}\}$ increases to one, then

$$C(B) = \lim_n C(K_n) = \lim_n \lim_m \int (f_m \hat{U})\, d\pi_{K_n}$$

$$= \lim_m \lim_n \langle f_m, U\pi_{K_n} \rangle$$

$$= \lim_m \langle f_m, U\pi_B \rangle = \pi_B(E)$$

where the interchange of limits is justified since $\int (f_m \hat{U})\, d\pi_{K_n} = \langle f_m, U\pi_{K_n} \rangle$

is increasing in both m and n. Let ν' be the restriction of π_B to B and $\nu = \pi_B - \nu'$. Let $\{G_n\}$ be a decreasing sequence of open sets containing B such that $(\hat{T}_{G_n} \wedge \hat{\zeta}) \uparrow (\hat{T}_B \wedge \hat{\zeta})$ almost surely $\hat{P}^\nu$. Then

$$\Phi_B = P_{G_n}\Phi_B = U\nu' + U\hat{P}_{G_n}\nu,$$

and, as in the proof of (2.12), $U\hat{P}_{G_n}\nu = P_{G_n}U\nu$ decreases to $P_B\,U\nu = U\hat{P}_B\,\nu$ a.e. Consequently $\Phi_B = U(\nu' + \hat{P}_B\,\nu)$ and so $\pi_B = \nu' + \hat{P}_B\,\nu$. Therefore π_B is concentrated on $B \cup {}^rB$. Whenever π_B exists it will be called the *capacitary measure* of B. If both π_B and $\hat{\pi}_B$ exist, then they must have the same mass—namely, $C(B)$.

We are now going to develop some of the implications of the following hypothesis:

(4.6) Every semipolar set is polar.

Obviously (4.6) is equivalent to the statement that if a compact set K is not polar, then some point of K must be regular for K. This last statement is hypothesis (H) of Hunt's memoir (Hunt [4], p. 193). Note that (4.6) is *not* satisfied by the process "uniform motion to the right in **R**," so that in one sense it is a strong restriction. On the other hand we shall see that (4.6) is satisfied by many of the familiar processes, in particular by Brownian motion and by the symmetric stable processes. Obviously X satisfies condition (4.6) if and only if for some $\alpha > 0$ the α-subprocess ((3.17) of Chapter III) satisfies (4.6). We will state explicitly when (4.6) is being assumed to hold. Let us begin with the following characterization of the capacitary measure π_K of a compact set K.

(4.7) PROPOSITION. Assume (4.6). If K is a compact subset of E, then π_K is the unique measure carried by K and having the following two properties: (i) $U\pi_K \le 1$ and (ii) $\{U\pi_K < 1\} \cap K$ is polar.

Proof. Since $U\pi_K(x) = \Phi_K(x) = P^x(T_K < \infty)$ and $\Phi_K = 1$ on K^r, it is clear that, under (4.6), π_K has properties (i) and (ii). Now let μ be any measure with support in K and satisfying (i) and (ii). If $x \notin K$ then $P_K(x, \cdot)$ charges no polar set and so, for such an x, $P_K U\mu(x) = P_K 1(x) = U\pi_K(x)$. Also $U\mu$ is bounded and so μ charges no polar set. But under (4.6), $K - {}^rK$ is polar and so $P_K U\mu = U\hat{P}_K\,\mu = U\mu$. Consequently $U\mu = U\pi_K$ except possibly on a polar set and hence $\mu = \pi_K$ proving (4.7).

We are now going to formulate some conditions which are equivalent to (4.6). A locally integrable α-excessive function f is said to be *regular* provided that almost surely the mapping $t \to f(X_t)$ is continuous wherever

$t \to X_t$ is continuous on $[0, \zeta)$. If f is finite this is clearly equivalent to the definition of regularity given in (5.1) of Chapter IV. The requirement that f is everywhere finite is not appropriate here—the local integrability of f is the appropriate finiteness condition. Of course, since f is locally integrable, $\{f = \infty\}$ is polar and so almost surely $t \to f(X_t)$ is finite on $(0, \infty]$. In particular the proof of (5.9) of Chapter IV remains valid when f is merely assumed to be locally integrable. As a result, a locally integrable f in $\mathscr{S}^\alpha$ is regular if and only if whenever $\{T_n\}$ is an increasing sequence of stopping times with limit T we have $f(X_{T_n}) \to f(X_T)$ almost surely on $\{T < \zeta\}$. We continue to state our results in the case $\alpha = 0$, but emphasize again that they carry over to positive α just under (2.1) and (4.1).

(4.8) PROPOSITION. A necessary and sufficient condition that all locally integrable excessive functions are regular is that $U\mu$ be regular whenever μ has compact support and $U\mu$ is bounded.

Proof. Only one of the implications requires proof. Suppose $f \in \mathscr{S}$ is locally integrable and that, for some $x \in E$ and some increasing sequence $\{T_n\}$ of stopping times with limit T, $P^x\{f(X_{T_n}) \nrightarrow f(X_T);\ T < \zeta\} > 0$. Then obviously there is a positive number k and an open set G with compact closure in E such that $P^x\{(f \wedge k)(X_{T_n}) \nrightarrow (f \wedge k)(X_T);\ T < T_{G^c}\} > 0$. Now $P_G(f \wedge k) = (f \wedge k)$ on G and $P_G(f \wedge k)$ is the bounded potential $U\mu$ of a measure μ carried on $\bar{G}$. Hence $P^x\{U\mu(X_{T_n}) \nrightarrow U\mu(X_T);\ T < \zeta\} > 0$, contrary to the hypothesis that every bounded potential of a measure with compact support is regular.

The next theorem, which is due to Hunt, is one of the most important results in this section.

(4.9) THEOREM. All locally integrable excessive functions are regular if and only if (4.6) holds.

Proof. Suppose (4.6) holds and $U\mu$ is a bounded potential. According to (3.6), $\mu = \mu_1 + \mu_2 + \mu_3$ where $\mu_1 + \mu_2$ is carried by a semipolar, and hence by a polar, set and μ_3 charges no semipolar set. Hence $\mu_1 + \mu_2 = 0$ because $U\mu$ is bounded. Consequently μ charges no semipolar set and so, by (3.5), $U\mu$ is a regular potential. It follows from (5.13) of Chapter IV then that $U\mu$ is regular and so one half of (4.9) follows from (4.8).

In proving the converse we first observe that if all locally integrable excessive functions are regular, then all locally integrable α-excessive functions are regular for each $\alpha > 0$. Indeed let μ be a measure with compact support such that $U^\alpha\mu$ is bounded. Clearly μ is finite since μ is in $\mathbf{M}^\alpha$ and so μ is in $\mathbf{M}$

also. Now $U\mu = U^\alpha \mu + \alpha U U^\alpha \mu$ and, since μ is in **M**, $U\mu$ and $UU^\alpha\mu$ are locally integrable. This and (4.8) plainly yield the assertion in the first sentence of this paragraph.

Now suppose there is a compact subset K of E which is thin but not polar. Let $\alpha > 0$ be fixed and let $\Phi_K^\alpha(x) = E^x(e^{-\alpha T_K})$. Then Φ_K^α is everywhere less than one and Φ_K^α is a regular α-excessive function. Since K is not polar Φ_K^α is not identically 0, and hence, since $\xi(K) = 0$, there is a point $x \notin K$ with $\Phi_K^\alpha(x) > 0$. Let $\{G_n\}$ be a decreasing sequence of open sets containing K with $\bigcap \bar{G}_n = K$. Let $T_n = T_{G_n}$. Then Proposition 2.9 and the fact that K is thin imply that almost surely $T_n < T_K$ on $\{T_K < \infty\}$. Clearly $\lim T_n = T_K$ almost surely P^x on $\{T_K < \infty\}$, and so, almost surely P^x, $\{T_K < \infty\} \subset \{T_n < T_K$ for all n, $\lim T_n = T_K < \infty\}$. But by (4.14) of Chapter IV, $\Phi_K^\alpha(X_{T_n}) \to 1$ almost surely on $\{T_n < T_K$ for all n, $\lim T_n = T_K < \infty\}$. Consequently the regularity of Φ_K^α and the fact that $\Phi_K^\alpha(X_{T_K}) < 1$ imply that $P^x(T_K < \infty) = 0$, proving (4.9).

It now follows easily from Theorems 3.4 and 3.5 and the results of Chapter IV that (4.6) holds if and only if every NAF with a finite potential (or a finite α-potential for some $\alpha > 0$) is actually continuous. A slightly more complicated argument shows that under (4.6) every NAF is continuous. See Exercise 4.17. In particular if X has continuous paths and (4.6) holds, then every (nonnegative, right continuous, continuous at ζ) additive functional of X is continuous. The most important example is, of course, Brownian motion.

We are next going to give some conditions which imply (4.6). In particular Proposition 4.10 will show that (4.6) is satisfied whenever $u^\alpha(x, y) = u^\alpha(y, x)$ for all x, y in E and $\alpha > 0$, that is, whenever X and $\hat{X}$ are equivalent.

(4.10) PROPOSITION. Condition (4.6) holds if and only if, for every finely closed $B \in \mathscr{E}$, B and the cofine closure of B differ by a polar set.

Proof. According to (1.25) the fine and cofine closures of a set in $\mathscr{E}$ differ by a semipolar set, so that if (4.6) holds then this difference is a polar set. Coming to the converse let K be a compact thin set and with $\alpha > 0$ fixed; let $\Phi_K^\alpha(x) = E^x\{e^{-\alpha T_K}\}$ and $B_n = \{\Phi_K^\alpha \geq 1 - 1/n\}$. Then each B_n is finely closed and $\bigcap B_n$ is empty. The last part of the proof of (4.9) shows that $\lim_{t \uparrow T_K} \Phi_K^\alpha(X_t) = 1$ almost surely P^x on $\{T_K < \infty\}$ for each x not in K. Therefore if $R_n = T_{B_n}$ and $x \in K^c$, then $R_n < T_K$ almost surely P^x on $\{T_K < \zeta\}$. Consequently

$$\begin{aligned} P_{B_n}^\alpha \Phi_K^\alpha(x) &= E^x\{\exp[-\alpha(R_n + T_K \circ \theta_{R_n})];\ T_K < \zeta\} \\ &= E^x\{e^{-\alpha T_K};\ T_K < \zeta\} = \Phi_K^\alpha(x), \end{aligned}$$

for $x \in K^c$, and, since K is thin, $P^\alpha_{B_n} \Phi^\alpha_K = \Phi^\alpha_K$ everywhere. Now $\Phi^\alpha_K = U^\alpha \pi^\alpha_K$ and so $\pi^\alpha_K = \hat{P}^\alpha_{B_n} \pi^\alpha_K$. Thus π^α_K is carried by $B_n \cup {}^r B_n$. But $B_n \cup {}^r B_n$ is the cofine closure of B_n and so, by hypothesis, it differs from B_n by a polar set. Therefore π^α_K must be carried by B_n for each n since π^α_K can not charge a polar set. But $\bigcap B_n$ is empty and so $\pi^\alpha_K = 0$. Consequently $\Phi^\alpha_K = 0$ and hence K is polar.

Of course when X and $\hat{X}$ are equivalent then the fine and cofine topologies coincide so the hypothesis of (4.10) is certainly satisfied in this case.

An interesting situation in which (4.6) obviously holds is when there are no semipolar sets except, of course, the empty set. Plainly this is equivalent to the condition that, for each x in E, x is regular for $\{x\}$. (Recall from Theorem 3.13 of Chapter V that this is also the condition that the local time for X at x exists.)

(4.11) PROPOSITION. Suppose that for some $\alpha > 0$ the function $x \to u^\alpha(x, x_0)$ is bounded and is continuous at $x = x_0$. Then x_0 is regular for $\{x_0\}$. In particular if this condition holds for all x_0 in E then (4.6) holds.

Proof. If μ is unit mass at x_0, then $U^\alpha \mu(x) = u^\alpha(x, x_0)$ and, since this is bounded by assumption, $\{x_0\}$ can not be polar. Let T be the hitting time of $\{x_0\}$ and suppose that x_0 is not regular for $\{x_0\}$. Then since $\{x_0\}$ is not polar there exists a $y \neq x_0$ such that $P^y(T < \infty) > 0$. Applying the argument in the third paragraph of the proof of Theorem 4.9 to the compact set $K = \{x_0\}$, we obtain an increasing sequence $\{T_n\}$ of hitting times such that almost surely P^y, $\{T_n\}$ increases to T strictly from below on $\{T < \infty\}$, and $\Phi^\alpha_K(X_{T_n}) \to 1$ as $n \to \infty$ on $\{T < \infty\}$. But $\Phi^\alpha_K = U^\alpha \pi^\alpha_K$ and, since $K = \{x_0\}$, $\pi^\alpha_K = C^\alpha(K)\, \varepsilon_{x_0}$. Therefore $\Phi^\alpha_K(x) = C^\alpha(K)\, u^\alpha(x, x_0)$ which is continuous at $x = x_0$, and so $\Phi^\alpha_K(X_{T_n}) \to \Phi^\alpha_K(X_T) = \Phi^\alpha_K(x_0)$ almost surely P^y on $\{T < \infty\}$. This is a contradiction since $\Phi^\alpha_K(x_0) < 1$ and hence Proposition 4.11 is proved.

The reader can easily check that the condition in (4.11) is satisfied for all x_0 by any stable process in $\mathbf{R}$ of index $\alpha > 1$. More generally, using the notation in (1.22), if X is any standard process in $\mathbf{R}$ with stationary independent increments such that $\int [\alpha + \psi_R(x)]^{-1}\, dx < \infty$ for some $\alpha > 0$, then the condition in (4.11) is satisfied at all points in $\mathbf{R}$. Thus (4.6) holds for these processes and, in addition, the local time L^x exists at all x. See the exercises for further information about these local times.

Under slightly stronger hypotheses, condition (4.6) is equivalent to two other familiar statements from classical potential theory. We are now going to formulate and prove these results following the development in Hunt [4, Sec. 20].

(4.12) THEOREM. Assume, in addition to (2.1), (2.2), (4.1), and (4.2), that $U^\alpha f$ is continuous on E whenever $\alpha > 0$ and $f \in b\mathscr{E}$. Then (4.6) is equivalent to each of the following statements:

(i) Let μ be a finite measure with compact support K. Then $U\mu$ is continuous if it is bounded and if its restriction to K is continuous.

(ii) If f is a locally integrable excessive function and $\varepsilon > 0$, then there is an open set G with $C(G) < \varepsilon$ such that the restriction of f to $E - G$ is finite and continuous.

Proof. First we show that (4.6) implies (i). Since $U\mu$ is bounded, μ does not charge polar sets and so, under (4.6), $\mu(K - {}^rK) = 0$. Therefore for any $\alpha \geq 0$, $P_K^\alpha U^\alpha \mu = U^\alpha \hat{P}_K^\alpha \mu = U^\alpha \mu$ and consequently $U^\alpha \mu$ is bounded on E by its maximum on K. Now if $\alpha > 0$ then $U\mu = U^\alpha \mu + \alpha U^\alpha U\mu$. But $U\mu$ is bounded so $U^\alpha U\mu$ is continuous by hypothesis, and hence the restriction of $U^\alpha \mu$ to K is continuous. Clearly $U^\alpha \mu$ decreases as $\alpha \to \infty$ and, since $\alpha U^\alpha U\mu \to U\mu$ as $\alpha \to \infty$, it follows that $U^\alpha \mu$ decreases to zero as $\alpha \to \infty$. By Dini's theorem $U^\alpha \mu$ decreases to zero uniformly on K and hence on E since $\|U^\alpha \mu\| = \sup_{x \in K} U^\alpha \mu(x)$. Therefore $\alpha U^\alpha U\mu \to U\mu$ as $\alpha \to \infty$ uniformly on E, and since $U^\alpha U\mu$ is continuous this yields Statement (i).

Next (i) implies (ii). Suppose first of all that f is the bounded potential $U\mu$ of a measure μ with compact support. Then μ is a finite measure which vanishes on polar sets and if ν is the restriction of μ to a compact set K then, by (i), $U\nu$ is continuous if its restriction to K is continuous. We will need a bound for the capacity of the (open) set B on which the potential $U\eta$ of a measure η in $\mathbf{M}$ exceeds a positive constant ρ. If D is a compact subset of B, then

$$\textbf{(4.13)} \qquad \rho\, C(D) \leq \int \hat{\pi}_D(dx)\, U\eta(x) = \int \hat{\Phi}_D\, d\eta \leq \eta(E)$$

and so $C(B) \leq \eta(E)/\rho$. Obviously this bound is useful only when η is a finite measure. Given $\delta > 0$ we apply Lusin's theorem to the finite measure μ to obtain an open set B such that $\mu(B) < \delta$ and the restriction of $U\mu$ to $E - B$ is continuous. Let ν_1 be the restriction of μ to $E - B$ and μ_1 the restriction of μ to B. The support of ν_1 is a compact subset of $E - B$ and the restriction of $U\nu_1$ to $E - B$ is continuous since $U\nu_1$ and $U\mu_1$ are lower semicontinuous on E and their sum $U\mu = U\nu_1 + U\mu_1$ has a continuous restriction to $E - B$. Therefore $U\nu_1$ is continuous on E. The set H on which $U\mu_1$ exceeds ρ is open and by (4.13), $C(H) < \delta/\rho$. Let $\delta_n = \varepsilon 4^{-n}$ and $\rho_n = 2^{-n}$ and apply the above argument. Then $U\mu = U\nu_n + U\mu_n$ where $U\nu_n$ is continuous and $U\mu - U\nu_n \leq 2^{-n}$ on $E - G_n$ where G_n is open and $C(G_n) < \varepsilon 2^{-n}$. If $G = \bigcup G_n$, then G is open, $C(G) < \varepsilon$, and $U\nu_n$ converges to $U\mu$ uniformly on $E - G$. Thus (ii) is proved when f is the bounded potential of a measure with compact support.

Next suppose that f is bounded. If D is an open set with compact closure, then $P_D f$ is the bounded potential of a finite measure with compact support and $P_D f = f$ on D. Consequently one can find an open set G of arbitrarily small capacity such that the restriction of f to $D - G$ is continuous. Clearly this yields (ii) for bounded f. If f is locally integrable then $\{f = \infty\}$ is polar and hence there exists an open set G_0 containing $\{f = \infty\}$ such that $C(G_0) < \varepsilon/2$. Let $f_n = f \wedge n$. By what has already been proved there exists for each $n \geq 1$ an open set G_n such that the restriction of f_n to $E - G_n$ is continuous and $C(G_n) < \varepsilon 2^{-(n+1)}$. Then $G = \bigcup_{n=0}^{\infty} G_n$ is open, $C(G) < \varepsilon$, and it is easy to see that the restriction of f to $E - G$ is finite and continuous. Therefore (i) implies (ii).

Finally we will show that (ii) implies (4.6) by using (4.8) and (4.9). Let f be a bounded excessive function and let $\nu = g\xi$ where g is a bounded non-negative function vanishing outside a compact set. The potential $g\hat{U}$ is bounded, say by M. If K is a compact subset of E, then

$$\int g(x)\, \Phi_K(x)\, dx = \int g(x)\, U\pi_K(x)\, dx$$

$$= \int g\hat{U}\, d\pi_K \leq M\, C(K).$$

Using (1.20) of Chapter V it now follows that, for any Borel set B, $\int g(x)\, \Phi_B(x)\, dx \leq M\, C(B)$. Let G be the set mentioned in (ii). Since $t \to f(X_t)$ is continuous wherever $t \to X_t$ is continuous on $(0, T_G)$ and since

$$P^{\nu}[T_G < \zeta] = \int g(x)\, \Phi_G(x)\, dx \leq M\, C(G) \leq M\varepsilon,$$

it follows, ε being arbitrary, that, almost surely P^{ν}, $t \to f(X_t)$ is continuous wherever $t \to X_t$ is continuous on $(0, \zeta)$. Given $\delta > 0$ let T be the smallest value of t such that $|f(X_t) - f(X_t)_-| > \delta$ and $X_t = X_{t-}$. Then T is a terminal time (see (4.16) of Chapter IV) and by what was proved above $P^{\nu}(T < \zeta) = 0$ whenever $\nu = g\xi$ with g as above. Therefore $P^x(T < \zeta) = 0$ a.e., and hence everywhere since $x \to P^x(T < \zeta)$ is excessive. But δ is arbitrary and so f must be regular. This completes the proof of Theorem 4.12.

REMARK. If X is such that $\|U\mu\| = \sup_{x \in K} U\mu(x)$ whenever μ is a measure with support K (by (1.26ii) this is the case whenever the fine and cofine topologies coincide), then one may replace the second sentence in (4.12i) by "Then $U\mu$ is continuous if its restriction to K is finite and continuous."

Exercises

In the exercises Conditions (2.1), (2.2), (4.1), and (4.2) are assumed to hold unless explicitly stated otherwise.

(4.14) (a) Show that the set function φ defined on compact subsets of E by $\varphi(K) = \pi_K(E)$ is a Choquet capacity. (b) Let φ be any Choquet capacity on the compact subsets of E with $\varphi(\emptyset) = 0$ and use φ also to denote the extension of this capacity to $\mathscr{E}$. Show that φ is countably subadditive on $\mathscr{E}$. [Hint: the first step is to show that $\varphi(A_1 \cup A_2) \leq \varphi(A_1) + \varphi(A_2)$ if A_1 and A_2 are open. For this, show first that if K is a compact subset of $A_1 \cup A_2$ then $K = K_1 \cup K_2$ where K_i is a compact subset of A_i $(i = 1, 2)$.]

(4.15) Let $\alpha \geq 0$ and let B be a Borel set for which π_B^α exists. If $\beta > \alpha$ show that π_B^β exists and that $\pi_B^\beta = \pi_B^\alpha + \hat{P}_B^\beta \nu_B$ where $\nu_B = (\beta - \alpha)\Phi_B^\alpha \xi$. [Hint: use the argument of (4.9) of Chapter V.]

(4.16) Let B be a Borel set with compact closure. Show that for every Borel set D, $\pi_B^\alpha(D)$ is a continuous nondecreasing function of α—in particular $C^\alpha(B)$ varies continuously with α. [Hint: use (4.15) and the fact that π_B^α is a finite measure to establish the assertion in the interval $(0, \infty)$. Next use the fact that $U\pi_B^\beta = \Phi_B^\beta + U(\beta\Phi_B^\beta)$ to obtain the continuity at 0.]

(4.17) Show that under (4.6) every NAF of X is continuous. [Hint: given $\varepsilon > 0$ let T be the smallest value of t such that $A_t - A_{t-} \geq \varepsilon$ and $X_t = X_{t-}$. Use (4.10) of Chapter IV to show that T is accessible on $\{T < \zeta\}$ and hence by (4.38) of Chapter IV, T is accessible since $T = \infty$ on $\{T \geq \zeta\}$. Let $\varphi(x) = E^x(e^{-T})$. Show that $\varphi \in \mathscr{S}^1$ and so $\varphi = U^1\mu + h$ by (2.11). Then adapt the argument of (4.10) to show that μ is carried by a polar set and hence $\mu = 0$. Now if K is a compact subset of E, $P_{K^c}^1\varphi = \varphi$. Show finally that this implies that $T \geq \zeta$.]

(4.18) Suppose that x_0 is regular for $\{x_0\}$ and let L be the local time at x_0. Let $K = \{x_0\}$ and let $T = T_K$. (a) Using the notation of (3.15) of Chapter V show that $u^\alpha(x) = c(1)\, u^\alpha(x, x_0)$ and that $u^\alpha(x_0) = c(1)/c(\alpha)$ where $c(\alpha) = C^\alpha(K)$ for $\alpha > 0$. (b) Use (4.15) to show that in the notation of (3.20) and (3.21) of Chapter V, $b = \lim_{\alpha \to \infty} g(\alpha)/\alpha = \xi(K)/c(1)$. Thus the corresponding subordinator Y has no linear term if $\xi(K) = 0$.

(4.19) Let X be a stable process in $\mathbf{R}$ with index $\alpha > 1$; that is, X is a Hunt process in $\mathbf{R}$ of the type described in (2.12) of Chapter I with the corresponding ψ of (2.13) of Chapter I given by

$$\psi(x) = |x|^\alpha\left[1 + i\beta \operatorname{sgn}(x) \tan \frac{\pi\alpha}{2}\right]$$

where β is a parameter with $-1 \leq \beta \leq 1$. As mentioned after (4.11) these processes satisfy the condition of (4.11) at all $x_0 \in \mathbf{R}$ and so a local time exists at each x_0. Let $p(t, x) = (2\pi)^{-1} \int_{-\infty}^{\infty} e^{-ixy} e^{-t\psi(y)} dy$ so that $f(t, x, y) = p(t, y - x)$ is the transition density for X with respect to Lebesgue measure and $u^\lambda(x, y) = \int_0^\infty e^{-\lambda t} p(t, y - x) dt$ is the potential kernel—here we use λ as the parameter in the potential kernel since α is reserved for the index of X. Show that if $a > 0$ then $p(at, x) = a^{-1/\alpha} p(t, a^{-1/\alpha} x)$ for all t and x. Use this to show that $u^\lambda(x, x) = p(1, 0) \Gamma(1 - 1/\alpha) \lambda^{(1/\alpha)-1}$ for all $\lambda > 0$ and all x. Using the notation of (4.18) show that $u^\lambda(x_0) = c(1)/c(\lambda) = \lambda^{(1/\alpha)-1}$ for each x_0. If $\tau(t)$ denotes the inverse of the local time at x_0, then use (3.21) of Chapter V to show that $(\tau(t), P^{x_0})$ is a stable subordinator of index $1 - 1/\alpha$. Compare this with (3.37) of Chapter V. Show that the hypothesis of Theorem 3.30 of Chapter V is satisfied with $h(x) = \min(1, |x|^{\alpha-1})$ so that the local time L_t^x is jointly continuous in t and x almost surely.

(4.20) Let X satisfy the hypothesis of Theorem 3.30 of Chapter V, and let L^x denote the local time at x. Let $h(x) = [C^1(\{x\})]^{-1} = u^1(x, x)$. Then show that if D is a Borel set, $\int_D h(x) L_t^x \, dx = \int_0^t I_D(X_u) \, du$ for all t almost surely. Compare with (3.41) of Chapter V.

(4.21) The assumptions and notation are as in (4.20). Let A be a CAF of X with a finite 1-potential. Show that there exists a measure ν such that $A_t = \int L_t^x \, \nu(dx)$ for all t almost surely. [Hint: show that $u_A^1 = U^1\mu$. Then show that $\nu = h\mu$ does the job.] Extend this result to arbitrary CAF's of X.

NOTES AND COMMENTS

Chapter 0

The reader will find a treatment of measure theory especially designed for probabilists in Neveu [1]. In particular the development given there of martingale theory is more than adequate for our purposes. For considering examples the reader will need a bit of Fourier transform theory. We refer him to Bochner [1] or Feller [1] for this.

The particular form of Theorem 2.2 given in the text appeared for the first time in Dynkin [1].

Chapter I

SECTION 2. We introduce the processes with stationary independent increments to provide examples illustrating various points in the general theory; also they enter into the discussion of local times (Chapter V). Of course these processes have an extensive theory of their own going back to the work of Kolmogorov and Lévy in the early 1930's. For an elegant treatment of the measure theoretic properties of these processes we refer the reader to Ito [2].

The result in (2.15) may be used to reduce the study of sample function properties in the general case to the temporally homogeneous case. It has not been used for much else however.

The operation of passing from the transition function P_t to Q_t in (2.16), see also (2.20), is called "subordination" by Bochner [1]. The term "subordinate to" is used in a much different sense in Chapter III.

SECTION 3. A definition of Markov process as general as the one given here is needed if one is to avoid endless additional qualifications in the future.

Even so it does not include some processes which arise in a natural manner. See, for example, (3.8) of Chapter III. A reader new to the subject may find the extensive axiomatization of this section annoying; he may, if he wishes, start with the function space representation described in Section 4, altering this later on to achieve the desired sample function regularities. However in some instances this is not the most natural representation of a process; for example, when discussing time changes in Chapter V.

SECTION 5. In some papers (including those of the present authors) $\mathscr{F}_t$ is defined as $\bigcap_\mu (\mathscr{F}_t^0)^{P^\mu}$, the intersection being over all finite measures μ on $\mathscr{E}_\Delta$. However this turns out not to be the appropriate completion—it is necessary to complete $\mathscr{F}_t^0$ in the larger σ-algebra $\mathscr{F}$ as we have done in the text.

Note that (5.17) doesn't say much unless $\mathscr{F}_0$ is considerably larger than $\mathscr{F}_0^0$ (as it will be in the cases of interest to us).

SECTION 6. We consider only those properties of stopping times which are immediately relevant to their use in the theory of Markov processes. For a much more thorough investigation see Chung and Doob [1].

SECTION 8. To the best of our knowledge the first formulation and proof of the strong Markov property for a class of Markov chains appeared in Doob [2]. Our main criterion for the strong Markov property (Theorem 8.11) was proved by Hunt [1] for processes with independent increments. The general case was obtained by Blumenthal [1] and by Dynkin and Yushekevich [1] independently. However, the use of the resolvent $\{U^\alpha\}$ rather than the semigroup $\{P_t\}$ in this condition seems to be due to Ito [1]. The difficult problem of finding more or less necessary conditions for the strong Markov property was first considered by Ray [1]. Two more references to related work are Ray [2] and Knight [1].

SECTION 9. The basic results concerning the absence of oscillatory discontinuities in the sample functions are due to Kinney [1] and Dynkin [3]. See also Blumenthal [1]. The idea of using martingale theory to study the trajectories of a Markov process goes back to Kinney [1] and to Doob [4] and [5]. The notion of quasi-left-continuity and the observation that it is the appropriate substitute for the continuity of the paths is due to Hunt. The fact that the conditions in Theorem 9.4 imply quasi-left-continuity appeared in Blumenthal [1]. The fact that Definition 9.2 delineates the appropriate class of processes on which to base a potential theory is due to Hunt. In fact what we call a Hunt process is just a process satisfying Hypothesis (A) of Hunt [2]. Also in [3] Hunt considered standard processes (implicitly if not explicitly). The terminology "standard process" is due to Dynkin. In [2]

Hunt points out that the equality $\mathscr{F}_{0+} = \mathscr{F}_0$ turns the zero-one law (5.17) into a useful result.

Theorem 9.4 on the existence of a Hunt process with a prescribed transition function is not the best result available. This is not important to us because in this book we assume that a standard process is given as the basic data. However in some other work an improvement of (9.4) is essential; see, for example, Ray [2] and Hansen [1].

Most of the familiar standard processes are actually Hunt processes. But even so, in passing to subprocesses the quasi-left-continuity on $[0, \infty)$ may be lost (see (9.16)) while the quasi-left-continuity on $[0, \zeta)$ is not. Assuming the process X to be standard rather than Hunt leads to an increase in the complexity of some proofs. From a technical point of view the possible failure of (11.3) of Chapter I and of (4.2) of Chapter IV for standard processes causes the most difficulty. However, (6.1) of Chapter III may be used in place of (11.3) of Chapter I in most situations in which (11.2) of Chapter I does not suffice.

SECTION 10. The idea of using Choquet's capacitability theorem to establish the measurability of hitting times, and, more important, the approximation theorems (10.16)–(10.19) is due to Hunt [2].

SECTION 12. Sometimes regular step processes are used as approximations to general standard processes; see, for example, T. Watanabe [1].

Chapter II

SECTIONS 1–4. With a few exceptions, most notably (3.6), the definitions and theorems in these sections are contained in Hunt [2]. However, some of the terminology and proofs differ a little from those given by Hunt. The probabilistic approach to the fine topology and to regular points is due to Doob [4] and [6] for Brownian motion and the heat process. The extension to general Markov processes was straightforward. Theorem 5.1 was announced in Dynkin [4] and a proof appeared in Dynkin [2].

Exercises 4.17–4.23 are designed to be a collection of the folklore concerning recurrence and the existence of nonconstant excessive functions. See Azéma *et al.* [1] for recent work in this direction.

SECTION 5. For results similar to those in (5.6), (5.9), and (5.11) see T. Watanabe [1] and Šur [2].

Chapter III

SECTIONS 1–4. Special types of MF's and some important analytic features of their relationship to processes have been studied (by physicists as well as mathematicians) for many years; see, for example, Kac [1] and Darling and Siegert [1]. In [3] Hunt used a rather wide class of MF's as the basis for the relative theory. Apparently Dynkin [1] was the first to axiomatize the properties of Definition 1.1 and systematically study the relationship between MF's and subprocesses. The construction of subprocesses and the discussion of their properties in Section 3 is due to Dynkin [1] and [2]. Most of the remainder of Sections 1–4 is due to Meyer [2].

Let M be a nonnegative MF satisfying $E^x(M_t) \leq 1$ for all x and t but which need not satisfy the stronger condition $M_t \leq 1$. Then (1.8) still defines a sub-Markov transition function and much work has been directed towards the construction of a process with this transition function. In this case the desired process can no longer be obtained simply by "killing" X at an appropriate time. We refer the reader to Dynkin [2], Ito and S. Watanabe [1], and Kunita and T. Watanabe [1] for different treatments of this interesting problem.

In [6] and [9] Meyer showed that if $\{M_t\}$ is a *complex-valued* MF, then $\{M_t\}$ has the strong Markov property if and only if the MF, $\{m_t\}$, defined by $m_t = 0$ if $M_t = 0$ and $m_t = 1$ if $M_t \neq 0$ has the strong Markov property. One can then easily obtain a criterion similar to (4.21) for a complex-valued MF to have the strong Markov property.

SECTION 5. Most of this material appears (in a slightly less general form) in the first half of Hunt [3].

SECTION 6. As mentioned in the text Theorem 6.12 is due to Hunt [2], at least for Hunt processes. In order to carry over Hunt's proof to the case of standard processes it seems to be necessary to have some result such as (6.1) available. Theorem 6.1 appeared in Blumenthal and Getoor [5].

Chapter IV

SECTIONS 1–3. The study of AF's of the form $t \to \int_0^t f(X_s)\, ds$ has about the same history as the study of MF's of the form $t \to \exp[-\int_0^t f(X_s)\, ds]$. Many interesting results about such functionals are given in Kac [1]. Blanc-Lapierre and Fortet [1] contains several sections devoted to the study of additive and multiplicative functionals. Darling [1] contains a bibliography of some of the earlier papers on this subject. Already in 1956 Ito and McKean had developed the theory of CAF's of a linear diffusion and the theory of time changes based on such AF's. See Sections 3 and 4 of Chapter V. The existence of local times for these processes enabled Ito and McKean to represent the additive functionals in question as integrals of the local time (see (4.21) of Chapter VI). Their work finally was published in Ito and McKean [1]. In the meantime many of their results were rediscovered by the Russian school. See especially, Volkonskii [1]. The study of the relationship between excessive functions and additive functionals for general standard processes was carried out by the Russian school and by P. A. Meyer in the late 1950's and early 1960's. The representation $f(x) = E^x(A_\infty)$ for a uniformly excessive f as the potential of a CAF was first given by Volkonskii [2]. Then Šur [1] made a clever extension of Volkonskii's method to treat general regular potentials. Theorem 3.8 is essentially Šur's result, and the proof given here is his with some modifications due to McKean and Meyer. Independently Meyer also extended Volkonskii's result to the general case in [2]. His method is used in the proofs of Theorem 2.1 of Chapter V and Theorem 3.5 of Chapter VI. In addition, in [2] Meyer introduced the concept of a natural additive functional, proved the uniqueness theorem (2.13), and extended the representation theorem to natural potentials (4.22). In the case of Brownian motion the basic representation theorem was found independently by McKean and Tanaka [1] and by Ventcel' [1]. The definition of the *characteristic* of an AF and the results contained in (2.19), (2.20), and (3.18) are due to Dynkin [2]. Motoo and S. Watanabe [1] contains an important extension of these results.

Usually one studies only AF's of (X, M) for M of the form $M_t = I_{[0,T)}(t)$ where T is an exact terminal time, the most important case being $T = \zeta$. The increased generality of Definition 1.1 does not seem to complicate the discussion in any significant way and is useful in some situations. The results and techniques of these sections have found their ultimate generalization in P. A. Meyer's work on the decomposition of a supermartingale into the sum of a martingale and a decreasing process. See Meyer [4], [5], and especially [1].

Section 4. The proof of Theorem 4.22 given in the text is due to Meyer [2]. We will now give an alternate proof based on Meyer's work on the decomposition of supermartingales. This proof *does not require* the assumption (4.1). For simplicity we consider the case $M_t = I_{[0,\zeta)}(t)$ and leave the straightforward extension to more general MF's to the reader. Thus let u be a natural potential of X (Definition 4.17). Let $\mathcal{U}^u$ denote the set of all finite measures μ on $\mathcal{E}_\Delta^*$ satisfying $\int u \, d\mu < \infty$. It follows immediately from the definition of a natural potential and (4.18), which is independent of (4.1), that u has the following two properties:

(1) The family $\{u(X_T);\ T$ a stopping time$\}$ is uniformly integrable with respect to P^μ for each $\mu \in \mathscr{U}^u$.
(2) For each $\mu \in \mathscr{U}^u$, $\lim_{t\to\infty} E^\mu\{u(X_t)\} = 0$.

We will use the terminology of Meyer [1] without special mention. Fix an element $\mu \in \mathscr{U}^u$ for the moment. Then (1) and (2) state that the supermartingale $\{u(X_t), \mathscr{F}_t, P^\mu\}$ is a right continuous potential of the class (D). Let $g_n = n(u - P_{1/n} u)$ and $A_t^n = \int_0^t g_n(X_s)\, ds$. Then according to Meyer's theorem [1, VII, T 29] there exists a unique natural increasing process $\{B_t^\mu\}$ such that for each stopping time T and $Y \in b\mathscr{F}$, $E^\mu(YA_T^n) \to E^\mu(YB_T^\mu)$ as $n \to \infty$, and such that $E^\mu(B_\infty^\mu) = \int u\, d\mu$. Now exactly as in the proof of Theorem 3.8 we can find a family $A = \{A_t;\ t \geq 0\}$ of nonnegative random variables that has all the properties of an AF of X (Definition 1.1) except that $t \to A_t$ may be discontinuous at ζ, and such that $t \to A_t$ and $t \to B_t^\mu$ are identical functions almost surely P^μ for each $\mu \in \mathscr{U}^u$. In particular we may assume that $t \to A_t$ is constant on $[\zeta, \infty]$ since each A^n has this property. Since $\varepsilon_x \in \mathscr{U}^u$ we see that $u(x) = E^x(A_\infty) = E^x(A_\zeta)$ for each x in E_Δ.

Next we will show that almost surely $t \to A_t$ and $t \to X_t$ have no common discontinuities on $[0, \zeta)$. If this were not the case there would exist an $\varepsilon > 0$ and an x in E such that if $T = \inf\{t\colon d(X_t, X_{t-}) > \varepsilon\}$, then

$$P^x(A_T - A_{T-} > 0;\ T < \zeta) > 0.$$

Here d is a metric for E_Δ. Let $R = T$ on $\{T < \zeta\}$ and $R = \infty$ on $\{T \geq \zeta\}$. The quasi-left-continuity of X implies that R is totally inaccessible (relative to P^x). But B^{ε_x}, and hence A, is a natural increasing process and so $A_R - A_{R-} = 0$ almost surely P^x (Meyer [1, VII, T 49]). Thus the assertion in the first sentence of this paragraph is established. A has all the properties of an NAF of X except that $t \to A_t$ may have a jump at ζ. Moreover $t \to A_t$ is a natural increasing process relative to P^μ for each $\mu \in \mathscr{U}^u$. For the purposes of the present discussion let us call such an A a *generalized* NAF of X. So far we have proved that a finite excessive function, u, satisfying (1) and (2) has a unique representation $u(x) = E^x(A_\infty)$ where A is a generalized NAF of X.

Finally we will show that if u is a natural potential, then we can modify the generalized NAF constructed above to obtain an NAF whose potential is u. To this end we need the following lemma whose proof is just the same as that of (4.37). Since we sketched this proof in the text we will omit the proof of the lemma.

LEMMA. Let μ be a fixed initial measure and let $\{R_n\}$ be an increasing sequence of stopping times with limit R. Suppose that $\Lambda \in \bigvee_n \mathscr{F}_{R_n}$ and let $\Gamma = \Lambda \cap \{R_n < R$ for all n, $R < \infty\}$. Then there exists an increasing sequence of stopping times $\{T_n\}$ such that P^μ almost surely $\lim T_n = R$ on Γ, $T_n < R$ for all n on Γ, and $T_n = \infty$ for all large n on Γ^c.

Let f be a finite excessive function satisfying (1) and (2) and let A be the generalized NAF such that $f(x) = E^x(A_\infty)$. Define $A_t^* = A_t$ if $t < \zeta$, $A_t^* = A_{\zeta-}$ if $t \geq \zeta$, and $J = A_\zeta - A_{\zeta-}$. Then $A^* = \{A_t^*\}$ is an NAF of X. Define

$$g_f(x) = E^x(A_\infty^*); \qquad h_f(x) = E^x(J).$$

It is easy to check that h_f is excessive and that it satisfies (1) and (2) since $h_f \leq f$. Thus $f = g_f + h_f$ where g_f is the potential of the NAF, A^*, and h_f again satisfies (1) and (2). Now for our given natural potential u define $u_1 = g_u$, $v_1 = h_u$ and $u_n = g_{v_{n-1}}$, $v_n = h_{v_{n-1}}$ for $n > 1$. Let $f_n = u_1 + \ldots + u_n$. Clearly $u = f_n + v_n$ for each n, and as $n \to \infty$, $\{f_n\}$ increases to an excessive function f while $\{v_n\}$ decreases to a super-mean-valued function v which is excessive since $u = f + v$. By construction for each n, $u_n(x) = E^x(A_\infty^n)$ where A^n is an NAF of X and so if $A_t = \sum_n A_t^n$, then $f(x) = E^x(A_\infty)$. It is easy to check that $A = \{A_t\}$ is an NAF of X. Thus in order to complete the proof of Theorem 4.22 it will suffice to show that $v = 0$.

To this end note that, for any j, $v_j - v = \lim_n (v_j - v_n) = \lim_n \sum_{j+1}^n u_\kappa$ is excessive. Both $v_j - v$ and v satisfy conditions (1) and (2) and since $v_j = (v_j - v) + v$ it follows from the uniqueness of the above representation that $h_{v_j} = h_{v_j - v} + h_v$. But $h_{v_j} = v_{j+1} \downarrow v$ and $h_{v_j - v} \leq v_j - v \to 0$ as $j \to \infty$. Therefore $v = h_v$ and this implies $v(x) = E^x(B_\zeta)$ where B is a generalized NAF with the property that $E^x(B_{\zeta-}) = 0$ for all x; that is, B is constant except for a possible jump at ζ. Fix x and let ζ_I and ζ_A denote the totally inaccessible part and the accessible part of ζ relative to P^x, respectively. Since $t \to B_t$ is a natural increasing process $B_\zeta = B_{\zeta-} = 0$ almost surely P^x on $\{\zeta_I < \infty\}$. Recall that $\zeta = \zeta_I$ on $\{\zeta_I < \infty\}$. But $\{\zeta_A < \infty\}$ is a countable union (up to a set of P^x measure zero) of sets of the form $K[(R_n)]$; here $\{R_n\}$ is an increasing sequence of stopping times with $R = \lim R_n \leq \zeta$ and

$$K[(R_n)] = \{R_n < R \text{ for all } n,\ R = \zeta < \infty\}.$$

See the remark following VII, T 44, of Meyer [1]. Since $\{\zeta < \infty\} = \{\zeta_I < \infty\} \cup \{\zeta_A < \infty\}$, in order to show that $v(x) = E^x(B_\zeta) = 0$ it suffices to show that $B_\zeta = 0$ almost surely P^x on $K[(R_n)]$. Let $\Lambda = \{B_R > 0\}$; then $\Lambda \in \bigvee_n \mathscr{F}_{R_n}$ (Meyer [1, VII, T 49]). Moreover $\Lambda \subset \{R = \zeta < \infty\}$. Let $\Gamma = \Lambda \cap \{R_n < R$ for all n, $R < \infty\}$. By the lemma there exists an increasing sequence $\{T_n\}$ of stopping times such that $\{T_n\}$ increases to R strictly from below on Γ and $\lim T_n = \infty$ on Γ^c almost surely P^x. Because $\Gamma \subset \{R = \zeta < \infty\}$, $\lim T_n \geq \zeta$ almost surely P^x. But v is a natural potential since $v \leq u$ and so $P_{T_n} v(x) \to 0$ as $n \to \infty$, while

$$P_{T_n} v(x) = E^x\{B_\zeta - B_{T_n};\ T_n < \zeta\}$$
$$\to E^x\{B_\zeta - B_{\zeta-};\ T_n < \zeta \text{ for all } n\}.$$

Consequently $B_\zeta = 0$ almost surely P^x on $\{T_n < \zeta$ for all $n\} \supset \Gamma$, and since $\Gamma = \{B_\zeta > 0\} \cap K[(R_n)]$ this complete the proof. Thus we have established Theorem 4.22 for arbitrary standard processes.

It is easy to give examples of excessive functions which satisfy (1) and (2) but which are not natural potentials. Such an excessive function, u, has the representation $u(x) = E^x(A_\infty)$ with A a generalized NAF but can not be represented as the potential of an NAF. Finally consider the following example. Let X be uniform motion to the right killed with probability $\frac{1}{2}$ as it passes through zero, i.e., the process discussed in (4.34) of Chapter IV, (3.18) of Chapter III, and (9.16) of Chapter I. Let $u(x) = 1$ if $x < 0$ and $u(x) = 0$ if $x \geq 0$. Then u is an excessive function which satisfies (1) and (2). Also it is not hard to see that u is a natural potential—however, this will become obvious in a moment. As in the first part of the above construction let $g_n = n(u - P_{1/n} u)$; then $g_n(x) = 0$ if $x \geq 0$ or if $x < -1/n$ and $g_n(x) = n$ if $-1/n \leq x < 0$. Consequently if T_0 is the hitting time of $\{0\}$, then $A_t^n = \int_0^t g_n(X_s)\, ds$ approaches A_t as $n \to \infty$ where $A_t = 0$ if $t < T_0 \wedge \zeta$ and $A_t = 1$ if $t \geq T_0 \wedge \zeta$. Clearly $A = \{A_t\}$ is the unique generalized NAF such that $u(x) = E^x(A_\infty)$. But if $x < 0$, $P^x(T_0 \wedge \zeta = \zeta) = \frac{1}{2}$ and so A has a jump at ζ. Define B_t by $B_t = 0$ if $t < T_0$ and $B_t = 2$ if $t \geq T_0$. It is easily verified that $B = \{B_t\}$ is an NAF such that $u(x) = E^x(B_\infty)$. Thus u is a natural potential and B is the unique NAF whose potential is u. The important point is that A and B are quite different functionals.

S. Watanabe [1] has studied the structure of general discontinuous additive functionals (not necessarily natural). See also Motoo and Watanabe [1].

SECTION 5. Most of the results of this section are due to Meyer [2].

Chapter V

SECTION 1. Meyer [2] was the first to introduce the condition (1.3) and to use it systematically in the study of Markov processes. It is somewhat surprising that such an innocuous regularity condition allows rather far-reaching simplifications in the theory. The results embodied in (1.5), (1.6), and (1.7) are due to Meyer [2]. Most of the remaining results in this section are due to Doob [3].

SECTION 2. Proposition 2.4 is due to Motoo [1]. The idea of using the inverse of a CAF as a "time change" in the process X has a long history. Already in 1956 it was used by Ito and McKean in discussing linear diffusions. See also

Section 15 of Hunt [3]. The results contained in (2.11) may be found in Volkonskii [1].

SECTION 3. The notion of the support of a CAF was introduced in Getoor [1]. These lectures also contained Theorem 3.8. The idea of a local time for the one-dimensional Brownian motion goes back to Lévy. See the discussion in Ito and McKean [1]. Trotter [1] gave a complete construction of the local time L_t^x for Brownian motion including the joint continuity in t and x. Additional properties of such local times may be found in Knight [2], McKean [1], and Ray [3]. Furthermore in their monograph [1] Ito and McKean develop many deep results about such local times and their inverses. Theorems 3.13, 3.17, and 3.21 appeared in Blumenthal and Getoor [3], although the definition of local time used in the text (3.12) appears here for the first time. As mentioned in the text, most of the results from (3.23) on are due to Meyer [8], although Theorem 3.30 is essentially due to Boylan [1]. Boylan's proof is much different from Meyer's proof, which is presented here. In [2] Kac used techniques not unrelated to those of Meyer.

SECTION 4. Most of the results of this section are due to Motoo [2], although some of the terminology and proofs differ a little from those given by Motoo.

SECTION 5. Theorem 5.1, at least for Hunt processes, appeared in Blumenthal *et al.* [1] and [2]. However in these papers the "piecing together" of the local time changes was somewhat slurred over and, in fact, the method outlined there is probably valid only under an assumption such as (1.3) which guarantees that every CAF is perfect. The procedure for carrying out this piecing together given in the text is new.

Chapter VI

SECTIONS 1 AND 2. The description of the dual process, or more precisely the demonstration of its relevance to the development of a strong potential theory, is due to Hunt [4]; nearly all the results of Sections 1, 2, and 4 are taken from that paper. We describe the relationship between X and $\hat{X}$ only analytically, although the interpretation of $\hat{X}$ as "X run backwards" suggests many theorems and proofs. Probabilistic descriptions of the relationship between the process and its dual are complicated; see, for example, Nagasawa [1].

The fundamental equality (1.17) more or less expresses the fact that the subprocesses (X, T_A) and $(\hat{X}, \hat{T}_A)$ are in duality also. Hunt establishes this fact for subprocesses corresponding to a much larger class of MF's—the proof is quite involved. The formulation and proof for general MF's has not yet been given.

Discussions similar to ours are given by Kunita and Watanabe [2] and by Meyer [3].

The equivalence of semipolar and cosemipolar sets was first pointed out by Meyer in [2].

SECTION 3. Theorems 3.1, 3.4, and 3.5 are due to Meyer [2]. The proofs of (3.1) and (3.4) given here are different from the original proofs of Meyer. An alternate proof of (3.5) in the same spirit as the proof of (3.4) is given in Blumenthal and Getoor [2]. Also that paper contains an extension of these results to certain classes of measures and CAF's which do not possess finite potentials.

SECTION 4. The results of this section are taken from Hunt [4]. One exception is (4.10) which appears here for the first time.

BIBLIOGRAPHY

IJM = Illinois J. Math.; *TAMS* = Trans. Am. Math. Soc.; *TV* = Teor. Veroyatnostei iee Primeneniya; *TP* = Theory of Probability and its Applications (translation of *TV*); *ZW* = Z. Wahrscheinlichkeitstheorie verw. Geb.

J. Azéma, M. Kaplan-Duflo, and D. Revuz:

[1] Récurrence fine des processus de Markov. *Ann. Inst. Henri Poincare* **B2**, 185–220 (1966).

H. Bauer:

[1] "Markoffsche Prozesse." Vorlesung an der Universität Hamburg (1963).

A. Blanc-Lapierre and R. Fortet:

[1] "Théorie des fonctions aléatoires." Masson, Paris, 1953.

R. M. Blumenthal:

[1] An extended Markov property. *TAMS* **85**, 52–72 (1957).

R. M. Blumenthal and R. K. Getoor:

[1] Sample functions of stochastic processes with stationary independent increments. *J. Math. Mech.* **10**, 493–516 (1961).

[2] Additive functionals of Markov processes in duality. *TAMS* **112**, 131–163 (1964).

[3] Local times for Markov processes. *ZW* **3**, 50–74 (1964).

[4] A theorem on stopping times. *Ann. Math. Statist.* **35**, 1348–1350 (1964).

[5] Standard processes and Hunt processes. *Proc. Symp. Markov Processes and Potential Theory, Madison, 1967*, p. 13–22. Wiley, New York, 1967.

R. M. Blumenthal, R. K. Getoor, and H P. McKean, Jr.:

[1] Markov processes with identical hitting distributions. *IJM* **6**, 402–420 (1962).

[2] A supplement to "Markov processes with identical hitting distributions." *IJM* **7**, 540–542 (1963).

S. Bochner:

[1] "Harmonic Analysis and the Theory of Probability." Univ. of California Press, Berkeley, California, 1955.

N. Bourbaki:

[1] "Élements de mathématique, Livre III, topologie générale," 2nd ed., Chapter 9. Hermann, Paris, 1956.

E. S. Boylan:

[1] Local times for a class of Markov processes. *IJM* **8**, 19–39 (1964).

M. Brelot:

[1] "Élements de la théorie classique du potentiel," 3rd ed. Centre de Documentation Universitaire, Paris, 1965.

K. L. Chung and J. L. Doob:

[1] Fields, optionality, and measurability. *Am. J. Math.* **87**, 397–424 (1965).

P. Courrège and P. Priouret:

[1] Temps d'arrêt d'une fonction aléatoire: Relations d'équivalence associées et propriétés de décomposition. *Publ. Inst. Statist. Univ. Paris* **14**, 245–274 (1965).

[2] Recollements de processus de Markov. *Publ. Inst. Statist. Univ. Paris* **14**, 275–377 (1965).

D. A. Darling:

[1] Étude des fonctionnelles additives des processus Markoviens. Le calcul des probabilités et ses applications. *Colloq. Intern. Centre Natl. Rech. Sci. (Paris)* **87**, 69–80 (1959).

D. A. Darling and A. J. F. Siegert:

[1] On the distribution of certain functionals of Markov chains and processes. *Proc. Natl. Acad. Sci. U.S.* **42**, 525–529 (1956).

J. L. Doob:

[1] "Stochastic Processes." Wiley, New York, 1953.

[2] Markoff chains-denumerable case. *TAMS* **58**, 455–473 (1945).

[3] Applications to analysis of a topological definition of smallness of a set. *Bull. Am. Math. Soc.* **72**, 579–600 (1966).

[4] Semimartingales and subharmonic functions. *TAMS* **77**, 86–121 (1954).

[5] Martingales and one dimensional diffusion. *TAMS* **78**, 168–208 (1955).

[6] A probability approach to the heat equation. *TAMS* **80**, 216–280 (1955).

E. B. Dynkin:

[1] "Foundations of the Theory of Markov Processes." Moscow, 1959 (in Russian). English translation: Prentice-Hall, Englewood Cliffs, New Jersey, 1961.

[2] "Markov Processes." Moscow, 1963. English translation (in two volumes): Springer, Berlin, 1965.

[3] Criteria of continuity and lack of discontinuities of the second kind for trajectories

of a Markov stochastic process (in Russian). *Izv. Akad. Nauk SSSR, Ser. Mat.* **16**, 563–572 (1952).

[4] Intrinsic topology and excessive functions connected with a Markov process (in Russian). *Dokl. Akad. Nauk SSSR* **127**, 17–19 (1959).

E. B. DYNKIN and A. A. YUSHEKEVICH:

[1] Strong Markov processes (in Russian). *TV* **1**, 149–155 (1956). English translation: *TP* **1**, 134–139 (1956).

W. FELLER:

[1] "An Introduction to Probability Theory and its Applications," Vol. II. Wiley, New York, 1966.

R. K. GETOOR:

[1] Additive functionals of a Markov process. Lecture notes, University of Hamburg (1964).

[2] Additive functionals and excessive functions. *Ann. Math. Statist.* **36**, 409–422 (1965).

[3] Continuous additive functionals of a Markov process with applications to processes with independent increments. *J. Math. Anal. Appl.* **13**, 132–153 (1966).

W. HANSEN:

[1] Konstruktion von Halbgruppen und Markoffschen Prozessen. *Invent. Math.* **3**, 179–214 (1967).

L. L. HELMS and G. JOHNSON:

[1] Class D supermartingales. *Bull. Am. Math. Soc.* **69**, 59–62 (1963).

E. HEWITT and K. STROMBERG:

[1] "Real and Abstract Analysis." Springer, Berlin, 1965.

G. A. HUNT:

[1] Some theorems concerning Brownian motion. *TAMS* **81**, 294–319 (1956).

[2] Markoff processes and potentials. I. *IJM* **1**, 44–93 (1957).

[3] Markoff processes and potentials. II. *IJM* **1**, 316–369 (1957).

[4] Markoff processes and potentials. III. *IJM* **2**, 151–213 (1958).

N. IKEDA, M. NAGASAWA, and S. WATANABE:

[1] A construction of Markov processes by piecing out. *Proc. Japan Acad.* **42**, 370–375 (1966).

K. ITO:

[1] "Stochastic Processes." Iwanami Shoten, Tokyo, 1957 (in Japanese). English translation of Chapters 4 and 5 by Y. Ito, Yale University, 1961.

[2] "Lectures on Stochastic Processes." Tata Institute of Fundamental Research, Bombay (1961).

K. ITO and H. P. MCKEAN, JR.:

[1] "Diffusion Processes and their Sample Paths." Springer, Berlin, 1965.

K. ITO and S. WATANABE:

[1] Transformation of Markov processes by multiplicative functionals. *Ann. Inst. Fourier* **15**, 13–30 (1965).

M. KAC:

[1] On some connections between probability theory and differential and integral equations. *Proc. 2nd. Symp. Math. Statist. Probability, Berkeley, 1950*, pp. 189–215. Univ. of California Press, Berkeley, California, 1951.

[2] Some remarks on stable processes. *Publ. Inst. Statist. Univ. Paris* **6**, 303–306 (1957).

J. R. KINNEY:

[1] Continuity properties of sample functions of Markov processes. *TAMS* **74**, 280–302 (1953).

F. B. KNIGHT:

[1] Markov processes on an entrance boundary. *IJM* **7**, 322–336 (1963).

[2] Random walks and a sojourn density process of Brownian motion. *TAMS* **109**, 56–86 (1963).

H. KUNITA and T. WATANABE:

[1] Notes on transformations of Markov processes connected with multiplicative functionals. *Mem. Fac. Sci., Kyushu Univ.* **A17**, 181–191 (1963).

[2] Markov processes and Martin boundaries. I. *IJM* **9**, 485–526 (1965).

J. LAMPERTI:

[1] An invariance principle in renewal theory. *Ann. Math. Statist.* **33**, 685–696 (1962).

P. LÉVY:

[1] "Théorie de l'addition des variables aléatoires," 2nd ed. Gauthier-Villars, Paris, 1954.

[2] "Processus stochastiques et mouvement Brownien," 2nd ed. Gauthier-Villars, Paris, 1965.

M. LOÈVE:

[1] "Probability Theory," 2nd ed. Van Nostrand, Princeton, New Jersey, 1960.

H. P. MCKEAN, JR.:

[1] A Hölder condition for Brownian local time. *J. Math., Kyoto Univ.* **1**, 195–201 (1962).

H. P. MCKEAN, JR. and H. TANAKA:

[1] Additive functionals of the Brownian path. *Mem. Coll. Sci., Univ. Kyoto* **A33**, 479–506 (1961).

P. A. MEYER:

[1] "Probability and Potentials." Ginn (Blaisdell), Boston, Massachusetts, 1966.

[2] Fonctionnelles multiplicatives et additives de Markov. *Ann. Inst. Fourier* **12**, 125–230 (1962).

[3] Semi-groupes en dualité. Séminaire de théorie du potentiel, directed by M. Brelot, G. Choquet, and J. Deny. Inst. Henri Poincaré, University of Paris, 5th year (1960/1961).

[4] A decomposition theorem for supermartingales. *IJM* **6**, 193–205 (1962).

[5] Decomposition of supermartingales: the uniqueness theorem. *IJM* **7**, 1–17 (1963).

[6] La propriété de Markov forte des fonctionnelles multiplicatives. *TV* **8**, 349–356 (1963). English translation: *TP* **8**, 328–334 (1963).

[7] Sur les relations entre diverses propriétés des processus de Markov. *Invent. Math.* **1**, 59–100 (1966).

[8] Sur les lois de certaines fonctionnelles additives: Applications aux temps locaux. *Publ. Inst. Statist. Univ. Paris* **15**, 295–310 (1966).

[9] Quelques resultats sur les processus. *Invent. Math.* **1**, 101–115 (1966).

[10] "Processus de Markov." Springer, Berlin, 1967.

M. MOTOO:

[1] Representations of a certain class of excessive functions and a generator of Markov process. *Sci. Papers Coll. Gen. Educ., Univ. Tokyo* **12**, 143–159 (1962).

[2] The sweeping-out of additive functionals and processes on the boundary. *Ann. Inst. Statist. Math.* **16**, 317–345 (1964).

M. MOTOO and S. WATANABE:

[1] On a class of additive functionals of Markov processes. *J. Math., Kyoto Univ.* **4**, 429–469 (1965).

M. NAGASAWA:

[1] Time reversions of Markov processes. *Nagoya Math. J.* **24**, 177–204 (1964).

J. NEVEU:

[1] "Mathematical Foundations of the Calculus of Probability." Holden-Day, San Francisco, California, 1965.

D. RAY:

[1] Stationary Markov processes with continuous paths. *TAMS* **82**, 452–493 (1956).

[2] Resolvents, transition functions, and strongly Markovian processes. *Ann. Math.* **70**, 43–72 (1959).

[3] Sojourn times of diffusion processes. *IJM* **7**, 615–630 (1963).

C. J. STONE:

[1] The set of zeros of a semi-stable process. *IJM* **7**, 631–637 (1963).

M. G. ŠUR:

[1] Continuous additive functionals of a Markov process (in Russian). *Dokl. Akad. Nauk SSSR* **137**, 800–803 (1961). English translation: *Soviet Math.* **2**, 365–368 (1961).

[2] A localization of the concept of an excessive function connected with a Markov process (in Russian). *TV* **7**, 191–196. English translation: *TP* **7**, 185–189 (1962).

H. F. TROTTER:

[1] A property of Brownian motion paths. *IJM* **2**, 425–433 (1958).

A. D. VENTCEL':

[1] Nonnegative additive functionals of Markov processes (in Russian). *Dokl. Akad. Nauk. SSSR* **137**, 17–20 (1961). English translation: *Soviet Math.* **2**, 218–221 (1961).

V. A. VOLKONSKII:

[1] Random time changes in strong Markov processes (in Russian). *TV* **3**, 332–350 (1958). English translation: *TP* **3**, 310–326 (1958).

[2] Additive functionals of Markov processes (in Russian). *Tr. Mosk. Mat. Obšč.* **9**, 143–189 (1960). English translation: *Selected translations in mathematical statistics and probability. Inst. Math. Statist. Am. Math. Soc.* **5**, 127–178 (1965).

S. WATANABE:

[1] On discontinuous additive functionals and Lévy measures of a Markov process. *Japan J. Math.* **34**, 53–70 (1964).

T. WATANABE:

[1] On the equivalence of excessive functions and superharmonic functions in the theory of Markov processes. I and II. *Proc. Japan Acad.* **38**, 397–401 and 402–407 (1962).

INDEX OF NOTATION

SUBJECT INDEX